D0955942

SCIENTIFIC BOOKS, LIBRARIES AND COLLECTORS

Galileo Galilei's *Dialogc . . . sopra id due massimi sistemi del mondo Tolemaico e Copernicano*, Florence, 1632.

(*Frontispiece*)

SCIENTIFIC BOOKS LIBRARIES AND COLLECTORS

A Study of Bibliography and the Book Trade in Relation to Science

by

JOHN L. THORNTON, F.L.A.

Librarian, St. Bartholomew's Hospital Medical College, London

and

R. I. J. TULLY, F.L.A.

Deputy-Librarian, University College of North Wales, Bangor

Third, Revised Edition

LONDON

THE LIBRARY ASSOCIATION

1971

THE LIBRARY ASSOCIATION
7 Ridgmount Street, London, WC1E 7AE

First edition, 1954
First edition, reprinted, 1956
Second edition, 1962
Third edition, 1971

SBN 85365 424 7

Made and printed in England by
STAPLES PRINTERS LIMITED
at their Rochester, Kent, establishment

Preface to the Third Edition

The second edition of this book has been out of print for several years, during which period we have attempted to keep up with the increasing output of literature on the subject. In bringing the book up to date we have omitted a few references, and added many new ones, and several individual scientists are included for the first time. We have recorded the publication of a number of reprints of classic writings, but have found it very difficult to trace bibliographical details of these. Many are announced but never published; some are published without being recorded in the appropriate reference books; and details of date of publication and publisher are often given inaccurately in advertisements. Occasionally more than one publisher has reprinted the same item, or items are advertised by booksellers without giving names of publishers and dates. We have sometimes had to record a reprint as 'announced', without providing fuller information.

In thanking reviewers of the previous editions, and the writers of personal letters, we are also grateful to those who sent us reprints of their papers, which have greatly facilitated the examination of the literature. Even librarians find it extremely difficult to gain access to all the material they require to consult, but very few items recorded in the Bibliography have not been seen by at least one of the authors.

The following have kindly given permission to use illustrations, which are the same as in the previous edition, and Messrs. Dawsons of Pall Mall also generously loaned the blocks for the frontispiece and Plates 1–2: Dr. C. P. Wendell-Smith (block for Plate 4); the Royal Society (Plate 13); and the Editors of *Nature* (Plate 15). The Royal College of Surgeons of England supplied photographs of Plates 13 and 15; the University of London made a print of Plate 14; Plates 6, 8–12 were made from photographs supplied by the Department of Medical Photography, St. Bartholomew's Hospital; and Mr. H. T. Davies of the Botany Department, University College of North Wales, Bangor, photographed the originals of Plates 3, 5, 7 and 16.

We are much indebted to Mr. L. M. Payne, Librarian of the Royal College of Physicians of London, and to the library staffs of the Library Association and of University College of North Wales, Bangor, for much help in tracing the literature.

J. L. T.
R. I. J. T.

Extract from the Preface to the First Edition

We have not attempted to make this book an exhaustive treatise on the bibliographical aspects of science, but rather an introductory history of the production, distribution and storage of scientific literature from the earliest times. Our aim has been the recording of information accessible only at the expense of much research, rather than the presentation of new material, and our selected bibliography guides readers to sources of additional information.

The history of science has received increasing attention in recent years, but in this book it is recorded only to illustrate the background of bibliographical development. Quite important epochs and individuals have been omitted, our chief object being emphasis on the bibliographical viewpoint. We have endeavoured to record the chief writings of every prominent scientific author, with details of the more important printed editions of his works, referring those interested to sources providing fuller details. While it is impossible to mention every book or edition published during the periods covered by the respective chapters, an attempt has been made to record the most important authorities and their writings.

The professional historian of science will find little new in these pages, but the student of the subject, and the scientist searching for 'bibliographical gaps', will find between two covers an accumulated wealth of material on the bibliography of science.

CONTENTS

ILLUSTRATIONS

LIST OF ABBREVIATIONS

The titles of periodicals have been so abbreviated that they can readily be recognized, and if necessary reconstructed. The following have been quoted by name of author or by initials.

Bib. Osler.	*Bibliotheca Osleriana: a catalogue of books illustrating the history of medicine and science, collected . . . by Sir William Osler,* [*etc.*], 1929.
Cole.	Cole, F. J. *A history of comparative anatomy from Aristotle to the eighteenth century*, 1944.
Ferguson.	Ferguson, John. *Bibliotheca chemica: a catalogue of the alchemical, chemical and pharmaceutical books in the collection of the late James Young of Kelly and Durris, Esq.*, 2 vols., Glasgow, 1906.
Ostwald's *Klassiker.*	Ostwald's *Klassiker der exakten Wissenschaften.*
Sarton.	Sarton, George. *Introduction to the history of science*, 3 vols. [in 5], Baltimore, 1927–48.
Science, Medicine and History.	*Science, Medicine and History: essays on the evolution of scientific thought and medical practice written in honour of Charles Singer. Collected and edited by E. Ashworth Underwood*, 2 vols., London, 1953.
S.T.C.	Pollard, A. W., and Redgrave, G. R. *A short title catalogue of books printed in England, Scotland and Ireland, and of English books printed abroad, 1475–1640,* [*etc.*], 1926.
Thornton.	Thornton, John L. *Medical books, libraries and collectors: a study of bibliography and the book trade in relation to the medical sciences. Second, revised edition*, 1966.

CHAPTER I

Scientific Literature Before the Invention of Printing

"So, too, we smile at the explanation of fossils as the earlier and clumsier attempts of an All-powerful Creator to produce the more perfect beings that we know ourselves to be. Yet such conceptions were legitimate stages in the development of modern geological theory, just as the scientific views of our own time are but stages in an agelong process that is leading to wider and more comprehensive conceptions of the nature of our world."

CHARLES SINGER

The development of science undoubtedly accelerated with its practical application to everyday problems, yet many epoch-making discoveries have evolved from inspired theories. Many outstanding figures in the history of science were philosophers rather than scientists, and one of the major problems of historians is the accurate interpretation of ideas expressed in the recorded versions of the thoughts of others, either from their own writings, or as conceived by their followers. Archaeological discoveries, scholarly investigation of findings, and re-surveys of materials existing in various collections have contributed to our knowledge of civilization in various countries. Just when we think we have discovered the origin of basic elements of civilization in a particular area, fresh evidence is uncovered tracing earlier signs of development, occasionally of higher standard than anticipated by the most ambitious archaeologist. The impression of one culture upon another by the literal building of city upon city presents particularly baffling problems. It is not just a question of studying the strata, as it were, but of sorting out mixtures of evidence, migrations of cults, and re-writing of history, with the strong possibility that fresh evidence will come to light upsetting our carefully thought-out theories.

The ancient civilizations of China, India, Egypt, Babylonia and Assyria were highly developed, but we still know comparatively little about their scientific knowledge. Probably similar prosperity existed elsewhere, notably in parts of Africa and South America, but we know even less about early civilization in these areas. Much of our knowledge is based upon written records, and the absence of these, or the inability to read them when found, considerably delays our appreciation of these civilizations, which we now survey merely as dumb statues, erections of stone and carved monoliths.

Some of the earliest writings discovered are devoted to science, but there is much research to be done upon these remains by those capable of dating the material, appreciating the science of the period, and above all, transcribing the texts. The literature is inscribed on clay tablets, on papyri and parchment, much being fragmentary remains of considerably larger texts, some of which are copied from material of greater antiquity. We may not possess the contributions of the best scientists of the period, and we cannot always be sure that the translations do not incorporate more recent developments read into the texts by the translators. Modern historians are rarely equipped with the thorough knowledge of ancient languages, a comprehensive understanding of the historical development of the subject concerned, and a completely unbiased mind, a combination which is eminently desirable for producing satisfactory results. However, the amateur in historical research has made important contributions to our knowledge of the development of science, and the solution would appear to be team-work by experts in the various spheres of knowledge involved.

Until recently we have had little knowledge of ancient civilization in China, but the researches of Joseph Needham and his assistants have done much to elucidate the problems of scientific development in that country. His *Science and civilisation in China*[1] presents a fully documented study of all branches of science, technology, industrial chemistry, biology, agriculture and medicine, while the final volume is planned to cover the social background, with unified bibliographies and indexes of all the previous volumes. It is impossible to attempt to abstract this monumental study, which adds so much to our knowledge of ancient Chinese science, and which might well serve as a model for similar studies of other civilizations.

Alchemy in China probably dates from about the same period as in Ancient Egypt, and was the product of Taoism, founded by Lao Tzu (*c.* 600 B.C.). Li Shao Chün, magician and alchemist, attempted to transmute metals, and his experiments are detailed in the *Shih Chi*, which was commenced by Ssŭ-Ma T'an and completed by his son, Ssŭ-Ma Ch'ien (145–87 B.C.).[2] Liu An or Huai-nan-tzŭ (died 122 B.C.) also studied alchemy, and wrote treatises upon it, while Wei Po-Yang, the "father of alchemy", wrote *Ts'an T'ung Ch'i* (*c.* A.D. 142), which is said to be the

[1] *Science and civilization in China. By Joseph Needham, with the research assistance of Wang Ling.* Vol. 1, Introductory orientations, Cambridge, 1954; Vol. 2, History of scientific thought 1956; Vol. 3, Mathematics and the sciences of the heavens and the earth, 1959; Vol. 4, Physics and physical technology Pt. 1, 1962, Vol. 4, Pt. 2, 1965; Vol. 4, Pt. 3, 1971. Each volume contains bibliographies of (*a*) Chinese books before 1800; (*b*) Chinese and Japanese books and journal articles since 1800; (*c*) Books and journal articles in Western languages. To be completed in seven volumes.

[2] Davis, Tenney L. The Chinese beginnings of alchemy. *Endeavour*, 2, 1943, pp. 154–160.

earliest known treatise on the subject. This has been translated into English.[1] Ko Hung (A.D. 4th century) wrote extensively on alchemy under the pseudonym Pao Pu Tzŭ. An alchemical poem by Kao Hsiang-Hsien, written not later than A.D. 1333, has been translated in recent times,[2] as has also an anonymous alchemical recipe entitled *Tai Shang Wei Ling Shên Hua Chiu Chuan Tan Sha Fa*, which is undated.[3] Additional information on Chinese alchemy is obtainable from an interesting book on the subject by Obed Simon Johnson.[4]

It is also of interest to record that China was the home of some of the earliest herbals, the first printed herbal in China appearing in A.D. 973. This was known as the *K'ai-pao hsiang-ting pên-ts'ao*, of which no copy is extant, although the Library of Congress houses numerous early herbals printed in China, Japan and Korea. These are mentioned in an article by A. W. Hummell,[5] who states that one of the rarest is the original edition of *Pên-ts'ao kang-mu* by Li Shih Chên, completed in 1578, and published about 1590, of which later editions appeared in 1603, 1640, 1655, 1714 (Japanese edition), 1846, 1872 and 1885. The same author gives details of the *Ch'ung-hsiu Chêng-ho Ching-shih chêng-lei pei-yung pên-ts'ao*, printed in P'ing-yang, Shansi, in A.D. 1249 (ten volumes), of which the Library of Congress possesses but thirteen of the original thirty sections. This originated in a work compiled by T'ang Shên-wei towards the end of the eleventh century, the first printed edition appearing in A.D. 1108.

Hideomi Tuge[6] has edited a general history of Japanese science and technology from about A.D. 50 up to the present time, a welcome contribution which is probably the first book on this subject in a western language.

Large numbers of Egyptian papyri and fragments thereof have been discovered, and certain of these are devoted to science. The Papyrus of Leyden, discovered in a tomb at Thebes in 1828, is preserved in the Leyden Museum.[7] A portion of the same papyrus was sent to Stockholm, and is

[1] An ancient Chinese treatise on alchemy entitled Ts'an T'ung Ch'i. Written by Wei Po-Yang about 142 A.D. Now translated from the Chinese into English by Lu-Ch'ing Wu. With an introduction and notes by Tenney L. Davis. *Isis*, 18, 1932, pp. 210–289.

[2] Davis, Tenney L., and Chao Yün Ts'ung. An alchemical poem by Kao Hsiang-Hsien. *Isis*, 30, 1939, pp. 236–240.

[3] Spooner, Roy C., and Wang, C. H. The Divine Nine Turn Tan Sha method: a Chinese alchemical recipe. *Isis*, 38, 1947–48 pp. 235–242.

[4] Johnson, Obed Simon. *A study of Chinese alchemy*, Shanghai, 1928; see also Waley, A. Notes on Chinese alchemy (supplementary to Johnson's A study of Chinese alchemy). *Bulletin School Oriental Studies; London University*, 6, 1930–32, pp. 1–24. [A criticism of Johnson's conclusions]; Dubs, Homer H. The beginnings of alchemy. *Isis*, 38, 1947–48, pp. 62–85; Dubs, Homer H. The origin of alchemy. *Ambix*, 9, 1961, pp. 23–36; Partington, J. R. An Ancient Chinese treatise on alchemy. *Nature*, 136, 1935, pp. 287–288; and Schramm, Gottfried. Über die Chemie im alten China. *Nova Acta Leopoldina*, N.F.27, 1963, pp. 145–166.

[5] Hummell, A. W. The printed herbal of 1249 A.D. *Isis*, 33, 1941–42, pp. 439–442; also in *Ann. Rep. Library of Congress for* 1940, 1941, pp. 155–157.

[6] Tuge, Hideomi, *ed. Historical development of science and technology in Japan*, Tokyo, 1961.

[7] Caley, Earle R. The Leyden Papyrus X. *J. chem. Educ.*, 3, 1926, pp. 1149–1166.

known as the Stockholm Papyrus.[1] The text is in Greek, and was probably written about A.D. 300, but reproduces material copied from much earlier sources. It deals with the production of imitations of an alloy of gold and silver, and of gems and dyes. The earliest discovery of metals goes back to at least 4000 B.C. in Egypt, and metals were worked before 3500 B.C. both in Egypt and Mesopotamia. Gold was probably the earliest known metal, followed by copper, silver, tin, bronze, brass and iron.[2]

Although the builders of the Pyramids obviously possessed a good knowledge of mathematics, we have no Egyptian mathematical texts prior to the Twelfth Dynasty. The Golenishcher Papyrus which is in Moscow, and the Rhind Papyrus (British Museum, Nos. 10057 and 10058), are representative of this type of literature, the former dating from the thirteenth century, while the Rhind Papyrus is a copy of a Twelfth Dynasty document.[3] Both contain mathematical problems.

Among the papyri discovered at Kahun in 1889 by Sir Flinders Petrie and now housed at University College, London, is the veterinary papyrus of Kahun. This is a fragment of a work apparently composed of case histories of diseases of animals, and dates from approximately 1900 B.C. This has been described and translated by R. E. Walker.[4]

Studies of Babylonian and Egyptian mathematics and astronomy, and of Hellenistic science, are contained in O. Neugebauer's *The exact sciences in antiquity*,[5] each chapter of which contains extensive notes and references. Neugebauer states that most Babylonian mathematical texts belong roughly to the period from 1800 to 1600 B.C., with a second group from the last three centuries B.C. He suggests that there are at least 500,000 Babylonian tablets in museums, which is but a small fraction of those still buried in ruins in Mesopotamia. Discoveries continue to be made; many mathematical tablets were unearthed in 1936 at Susa, but the texts remain unpublished, and many tablets are disintegrating before being recorded. However, a number of cuneiform texts devoted to mathematics have been studied by O. Neugebauer and A. Sachs.[6] These

[1] Translations of both papyri are contained in Berthelot, Marcelin. *Introduction a l'étude de la chimie des anciens et du moyen âge*, [etc.], Paris, 1889; see also Caley, Earle R. The Stockholm Papyrus, [etc.]. *J. chem. Educ.*, 4, 1927, pp. 979–1002.

[2] See Partington, J. R. Chemistry in the ancient world. In *Science, medicine and history*, Vol. 1, 1953, pp. 35–46.

[3] See Archibald, Raymond Clare. *Bibliography of Egyptian mathematics with special references to the Rhind Mathematical Papyrus and sources in its study*, Oberlin, 1927.

[4] Walker, R. E. The veterinary papyrus of Kahun. A revised translation and interpretation of the ancient Egyptian treatise known as the veterinary papyrus of Kahun. *Vet. Rec.*, 76, 1964, pp. 198–200.

[5] Neugebauer, O. *The exact sciences in antiquity . . . Second edition*. Providence, Rhode Island, 1957; see also Neugebauer, O., and Parker, Richard A. *Egyptian astronomical texts. I. The early Decans*, Providence, R. I., London, 1960.

[6] Neugebauer, O., and Sachs, A. *Mathematical cuneiform texts*. (Amer. Oriental Series, Vol. 29), New Haven, Conn., 1945; see also Neugebauer, O., *Astronomical cuneiform texts*, 3 vols., Lund, 1955.

texts are Babylonian in origin, and date from about 1800–300 B.C. Excavation of the Library of Ashurbanipal (668–626 B.C.) has resulted in the discovery of about twenty thousand fragments of cuneiform tablets, more than one-third of which are devoted to science. They deal with herbs, animals, birds, fishes, and insects, and are now housed in the British Museum. One of the tablets (B.M. No. 120960) contains the earliest record of formulae for making glaze. The Assyrians recognized iron as early as 2700 B.C. and also copper, bronze and other metals. They minted silver coins, made glass from various compounds, and used gold to give glass the colour of coral.[1,2]

Hindu science requires intensive investigation, but a study of chemistry in ancient and mediaeval India has been edited by Priyadaranjan Rây,[3] who includes several Sanskrit and Tibetan texts, with English translations, in his comprehensive history. The Harappan Culture flourished between 2500 and 1800 B.C., and the ruins of this ancient civilization have been unearthed at Mohenjo-daro, revealing much information regarding the chemical knowledge of the period. After the decline of this civilization there followed a bleak period in India until the advent of the Aryans about 1500 B.C. David Pingree[4] has published a survey of the influence of foreign ideas on early Indian science, and gives useful references to recent work on the subject. He has also described the contents of eighty Sanskrit astronomical tables, and many more fragments, housed in the United States.[5] A bibliography of Sanskrit works on astronomy and mathematics has been published by S. N. Sen.[6]

The earliest scientific documents that can be termed complete in any sense are in Greek, and were composed about 500 B.C. These were largely influenced by the civilizations of Egypt and Mesopotamia, but their ideas were vastly improved upon. Thales of Miletus, an Ionian Greek, was born about 624 B.C., and has been described as the founder of Greek science and philosophy. He was a merchant who travelled extensively in Egypt and Mesopotamia, and predicted the solar eclipse which was visible at Miletus in 585 B.C. Thales formulated various geometrical

[1] See Thompson, Reginald Campbell. *A dictionary of Assyrian chemistry and geology*. Oxford 1936; Thompson, Reginald Campbell. *The Assyrian herbal*, [*etc.*], 1924; and Thompson, Reginald Campbell. *A dictionary of Assyrian botany*, 1940.

[2] See also, Levey, Martin. Research sources in ancient Mesopotamian chemistry. *Ambix* 6, 1958, pp. 149–154; and Levey, Martin. Evidence of ancient distillation, sublimation and extraction in Mesopotamia. *Centaurus*, 4, 1955–56, pp. 23–33.

[3] Rây, Priyadaranjan, *ed. History of chemistry in ancient and mediaeval India, incorporating the History of Hindu chemistry by Acharya Prafulla Chandra Rây. Edited by P. Rây*, Calcutta, 1956.

[4] Pingree, David. Astronomy and astrology in India and Iran. *Isis*, 54, 1963, pp. 229–246.

[5] Pingree, David. Sanskrit astronomical tables in the United States. *Trans. Amer. Philos. Soc.*, 58, iii, 1968.

[6] Sen, S. N. *A bibliography of Sanskrit works on astronomy and mathematics. Part I. Manuscripts texts, translations and studies*, Delhi, 1966.

propositions, and regarded water as the essence of all things. None of his writings has come down to us.[1]

A pupil of Thales, Anaximander of Miletus (611–547 B.C.) was interested in geography and was the first of the Greeks to indicate on maps the surface of the earth. He introduced the sun-dial from Babylon, evolved a theory of organic evolution, and wrote a book on natural philosophy, among other accomplishments, but none of his writings remains.[2] His pupil, Anaximenes of Miletus (*c.* 570–*c.* 528 B.C.), extended the ideas of his master, particularly in astronomy, and regarded air as the principal of all things.[3]

Hecataeus (*c.* 550–*c.* 475 B.C.) also of Miletus, was a great explorer, and has been called the father of geography. He explored Egypt, the Persian Empire, Thrace, Lydia, the Dardanelles and the coast of the Black Sea, and reached Gibraltar. He compiled a geographical handbook entitled *Circuit of the earth*, only fragments of which have come down to us.[4]

Although better known as a poet and philosopher than as a scientist, Heraclitus of Ephesus (*c.* 540–475 B.C.) keenly influenced the development of science, and is particularly remembered for the ideas that "fire is the origin of all things", and that "everything flows".[5] Little is known of Leucippus (*fl. c.* 475 B.C.),[6] who has been described as the founder of the atomic doctrine of matter, for he was overshadowed by his pupil Democritus of Abdera (*c.* 460–370 B.C.).[7] Mere fragments of the writings of Leucippus remain, but the views of the pupil reproduce those of the master. Democritus taught that all things are made of solid atoms, with space between them. This idea was not based upon experiment, but solely upon hypotheses. He studied mathematics, physiology, the anatomy of the chameleon, and contributed to our knowledge of psychotherapy and medicine.[8] His followers were known as Epicureans, after his disciple Epicurus of Samos (341–270 B.C.),[9] who founded a new school

[1] See Sarton, Vol. I, p. 72. Reference is frequently made to George Sarton's *Introduction to the history of science*, where he gives fuller information, additional references, details of further publications, translations, etc., and also to references in periodical literature.

[2] See Kahn, Charles H. *Anaximander and the origins of Greek cosmology*, New York, 1960; and Seligmen, Paul. *The 'Apeiron' of Anaximander. A study in the origin and function of metaphysical ideas*, London, 1962.

[3] See Sarton, Vol. I, p. 73.

[4] See Sarton, Vol. I, p. 78.

[5] See *Heraclitus: the cosmic fragments, edited with an introduction and commentaries by Geoffrey Stephen Kirk*, Cambridge, 1954; Wheelwright, Philip. *Heraclitus*, Princeton, London, 1959; and Sarton, Vol. I, p. 85.

[6] See Sarton, Vol. I, p. 88.

[7] See Sarton, Vol. I, pp. 88–89; Melsen, Andrew G. van. *From atomos to atom. The history of the concept Atom. . . . Translated by Henry J. Koren*, Pittsburgh, 1952; and Sambursky, S. Conceptual developments in Greek atomism. *Arch. int. Hist. Sci.*, 11, 1958, pp. 251–261.

[8] See Siegel, R. E. Theories of vision and color perception of Empedocles and Democritus: some similarities to the modern approach. *Bull. Hist. Med.*, 33, 1959, pp. 145–159.

[9] See Sarton, Vol. I, p. 137.

of philosophy in Athens, and was responsible for perpetuating the atomic theory. The intricate features of Epicureanism have been explained by David J. Furley[1] in a preliminary study contributing towards an exhaustive history of the struggle between the Atomists and the Aristotelians which waged at intervals, particularly in the sixteenth and seventeenth centuries.

Hellenic chemistry was based upon scientific research, as distinct from alchemy with its Arabic and Egyptian mysticism, a viewpoint which is upheld by P. D. Zacharias,[2] while Greek physics has been investigated by S. Sambursky.[3] In a later study[4] he deals particularly with the Stoics whose physics is based on the continuum theory, developed mainly by Zeno and Chryssipos in the third and fourth centuries B.C.

Herodotus of Halicarnassos (*c.* 484–*c.* 425 B.C.) has been called the father of history and of anthropology. He travelled extensively in Europe, Asia and Africa, and finally joined Greek immigrants settling in Italy. There he spent the remainder of his life writing the nine books of his *History*. His accounts of people and their countries are remarkably accurate, and are not without significance in assisting us to check the early records of the history of science. The *History* was first printed at Venice by Aldus, 1502, a critical edition by Heinrich Stein was published in two volumes, Berlin, 1869–71, and a Greek-English edition by A. D. Godley has been issued in the Loeb Library, four volumes, 1921–24.[5]

Although Pythagoras of Samos (*fl. c.* 532 B.C.)[6] left nothing in writing, he founded an important scientific school devoted to the subject of numbers, which were regarded as the element of all things. The Pythagorean school lasted for many years, and treatises describing its mathematical doctrines were written, the first being composed by Philolaus of Tarentum (*c.* 480–400 B.C.). This was used by Plato during his composition of the *Timaeus*, and Philolaus also advanced the thesis that the earth and other planets revolve round a central fire.[7]

Alcmaeon of Crotona (*c.* 500 B.C.), another disciple of Pythagoras, was the most distinguished Greek physician preceding Hippocrates. He practised scientific dissection, described the nerves of the eye, and the tube connecting the mouth and the ear, investigated by Eustachius (1520–74) many centuries later, and known as the Eustachian tube.[8]

The belief that the blood is the seat of the soul, and that the heart is

[1] Furley, David J. *Two studies in the Greek atomists: Study I. Indivisible magnitudes. Study II. Aristotle and Epicurius on voluntary action*, Princeton, N.J., 1967.

[2] Zacharias, Procopios D. Chymeutike. The real Hellenic chemistry. *Ambix*, 5, 1956, pp. 116–118.

[3] Sambursky, S. *The physical world of the Greeks*, 1956; see also essay review of this by Gillispie, Charles Coulston. A physicist looks at Greek science. *Amer. Scient.*, 46, 1958, pp. 62–74.

[4] Sambrusky, S. *Physics of the Stoics*, 1959.

[5] See Sarton, Vol. 1, pp. 105–106.

[6] See Sarton, Vol. 1, pp. 73–75.

[7] See Sarton, Vol. 1, pp. 93–94.

[8] See Sarton, Vol. 1, p. 77.

the special seat of life, originated with Empedocles of Acragas (Sicily) (*c.* 490–*c.* 435 B.C.)[1] He was also a physicist and a physician, formulating theories of the flux and reflux of the blood to and from the heart, and also of respiration. He founded the medical school in Sicily, and influenced Diogenes of Apollonia (Crete) (*c.* 430 B.C.), who wrote a book on nature. Diogenes gave the first coherent description of the vascular system, and recognized the pulse.[2]

Another book on nature was written by Anaxagoras of Clazomenae (488–428 B.C.), who went to Athens in 464 B.C. He was a keen scientist, being interested in anatomy, medicine and astronomy, giving scientific explanations of eclipses, meteors, rainbows and other natural phenomena. He dissected animals, investigated the brain, and believed that disease was caused by black or yellow bile permeating the blood and organs.[3]

Hippocrates of Cos (*c.* 460–*c.* 370 B.C.), the greatest figure in Greek medicine, and the first to separate medicine from superstition, is dealt with elsewhere.[4] He must not be confused with Hippocrates of Chios (*fl.* 450), the mathematician. The latter was one of the greatest of Greek mathematicians, and composed the first textbook of geometry. He is possibly the first to have used letters in geometrical figures, and proved the theorem that circles are to one another as the squares of their diameters.[5]

The Academy of Plato existed for nearly nine centuries, and was mainly concerned with philosophy, although over the school were inscribed the words, "Let none who has not learnt mathematics enter here". Plato (427–347 B.C.), a disciple of Socrates, was also interested in astronomy, and has been called the founder of jurisprudence. His work the *Timaeus*, a mythical theory of the universe of phenomena, greatly influenced mediaeval thought, and it has been translated by Francis M. Cornford.[6] Thomas S. Hall[7] has contributed an interesting paper on the biology of the *Timaeus*, with particular reference to physiology, generation, nutrition, respiration, disease, perception, locomotion and classification, relating these to the physiology of later periods. Plato also wrote the *Meno* and the *Phaedo*.[8] His works were printed at Florence [*c.* 1485]; Venice, 1491; *editio princeps*, Venice, 1513; later Greek edition by Henri Estienne with Latin translation by Serranus, three volumes, Paris, 1578;

[1] See Sarton, Vol. 1, p. 87.

[2] See Sarton, Vol. 1, p. 96.

[3] See Bicknell, P. J. Did Anaxagoras observe a sunspot in 467 B.C.? *Isis*, 59, 1968, pp. 87–90; Gershenson, Daniel E., and Greenberg, Daniel A. *Anaxagoras and the birth of physics*, New York, 1964; and Sarton, Vol. 1, p. 86.

[4] Thornton, pp. 5–7; see also Sarton, Vol. 1, pp. 96–102.

[5] See Sarton, Vol. 1, pp. 91–92.

[6] *Plato's Cosmology. The* Timaeus *of Plato translated with a running commentary by Francis M. Cornford*, 1956.

[7] Hall, Thomas S. The biology of the 'Timaeus' in historical perspective. *Arion*, 4, 1965 pp. 109–122.

[8] See *Plato's Phaedo. Translated by R. Hackforth*, Cambridge, 1955.

best critical edition by John Burnet, five volumes, Oxford, 1899–1906; English translation by Benjamin Jowett, third edition, five volumes, Oxford, 1892.[1] Plato's chief contribution to mathematics was his encouragement of its study in the Academy, and an associate of his pupils, Eudoxus of Cnidos (391–338 B.C.) studied medicine, but became renowned as an astronomer and mathematician. He founded a school at Cyzicus, where he wrote *The Phaenomena*, describing the constellations, and probably also his major work *On Speeds*, which analyses the movements of the sun, moon and planets. He revised *The Phaenomena* under the title *The Mirror*, and also wrote the seven books of *Tour of the earth*. He possibly wrote another book on astronomy with the title *Sphaerics*. Eudoxus was the founder of idealistic cosmology, and estimated the solar year as 365 days and 6 hours. The scattered material ascribed to Eudoxus has been collected together by François Lassere as *Die Fragmente des Eudoxus von Knidos. Herausgegeben, übersetzt und kommentiert von François Lassere.* (*Texte und Kommentare-eine altertumswissenschaftliche Reihe*, Bd. 4), Berlin, 1966.[2]

Yet another pupil of Plato, Heraclides of Pontos (*c.* 388–315 B.C.), suggested that the earth rotates on its own axis once in twenty-four hours, that Mercury and Venus revolve round the sun, and that the sun, moon and certain other planets revolve round the earth. He wrote extensively, but only fragments remain.[3]

Aristotle of Stagira (384–322 B.C.) was a pupil of Plato, and founded the Peripatetic school in the Lyceum at Athens. The writings of Aristotle cover the entire field of knowledge, but he is best known as a naturalist and biologist, his physical and astronomical observations being now considered as inferior to his biological investigations.[4] Aristotle studied the development of the chick, and embryology in general. He experimented on octopuses and squids, investigated the heart and circulation, and introduced diagrams (now all lost) to illustrate his anatomical writings. A catalogue of Latin manuscripts of his works and of pseudo-Aristotelian writings, made before the fourteenth century, has been published.[5] Many of the writings ascribed to him are apocryphal, but he made a considerable contribution to our knowledge of botany, zoology, and

[1] See Sarton, Vol. 1, p. 113; Wedberg, Anders. *Plato's philosophy of mathematics*, Stockholm, 1955; and Lodge, Rupert C. *The philosophy of Plato*, 1956.

[2] See also Sarton, Vol. 1, pp. 117–118; see also Duhem, Pierre. *Le système du monde. Histoire des doctrines cosmologiques de Platon à Copernic*, 10 vols., Paris, 1954–58; and Lasserre, François. *The birth of mathematics in the age of Plato*, 1964.

[3] See Sarton, Vol. 1, p. 141.

[4] See Coonen, L. P. Aristotle on biology. *Thomist Reader*, 1958, pp. 35–71. This study of Aristotle's work in various biological fields suggests that he was the greatest biologist of all time.

[5] *Aristoteles Latinus; codices descripserunt G. Lacombe, A. Birkenmajer, Ae. Franceschini, M. Dulong; supplementis indicibusque instruxit L. Minio-Paluello, Pars prior*, Rome, 1939, and Bruges & Paris, 1957; *Pars posterior*, Cambridge, 1955.

anatomy. He formulated theories of generation and heredity, although unfortunately he made errors which misled scientists for many centuries. Aristotle made many references to mathematics throughout his writings, but wrote no separate work on the subject as far as we know. His work on the philosophy of mathematics has been the subject of a study by Hippocrates George Apostle.[1] Friedrich Solmsen[2] has examined in considerable detail Aristotle's writings on physics, and the failure of the Greeks to found the science of chemistry has been the subject of a paper by R. A. Horne,[3] with particular reference to the lack of chemistry in Aristotelian writings.

The editions of Aristotle's writings are very extensive and are listed elsewhere,[4] but the following represent the more important issues of his significant writings.[5] *De anima*, Padua, 1472, Cologne, 1491 and 1497, and [Leipzig, 1492–96] which has been edited and translated by Sir David Ross,[6] by D. W. Hamlyn,[7] and by Walter Stanley Hett;[8] *De animalibus*, Venice, 1476, 1492, 1495 and 1498; *De coelo et mundo*, Padua, 1473, Venice, 1495, and [Leipzig, 1492–1496]; *De generatione et corruptione*, Padua, 1474, and Leipzig [1498], of which there was a French edition by C. Mugler in 1966; *Metaphysica*, Padua, 1473, Venice, 1496, and Leipzig, 1499; *Meteororum libri IV*, Padua, 1474, Venice, 1491 and [1496],[9] *Parva naturalia*, [Padua, 1473] and 1493, Cologne, 1491,[10] [Leipzig, 1495 and 1496], and Cologne, 1498; *Physica sive de physico auditu*, [Padua, 1472], [Louvain, 1475], [Rome, 1481], Venice, 1495, and Leipzig, [1495]; *Problemata*, Mantua, [1473], Rome, 1475, and many other editions. An edition of *De generatione animalium* by H. J. Drossart Lulofs was published at Oxford in 1965, and a translation of *Historia animalium* by A. L. Peck, 3 vols., London, 1969. Aristotle's *Opera*, with a commentary by Averroës, was printed at Padua, 1472–74; Venice, 1483, 1489 and 1495–96. The

[1] Apostle, Hippocrates George. *Aristotle's philosophy of mathematics*, Chicago, 1952.

[2] Solmsen, Friedrich. *Aristotle's system of the physical world. A comparison with his predecessors*, Ithaca, [*etc.*], 1960.

[3] Horne, R. A. Aristotelian chemistry. *Chymia*, 11, 1966, pp. 21–27.

[4] See Sarton, Vol. 1, pp. 127–136; this includes a bibliography divided into twenty-five sections, covering seven pages.

[5] Others are in Chapter II, pp. 47–50.

[6] Aristotle. *De anima. Edited, with introduction and commentary by Sir David Ross*, Oxford London, 1961. The same editor was responsible for *Aristotle's Physics. A revised text with introduction and commentary by William David Ross*, Oxford, 1936. See also St. Thomas Aquinas. *Commentary on Aristotle's 'Physics'. Translated by Richard J. Blackwell, Richard J. Spath and W. Edmund Thirkell*, 1963.

[7] Aristotle. *De anima. Books II and III (with certain passages from Book I.) Translated with introduction and notes by D. W. Hamlyn*, Oxford, 1968.

[8] See Hett, Walter Stanley. *Aristotle, On the soul, Parva naturalia, On breath. With an English translation*, Cambridge, Mass., 1935; and Aristotle. *Categories, and De interpretatione. Translated with notes by J. R. Ackrill*, London, 1963.

[9] Translated in the Loeb Classical Library by H. D. P. Lee, 1952.

[10] See Aristotle. *Parva naturalia. A revised text with introduction and commentary by Sir David Ross*, Oxford, London, 1955.

first edition of the *Opera* without Averroës' commentary was printed in Venice, 1482, 1496, and in Cologne, 1497. A Greek edition was printed in Venice, five volumes, 1495–98; a Greek-Latin bilingual text, five volumes, Berlin, 1851–70; and the *Works. . . . Edited by J. A. Smith and W. D. Ross*, twelve volumes, Oxford, 1908–52.[1]

Theophrastus of Eresus (*c.* 370–286 B.C.) succeeded Aristotle as head of the Lyceum, and also inherited his master's library and botanic garden. Theophrastus wrote on many scientific subjects, including a book on rocks and minerals, on odours, on winds, on weather signs, and on the history of geometry, astronomy and arithmetic among 227 treatises on various subjects, most of which are now lost, but it is as a botanist that he is chiefly remembered.[2] His writings on that subject are of vital interest, and consist of *De causis plantarum* and *De historia plantarum*, the latter being a collection of works. These two represent the only complete writings of Theophrastus that have come down to us. They were first printed as *De historia et causis plantarum Latine, Theodoro Gaza interprete*, Treviso, 1483, Greek editions appearing from the press of Aldus, 1495–98; with an edition by Joannes Bodaeus à Stapel, Amsterdam, 1644. Also under the title *Opera*, Basle, 1541. A Greek-English bilingual edition, edited by Sir Arthur Hort, was published in the Loeb Classical Library, two volumes, 1916. Other editions are recorded elsewhere.[3] *De lapidibus* was edited with an introduction, translation and commentary by D. E. Eichholz in 1965. Theophrastus introduced numerous technical botanical terms which survive in modern botany. He distinguished between monocotyledons and dicotyledons, and attempted to distinguish sex in plants. Scientific botany originated with his painstaking observations, and Theophrastus is recognized as the founder of that subject.

A contemporary of Euclid, Autolycus of Pitane (*c.* 360–*c.* 300 B.C.), was an astronomer and mathematician who wrote *On the moving sphere* and *On risings and settings*, collected together with the writings of other astronomers in the *Little astronomy*. He suggested that the relative sizes of the sun and moon as apparent to the eye was due to variations in their distances from the earth.[4]

[1] See also Ross, Sir W. D. *Aristotle. Fifth edition, revised*, 1949 (reprinted 1960); Taylor, A. E., *Aristotle*, (Nelson's Discussion Books, No. 77), London, [*etc.*], 1943, (brief bibliography, pp. 158–159); Jaeger, Werner. *Aristotle, Fundamentals of the history of his development: translated with the author's corrections and additions by Richard Robinson, Second edition*, Oxford, 1948; Lewes, George Henry. *Aristotle: a chapter from the history of science, including analyses of Aristotle's scientific writings*, 1864; Wingate, S. D. *The mediaeval Latin versions of the Aristotelian scientific corpus, with special reference to the biological works*, London, Leamington Spa, 1931; and a useful survey intended for the undergraduate by Lloyd, G. E. R. *Aristotle: the growth and structure of his thought*, Cambridge, 1968.

[2] See Coonan, L. P. Theophrastus revisited. *Centennial Rev.*, 1, 1957, pp. 404–418; this evaluates the work of Theophrastus as a botanist, summarizing his achievements as "the first great botanist".

[3] See Sarton, Vol. 1, pp. 143–144.

[4] See Sarton, Vol. 1, pp. 141–142.

A pupil of Aristotle, Dicaearchus of Messina (*c.* 355–*c.* 285 B.C.), wrote extensively on many subjects, but most of his work is lost to us. He is mainly remembered as a geographer, writing on physical geography, and including a map. He was the first to draw a parallel of latitude across a map, to measure the height of mountains, and he also estimated the circumference of the earth.[1] Another geographer, Pytheas of Marseilles (*c.* 360–*c.* 290 B.C.), explored extensively the coast from Marseilles to Spain, rounded Great Britain, sailed to the mouth of the Elbe, and up to Trondheim. Pytheas was also an astronomer, and appreciated the influences of the moon upon the tides. Only fragments of his writings remain.[2]

The development of Alexandria as a seat of learning was promoted by the characters of the scholars who taught there. Attracted by the incomparable library,[3] the learned of the civilized world congregated at the Academy, which flourished for over two hundred years, and although languishing after 100 B.C., survived for several more centuries. Euclid (*c.* 330–260 B.C.) was undoubtedly the most distinguished person associated with Alexandria, but we know nothing of his life beyond the fact that he was teaching mathematics there as early as the reign of Ptolemy I. Euclid's *Elements of geometry*, in thirteen books, was his major contribution to scientific literature, but he also wrote the *Data*, intended to facilitate the analytical treatment of theorems; the *Porisms*, unfortunately lost, as was also the *Pseudaria*, on fallacious solutions; *On the division of figures*, which survives in an Arabic edition; and students' texts on optics (i.e., perspective), the mathematical theory of pitch, and on mathematical astronomy. Innumerable editions of Euclid's *Elements* have been published, and Piètro Riccardi[4] and Charles Thomas-Stanford[5] have prepared bibliographies of this work. The latter records, with full bibliographical descriptions, editions published between 1482 and 1600, and the following details are gleaned from that work. The *Elements* was first printed by Ratdolt at Venice in 1482, and the same year a variant edition appeared with certain corrections, and with leaves 1–9 reset. A close reprint of Ratdolt's edition, but omitting the dedication, was printed at Vicenza, 1491, to be followed by editions from Venice in 1505, 1509, and 1510, the latter being a reissue of the sheets of the 1505 edition up to quire O, the remainder being reset. Then appeared the following editions: Paris,

[1] See Sarton, Vol. 1, p. 145.

[2] See Sarton, Vol. 1, pp. 144–145.

[3] See Parsons, Edward Alexander. *The Alexandrian Library, glory of the Hellenic world: its rise, antiquities, and destructions*, 1952; reprinted Amsterdam, 1967.

[4] Riccardi, Piètro. *Saggio di una bibliografia Euclidea*, 5 parts, Bologna, 1887–93; (2,500 entries).

[5] Thomas-Stanford, Charles. *Early editions of Euclid's Elements*, 1926, (Bibliographical Society, Illustrated Monographs, No. XX; 12 plates); see also Sarton, Vol. 1, pp. 153–156.

[1516]; Basle, 1533, the first edition in Greek; Paris, 1536, 1544 and 1551; Basle, 1537, which was reprinted on several occasions; Lyons, 1557; Strassburg, 1566; Paris, 1566, reprinted 1578; Pesaro, 1572; Rome, 1574; Leipzig, 1577; Paris, 1578; Rome, two volumes, 1589; Cologne, 1591, also in two volumes; Paris, 1598; and Cologne, 1600. There were also editions in Greek and Latin of the enunciations only, and translations into current European languages and into Arabic. The first translation into any modern language was the Italian version first printed at Venice in 1543, and reprinted in 1544, 1545, 1565, 1569, 1585 and 1586. A German translation was published at Basle, 1562, a French version at Paris, 1564, and in 1570 John Day published in London the first English translation [S.T.C. 10560]. A Spanish text was published at Seville in 1576. Among more modern editions of Euclid's works the following are noteworthy: the Greek text, with Latin and French translations by F. Peyrard, three volumes, Paris, 1814–18; the edition by J. L. Heiberg and H. M. Menge, eight volumes, Leipzig, 1883–1916, with a supplement dated 1899; and a translation from Heiberg's text by Sir Thomas L. Heath of the *Elements*, three volumes, Cambridge, 1908, and second edition, 1925; reprinted 1956. Euclid's *Optics* has recently been translated into English for the first time.[1,2] A translation from the Greek into French by Paul ver Eecke was published as *L'Optique et la Canoptrique*, Paris, 1959. William Austin (1754–93) was the author of an early critical evaluation of Euclid in English, but his book, entitled *An examination of the first six books of Euclid's Elements*, Oxford, 1781, disappeared. A copy was discovered in the De Golyer Collection in the University of Oklahoma Libraries, and has been described by George A. Summent.[3,4] An outline study of the English editions of Euclid up to the end of the seventeenth century, including re-issues and new editions of these versions into the eighteenth century, has been published by Diana M. Simpkins.[5] This also provides considerable biographical information, and is a very useful supplement to the bibliographical studies by Riccardi and by Thomas-Stanford.

Another prominent mathematician at Alexandria was Aristarchus of Samos (*c.* 310–230 B.C.). He made the first scientific attempt to measure the distances of the sun and moon from the earth, and also to estimate

[1] The Optics of Euclid. Translated by Harry Edwin Burton. *J. Optical Soc. America*, 35, 1945, pp. 357–372.

[2] It is probable that Adelard of Bath first brought Euclid to England, and there is no proved reference to it here before his period; see Yeldham, Florence A. The alleged early English version of Euclid. *Isis*, 9, 1927, pp. 234–238.

[3] Summent, George A. A lost work on Euclid rediscovered. *Isis*, 48, 1957, pp. 66–68.

[4] See also Hauser, Gaston. *Geometrie der Griecher von Thales bis Euklid, mit einem einleitenden Abschnitt über die vorgrieche Geometrie*, Lucerne, 1955.

[5] Simpkins, Diana M. Early editions of Euclid in England. *Ann. Sci.*, 22, 1966 [1967], pp. 225–249; see also Wallis, P. J. *A check list of British Euclids up to 1950*, Newcastle, 1970. [Typescript.]

their relative sizes. Aristarchus wrote on vision, light and colours, and his *De magnitudinibus et distantiis solis et lunae* first appeared in Georgio Valla's *Collectio. Nicephoros, logica,* [*etc.*], 1488 (reprinted 1498). The first independent Latin edition, by Federigo Commandino, was printed at Pesaro, 1572, and the first edition of the Greek text, with the Latin translation, by John Wallis, Oxford, 1688, was reprinted in Wallis's *Opera*, Vol. 3, 1695–99, pp. 565–594. A new Greek text, with a translation and notes by Sir Thomas L. Heath, appears in his *Aristarchus of Samos, the ancient Copernicus,* [*etc.*], Oxford, 1913.[1]

Herophilus of Chalcedon (*fl. c.* 300 B.C.) was a contemporary of Euclid at Alexandria, and was the founder of scientific anatomy. He distinguished between arteries and veins, and compared the anatomy of man with that of animals. Herophilus wrote treatises on anatomy, on dietetics, on the eyes and on midwifery. Erasistratus of Iulis (*c.* 310–250 B.C.), a younger contemporary of Herophilus at Alexandria, was an anatomist, physiologist and physician, and has been called the father of physiology. He furthered the work of Herophilus on the brain, dissected extensively both human and animal bodies, wrote two works on anatomy and also several treatises on clinical subjects. It has been suggested that Erasistratus came near to discovering the circulation of the blood, for he recognized that the heart is the origin of venous as well as arterial blood. Aristotle held similar views, but failed to distinguish clearly between veins and arteries, and I. M. Lonie[2] has discussed their respective contributions.[3]

The title of "the greatest mathematician, physicist, and engineer of antiquity" has been accorded to Archimedes of Syracuse (287–212 B.C.), and his work influenced every department of science. He invented the Archimedean screw for raising water, which is still in use in Egypt, improved our knowledge of the mathematical principles of levers, while his conception of specific gravity is set forth in *On floating bodies*. The principle of the steelyard is contained in *On plane equilibrium*, and he also wrote *Quadrature of the parabola; On method; Elements of mechanics; On the sphere and the cylinder; The measurement of the circle; On parallelepipeds and cylinders; On spirals; Catoptrics;* and a treatise entitled the *Sand-reckoner*. Editions of his writings are numerous, and the following are of special importance: the first (partial) Latin edition, Venice, 1543 attributed to Tartaglia, but actually by William of Moerbeke; the first Greek edition by Thomas Gechauff, Basle, 1544; a Latin translation by F. Commandino,

[1] See Sarton, Vol. 1, pp. 156–157.

[2] Lonie, I. M. Erasistratus, the Erasistrateans, and Aristotle. *Bull. Hist. Med.*, 38, 1964, pp. 426–443.

[3] See also Sarton, Vol. 1, pp. 159–160; also Brown, John R., and Thornton, John L. Physiology before William Harvey. *St. Bart's Hosp. J.*, 63, 1959, pp. 116–124; and O'Malley, C. D. The evolution of physiology. *J. Int. Coll. Surg.*, 30, 1958, pp. 115–129.

Venice, 1558; François Peyrard's *Oeuvres d'Archimède* [*etc.*] Paris, 1807; the edition by J. L. Heiberg, three volumes, Leipzig, 1910–15, first published in 1880–81, which is the best edition, and contains a Latin translation; Sir Thomas L. Heath's translation based on Heiberg's earlier version, Cambridge, 1897; and Paul ver Eecke's French translation, based on Heiberg's second edition, Paris and Brussels, 1921. A two-volume edition of this was published in Paris in 1961. A manuscript of Archimedes' *Method* was discovered by J. L. Heiberg in 1906, and was first published by him in 1907,[1] while a German translation of it by Heiberg and H. G. Zeuthen was published the same year.[2] An English translation of the *Method* was published by Sir Thomas L. Heath as a supplement to his *Works of Archimedes*, 1897, at Cambridge, 1912, and also 1951.[3,4] Marshall Clagett[5] has published an annotated collection of texts based on an analysis of numerous manuscripts, many of which have not previously been studied.

A successor of Archimedes at Alexandria was Apollonius of Perga (260–200 B.C.), author of a monumental treatise *On conic sections*, one of the few works of his that have survived. He also wrote *On unordered irrationals;* a work on the cylindrical helix entitled *On the cochlias;* and another, *Quick delivery*. The original Greek texts extant have been edited and translated into Latin by J. L. Heiberg, two volumes, Leipzig, 1891–93. The work *On conic sections* was issued in a Latin translation, Venice, 1537, with an improved version from Bologna, 1566, but these versions contained only Books I–IV. Books V–VII were first printed in a Latin translation at Florence, 1661. Edmond Halley edited a folio edition from Oxford, 1710, and Sir Thomas L. Heath based his edition of this work on those of both Heiberg and Halley. It was published at Cambridge in 1896, and reprinted in 1961. A French translation by Paul ver Eecke of *The Conics* was published at Bruges, 1924.[6]

Eratosthenes (*c.* 276–*c.* 194 B.C.) was a mathematician, astronomer, geographer, philologist and librarian at Alexandria. He measured the globe of the earth, advanced our knowledge of prime numbers, and wrote a geographical treatise in three books; a mathematical treatise *On*

[1] *Hermes*, 42, 1907, pp. 235–303.
[2] *Bibliotheca Mathematica*, 7, 1907, pp. 321–363.
[3] See Sarton, Vol. 1, pp. 169–172.
[4] See also Dijksterhuis, E. J. *Archimedes*, Copenhagen, 1957. (Acta Historica Scientiorum Naturalium et Medicinalium, Vol. 12); Kliem, Fritz, and Wolff, Georg. *Archimedes*, Berlin, 1927; Clagett, Marshall. The impact of Archimedes on mediaeval science. *Isis*, 50, 1959, pp. 419–429; and Middleton, W. E. Knowles. Archimedes, Kircher, Buffon and the burning mirrors. *Isis*, 52, 1961, pp. 533–543.
[5] Clagett, Marshall. *Archimedes in the Middle Ages. Volume I. The Arabo-Latin tradition*, Madison, 1964. This contains the original Latin texts with English translations, with excellent commentaries and a bibliography (pp. 684–690).
[6] See Sarton, Vol. 1, pp. 173–175.

means, and a book on stars is attributed to him, but only fragments of his writings remain.[1]

One of the greatest astronomers of antiquity, Hipparchos of Nicaea (*c.* 190–*c.* 126 B.C.), erected an observatory at Rhodes, and contributed usefully to astronomy, mathematics and geography. He collated the works of earlier astronomers, made a catalogue of the stars and studied the activities of the planets. Hipparchos probably constructed the first celestial globe, invented numerous astronomical instruments, and is recognized as the founder of trigonometry. Of his extensive writings the only complete surviving text is that of an early immature work, but the fragments of his geographical writings have been edited and translated.[2]

Crateuas (*fl.* 80 B.C.) was physician to Mithridates, and wrote a herbal that was illustrated with drawings. This is probably the earliest example of botanical illustration. His plants were all of medicinal interest, and Crateuas also wrote a materia medica.[3]

Most previous mathematicians had been interested solely in the theoretical aspect of the subject, but Hero of Alexandria (*c.* first century B.C.) was more concerned with its practical application. He invented numerous contrivances, including automatic machines, siphons, a fire engine, and also contributed to the science of optics. Hero wrote a commentary on Euclid's *Elements*; a work entitled *Pneumatica*; *Belopoeiica; Catoptrica;* and *Metrica*, which contains examples of the use of the cog wheel, rack and pinion and multiple pulleys. A Latin translation of the *Pneumatica* was published at Urbino, 1575, while the first edition of the *Belopoeiica* was printed at Augsburg in 1616. Wilhelm Schmidt edited a critical edition of Hero, with a German translation, in five volumes, Leipzig, 1899–1914.[4] It is of interest to record that Hero's *Pneumatica* contains much of the work of Strato, a native of Lampsacus, who became head of the Lyceum at Athens. Strato was a student of natural philosophy, and the author of about forty works, none of which has come down to us.

Diophantus of Alexandria (*c.* A.D. 180) has been described as the greatest Greek writer on algebra. His main work was the *Arithmetica*, of which six of the original thirteen books have survived. This appeared in a Latin translation at Basle, 1575, with a Greek text by Bachet de Méziriac, Paris, 1621. An important edition by Paul Tannery, containing Greek commentaries and a Latin translation, was published in two volumes, Leipzig, 1893–95, and the first English translation, with a commentary by Sir Thomas L. Heath was published at Cambridge in 1885. This appeared

[1] See Sarton, Vol. 1, pp. 172–173; also *La géographie d'Eratosthène*, Paris, 1921; and Thalamas A. *Etude bibliographique de La géographie d'Eratosthène*, Paris, 1921.

[2] See *The geographical fragments of Hipparchus. Edited and annotated by D. R. Dicks*, 1960; see also Sarton, Vol. 1, pp. 193–195.

[3] See Sarton, Vol. 1, p. 213.

[4] See Sarton, Vol. 1, pp. 208–211.

in a new enlarged edition in 1910, which is the most important study of Diophantus. A French translation by Paul ver Eecke was published as *Les six livres arithmétiques et le livre des nombres polygones*, Paris, 1959. A German translation by G. Wertheim of all the works of Diophantus was published in Leipzig in 1890.[1]

The writings of Claudius Ptolemaeus of Alexandria (*fl.* A.D. 170) covered all branches of science, and he is recognized particularly as astronomer, mathematician, geographer and physicist. His influence extended into the middle of the sixteenth century. His main works are the *Mathematical treatise*, usually known as the *Almagest;* the *Geographical treatise*;[2] the *Optics*; the *Tetrabiblos*, a standard work on astrology; *On planetary hypotheses*; and the *Hand-tables*. The complete writings of Ptolemy in Greek, with a French translation by the Abbé Nicolas B. Halma, were published in Paris, 1816–20, and a further complete edition began publication at Leipzig, three volumes, 1898–1907. Edited by J. L. Heiberg, this includes Ptolemy's minor astronomical works, but Heiberg's translation of *Planetary hypotheses* is incomplete, excluding a section on planetary sizes and distances. Bernard R. Goldstein[3] has published a translation and commentary on this unpublished part, printing the Arabic text, which was also previously unpublished. Goldstein reproduces the British Museum text (B.M. MS. arab. 426 (Add. 7473), dated A.D. 1242), and includes variant readings.

The *Almagest* was first printed at Venice, 1496 (epitome only), with a complete Latin translation by P. Liechtenstein from the Arabic, Venice, 1515. A Greek and French text by Halma was published in two volumes, Paris, 1813–16.[4] The *Geographical treatise* was first printed in a Latin text in Vicenza, 1475, with later editions from Bologna, 1462 [1477]; Florence, 1482; and Rome, 1478. The first Greek edition by Erasmus was published at Basle, 1533.[5] An excellent Greek edition by C. F. A. Noble was published in three volumes, Leipzig, 1843–45; and a Greek and Latin edition by Charles Müller and C. Th. Fischer was published at Paris, 1883–1901. The *Tetrabiblon*, or *Quadripartitum*, was printed at Venice in

[1] See Sarton, Vol. 1, pp. 336–337.

[2] See Schütte, Gudmund. *Ptolemy's maps of northern Europe. A reconstruction of the prototypes*, Copenhagen, 1917; and Neugebauer, O. Ptolemy's Geography, Book VII, Chapters 6 and 7. *Isis*, 50, 1959, pp. 22–29, which provides a translation of these sections.

[3] Goldstein, Bernard R. The Arabic version of Ptolemy's *Planetary hypotheses. Trans. Amer. Philos. Soc.*, N.S.57, iv, 1967.

[4] A facsimile edition of this was printed in 1927. See also Peters, Christian H. F., and Knobel, Edward Ball. *Ptolemy's catalogue of stars. A revision of the Almagest*, Washington, Carnegie Institution, Publication No. 86, 1915.

[5] See also Winsor, Justin. *A bibliography of Ptolemy's Geography*, (Library of Harvard University, Bib. Contrib.), Cambridge, Mass., 1884. [Lists 76 items]; [Eames, Wilberforce]. *A list of editions of Ptolemy's Geography, 1475–1730*, New York, 1886. [Lists 40 items]; Stevens, Henry N. *Ptolemy's Geography. A brief account of all the printed editions down to 1730. . . . Second edition*, 1908. [Lists 60 items.]

1484 and 1493, and in a Greek and Latin edition by Camerarius, Nuremberg, 1535, and also by Melanchthon, Basle, 1533. It was translated into English by J. M. Ashmand, London, 1822. A facsimile reprint of this was printed in Chicago by the Aries Press in 1936. Another translation into English, by F. E. Robbins, is included in the Loeb Classical Library, and was published in 1940. A German translation of Books III–IV was published at Berlin in 1923.[1]

Pedacius Dioscorides, a Greek army surgeon of the first century A.D., studied the flora of the countries visited during the course of his military duties, and was author of the first medical botany. His *De materia medica* influenced the development of the subject for several centuries, and much of his nomenclature is still employed by botanists. The work describes about six hundred plants, and is of interest as a contribution to pharmacy and chemistry. There have been many editions and translations of *De materia medica*. The first Latin translation by Petrus Paduanensis was published at Colle in 1478, to be followed by a Greek edition from Venice in 1499, with a second Greek edition in 1518. An edition by Max Wellmann in three volumes was published at Berlin, 1907–14. A fine manuscript dating from about 512 is beautifully illustrated, and is known as the *Codex Vindobonensis*. It is preserved in the National Library at Vienna, and contains 479 illustrations. The restoration of this has been described by Otto Wächter,[2] and the manuscript has been reproduced in an English translation.[3,4] A Spanish version by Andrés de Laguna (1499–1560), printed in Antwerp, 1555; Salamanca, 1563, 1566, 1570, 1594; Valencia, 1626, 1635, 1636, 1651 and 1695, has been described in a paper by V. Peset,[5] who also records numerous other Spanish editions.

The work of Galen (A.D. 130–200) as anatomist and physician is dealt with elsewhere,[6] but he must be mentioned for his experiments in the fields of comparative anatomy, physiology and embryology. Galen investigated the functions of the arteries, the kidneys, the cerebrum and the spinal cord, and was interested in respiration and pulsation. He is represented by numerous MSS. and many incunabula, and the first Latin edition of his writings was printed in two volumes in Venice,

[1] See Sarton, Vol. 1, pp. 272–278; also Zinner, Ernst. Cl. Ptolemaeus und das Astrolab. *Isis*, 41, 1950, pp. 286–287.

[2] Wächter, Otto. The 'Vienna Dioskurides' and its restoration. *Libri*, 13, 1963, pp. 107–111.

[3] *The Greek herbal of Dioscorides, illustrated by a Byzantine A.D. 512. Englished by John Goodyer A.D. 1655, edited and first printed A.D. 1933 by Robert T. Gunther*, [*etc.*], Oxford, *for the author*, 1934, reprinted in 1959, (396 illustrations).

[4] See Sarton, Vol. 1, pp. 258–260; see also, Blunt, Wilfrid. *The art of botanical illustration*, [*etc.*], 1950, pp. 9–12.

[5] Peset, V. Spanish version of Dioscorides' "Materia medica". *J. Hist. Med.*, 9, 1954, pp. 49–58.

[6] Thornton, pp. 10–13.

1490, and the first edition of his *Opera* in Greek was printed by Aldus in five volumes in 1525.[1] The most useful text is the Greek-Latin edition by Carolus Gottlob Kühn, published in Leipzig, twenty volumes in twenty-two, between 1821 and 1833, and reprinted in Hildesheim, 1964–65. Volume one contains a bibliography of Galen's writings up to the beginning of the nineteenth century, compiled by J. C. G. Ackermann (1756–1801). There is a Greek-French edition by Charles Daremberg in two volumes, published in Paris between 1854 and 1856 with the title *Oeuvres anatomiques, physiologiques et médicales*. Critical editions of the Greek text of twelve treatises were published in *Scriptora minora*, 3 volumes Leipzig, 1884–93. Among Galen's minor writings the following are of particular interest: *De anatomicis administrationibus*, a guide to the dissection of animals, chiefly monkeys, in fifteen books. Galen began this work in A.D. 177, but books xii–xv were destroyed by fire in A.D. 192, and they were rewritten by Galen. Greek manuscripts of only the early books survived, but complete Arabic versions are extant. The first eight-and-a-half books were translated into English from the original Greek, by Charles Singer.[2] A German translation from an Arabic manuscript of the complete work was made by Max Simon (Leipzig, 1906), and W. H. L. Duckworth translated the later books into English from this German text, although his text was revised by editors from the Arabic text.[3] Also, *De venorum arteriorumque dissectione: De nervorum dissectione; De musculorum dissectione ad tirones;* and *De ossibus ad tirones*, this last named being the only anatomical treatise dealing with human material.[4] Several translations and studies of Galen's anatomical writings have been contributed by Charles Mayo Goss,[5] who stresses the accuracy of Galen's anatomy, and Margaret Tallmadge May[6] has translated *On the usefulness of the parts of the body*. George Sarton's lectures devoted to Galen contain a chapter on his writings (pp. 25–29), a chronological summary, information on Galenic treatises translated from the Arabic, and on Galenic

[1] Mani, Nikolaus. Die griechische Editio princeps des Galenos (1525), ihre Entstehung und ihre Wirkung. *Gesnerus*, 13, 1956, pp. 29–52.

[2] *Galen on anatomical procedures. . . . De anatomicis administrationibus. Translation of the surviving books with introduction and notes by Charles Singer*, London, [*etc.*], 1956. (Publications of the Wellcome Historical Medical Museum, N.S.7.)

[3] *Galen on anatomical procedures. The later books. A translation by W. H. L. Duckworth. Edited by M. C. Lyons and Bernard Towers*, Cambridge, 1962.

[4] See Singer, Charles. Galen's Elementary course on bones. *Proc. Roy. Soc. Med.*, 45, 1952, pp. 767–776.

[5] Goss, Charles Mayo. On anatomy of veins and arteries, by Galen of Pergamos. *Anat. Rec.*, 141, 1961, pp. 355–366; On the anatomy of muscles for beginners, by Galen of Pergamon. *Anat. Rec.*, 145, 1963, pp. 477–501; The precision of Galen's anatomical descriptions compared with Galenism. *Anat. Rec.*, 152, 1965, pp. 376–380; On anatomy of nerves, by Galen of Pergamon. *Amer. J. Anat.*, 118, 1966, pp. 327–335.

[6] Galen. *On the usefulness of the parts of the body. Translated from the Greek with an introduction and commentary by Margaret Tallmadge May*, 2 vols., Ithaca, 1968.

texts available in English translations.[1] The misconceptions by historians regarding Galen's ideas on physiology have been enumerated by Donald Fleming,[2] who suggests that Galen's views on the motion of the blood in the heart and lungs did not obstruct the work of Harvey, who was not influenced by them in any way. Rudolph E. Siegel[3] has also investigated Galen's physiological ideas, confirming the view that he made many useful observations, and providing a thoroughly documented survey of his treatises, and a useful biography. Richard J. Durling[4] has supplemented, and occasionally corrected, the bibliography of Galen's writings printed up to the beginning of the nineteenth century, compiled by J. C. G. Ackermann (1756–1801) and printed in the first volume of Kühn's Greek-Latin edition of Galen's writings. Durling lists 660 editions and translations.[5] Lynn Thorndike[6] has examined two Latin translations from the Arabic of a spurious work attributed to Galen, on forty-six plants and occult medicine. Two are fifteenth-century manuscripts translated into Latin by Grumerus, one in the Vatican, and the other in the Laurentian Library, Florence; two others are translations by Jacobus Albensis of Lombardy, both in the Vatican.

A follower of Epicurus, Titus Lucretius Carus (*c.* 95–55 B.C.), was a Roman philosopher and scientist who expounded the atomic theory, and taught a theory of evolution vaguely similar to that of natural selection. Lucretius has been described as "the only Latin writer who gives us a complete and coherent scheme of natural knowledge", and his work *De natura rerum* was printed several times during the fifteenth century, e.g., [Brescia, 1473]; Verona, 1486; Venice, 1495 and 1500; and Brescia, 1496. An edition containing an English translation by H. A. J. Munro was published in a fourth edition in three volumes, London, 1905–10, while an edition by William A. Merrill was published at New York in 1907.[7] A translation by Ronald Latham was published as *The nature*

[1] Sarton, George. *Galen of Pergamon*, Lawrence, Kansas, 1954. (Logan Clendening Lectures on the History and Philosophy of Medicine, 3rd series.)

[2] Fleming, Donald. Galen on the motions of the blood in the heart and lungs. *Isis*, 46, 1955, pp. 14–21.

[3] Siegel, Rudolph E. *Galen's system of physiology and medicine. An analysis of his doctrines and observations on bloodflow, respiration, humors and internal diseases*, Basle, New York, 1968; Siegel, Rudolph E. Why Galen and Harvey did not compare the heart to a pump. *Amer. J. Cardiol.*, 20, 1967, pp. 117–121; see also Taylor, D. W. Galen's physiology. *N.Z. med. J.*, 66, 1967, pp. 176–181.

[4] Durling, Richard J. A chronological census of Renaissance editions and translations of Galen. *J. Warburg & Courtauld Institutes*, 24, 1961, pp. 230–305.

[5] See also Wilson, Leonard G. Erasistratus, Galen and the *Pneuma. Bull. Hist. Med.*, 33, 1959, pp. 293–314; Kilgour, Frederick. Galen. *Sci. Amer.*, 196, 1957, pp. 105–114; Thorndike, Lynn. Translations of works of Galen from the Greek by Peter of Abano. *Isis*, 33, 1941–42, pp. 649–653; and Major, Ralph H. Cl. Galen. *Int. Rec. Med.*, 172, 1959, pp. 37–43.

[6] Thorndike, Lynn. The Pseudo-Galen *De plantis* (with Latin text of chapters on stones and those of chemical interest). *Ambix*, 11, 1963, pp. 87–96.

[7] See Sarton, Vol. 1, pp. 205–206.

of the universe in the Penguin Classics series; an edition by D. R. Dudley was published in London in 1965; and the tenth edition of a French version edited by A. Ernout, was published in two volumes, Paris, 1966. G. D. Hadzsits[1] and A. D. Winspear[2] provide information on the life and writings of Lucretius, stressing his originality and influence on later scientific thought.

Only two of the numerous works of Marcus Terentius Varro (116–27 B.C.) have come down to us, *De lingua latina* and *Res rusticae*. He was a country gentleman, wrote an encyclopaedia of science, and books on mensuration, arithmetic and geometry. His *Res rusticae* deals with botany and animal diseases, among other subjects. This work was transcribed by an unknown person in 1329, and later added to other writings to form *Scriptores rei rusticae*, which was first printed by Jenson at Venice in 1472. An English translation by Lloyd Storr-Best was published in London, 1912, and one by Fairfax Harrison in New York, 1913.[3]

Gaius Plinius Secundus (Pliny the Elder) (A.D. 23–79) was the author of *Historia naturalis*, a scientific encyclopaedia that was widely read for many centuries. Pliny mentions his authorities but was uncritical, and the book has been described by Charles Singer as "scientifically worthless".[4] Nevertheless, it is not without merit, and contains a wealth of information not available elsewhere. The work is in thirty-seven books, and was dedicated to the Emperor Titus. It was first printed at Venice, 1469, to be followed by printings at Rome, 1470; Venice, 1472; Rome, 1473; Venice, 1476; Parma, 1476; Treviso, 1479; Parma, 1480, 1481; and numerous other editions. An Italian translation by Christophorus Landinus, *Storia naturale*, was printed at Venice in 1476. At least eighty-nine editions appeared in the sixteenth century, forty-three editions in the seventeenth century, nineteen editions in the eighteenth century, and thirteen editions in the nineteenth century. These figures are provided in a paper by E. W. Gudger,[5] in which he mentions 222 editions. Arnold C. Klebs[6] has criticized Gudger's total of thirty-nine editions published between 1469 and 1499, and lists only eighteen, and it is probable that this figure is the more reliable. A Latin-French bilingual edition appeared in twenty volumes, Paris, 1829–33, while an English translation by John Bostock and H. T. Riley, in six volumes, was published 1855–57. The first English version by Philemon Holland, London, 1601, was issued in

[1] Hadzsits, G. D. *Lucretius and his influence*, New York, 1963.

[2] Winspear, A. D. *Lucretius and scientific thought*, Montreal, 1963.

[3] See Sarton, Vol. 1, pp. 225–226; see also Boissier, Gaston, *Étude sur la vie et les ouvrages de Varron*, Paris, 1861.

[4] Singer, Charles. *The evolution of anatomy*, [*etc.*], 1925, p. 40.

[5] Gudger, E. W. Pliny's Historia naturalis; the most popular natural history ever published. *Isis*, 6, 1924, pp. 269–281.

[6] Klebs, Arnold C. Incunable editions of Pliny's Historia naturalis. *Isis* 24, 1935–36, pp. 120–121.

two volumes [S.T.C. 20029], and appeared in several later editions. Selections from this translation, edited by J. Newsome, were published by the Clarendon Press in 1964. An English translation in ten volumes is in course of publication in the Loeb Classical Library, the first volume appearing in 1938.[1] As a further indication of its popularity, Lynn Thorndike[2] mentions the epitomes of the *Natural history* appearing in fifteenth-century manuscripts. Pliny's book is of interest to the astronomer, chemist,[3] geologist, botanist, zoologist and anthropologist, and he died of a heart attack at Stabiae, some nine miles from Vesuvius, during the eruption that destroyed Pompeii.[4]

Lucius Annaeus Seneca (4 B.C.–A.D. 65), a native of Cordova, went to Rome and became the leader of Stoicism. He was a brilliant rhetorician, and was more philosophical than scientific. His *De quaestionibus naturalibus* contains material on astronomy, geography, geology, meteorology and physics, but is mainly derived from other authorities. Seneca was keenly interested in earthquakes, on which he wrote a treatise which has not survived. His *De quaestionibus naturalibus* was printed at Naples, 1475, and Leipzig, [1494–95], while a critical edition was published in four volumes, Leipzig, 1898–1907. An English translation by John Clarke, annotated by Sir Archibald Geikie, was published in London, 1910.[5]

Although he is remembered chiefly as an historian and a geographer, the writings of Strabo (born *c.* 63 B.C.) contain much of interest to the general scientist. He contributed usefully to botany, geology, biology and mathematics. Editions of his *Geographia* were printed at Rome, [1496]; [Venice], 1472; [Treviso], 1480; [Venice], 1494 and 1494[5]. The first Greek edition was published at Venice in 1516, while scientific critical editions were published by G. Kramer in three volumes, Berlin, 1844–52, and by A. Meineke, three volumes, Leipzig, 1852–53, with a new edition, 1866–77. A Greek and Latin edition by C. Müller and F. Dübner was published in Paris, 1853–58, while English translations by H. C. Hamilton and W. Falconer, three volumes, London, 1854–57, and by Horace Leonard Jones and J. R. S. Sterrett, with the Greek text, were published in the Loeb Classical Library, eight volumes, London, 1917–32.[6]

Anicius Manilius Severinus Boetius (Boëthius) (A.D. 480–524), a Roman, compiled elementary manuals summarizing the mathematical knowledge of Latin authors, those on arithmetic and music being extant. He trans-

[1] See Sarton, Vol. 1, pp. 249–251.

[2] Thorndike, Lynn. Epitomes of Pliny's Natural history in the fifteenth century. *Isis*, 26, 1936, p. 39.

[3] The sections from Pliny's *Natural history* devoted to chemistry have been reproduced in *The Elder Pliny's chapters on chemical subjects.... Edited, with translations and notes, by Kenneth C. Bailey*, 2 vols., 1929–32.

[4] See Zirkle, Conway. The death of Gaius Plinius Secundus (23–79 A.D.). *Isis*, 58, 1967, pp. 553–559.

[5] See Sarton, Vol. 1, pp. 247–249.

[6] See Sarton, Vol. 1, pp. 227–229.

lated some of Aristotle's writings, and wrote *De consolatione philosophiae*, a popular treatise that was widely printed and translated, and which was printed by Caxton about 1479. An English translation was published in the Loeb Classical Library in 1926, and there were numerous earlier editions. His *Arithmetica* was printed at Augsburg, 1488, and his *Opera* at Venice, 1491–92, and also 1497[8]–1499.[1]

A comprehensive survey of Roman science has been published by William H. Stahl,[2] covering six hundred years from Cato the Elder and eight hundred years of the Middle Ages.

An elementary encyclopaedia was compiled by Martianus Mineus Felix Capella [*c.* A.D. 500]. Entitled *Satyricon*, or *De nuptiis philologiae et mercurii et de septem artibus liberalibus*, it deals with the seven arts, including geometry, arithmetic and astronomy. It was published at Vicenza, 1499, Modena, 1500, with an edition by Franz Eyssenhardt, Leipzig, 1866.[3] Bishop Isidore of Seville (*c.* A.D. 560–636) was another encyclopaedist, and from his numerous writings on religion, philology, etc., we must mention his *De natura rerum*, devoted to astronomy, cosmography and meteorology. This was separately printed at Berlin, 1857, with a complete edition by F. Arevala, seven volumes, Rome, 1797–1803.[4]

The Armenians regard Ananias of Širak (born between A.D. 595 and 600) as their most outstanding mediaeval scientist, and he has been called "father of the exact sciences in Armenia". He has been neglected chiefly because his writings have only recently become available, but an article by Robert H. Hewsen[5] evaluates his contributions to science, and gives references to modern editions of his works and to writings about him in Russian and Armenian. Ananias was born in the village of Anonia, the son of John of Širak. He gained a knowledge of mathematics under the teacher Tychicus, and opened a school of his own in Armenia. Conducting research into astronomy, chronology, geography and mathematics, Ananias taught that the world is a sphere, and that when it was day on one side it was night on the other; he described the Milky Way, and appreciated the fact that the moon reflects the light of the sun. Among the numerous writings of Ananias were *Cosmography and the calendar; Table of the motions of the moon; On the course of the sun;* a translation of Paul of Alexandria's *Introduction to astronomy; Problems and solutions; Book of arithmetic; Arithmetic; On weights and measures;* and *Geography.*

The Venerable Bede (A.D. 673–735) is honoured as the father of British history, but he was also interested in science. His *De natura rerum*, based

[1] See Sarton, Vol. 1, pp. 424–426.
[2] Stahl, William H. *Roman science. Origins, development and influence to the later Middle Ages*, Madison, 1962.
[3] See Sarton, Vol. 1, pp. 407–408.
[4] See Sarton, Vol. 1, pp. 471–472.
[5] Hewsen, Robert H. Science in seventh-century Armenia: Ananias of Širak. *Isis*, 59, 1968 pp. 32–45.

on Pliny and Isidore of Seville, deals with natural phenomena, and he also wrote on arithmetic. An English edition of Bede's works by John Allen Giles, was published in twelve volumes, London, 1843–44, volume 6 containing scientific material.[1] Alcuin of York (A.D. 735–804), the eminent educationalist and theologian, wrote numerous works, including *Propositiones ad acuendos invenes*. His writings were collected together and printed in Paris, 1617, and in two volumes, Ratisbon, 1777.[2]

Hrabanus Maurus, or Rabanus, Archbishop of Mainz (*c.* 776–856), a pupil of Alcuin, was the author of the first encyclopaedia to be printed. Written about A.D. 820, *De sermonum proprietate, seu de universo* was printed at Strassburg about 1467. It has been termed the first printed treatise on medicine, but only one chapter is devoted to that subject. Book VII is devoted to animals, fishes and serpents.[3,4]

The reintroduction of the abacus was due to Gerbert, Pope Sylvester II (*c.* 930–1003), who derived his knowledge of arithmetic from Boëthius. He also wrote on the astrolabe and on other scientific subjects. A complete edition of his works was published at Clermont, 1867, and his *Opera mathematica* (972–1003), [*etc.*], was printed in Rome in 1899, and reprinted at Hildesheim in 1963.[5,6]

Greek learning was transmitted through Syriac translations (750–850) and Arabic translations (850–950), much of it coming down to us solely through Arabic versions.[7] Thousands of Arabic MSS. still remain unexamined in the great libraries of Europe, Egypt, Turkey, India and Pakistan, and little is known of Muslim contributions to alchemy. Many of these writings in Arabic were actually the work of Persians and others. The Muslim conquerors set up academies and observatories, and the main Greek writings on astronomy, mathematics, medicine and other sciences were translated into Arabic.[8] The Caliph Al-Mamum (813–833) established a school for translation in Baghdad, and Honain ibn Ishaq (809–

[1] See Sarton, Vol. I, pp. 510–511; also Jones, Charles W. *Bedae Pseudoepigraphia: scientific writings falsely attributed to Bede*, Ithaca, New York, 1939; Browne, G. F. *The Venerable Bede, his life and writings*, 1919; and Jones, Charles W. Manuscripts of Bede's De natura rerum. *Isis*, 27, 1937, pp. 430–440; [lists 66 MSS.].

[2] See Sarton, Vol. I, pp. 528–529.

[3] See Sarton, Vol. I, p. 555.

[4] See Jessup, Everett Colgate. Rabanus Maurus: "De sermonum proprietate, seu de universo". *Ann. med. Hist.*, N.S.6, 1934, pp. 35–41.

[5] See Sarton, Vol. I, pp. 669–671.

[6] See also Clagett, Marshall. *Greek science in antiquity*, New York, 1955; and Sarton, George. *The appreciation of ancient and mediaeval science during the Renaissance (1450–1600)*, Philadelphia, 1955. (The A. S. W. Rosenbach Fellowship in Bibliography.) The latter traces the influence of the ancients on scientists of the Renaissance.

[7] See Mieli, A. *La science arabe et son rôle dans l'évolution scientifique mondiale. Reimpression anastatique augmentée d'une bibliographie avec index analytique par A. Mazahéri*, Leiden, 1966; Farrukh, O. A. *Arab genius in science and philosophy*, Washington, 1954; O'Leary, De Lacy. *How Greek science passed to the Arabs*, (1948) [1949], [bibliography pp. 189–192]; Winter, H. J. J. The Arabic achievement in physics. *Endeavour*, 9, 1950, pp. 76–79; and Youschkevitch, A. P. Recherches sur l'histoire des mathématiques au moyen-âge dans les pays d'orient: bilans et perspectives. *History of Science*, 6, 1967, pp. 41–58.

[8] Holmyard, E. J. Alchemy in medieval Islam. *Endeavour*, 14, 1955, pp. 117–125.

877) translated most of the Galenic writings into Arabic, and also commenced the translation of Ptolemy's *Almagest*, and Aristotle's works. With his pupils he prepared versions of the Hippocratic writings, and of several mathematical and astronomical works. It was at Baghdad that many Greek alchemical writings,[1] and also texts from India and Persia were made available in Arabic. The outstanding Arabic alchemist, Jābir ibn Hāyyan (*c.* A.D. 722–803 or 813?), the Geber of the Middle Ages, was probably born at Tus, and spent much of his life at Kufa. E. J. Holmyard[2] has given an account of his background and the part he played in the development of alchemy in the light of recent research, although much is still conjectural. It is clear that many of the works attributed to Jābir cannot have been written by him in the form they have reached us, although some of the Arabic works exist in mediaeval Latin translations. He wrote books on medicine, a commentary on Euclid, on mirrors, logic and poetry, but his chemical investigations are of first importance. Among these are the *Book of properties*; the *Sum of perfection*; and the *Chest of wisdom*, which contains the earliest known recipe for the preparation of nitric acid. Early editions of his chemical writings include printings at [Rome, 1473? and about 1520]; Strassburg, 1528, 1529, [1530?], 1531; Nuremberg, 1541, 1545; Venice, 1542; Nuremberg, 1545, [1562?]; Basle, 1572; Strassburg, 1598, [1649 and 1670?]. An Arabic text with German translation of his *Book of poisons* has been edited by Alfred Siggel.[3] English versions of Jābir's works were published by William Salmon, London, 1692, and by Richard Russell, London 1678 and 1686, of which latter translation a new edition with an introduction by E. J. Holmyard was published in London, [*etc.*], 1928.[4,5]

Al-Rāzī, or Rhazes (860–932), is chiefly known for his medical

[1] Greek alchemy, and the transmission of ancient science to the Middle Ages, have been the subject of studies by Marcelin Berthelot, *Les origines de l'alchimie*, Paris, 1885; *Introduction à l'étude de la chimie des anciens et du moyen âge*, [*etc.*], Paris, 1889; *Histoire des sciences. La chimie au moyen âge*, [*etc.*], 3 vols., Paris, 1893, reprinted Amsterdam, 1967; *Collection des anciens alchimistes grecs*, [*etc.*], 3 vols., Paris, 1887–88, reprinted 1963. See also Taylor, F. Sherwood. A survey of Greek alchemy. *J. Hellenic Stud.*, 15, 1930, pp. 109–139; and Wilson, William Jerome. The origin and development of Greco-Egyptian alchemy, Special number of *Ciba Symposia*, 3, August, 1941.

[2] Holmyard, E. J. *Alchemy*, Harmondsworth, 1957, pp. 66–80, 131–138, *etc.* Pelican Books.

[3] Das Buch der Gifte des Ǧābir ibn Hayyān. Arabischer Texte in Faksimile. . . . Übersetzt und erläutert von Alfred Siggel. *Veröffentlichungen der orientalischen Kommission der Akademie der Wissenschaften und der Literatur*, Bd. 12, 1958.

[4] See also Darmstaedter, Ernst. *Die Alchemie des Geber*, [*etc.*], Berlin, 1922; and Kraus, Paul. *Jābir ibn Hāyyan. Contribution a l'histoire des idées scientifiques dans l'Islam*. . . . Vol. 1. *Le corpus des écrits jabiriens*. (Vol. 2. *Jābir et la science grecque*.) (*Mémoires presentés à l'Institut d'Égypte*, Tomes 44–45), Cairo, 1942–43; Ruska, J. *Arabische Alchemisten*, Vol. 1, Heidelberg, 1924; Haschmi, Mohamed Yahia. The beginnings of Arab alchemy. *Ambix*, 9, 1961, pp. 155–161, which suggests an earlier date than advanced by Kraus and by Ruska; and Stapleton, H. E. The antiquity of alchemy. *Ambix*, 5, 1953, pp. 1–43. This includes a discussion of possible connections between Chinese and Jabirian alchemy.

[5] See Ferguson, Vol. 1, pp. 209–304; Sarton, Vol. 1, pp. 532–533; Holmyard, E. J. Jābir

writings,[1] but he also wrote on astronomy, chemistry, mathematics, philosophy and physics. Rhazes divided mineral bodies into six classes, and his book on alchemy, *Secretum secretorum*, was translated by Julius Ruska, Berlin, 1937. A hitherto unknown text was discovered in Tashkent in the Institute of Oriental Studies of the Academy of Sciences of Uzbek SSR in 1950. This was a transcript, completed on 7th June, 1506, of an authentic manuscript of Al-Rāzī, "The Book of the Secret of Secrets" (*Kitāb sirr al-asrār*), of which the "Book of Secrets" (*Kitāb al-asrār*) is an abridged version. This has been published with a Russian translation by U. I. Karimov, and a facsimile of the Arabic manuscript.[2] Rhazes' *De alimunibus et salibus*, also known under the title *De spiritibus et corporibus*, was translated by Gerard of Cremona,[3] and Rhazes is said to have been the author of over two hundred works, few of which have appeared in print.

Abū Kāmil Shujāᶜ ibn Aslam (850–930) was the author of works on integral solutions of differential equations, on the pentagon and decagon, and an outstanding book on algebra, which has been translated into English by Martin Levey[4] and published together with the Hebrew text. This is based on four extant manuscripts in Latin, Hebrew (2) and Arabic. The earliest known arithmetic extant in Arabic has come down to us in a unique copy preserved in the Yeni Gami Library, Istanbul, and was written in Damascus in 952–3 A.D. The author, Abu al-Hasan, Ahmad ibn Ibrāhīm al-Uqlīdisī provides information on early Hindu-Arabic arithmetic, on the cube and the cube root, and he grasped the decimal idea long before Stevin. Al-Uqlīdisī used a stroke (/) as a decimal point, and wrote on decimal fractions.[5]

Another physician, Abū 'Alī al-Husain ibn 'Abd-Allāh ibn Hasan ibn 'Alī ibn Sīnā, generally known as Avicenna (980–1037), was born at Kharmaithan, a large village near Bokhara, and is best remembered for

ibn Hayyān. *Proc. Roy. Soc. Med.*, 16, Parts 1–2, 1923, Sect. Hist. Med., pp. 46–57; and Iskandar, A. Z. *A catalogue of Arabic manuscripts on medicine and science in the Wellcome Historical Medical Library*, 1967.

[1] See Thornton, p. 16.

[2] [*An unknown work of Al-Rāzī, "The Book of the Secret of Secrets". By U. I. Karimov. Publication of the Academy of Sciences of the Uzbek SSR.* Tashkent, 1957.] (In Russian.) Reviewed in *Ambix*, 10, 1962, pp. 146–149.

[3] See Steele, Robert. Practical chemistry in the twelfth century. Rasis de aluminibus et salibus. Translated by Gerard of Cremona. *Isis*, 12, 1929, pp. 10–46; Ruska, Julius. Pseudepigraphe Rasis-Schriften. *Osiris*, 7, 1939, pp. 31–94; and Leibowitz, J. O. Manuscript notes in a Rhazes-Maimonides incunable, 1497. An appreciation of Rhazes and an exposition of the anatomy of the heart. *Bull. Hist. Med.*, 39, 1965, pp. 424–434.

[4] Levey, Martin, *ed. & trans. The algebra of Abū Kāmil, Kitāb fī al-jābr wa'l-mugābala, in a commentary by Mordecai Finzi*, [*etc.*], Madison, 1966.

[5] See Saidan, A. S. The earliest extant arithmetic. *Kitāb al-Fusūl fī al Hisab al-Hindī* of Abū al-Hasan, Ahmad ibn Ibrāhīm al-Uqlīdisī. *Isis*, 57, 1966, pp. 475–490.

his medical writings,[1] but was the author of numerous books devoted to astronomy, mathematics and philosophy, some being in Persian and the others in Arabic. There are two bibliographies of his writings,[2] and a biography by Soheil M. Afnan[3] also contains lists of writings on aspects of his work, of source-books on his life, of MS. texts, printed Arabic and Persian texts, etc. Avicenna was the author of an important philosophical encyclopaedia, *Kitāb al-shifā*,[4] a work on logic entitled *Kitab al-ishārāt wal-tanbīhāt*; and numerous other writings. He was one of the most eminent philosophers of his period, and his writings have appeared in numerous editions and translations.[5] The encyclopaedia, or portions thereof, have been frequently printed, and the mineralogy portion of this was translated into English in 1927.[6] His collected writings were printed as *Opera*, Venice, 1508, of which a reprint was published in 1961, and a German translation of his work on metaphysics was published by M. Horten in 1907 as *Die Metaphysik Avicennas. Das Buch der Genesung der Seele. Eine philosophische Enzyklopädie Avicennas. Die Metaphysik, Theologie, Kosmologie und Ethnik*, a reprint of which was announced in 1969. Two alchemical treatises attributed to Avicenna have been translated into English,[7] and possibly represent his views before he wrote the *Shifā*.[8]

A Persian, Muhammad ibn Mūsā al-Khowārizmī, or Al-Kwarizmi (*c.* A.D. 830), wrote an *Arithmetic* that was of primary significance, an *Algebra* that led to the use of this term in the mathematical sense, and also compiled astronomical tables.[9] His *Algebra* was the earliest Arabic book on mathematics, and in the twelfth century was translated into Latin by Gerard of Cremona and by Robert of Chester. There is no trace of Euclid in his geometry, but S. Gandz,[10] in a study of Al-Khowārizmī's

[1] See Thornton, p. 17, for his medical works.

[2] See Anawati, G. C. *Essai de bibliographie avicennienne*, Cairo, 1950.

[3] Afnan, Soheil M. *Avicenna: his life and works*, 1958.

[4] The psychological section of this has been critically edited by Fazlur Rahman as *Avicenna's De anima*, [*etc.*], (University of Durham publications), Oxford, 1959.

[5] See Avicenna. *Le livre de science. Tome 1. Logique; metaphysique.* [*Tome 2. Physique; Mathématique*], [*etc.*], Paris, 1955–58.

[6] *De congelatione et conglutinatione lapidum. Being sections of the Kitab al-shifa. The Latin and Arabic texts edited with an English translation of the latter and with critical notes by E. J. Holmyard and D. C. Mandeville*, Paris, 1927.

[7] See Stapleton, H. E., Azo, R. F., Husain, M. Hidāyat, and Lewis, G. L. Two alchemical treatises attributed to Avicenna. *Ambix*, 10, 1962, pp. 41–82.

[8] See also Ruska, Julius. Die Alchemie des Avicenna. *Isis*, 21, 1934, pp. 14–51; and Soubiran, André. *Avicenne, prince des médecins; sa vie et sa doctrine*, Paris, 1935.

[9] *The astronomical tables of Al-Khwārizmī. Translation with commentaries of the Latin version by H. Suter supplemented by Corpus Christi College MS.* 283, *by O. Neugebauer*, Copenhagen, 1962. (*Hist. Filos, Skr. Dan. Vid. Selsk.*, 4, ii); see also Millás-Vallicrosa, J.-M. La autenticidad de comentario a las Tablas astronómicas de Al-jwárizmi por Ahmad ibn al-muttannā. *Isis*, 54, 1963, pp. 114–119.

[10] Gandz, S. The sources of al-Khowarizmi's algebra. *Osiris*, 1, 1936, pp. 263–277; see also Ruska, Julius. Zur ältesten arabischen Algebra und Rechenkunst. *Sitzungsberichte der Heidelberger Akademie der Wissenschaften, Philosophisch-historische Klasse*, 1917, pp. 1–125.

sources, considers his algebra to be "the foundation and cornerstone of the science". He was also the author of an encyclopaedia of the sciences, *Keys of the sciences*, which has been investigated and summarized by C. E. Bosworth.[1] Messahala, or Māshāllāh (770–820), flourished in Baghdad, and wrote on astronomy and astrology, but none of his works is preserved in the original Arabic. There are numerous Latin translations under various titles, and extracts under new titles, and his astrological writings in these forms have been investigated by Lynn Thorndike.[2] Messahala's *De scientia motus orbis* was published in Nuremberg in 1504 and 1549, the second edition being entitled *De elementis et orbibus coelestibus*.[3] The greatest astronomer of his period, Al-Battani, or Albategnius (*c.* 858–929), made numerous important discoveries, compiled a catalogue of fixed stars, and was the author of several astronomical books. The major work was *De scientia stellarum; De numeris stellarum et motibus*. His astronomical writings were translated into Latin by Plato of Tivoli, and published at Nuremberg, 1537, while the Arabic text, with a Latin translation and notes by C. A. Nallino, was published in three volumes, Milan, 1899–1907; and reprinted in 1967.[4]

Abū Yūsuf Ya'qūb ibn Ishāq al-Kindī, generally known as Al-Kindī (*c.* 800–*c.* 873) was born in the early years of the ninth century at Kufai now in Iraq, and flourished in Baghdad. He was the earliest Arabic writer on physics, and has been described as the "first philosopher of the Arabs". His writings number about two hundred and sixty-five, many of which are lost, and cover astrology, mathematics, medicine, pharmacy, physics and other subjects. The most important was the work on optics and the reflection of light, *De aspectibus*.[5] Sami Hamarneh[6] gives the locations of some Arabic manuscripts of his writings, and one previously unknown was discovered among hundreds of uncatalogued scientific manuscripts in Istanbul in the Aya Sofya Kutubkhane (MS. 3603, fols. 91b–139a). Entitled *The Aqrābādhīn*, this deals with electuaries, poultices, gargles, clysters, ointments, etc., and has been described by Martin Levey.[7] Commentaries on and translations of two treatises by Al-Kindī,

[1] Bosworth, C. E. A pioneer Arabic encyclopedia of the sciences: al-Khwārizmī's Keys of the sciences. *Isis*, 54, 1963, pp. 97–111.
[2] Thorndike, Lynn. The Latin translations of astrological works of Messahala. *Osiris*, 12, 1956, pp. 49–72.
[3] See also Sarton, Vol. 1, p. 531.
[4] See Sarton, Vol. 1, pp. 602–603.
[5] See Sarton, Vol. 1, pp. 559–560.
[6] Hamarneh, Sami. Al Kindī, a ninth-century physician, philosopher and scholar. *Med. Hist.*, 9, 1965, pp. 328–342.
[7] Levey, Martin. The Aqrābādhīn of al-Kindi and early Arabic chemistry. *Chymia*, 8, 1962, pp. 11–20.

Epistle on the concentric structure of the universe,[1] and *Epistle on the finitude of the universe*,[2] have been published.

Al-Bitrūjī (Alpetragius) was the author of a book on the configuration of the heavenly bodies, which was translated into Latin by Michael Scot in 1217, into Hebrew by Moses ibn Tibbon in 1259, this latter version having been printed with other works at Venice in 1531. Michael Scot's translation, edited by Francis J. Carmody, was published in 1952.[3] This work was composed about 1185, but little else is known about Al-Bitrūjī.

An Arabian of Turkish stock, Al-Fārābī (870–960) studied mathematics, medicine and philosophy in Baghdad. He wrote on the fundamental principles of science, on music, and a book on gems, which was translated in 1906.

Abū Hāyyan al-Tauhīdī (died *c.* 1009) was the author of many writings now lost, while others are preserved in unique manuscripts. His *Kitāb al Imta' wal-Mu'ānasa*, "The Book of Enjoyment and Entertainment", is divided into forty chapters or 'nights', corresponding to meetings. The complete work was published in Cairo in three volumes, 1939–44, and the zoological chapter has been translated into English.[4]

Of Persian parentage, Al-Bīrūnī (973–1048) travelled extensively in India, and finally settled in Afghanistan. He studied medicine, and was one of the greatest scientists of his period. Al-Bīrūnī became deeply interested in Sanskrit literature, and wrote a *Chronology of ancient nations*, translated into English in 1879, and a history of religion, philosophy, astronomy, astrology, geography, etc., of India, written about 1030, and translated into English in 1888. He was the author of an encyclopaedia of astronomy and other works, some of which have been published as *Rasa'ilu'l-Biruni. Containing four tracts based on the unique compendium of mathematical and astronomical treasures in the Oriental Public Library, Bankipore*, Hyderabad, 1948.

Martin Levey[5] has provided a translation of an interesting work by Al-Muᶜizz ibn Bādīs (*c.* 1025) entitled "Staff of the scribes and implements of the discerning with a description of the line, the pens, soot inks, *liq*,

[1] Khatchadourian, Haig, and Rescher, Nicholas. Al-Kindī's Epistle on the concentric structure of the universe. *Isis*, 56, 1965, pp. 190–195.

[2] Rescher, Nicholas, and Khatchadourian, Haig. Al-Kindi's Epistle on the finitude of the universe. *Isis*, 56, 1965, pp. 426–433; see also Rescher, Nicholas. *Al-Kindī: an annotated bibliography*, Pittsburgh, 1965.

[3] *De motibus coelorum. Critical edition of the Latin translation of Michael Scot. Edited by Francis J. Carmody*, Berkeley, [*etc.*], 1952.

[4] Kopf, L. The zoological chapter of the *Kitāb al-Imta' wal-Mu'ānasa* of Abū Hayyān al-Tauhīdī (10th century). (Translated from the Arabic and annotated.) *Osiris*, 12, 1956, pp. 390–466.

[5] Levey, Martin. Mediaeval Arabic bookmaking and its relation to early chemistry and pharmacology. *Trans. Amer. Philos. Soc.*, 52, iv, 1962.

gall inks, dyeing, and details of bookbinding", with an introduction on inks, dyes, glues and paper.

The work of Job of Edessa, or Ayyūb ar-Ruhāwi (*c.* A.D. 760–*c.* A.D. 835) must also be mentioned, as a modern translation of his writings summarizes all natural and philosophical science as known and taught in Baghdad during his times. He translated many Greek works, and was a prolific writer both in Arabic and Syriac, but only the latter survive. His *Book of treasures*[1] was written about A.D. 817, and deals with astronomy, chemistry, mathematics and physics, among other subjects.

The golden age of Arabic literature extended through the tenth century and into the early part of the eleventh. Alhazen, Ibn Al-Hasan, Ibn al-Haytham, or abu-'Ali al-Hasan ibn al-Haythan (965–1038), a native of Basra, was pre-eminent as a physicist and student of optics, but also contributed to mathematics, astronomy and medicine. He wrote commentaries on Aristotle and Galen, a book *On the burning-sphere* containing the first mention of the camera obscura, and his magnum opus, *Optics*. His book on twilight, *De crepusculis et nubium ascensionibus*, was printed in Lisbon, 1542, with a work by another author, and republished with the *Optics* at Basle, 1572.[2] Matthias Schramm[3] has made a thorough study of Ibn al-Haytham's work in physics, suggesting that he was probably the greatest of Arab scientists, and recording later physicists who interpreted and popularized his findings. The most significant of these was Witelo (Vitellius) (1220–80), a native of Silesia who was educated in Paris, Padua and Viterlo, and who derived his optics largely from Ibn al-Haytham. Witelo was the author of *De natura daemonum; De intelligentiis; Optics*, or *Perspectivae;* and *De primaria causa paenitentiae*. He died in the Monastery of Witow, near Piotrkow, Poland. It has been suggested by A. I. Sabra[4] that *De crepusculis* is in fact the work of an Andalusian mathematician, Abū 'Abd Allāh Muhammad ibn Mu'ādh (born A.D. 989–90), who was the author of several important mathematical works. The evidence presented by Sabra is convincing.

The Brethren of Purity were established at Basra about A.D. 983 as a secret association, and were responsible for fifty-two treatises, on mathematics, logic, natural sciences, metaphysics, etc. This encyclopaedia was widely read, and its influence spread as far as Spain. It was printed at

[1] Job of Edessa. *Encyclopaedia of philosophical and natural sciences as taught in Baghdad about A.D.* 817; *or, Book of treasures. . . . Syriac text edited and translated with a critical apparatus by A. Mingana*, [*etc.*], Cambridge, 1935. Contains translation and also facsimile of Syriac text.

[2] See Sarton, Vol. 1, pp. 721–723.

[3] Schramm, Matthias. *Ibn Al-Haythan's Weg zur Physik*, Wiesbaden, 1963. (Boethius. Texte und Abhandlungen zur Geschichte der exakten Wissenschaften, Bd. 1); see also Sarton, Vol. 1, pp. 721–723; and Lindberg, David C. Alhazen's theory of vision and its reception in the West. *Isis*, 58, 1967, pp. 321–341.

[4] Sabra, A. I. The authorship of *Liber de crepusculis*, an eleventh century work on atmospheric refraction. *Isis*, 58, 1957, pp. 77–85.

Leipzig, 1886, and portions of it have been translated, but there is no English version.[1]

Hasdai ibn Shaprut (*c.* 915–*c.* 990) was a Jewish physician who greatly assisted in the development of science in Spain. He flourished at Cordova, translated into Arabic a MS. of Dioscorides, and extended the influence of Jewish science. A library and academy were founded at Cordova in 970, and these were also formed at Toledo and other centres. Arzachel, or Al-Zarquālī (*c.* 1029–*c.* 1087), of Cordova, worked mainly at Toledo, and was the most prominent astronomer of his period. He invented an improved astrolabe, and edited the Toledan Tables, which were immensely popular.[2]

Although an eminent physician, Moses Maimonides, or Rabbi Moses ben Maimon (1135–1204), was also a philosopher, theologian and astronomer. He wrote a commentary on the astronomy of Jābir ibn Aflah, and a treatise on the Jewish calendar.[3] Maimonides was the author of a treatise on poisons, which contains much interesting zoological information.[4] His writings were originally in Arabic, and early translations were into Latin and ancient Hebrew, but French, German, Italian and English versions are becoming available.[5]

Another physician, Averroës, or ibn Rushd (1126–98), was one of the greatest of Muslim philosophers. He wrote commentaries on Aristotle's works, a medical encyclopaedia, a treatise on the motion of the sphere, and a summary of the Almagest.[6] His commentaries on Aristotle appeared in the latter's *Opera*, published in Padua, 1472–74; and Venice, 1483, 1489 and 1495–96.[7]

Hermann the Lame, or Hermann of Reichenau (1013–54), where he was Abbot of the Benedictine monastery, wrote several mathematical and astrological works that were widely used in the following century. These included a treatise on the abacus, a work on the arithmetical game called rithmomachia, two books on the astrolable, *De mensura astrolabii* and *De utilitatibus astrolabii*,[8] and a treatise on music.[9]

The fascinating early mediaeval bestiaries were immensely popular, manuscript copies, translations and printed versions having been in circulation for centuries. The *Physiologus* by Bishop Theobald, Abbot

[1] See Sarton, Vol. 1, pp. 660–661.
[2] See Sarton, Vol. 1, pp. 758–759.
[3] See Sarton, Vol. 2, Part 1, pp. 369–380.
[4] See Théodoridès, Jean. Les sciences naturelles et particulièrement la zoologie dans le "Traité des poisons" de Maimonide. *Rev. Hist. Méd. hébr.*, No. 31, 1956, pp. 87–104.
[5] See Thornton, pp. 18–19; see also Rosner, Fred. Moses Maimonides (1135 to 1204). *Ann. intern. Med.*, 62, 1965, pp. 372–375.
[6] A list of his writings is contained in Renan, Ernest. *Averroës et l'averroisme*, 3rd ed., 1869.
[7] See Sarton, Vol. 2, Part 1, pp. 355–361.
[8] See Drecker, J. Hermannus Contractus Über das Astrolab. *Isis*, 16, 1931, pp. 200–219.
[9] See Sarton, Vol. 1, p. 757.

of Monte Cassino from A.D. 1022–35, was the immediate ancestor of the bestiary, and it went into innumerable editions, including printings from Cologne, [before 1489], 1492 and 1502. A translation of the 1492 edition was published by Alan Wood Rendell, London, 1928. Another form of the *Physiologus*, the *Dialogus creaturarum*, was first printed at Gouda, in Holland, in 1480, with a modern edition published in Munich as *Die Zweisprach der Tiere* in 1923. A Latin bestiary of the twelfth century was presented to Cambridge University Library[1] by Osbert Fowler in 1655, and was edited for the Roxburghe Club in 1928 by M. R. James. This has also been translated and edited by T. H. White.[2] The Dutch work *Der dieren palleys*, Antwerp, 1520, is derived from the *Liber bestiarum*, and an English translation of the former by Laurence Andrewe was published as *The noble lyfe & natures of man, of bestes, serpentys, fowles & fisshes*, Antwerp, [*c.* 1521]. This translation is very rare,[3] and in publishing a facsimile, Noel Hudson[4] has proved its association with the *Hortus sanitatis*, which has never been completely translated into English, and has not been printed for over four hundred years. A facsimile reproduction of an early sixteenth-century bestiary, the *Libellus de natura animalium*, Mondovi, 1508, has also recently been published.[5] The original of this is also rare, and it is believed that only three copies exist.[6]

A prominent translator from Arabic into Latin was an Englishman, Adelard of Bath (*c.* 1090–*c.* 1150). He was a philosopher, mathematician and scientist, and travelled extensively in France, Spain, Italy, Sicily and the Near East. He translated a treatise on the abacus, the *Arithmetic* of Al-Kwarizmi, and Euclid. Adelard was the author of *De eodem et diverso*, and *Quaestiones naturales*, which represents a compendium of Arabic science. Also, a treatise on the abacus, *Regule abaci*, a book on the astrolabe, *De opere astrolapsus*, and the earliest Latin treatise on falconry. The *Quaestiones naturales* was printed at [Louvain, 1475], 1484 and 1490, and an English version, edited by Hermann Gollancz, was printed at Oxford, 1920.[7]

[1] Cambridge University Library, MS. li.4.26.

[2] *The book of beasts. Being a translation from a Latin bestiary of the twelfth century, made and edited by T. H. White*, 1954.

[3] It is believed that copies exist only in Cambridge University Library and the Wellcome Historical Medical Library.

[4] *An early English version of Hortus sanitatis: a recent bibliographical discovery by Noel Hudson*, 1954 [1955].

[5] *Libellus de natura animalium, Mondovi, 1508. A book of the nature of animals. A facsimile reproduction of an early sixteenth-century Italian bestiary illustrated with woodcuts. With an introduction by J. I. Davis*, 1958.

[6] One is in the Bodleian Library, Oxford.

[7] See Haskins, Charles Homer. *Studies in the history of mediaeval science. . . . Second edition*, Cambridge, Mass., Chapter II, Adelard of Bath, pp. 20–42, *etc.*; also Sarton, Vol. 2, Part 1, pp. 167–169.

Another Englishman, Robert of Chester, or Robert the Englishman (*c.* 1110–*c.* 1160), also translated from Arabic into Latin, and was noted as an alchemist, astronomer and mathematician. For some years he lived in Spain, but returned to live in London in 1147. Robert of Chester translated Al-Kindi's *Judicia*; the Koran; a treatise on alchemy, *Liber de compositione alchemiae*, possibly the first alchemical work to appear in Latin; the *Algebra* of Al-Khwarizmi,[1] a treatise on the astrolabe; and compiled astronomical tables for the longitude of London, 1149–50, based on Albategnius and Al-Khwarizmi, and for the latitude of London, based on the tables of Al-Khwarizmi, translated by Adelard of Bath. The translation of the Koran was first printed in three volumes at Basle, 1543; *De compositione alchemie* at Basle, 1559 and 1593, and Paris, 1564.[2]

John of Seville (*fl.* 1139–55) worked at Toledo, and made many translations from Arabic into Latin, several in collaboration with Domingo Gundisalvo or González (*fl.* 1140). They translated numerous Arabic treatises on astronomy, astrology, mathematics, medicine and philosophy, and most have been printed.[3] But the greatest of translators from Arabic into Latin was undoubtedly Gerard of Cremona or Gherardo Cremonese (1114–87). He studied Arabic at Toledo, and is said to have translated over ninety Arabic works, including the *Almagest* of Ptolemy, Avicenna's *Canon*, Archimedes's *On the quadrature of the circle*, many of Aristotle's works, Euclid's *Elements*, Galen's medical writings, and those of Hippocrates, Rhazes and Albucasis, together with the works of Geber, Al-Kindī, Alfargani, and many others. George Sarton provides a complete list of these translations.[4]

Little is known about the development of early science in Russia, which had contacts with Byzantium and the east from early times and with western Europe from the sixteenth century. A book by Alexander Vucinich[5] reveals something of the history of science as a social activity, suggesting that much remains to be investigated, particularly manuscript material.

Early in the thirteenth century two religious orders were founded that influenced the development of the universities, and provided the majority of the great university teachers. The Dominicans, or Black Friars, were founded at Toulouse in 1215 by Dominic (1170–1221), while the

[1] See Karpinski, Louis Charles. *Robert of Chester's Latin translation of the Algebra of Al-Khowarizmi*, [*etc.*], New York, 1915. (University of Michigan Studies, Humanistic Series, Vol. IX.)

[2] See Sarton, Vol. 2, Part 1, pp. 175–177. Confusion between the individuals known as Robert of Chester, Robert Anglicus, and others with similar names is mentioned in Lynn Thorndike's Robertus Anglicus. *Isis*, 34, 1942–43, pp. 467–469.

[3] See Sarton, Vol. 2, Part 1, pp. 169–173.

[4] See Sarton, Vol. 2, Part 1, pp. 338–344.

[5] Vucinich, Alexander. *Science in Russian culture. A history to 1860*, 1965; see also Ryan, W. F. Science in medieval Russia: some reflections on a recent book. *History of Science*, 5, 1966, pp. 52–61.

Franciscans, or Grey Friars, were formed in 1209 by Francis of Assisi. Alexander of Hales, Robert Grosseteste, Roger Bacon and Bartholomew the Englishman were Franciscans, while Albertus Magnus, St. Thomas Aquinas, William of Moerbeke and Vincent of Beauvais belonged to the Dominicans.

Michael Scot (*c.* 1175–*c.* 1234) was born in Scotland, but spent much of his time at Toledo, Padua, Bologna and Rome, and for a considerable period was astrologer and translator to Frederick II. His name became associated with sorcery and black magic, but he translated several important works from Arabic into Latin. These included the writings of Alpetragius, Averroës, Aristotle, Avicenna and others, while many spurious works are attributed to him. Michael Scot was the author of *Liber introductorius*, and *Liber particularis*, both on astrology; *Physionomia*, also known as *Liber physiognomiae*, or *De secretis naturis;* and probably two works on alchemy. The *Liber physiognomiae* was printed at [Venice], 1477; [Lyons, 1480]; Louvain, [1484?]; [Toulouse, 1485]; [Cologne, 1485]; together with numerous other editions; his *Expositio super auctoren sphaerae cum quaestionibus*, Bologno, 1495 and 1518. S. Harrison Thomson[1] has reproduced three alchemical texts attributed to Michael Scot, a list of his MSS. preserved in this country is provided in an article by John D. Comrie,[2] Charles H. Haskins[3] has provided information on Michael Scot's relationship with Frederick II, while Sarton[4] and Ferguson[5] give extensive lists of his works. A biographical study by Lynn Thorndike[6] deals exhaustively with Scot's life, times and beliefs, and is based on an extensive knowledge of the period.

Frederick II of Hohenstaufen (1194–1250), Holy Roman Emperor, King of Sicily and Jerusalem, was a keen patron of learning, and himself studied mathematics, natural history and philosophy. He founded the University of Naples in A.D. 1224. Frederick II's book on falconry was the earliest scientific treatise on zoology, and represents an important treatise on comparative ornithology and avian anatomy. *De arte venandi cum avibus* was composed between 1244 and 1250, and frequently cites Aristotle, although the book is mainly based on personal observation and experience. It contains six books, only two of which appeared in

[1] Thomson, S. Harrison. The texts of Michael Scot's Ars Alchemie. *Osiris*, 5, 1938, pp. 523–559; see also Haskins, Charles H. The "Alchemy" ascribed to Michael Scot. *Isis*, 10, 1928, pp. 350–359.

[2] Comrie, John D. Michael Scot: a thirteenth-century scientist and physician. *Edinb. med. J.*, N.S.25 1920, pp. 50–60.

[3] Haskins, Charles H. Michael Scot and Frederick II. *Isis*, 4, 1921–22, pp. 250–275; revised version in his *Studies in the history of mediaeval science. . . . Second edition*, Cambridge, Mass., 1927, Chapter XIII, pp. 272–298.

[4] Sarton, Vol. 2, Part II, pp. 579–582.

[5] Ferguson, Vol. 2, pp. 355–360; also Ferguson, John. *A short biography and bibliography of Michael Scotus*, Glasgow, *for Glasgow Bib. Soc.*, 1931.

[6] Thorndike, Lynn. *Michael Scot*, London, [*etc.*], 1965.

print until the complete work was translated and edited by Casey A. Wood and F. Marjorie Fyfe[1] in 1943. M. F. M. Meiklejohn[2] suggests that this edition "contains possibly more errors than any book ever published on birds", and that the work deserves an authoritative edition as "the best book written about birds between Aristotle's *Historia animalium* and Gilbert White's *Natural history of Selbourne*". Wood and Fyfe's edition lists and describes six six-book manuscripts and six two-book manuscripts, in addition to the four printed texts. The first printed edition, edited by Johann Velser and published at Augsburg, 1596, was reprinted with a commentary by Johann Gottlieb Schneider, two folio volumes, Leipzig, 1788–89. German translations have been prepared by Johann Erhard Pacius, Onolzbach, 1756, and by H. Schöpffer, Berlin, 1896. Charles H. Haskins[3] has contributed an important paper on Frederick II and his book.

The writings of Michael Scot were rivalled in popularity by those of Albertus Magnus, or Albrecht von Bollstädt (1193–1280). The latter was a great Dominican teacher, philosopher and scientist, who wrote numerous treatises that were later published in twenty-one folio volumes at Leyden, 1651, while the first volume of a new complete critical edition in forty volumes, which will include a number of works previously unpublished, commenced publication in 1951. This edition is the work of the Albertus Magnus Institute at Cologne, under the direction of Bernhard Geyer. Albertus was undoubtedly one of the greatest teachers of the Middle Ages, and the popularity of his writings survived long after his death. His scientific knowledge was based on Aristotle, and he contributed to zoology, botany, astronomy, chemistry and related subjects, while several treatises are falsely attributed to him. A brief list of some of the editions of certain of his works published in the fifteenth century indicates the popularity of Albertus, and more comprehensive details are available in Sarton[4] and in Ferguson.[5] Albertus' *De anima* was first printed at Venice in 1481, and again in 1494; *De animalibus* was published in Rome, 1478, Mantua, 1479, and Venice, 1495;[6] *De coelo et*

[1] *The art of falconry: being the De arte venandi cum avibus of Frederick II of Hobenstaufen. Translated and edited by Casey A. Wood and F. Marjorie Fyfe*, Stanford Univ., London, 1932. Translation mainly follows the text of the Bologna University MS. Lat. 419 (717); pp. lvii–lxxxvii are devoted to "Manuscripts and editions of the 'De arte venandi cum avibus'"; the work also contains "An annotated bibliography of ancient, mediaeval and modern falconry", pp. 559–609.

[2] Meiklejohn, M. F. M. The birds of Dante. *Ann. Sci.*, 10, 1943, pp. 33–44.

[3] Haskins, Charles H. The "De arte venandi cum avibus" of the Emperor Frederick II. *Engl. hist. Rev.*, 36, 1921, pp. 334–335.

[4] Sarton, Vol. 2, Part II, pp. 934–944.

[5] Ferguson, Vol. 1, pp. 15–17.

[6] Also, Albertus Magnus. *De animalibus libri XXVI nach der Cölner Urschrift hrg. von Hermann Stadler.* (*Beitr. zur Gesch. d. Philos. d. Mittelalters, Bd.* 15–16), Münster, 1816–20; see also his *De vegetabilibus libri VII. Editionem criticam ab Ernesto Meyero coeptam absobrit Carolus Jessen*, Berlin, 1867.

mundo appeared in two editions at Venice, 1490 and 1495; *De generatione et corruptione*, Venice, 1495; *De intellectu et intelligibili*, [Venice, 1472], and [Leipzig, 1492]. His *Liber aggregationis* went into innumerable editions and translations, including printings at [Cologne, 1499–1500], Strassburg, 1478; Reutlingen, 1483; [Ferrara, 1477]; [Bologna], 1478; [Rome, 1480 and 1481]; [Paris, 1483, 1493? and 1500]; and London, [1485] [S.T.C. 258]. The *Metaphysica* was printed at Venice, 1494; *Meteororum libri IV* at Venice 1488 and 1494; *De mineralibus* at [Padua], 1476, Pavia, 1491, Venice, 1495, and [Cologne, 1499]; *De mirabilibus mundi*, [Venice, 1472 (two printings), and 1478], and Colle, 1478; *Philosophia pauperum* Barcelona, 1482, [Lerida, 1485], and [Toulouse, 1485]; *Physica sive de physico auditu*, Venice, 1488 and 1494; *Secreta mulierum et virorum cum commento*, [Perugia, 1476], [Bologna or Venice], 1478, Augsburg, 1489, Vienna, [1500], "Rome, 1499" [Venice, 1500?], [London, 1485] [S.T.C. 273], and numerous other editions from the presses of Germany, Holland and France; *Summa de creaturis et de homine*, Venice, 1498. A large number of alchemical writings have been ascribed to Albertus Magnus, and some of these have been studied by Pearl Kibre,[1] who has translated the *De occultis nature*,[2] following a paper suggesting that it is falsely attributed to Albertus Magnus.[3] The *Libellus de alchimia*, also attributed to Albertus Magnus, has been translated into English,[4] and certain of his zoological writings have been the subject of an article by Heinrich Balss.[5] Bernhard Geyer is editing the *Opera omnia* of Albertus Magnus which is in course of publication, and Dorothy Wyckoff[6] has translated *De mineralibus*.

Robert Grosseteste (*c.* 1168–1253), Bishop of Lincoln, was born at Stradbrook, Suffolk, and became first Chancellor of the University of Oxford, wheie he lectured. He was an astronomer, mathematician, philosopher and physicist, and translated several works from Greek into Latin. The influence of Robert Grosseteste was extended not only through his writings, but through his pupils, notably Roger Bacon. He wrote on the compotus; the *Compendium sphaerae*, which was based on Sacro-

[1] Kibre, Pearl. Alchemical writings ascribed to Albertus Magnus. *Speculum*, 17, 1942, pp. 499–518; Kibre, Pearl. An alchemical tract attributed to Albertus Magnus. *Isis*, 35, 1944, pp. 303–316; Kibre, Pearl. The *Alkimia minor* ascribed to Albertus Magnus. *Isis*, 32, 1940, pp. 267–300; and Further manuscripts containing alchemical tracts attributed to Albertus Magnus. *Speculum*, 34, 1959, pp. 238–247.

[2] Kibre, Pearl. Albertus Magnus, *De occultis nature*. *Osiris*, 13, 1958, 157–183.

[3] Kibre, Pearl. The *De occultis naturae* attributed to Albertus Magnus. *Osiris*, 11, 1954, pp. 23–39.

[4] *Libellus de alchemia, ascribed to Albertus Magnus. Translated from the Borgnet Latin edition; introduction and notes by Sister Virginia Heines. With a foreword by Pearl Kibre*, Berkeley, Los Angeles, 1958.

[5] Balss, Heinrich. Die Tausendfüssler, Insekten und Spinnen bei Albertus Magnus. *Sudhoffs Arch. Gesch. Med.*, 38, 1954, pp. 303–322.

[6] Albertus Magnus. *Book of minerals. Translated by Dorothy Wyckoff*, Oxford, 1967. Selections are also contained in Wyckoff, Dorothy. Albertus Magnus on ore deposits. *Isis*, 49, 1958, pp. 109–122.

bosco; *De generatione stellarum; Hexaemeron; De luce seu de inchoatione formarum; De colore; De impressionibus aeris seu de prognosticatione*, and others. A Latin version of *De cometis*, from a manuscript in the Biblioteca Manicelliana, has been published by S. Harrison Thomson.[1] *De iride*, a treatise on refracted light, was composed about 1235, and includes an explanation of the rainbow based on refraction. This has been the subject of a paper by Bruce S. Eastwood.[2] Most of Grosseteste's scientific writings have been published by Ludwig Baur,[3] while early editions of separate works include *Compendium sphaerae*, Venice, 1508, 1513 or 1514, 1518 and 1531; *Commentarius in analytica posteriora*, [1475?], Venice, 1494, 1497, 1499, 1504 and 1537; *Super libros physiorum*, Venice, 1506; *Opuscula quaedam philosophica*, Venice, 1514; and *Summa super physica Aristotelis*, Venice, 1500. Richard C. Dales[4] questions the attribution of *Summa in octo libros physicorum Aristotelis* to Grosseteste, and the same book has been the subject of an article by S. Harrison Thomson.[5] The latter has provided a scholarly study of the manuscripts of Grosseteste's writings, and published the texts of *Questio de calore* and *De operacionibus solis* for the first time, from MS.3314 of the Biblioteca Nacional, Madrid.[6]

Robert Grosseteste experimented with lenses, and the action of the mirror.[7] He advocated the study of Greek and Hebrew, and his influence extended over the two hundred yeaıs following his death.[8] There has been much controversy over the attribution of certain writings to Grosseteste, particularly *Questio de fluxuet reflexus maris*, but Richard C. Dales[9] suggests that this should be included among Grosseteste's authentic writings, and presents much evidence to support this following a careful

[1] Thomson, S. Harrison. The text of Grosseteste's De cometis. *Isis*, 19, 1933, pp. 19–25.

[2] Eastwood, Bruce S. Grosseteste's 'quantitative' law of refraction: a chapter in the history of non-experimental science. *J. Hist. Ideas*, 28, 1967, pp. 403–414.

[3] Baur, Ludwig. *Die philosophischen Werke des Robert Grosseteste.* (*Beiträge zur Geschichte der Philosophie des Mittelalters: Bd.* 9, 1912); see also Baur, Ludwig. *Die Philosophie des Robert Grosseteste, Bischofs von Lincoln* (†1253). (*Beiträge zur Geschichte des Mittelalters*), 1917.

[4] Dales, Richard C. The authorship of the Summa in physica attributed to Robert Grosseteste *Isis*, 55, 1964, pp. 70–74.

[5] Thomson, S. Harrison. *The Summa in VIII libros physicorum* of Grosseteste. *Isis*, 22, 1934–35, pp. 12–18.

[6] Thomson, S. Harrison. *The writings of Robert Grosseteste, Bishop of Lincoln, 1235–1253*, Cambridge, 1940; and Grosseteste's *Questio de calore, De cometis* and *De operationibus solis. Medievalia et Humanistica*, 11, 1957, pp. 34–43.

[7] See Turbane, C. M. Grosseteste and an ancient optical principle. *Isis*, 50, 1959, pp. 467–472.

[8] See Sarton, Vol. 2, Part II, pp. 583–586. See also Boulter, Ben Consitt. *Robert Grossetête. The defender of our Church and our liberties*, 1936; Crombie, A. C. *Robert Grosseteste and the origins of experimental science, 1100–1700*, Oxford, 1953; and Callus, Daniel Angelo, *ed. Robert Grosseteste: scholar and bishop. Essays in commemoration of the seventh centenary of his death*, Oxford, London, 1955; Dales, Richard C. Robert Grosseteste's scientific work. *Isis* 52, 1961, pp. 381–402; and Dales, Richard. Robert Grosseteste's views on astrology. *Mediaeval Studies*, 29, 1967, pp. 357–363.

[9] Dales, Richard C. The authorship of the *Questio de fluxuet reflexus maris* attributed to Robert Grosseteste. *Speculum*, 37, 1962, pp. 582–588.

examination of the alternative claims. Dales has also contributed a further study on Grosseteste's short work on the tides, *Questio de fluxu et refluxu maris*, providing an English translation,[1] and on *Commentarius in octo libros Physicorum Aristotelis*,[2] which exists in only three manuscripts, and which Richard C. Dales has edited.[3]

Another Franciscan, John Peckam (*c.* 1220–92) was a native of Sussex, studied at Oxford, and became Archbishop of Canterbury in 1279. He wrote much on theology and also four scientific treatises. The *Perspectiva communis* was devoted to optics, and was printed at Milan, 1482, Leipzig, 1504, Venice, 1504 and [1505?], Cologne, 1508, 1542 and 1627, Nuremberg, 1542, with an Italian translation published at Venice, 1593. The other scientific writings were entitled *Theorica planetarum*, *Tractatus sphaerae*, and *De numeris*, but it is as a writer on optics that he is remembered.[4]

The greatest figure in mediaeval science was undoubtedly Roger Bacon (1214–94), also a Franciscan, who has been described as "the first man of science in the modern sense". He was born at Ilchester, Suffolk, and studied under Robert Grosseteste at Oxford before visiting Paris and Italy. Roger Bacon was an encyclopaedist. He wrote a textbook on optics that was in use for the next two centuries, constructed astronomical tables, made suggestions for the mechanical propulsion of vehicles and ships, and also planned a flying machine. He was also a geographer, described the composition and manufacture of gunpowder, and was a mathematician of no small repute. Roger Bacon was not an experimental scientist, but a philosopher, and foresaw many scientific developments. His writings are bibliographically complex, for he often rewrote his treatises under different titles. None of his works was printed before 1500, and editions of his separate works were few. A bibliography of his writings was published in 1914 by A. G. Little.[5] The *Speculum alchemiae* was first printed at Nuremberg, 1541, with a French translation in 1557, 1612 and 1627, while the first edition in English was entitled *The mirror of alchemy*, London, 1597 [S.T.C. 1182].[6] Bacon's *Epistola de secretis* was

[1] Dales, Richard C. The text of Robert Grosseteste's *Questio de fluxu et refluxu maris* with an English translation. *Isis*, 57, 1966, pp. 455–474.

[2] Dales, Richard C. Robert Grosseteste's *Commentarius in octo libros Physicorum Aristotelis. Medievalia et Humanistica*, 11, 1957, pp. 10–33.

[3] Grosseteste, Robert. *Commentarius in VIII libros Physicorum Aristoteles, Edited by Richard C. Dales*, Boulder, 1963.

[4] See Sarton, Vol. 2, Part II, pp. 1028–1030.

[5] *Roger Bacon: essays contributed by various writers on the commemoration of the seventh centenary of his birth. Collected and edited by A. G. Little*, Oxford, 1914; Appendix, pp. 375–426, "Roger Bacon's works with references to the MSS. and printed editions".

[6] A modern English translation is contained in, The mirror of alchemy of Roger Bacon. Translated into English by Tenney L. Davis. *J. chem. Educ.*, 8, 1931, pp. 1945–1953. See also Singer, Dorothea Waley. Alchemical writings attributed to Roger Bacon. *Speculum*, 7, 1932, pp. 80–86.

printed in Paris, 1542, Oxford, 1594, and Hamburg, 1618, with French translations from Lyons, 1557, and Paris, 1558, and English translations from London, 1597 and 1659, and a new English translation by Tenney L. Davis, Easton, Pa., 1923. *De consideratione quintae essentiae* was published at Basle in 1561 and 1597, while several others were included in collections. Bacon's works were also published as *Opus majus*,[1] *Opus minus* and *Opus tertium*. The first edition of *Opus majus*, by Samuel Jebb, was printed in London, 1733, Venice, 1750, with a new edition by John Henry Bridges, two volumes, Oxford, 1897, with a third volume, London, 1900.[2] An English translation of this by Robert Belle Burke was published in two volumes, Philadelphia, 1928. John Sherren Brewer published Bacon's *Opera quaedam hactenus inedita*, London, 1859, but there is no complete edition of Roger Bacon's writings, although separate works have been published by the Oxford University Press, mainly edited by Robert Steele, 1905–40. Bacon's work on optics has been the subject of a paper by R. R. James,[3] and further references are provided by Sarton[4] and by Ferguson.[5] A transcript of a manuscript in the Biblioteca Nacional, Madrid (MS. 3314), published by S. Harrison Thomson,[6] has revealed a previously unknown work by Roger Bacon, and David C. Lindberg[7] has compared Bacon's theory of the rainbow with that of Grosseteste, in favour of Bacon.

King Alfonso the Wise (1223–84) of Castile gathered together a group of scholars to translate Arabic writings into Spanish and fostered scientific research. He attracted to Toledo a number of scholars who calculated the set of astronomical tables known as the *Alfonsine tables*, which attained great popularity, and he also initiated an enormous encyclopaedia of astronomical knowledge, chiefly from Arabic sources, known as the *Libros del saber de astronomia*. This was published in five folio volumes at Madrid 1863–67, the final volume not being completed. The *Tabulae*

[1] The seventh part, on moral philosophy, of the *Opus majus* has hitherto been left incomplete, but in 1919 a fuller MS. was discovered in the Vatican Library. Corrected by Roger Bacon himself, it was included in a new edition of the seventh part of the *Opus*, and published as *Rogeri Baconis Moralis philosophia. Post Ferdinand Delarme critice instruxit et edidit Eugenio Massa*, Zurich, 1954; see also Crombie, A. C. The *Opus maius* of Roger Bacon. *Endeavour*, 8, 1949, pp. 163–166.

[2] See also Bridges, John Henry. *The life and work of Roger Bacon, with additions and notes by H. Gordon Jones*, 1914; and Easton, Stewart C. *Roger Bacon and his search for a universal science*, [*etc.*], Oxford, 1952.

[3] James, R. R. The father of British optics. Roger Bacon, *c.* 1214–94. *Brit. J. Ophthal.*, 12, 1928, pp. 1–14.

[4] Sarton, Vol. 2, Part II, pp. 952–967.

[5] Ferguson, Vol. 1, pp. 64–66.

[6] Thomson, S. Harrison. An unnoticed treatise of Roger Bacon on time and motion. *Isis*, 27, 1937, pp. 219–224.

[7] Lindberg, David C. Roger Bacon's theory of the rainbow: progress or regress? *Isis*, 57, 1966, pp. 235–248.

alphonsinae was first printed at [Venice], 1483, and Venice, 1492, 1518, 1521, 1524; Paris, 1545, 1553; and Madrid, 1641.[1]

John Holywood (John of Halifax, or Holywood) is more generally known as Sacrobosco (died 1256). Although probably born at Halifax, Yorkshire, and educated at Oxford, he spent most of his life in Paris. He wrote a popular textbook on astronomy that exists in numerous manuscripts, and also an elementary treatise on arithmetic, entitled *Algorismus*. The *Tractatus de sphaera*, or *Sphœra mundi*, was printed at Ferrara, 1472; [Venice, 1472 and 1476]; Bologna, 1477 and 1480; Milan, 1478; Venice, 1478 (twice) and 1482; [Venice], 1485, and numerous other editions and translations. In fact, there were at least thirty printed editions before 1501, and at least two hundred more between 1501 and 1600. Although written in the early thirteenth century, it was still the most popular text on the subject in the sixteenth. The 1485 edition, printed by Erhard Ratdolt, was the first printed book in which more than two colours were employed, black, red, yellow, brown and grey being used. A modern critical edition by Lynn Thorndike[2] was published in 1948, but omits the mediaeval diagrams. The *Algorismus* was first printed with another work at Strassburg in 1488, and numerous subsequent editions have appeared.[3]

Nicole Oresme (*c.* 1320–82) was the author of a work on fractional exponents, *De proportionibus proportionum*, and on the kinematics of circular motion, *Ad pauca respicientes*, both of which have been translated by Edward Grant[4] in a critical study. Grant[5] has also provided a translation of Part I of Oresme's *Algorismus proportionum*, a systematic attempt to present rules for the multiplication and division of ratios involving integral and fractional exponents.

That Geoffrey Chaucer (*c.* 1340–1400), as a translator of Boëthius' *De consolatione*, would be keenly interested in science is to be expected. This translation, and a short unfinished *Treatise on the astrolabe*, are the only two prose works attributed to Chaucer, and the latter is probably the best-known Middle English scientific work. It was first edited in Chaucer's works by William Thynne in 1532, and there is a modern English version by R. T. Gunther.[6] Compiled from Messahala and Sacrobosco, and intended for the use of Chaucer's son Lewis, it contains worked examples for the year 1391, and is an excellent example of

[1] See Sarton, Vol. 2, Part II, pp. 834–842.

[2] Thorndike, Lynn. *The Sphere of Sacrobosco and its commentators*, Chicago, 1949.

[3] See Sarton, Vol. 2, Part II, pp. 617–619.

[4] Oresme, Nicole. *De proportionibus proportionum and Ad pauca respicientes. Edited with introductions, English translations, and critical notes by Edward Grant*, Madison, 1966.

[5] Part I of Nicole Oresme's *Algorismus proportionum*. Translated and annotated by Edward Grant. *Isis*, 56, 1965, pp. 327–341.

[6] Gunther, E. T. *Early science in Oxford*, Vol. 5, Oxford, 1929.

scientific writing. E. J. Holmyard[1] has demonstrated that Chaucer showed a wide knowledge of alchemy in his study of "The Canon's Yeoman's tale" in the *Canterbury tales*, and Derek J. Price[2] brings strong evidence to prove that Peterhouse MS. 75. I at Cambridge, *The equatorie of the planetis*, is from Chaucer's hand. This evidence is given with a transcript of the MS., a translation and much other relevant information regarding astronomy at that period.[3] Another manuscript ascribed to Chaucer is in Trinity College, Dublin (MS. D.2.8), and was probably from the library of John Dee. This throws new light on the tradition that Chaucer was a master of the hemetic art and was the subject of a paper by Gareth W. Dunleavy.[4]

Leonardo of Pisa, or Leonardo Fibonacci (*c.* 1170–*c.* 1245), travelled widely in the East and on the Barbary Coast, and made an extensive study of mathematics. He has been called the "greatest Christian mathematician of the Middle Ages", and in 1202 wrote the *Liber abaci*, his *Practica geometriae* following in 1220. He also wrote smaller works entitled *Flos super solutionibus quarundam questentium ad numerum et ad geometriam vel ad utrumque pertinentium* and *Liber quadratorum*. The smaller writings were published in Florence, 1854 and 1856, with a complete edition in two volumes, Rome, 1857–62.[5]

A botanist of the thirteenth century, Rufinus, was the author of a *Herbal* written about A.D. 1287, and probably based on a text of Dioscorides. The work of Rufinus was apparently unknown to sixteenth-century writers but a thirteenth-century copy of his *Herbal* (MS. Ashburnham 189 of the Laurentian Library, Florence) has been edited by Lynn Thorndike,[6] who has also provided us with additional information on Rufinus.[7]

Before passing from the manuscript era to that of the printed book, we must pause to consider the nature of the vehicle for disseminating learning. Manuscripts were copied in the *scriptoria* of monasteries and by scribes attached to publishing houses. In university towns, publishing was a flourishing business, and popular texts were frequently issued.

[1] Holmyard, E. J. *Alchemy*, Harmondsworth, 1957.

[2] *The equatorie of the planetis. Edited from Peterhouse MS. 75.I by Derek J. Price, with a linguistic analysis by R. M. Wilson*, Cambridge, 1955; see also Price, Derek J. Chaucer's astronomy. *Nature*, 170, 1952, pp. 474–475.

[3] See also Herdan, G. Chaucer's authorship of *The equatorie of the planetis:* the use of Romance vocabulary as evidence. *Language*, 32, 1956, pp. 254–259; and Curry, Walter Clyde, *Chaucer and the mediaeval sciences. Revised and enlarged edition*, 1960. Selected bibliography, pp. 353–367.

[4] Dunleavy, Gareth W. The Chaucer ascription in Trinity College, Dublin MS. D.2.8. *Ambix*, 13, 1965, pp. 2–21.

[5] See Sarton, Vol. 2, Part II, pp. 611–613.

[6] *The Herbal of Rufinus. Edited from the unique manuscript by Lynn Thorndike, assisted by Francis S. Benjamin, Jr.*, Chicago, 1946. Transcribes Latin text with annotations and indexes.

[7] Thorndike, Lynn. Rufinus: a forgotten botanist of the thirteenth century. *Isis*, 18, 1932, pp. 63–76.

Manuscripts were sometimes hired out to students, often a few pages at a time, and complete texts were very costly. Mediaeval manuscripts lack title-pages and indexes, and were not paginated, but the colophon at the end of the text usually provides details of the author, title, and perhaps a date, together with the name of the scribe. In certain manuscripts the scribes left spaces for the rubrication of initial letters, and for other illustrations. Printing did not kill the manuscript trade immediately, for it lingered until the middle of the fifteenth century, and it is also of interest to remember that the early printed books closely resembled manuscripts. Their main advantage was the fact that they were more readily duplicated, contained fewer copyist errors, and were infinitely cheaper.

Locating scientific manuscripts is a difficult problem, but most of the older university and national libraries contain this material. Their catalogues are useful, but personal research frequently reveals hitherto unknown items. The catalogues of alchemical manuscripts prepared by Dorothea Waley Singer,[1] and by Joseph Bidez and others,[2] are examples of invaluable bibliographical tools that render useful assistance to the research worker. A similar catalogue devoted to alchemical manuscripts in America describes seventy-nine MSS. in thirty-one different libraries, and was compiled by William Jerome Wilson.[3] The latter assisted Seymour De Ricci in compiling his *Census of mediaeval and renaissance manuscripts in the United States and Canada.*[4] Lynn Thorndike, who spent much of his life investigating manuscripts as sources of knowledge, contributed two useful articles on the Latin scientific manuscripts in the Vatican,[5] and on manuscripts at Bologna, Florence, Milan and Venice.[6] Alfred Siggel[7] has compiled, under the auspices of the Union Aca-

[1] Singer, Dorothea Waley. *Catalogue of Latin and vernacular alchemical manuscripts in Great Britain and Ireland dating from before the XVth century, by Dorothea Waley Singer assisted by Annie Anderson*, 3 vols., Brussels, 1928–31. This is supplemented by Robbins, Rossell Hope. Alchemical texts in middle English verse: corrigenda and addenda. *Ambix*, 13, 1966, pp. 62–73.

[2] Bidez, Joseph, and others. *Catalogue des manuscrits alchimiques grecs*, 8 vols., Brussels, 1924–32.

[3] Wilson, William Jerome. Catalogue of Latin and vernacular alchemical manuscripts in the United States and Canada. *Osiris*, 7, 1939.

[4] Two vols., 1935–37, index 1940; supplement in preparation.

[5] Thorndike, Lynn. Vatican Latin manuscripts in the history of science and medicine; *Isis*, 13, 1929–30, pp. 51–102; Thorndike, Lynn. Notes on some mediaeval Latin astronomical, astrological and mathematical manuscripts at the Vatican. Part I [-II]. *Isis*, 47, 1956, pp. 391–404; 49, 1958, pp. 34–49.

[6] Thorndike, Lynn. Some alchemical manuscripts at Bologna and Florence. *Ambix*, 5, 1956, pp. 85–110; and Thorndike, Lynn. Notes on some medieval astronomical, astrological and mathematical manuscripts at Florence, Milan, Bologna and Venice. *Isis*, 50, 1959, pp. 33–50.

[7] Union Académique Internationale. *Katalog der arabischen alchimistischen Handschriften Deutschlands: Handschriften der Öffentlichen Wissenschaftlichen Bibliothek. Bearbeitet von Alfred Siggel*, Berlin, 1949; and *Handschriften der Ehemals Herzoglichen Bibliothek zu Gotha. Bearbeitet von Alfred Siggel*, Berlin, 1950.

démique Internationale, two catalogues of Arabian alchemical manuscripts in German libraries. These catalogues are of particular value as all important material for the historian is translated into German. Theodore Silverstein[1] has published a provisional catalogue of mediaeval Latin scientific writings in the Barberini Collection in the Vatican, and Lynn Thorndike and Pearl Kibre[2] have studied the incipits of mediaeval scientific writings in Latin.

The increase in the numbers of schools and universities, and the introduction of printing, initiated a new era in the development of science and the diffusion of scientific literature. Not only were the older texts re-edited and made generally available, but scientists were encouraged to produce new works, and their ideas and results of experimental enquiry were quickly made available throughout the civilized world. Science was beginning to gain the impetus that is still carrying it steadily forwards.

[1] Silverstein, Theodore. *Medieval Latin scientific writings in the Barberini Collection: a provisional catalogue*, Chicago, 1957.
[2] Thorndike, Lynn, and Kibre, Pearl. *Incipits of mediaeval writings in Latin*, London, 1963.

CHAPTER II

Scientific Incunabula

"Science is very old, as old as tool-using man, but the meticulous precision with which it advanced to our day dates from the time when thought could as sharply be cut and cast and composed into clear phrasing and formulation, as a printer's form. Nowhere can the new mechanically orientated mentality be better studied than in embryo, in these very incunabula with their clear beginning in time, and still bearing the traces of vivid spontaneity, only too quickly lost in subsequent routine."

ARNOLD C. KLEBS

The diffusion of knowledge was a slow process before the invention of printing. Relying upon lectures, discussion between scholars, letters transported by laborious means across the continent, and travels by seekers after truth, advancement was by development in isolated communities, with but occasional contact with the outside world. Undoubtedly the most significant event in the revival of learning was the invention of printing from movable type in the middle of the fifteenth century in Europe. The success of the Renaissance depended upon the rapid diffusion of knowledge throughout the civilized world, and the multiplication of copies of learned works was essential for this purpose. Paper was in general use in China during the early centuries of our era and the date A.D. 105 is usually given as the date of its invention by Ts'ai Lun. The secret of its manufacture passed from the Chinese to the Arabs in the eighth century. In turn the Moors imparted this knowledge to their Spanish conquerors in the twelfth and thirteenth centuries. The study of the history of printing in China by T. F. Carter[1] provides a scholarly survey of the subject, producing evidence of a block-printed book from China in the ninth century. This is *The Diamond Sutra*, dated A.D. 868, which was discovered by Sir Aurel Stein in Tun-huang, West China, and represents the oldest extant printed book. It is now housed in the British Museum. It is stated that movable type was invented by Pi Sheng (*c.* 1041–48), as described by Shen Kua in his *Meng ch'i pi t'an*. A detailed description of the process by Wang Chen is contained in his *Nung shu*. The earliest

[1] Carter, Thomas Francis. *The invention of printing in China and its spread westward. . . . Revised by L. Carrington Goodrich. Second edition*, New York, 1955.

extant book printed from movable type is the Korean *Sonja sail kaju*, dated 1409.

Historical evidence points to Mainz as the city whence printing rapidly spread throughout Europe. About the year 1450 Johann Gutenberg perfected the art of printing from movable type, and from Mainz, where Gutenberg experimented with his invention, the craft was introduced to other centres of learning. By 1458 Strassburg had a press, Bamberg followed in 1461, and in 1465 Ulrich Zel set up his press at Cologne. Three years later printers were working at Augsburg, in 1470 the craft had spread to Nuremberg, and by the year 1500 fifty-one German towns housed printing presses.[1]

The well-known partners Conrad Sweynheym and Arnold Pannartz carried the art to Rome in 1467, having paused on their journey in 1465 to set up a press at Subiaco. By 1469 Venice had its press, and that city rapidly became the main centre of book production in Italy, numbering among its printers the distinguished Nicolas Jenson, Erhard Ratdolt and Aldus Manutius. Milan, Florence and Naples had printing presses in 1471, and it is significant to note that printing houses were usually found in university towns or centres of learning.

Martin Crantz, Ulrich Gering and Michael Friburger introduced printing at the Sorbonne in 1470, and Lyons and Rouen became additional centres of the book trade in France. Switzerland had a printing press at Basle in 1468, Holland its first printer at Utrecht in 1470, while four years later a press was set up at Valencia, in Spain.

The Englishman, William Caxton, gained his knowledge of the art of printing on the Continent, and his *Recuyell of the historyes of Troye*, printed at Bruges in 1474, was the first book to be printed in English. Two years later Caxton returned to England, setting up a press at Westminster, his earliest dated book being *Dictes or Sayengis of the philosophres*, which was completed on 18th November, 1477. Caxton printed over one hundred books, and was a scholar who edited and translated some of his publications.[2] Other early printers in England include Theodoric Rood, who worked at Oxford from 1478 to 1485, and an unnamed craftsman who had printed eight books at St. Albans by 1485.

The first printing types closely resembled the script of manuscripts copied by scribes, and the early printed books had initial letters left blank for rubrication, and similar spaces for ornamentation, as in the manuscripts of the period. At the end of the work appeared what is known as the colophon, where the printer added the title, possibly the

[1] Aldis, Harry G. *The printed book. . . . Revised and brought up to date by John Carter and Brooke Crutchley. Third edition*, Cambridge, 1951, p. 9.

[2] See Bennett, H. S. *English books and readers 1475 to 1557: being a study in the history of the book trade from Caxton to the incorporation of the Stationers' Company*, Cambridge, 1952.

name of the author, and less frequently his own name, together with the date. The printer's device or trade-mark was also printed here, famous examples of these devices being the anchor and dolphin of Aldus of Venice, which first appeared in 1502, and Froben's twined serpents. Occasionally these devices incorporated portraits of the printers. Title-pages did not come into general use until about 1480, and the information now incorporated therein was confined to the colophon.

It has been estimated that 38,000 editions were printed in the fifteenth century, of which one-third was issued in Italy, one-third in Germany, and the remainder in other countries.[1] Yet another estimate, by Curt F. Bühler,[2] based on Klebs' list,[3] gives the following totals for scientific incunabula printed in various countries: Italy, 1,445; Germany, 1,003; France, 387; and England, 23. The same author provides tables for certain towns in these countries.

The first illustrated printed book was issued about 1461, and contained woodcuts, but this ornamentation did not become general for another ten years. Two of Caxton's books printed about 1481 are the first volumes in English to be illustrated with woodcuts. Probably the first example of colour printing was in the second printed edition of Sacroboscos' *Sphaera mundi*, printed in 1485 by Erhard Ratdolt at Venice.

Arnold C. Klebs[4] has listed alphabetically by names of authors over 3,000 incunabula devoted to medicine and science, of which about 850 items are devoted to medicine; roughly 400 editions are on natural philosophy (botany, zoology, etc.); 300 deal with arithmetic, alchemy,[5] agriculture, etc.; astronomy and astrology provide roughly 200 editions; while mathematics, geometry and physics are represented by 100 entries. Details of some of the incunabula of science are here provided in roughly chronological order, information being taken from Klebs' list, which should be consulted for additional material. Some of the printers of these books are mentioned in Chapter XI.

An edition of Ptolemy's *Cosmographia* was printed at Bologna and dated 1462, but Klebs gives [1477] as the correct year of issue.[6] Apparently only two scientific books were printed before 1470, these being Pliny the Elder's *Historia naturalis*, printed at Venice in 1469, and the *Geographia*

[1] Aldis, Harry G. *Op. cit.*, p. 21.
[2] Bühler, Curt F. The statistics of scientific incunabula. *Isis*, 39, 1949, pp. 163–168.
[3] See below.
[4] Klebs, Arnold C. Incunabula scientifica et medica. *Osiris*, 4, i, 1938, pp. 1–359; reprinted Hildersheim, 1963.
[5] An alphabetical list of ninety-eight writers on alchemy, with titles and dates of their works printed up to 1536 is provided by Hirsch, Rudolf. The invention of printing and the diffusion of alchemical and chemical knowledge. *Chymia*, 3, 1950, pp. 115–141.
[6] See also Lynam, Edward. *The first engraved atlas of the world. The* Cosmographia *of Claudius Tolemaeus, Bologna*, 1477, Jenkintown, 1941. (The George H. Beans Library, Publication No. 16.)

of Strabo, printed at Rome, [1469], by Sweynheym and Pannartz. These were followed by *De proprietatibus rerum*, by Bartholomew the Englishman, [Basle, 1470?], and another edition of Pliny's *Historia naturalis*, this time from Rome, 1470. In 1472 we find printings of Albertus Magnus' *De intellectu et intelligibili*, [Venice, 1472]; the same author's *De mirabilibus mundi*, Venice (two printings); Aristotle's *Opera*, with commentary by Averroës, Padua, 1472–74; his *De anima*, Padua, 1472; his *Physica sive de physico auditu* [Padua, 1472]; and his *Secreta secretorum*, [Cologne, 1472]; another edition of Bartholomew the Englishman's *De proprietatibus rerum*, [Cologne, 1472?]; the *Vitae et sententiae philosophorum* of Diogenes Laërtius, [Rome, 1472]; Isidore of Seville's *De responsione mundi*, Augsburg 1472; an edition of Macrobius' *In somnium Scipionis expositio Saturnalia*, Venice, 1472; a further printing of Pliny the Elder's *Historia naturalis*, Venice, 1472; *Sphaera mundi*, by Johannes de Sacrobosco, Ferrara, 1472, with another edition, [Venice, 1472]; Thomas of Aquinas' *Quaestiones de anima*, [Venice,] 1472; and *Scriptores rei rusticae*, Venice, 1472.

During the following year several other important works appeared, including Alchabitius' *Liber isagogicus ad scientiam indicialem astronomiae*, [Bologna, 1473]; Aristotle's *De coela et mundo*, Padua, 1473; his *Lapidarius*, Merseburg, 1473; his *Metaphysica*, Padua, 1473; his *Parva naturalia*, [Padua, 1473]; and the same author's *Problemata*, Mantova, [1473]; two editions of Arnaldo de Villanova's *De arte cognoscendi venena*, [Mantova, 1473], and [Padua, 1473]; Lucretius' *De rerum natura*, [Brescia, 1473]; and Pliny's *Historia naturalis*, Rome, 1473.

In 1474 appeared an edition of Aristotle's *De generatione et corruptione*, Padua, the same author's *Meteororum libri IV*, Padua, and his *De pomo et morte*, [Cologne, 1474]; Arnold de Villanova's *De arte cognoscendi venena*, [Turin or Piedmont, 1474]; and his *Regimen sanitatis ad regem Aragonum*, [Turin or Piedmont, 1474].

Adelard of Bath's *Quaestiones naturales*, [Louvain, 1475] (two printings) appeared the next year. Also, *De mirabilibus mundi*, by Albertus Magnus, [Santorso?, 1475]; Antonio Andrea's *Quaestiones super metaphysicam Aristotelis*, Naples, 1475, and the same author's *De tribus principiis rerum naturalium*, Padua, 1475; Aristotle's *Physica sive de physico auditu*, [Louvain, 1475], and his *Problemata*, Rome, 1475; Arnaldo de Villanova's *De arte cognoscendi venena*, [Milan, 1475], and [Rome, 1475], the *Vitae et sententiae philosophorum* of Diogenes Laërtius, Venice, 1475; and Ptolemy's *Cosmographia*, Vicenza, 1475, the first edition, without maps.

Albertus Magnus' *De mineralibus*, [Padua], was printed in 1476, as was his *Secreta mulierum et virorum cum commento*, [Perugia]; also Aristotle's *De animalibus*, Venice; Pliny's *Historia naturalis*, Parma, and an edition in Italian at Venice; and Johannes de Sacrobosco's *Sphaera mundi*, [Venice, 1476].

Another edition of Albertus Magnus' *Liber aggregationis*, with *De mirabilibus mundi*, was printed in Italy, [Ferrara, 1477]; also Albert von Sachsen's *De proportionibus*, [Padua, 1477]; Antonio Andrea's *Quaestiones super metaphysicam Aristotelis*, Vicenza, 1477; Arnaldo de Villanova's *De aqua vitae*, [Venice, 1477]; Plutarch's *Problemata*, [Venice, 1477], and [Ferrara, 1477?]; Ptolemy's *Cosmographia*,[1] Bologna, 14[77], the first edition to contain copper engraved maps; Robert the Englishman's *De astrolabio canones*, Perugia, [1477]; Johannes de Sacrobosco's *Sphaera mundi*, Bologna, 1477, and Michael Scot's *Physiognomia*, [Venice], 1477.

The year 1478 saw the publication of Albertus Magnus' *De animalibus*, Rome; his *Liber aggregationis*, with *De mirabilibus*, [Strassburg, 1478], and [Bologna]; a separate edition of *De mirabilibus mundi*, [Venice, 1478]; and the same author's *Secreta mulierum et virorum cum commento*, [Bologna or Venice]; Aristotle's *Secreta secretorum* in French, [Lyons, 1478]; the first printed arithmetic, *Arte dell' abbaco*, Treviso; Dioscorides' *De materia medica*, Colle; the *Cosmographia* of Ptolemy, Rome, containing engraved maps; and three editions of Johannes de Sacrobosco's *Sphaera mundi*, Milan, and Venice (two printings).

The *De animalibus* of Albertus Magnus was also printed at Mantova in 1479, and in the same year appeared Aristotle's *Opera*, Augsburg; also a German version of Arnaldo de Villanova's *De vinis*, Augsburg.

Another edition of *Liber aggregationis*, with *De mirabilibus mundi*, [Rome, 1480], appears among the numerous editions of the writings of Albertus Magnus, and Bartholomew the Englishman's *De proprietatibus rerum*, [Lyons], was also printed in 1480, as was a Hebrew edition of Maimonides' *Liber scientiae, de amore, de temporibus*, [Italy, 1480?]; Pliny's *Historia naturalis*, Parma; Ptolemy's *Cosmographia*, Florence, [1480]; the *Physiognomia* of Michael Scot, [Lyons, 1480]; Johannes de Sacrobosco's *Sphaera mundi*, Bologna; Thomas of Aquinas' *Super physica Arist.*, [N.P.], and the same author's *In metaphysicam Arist.*, Pavia.

Among the publications of 1481 must be mentioned Aristotle's *Physica sive de physico auditu*, [Rome, 1481]; Apuleius Platonicus, *Herbarium*, Rome; Johannes Philippus' *De Lignamine*, the first printed book in which engravings of plants appeared; *De proprietatibus rerum*, of Bartholomew the Englishman, Cologne; Maimonides' *De regimine sanitatis*, Florence, [1481]; Pliny's *Historia naturalis*, Parma; and Thomas of Aquinas' *Commentarius in Arist. de anima*, Venice. The following year appeared, among others, Albert von Sachsen's *De proportionibus*, Padua, 1482; and his *Quaestiones de coelo et mundo Aristotelis*, Pavia, 1482; Antonio Andrea's *Scriptum super metaphysicam Aristotelis*, Venice, 1482; Aristotle's *Opera*, Venice, 1482; a German version of Arnaldo de Villanova's *De vinis*,

[1] The 14[77], 1478 and 1482 editions have all been reprinted, with a bibliographical note by R. A. Skelton, Amsterdam, 1967, in the facsimile series *Theatrum Orbis Terrarum*.

Augsburg, 1482; Bartholomew the Englishman's *De proprietatibus rerum* was printed at [Lyons] (three editions), 1482; Euclid's *Elementa geometriae*, Venice, 1482, the first appearance in print of this author, and containing numerous geometrical illustrations; Ptolemy's *Cosmographia*, Ulm, 1482, the first to have woodcut maps; and Sacrobosco's *Sphaera mundi*, Venice, 1482.

Publications of 1483 include Alfonso X el Sabio's *Tabulae astronomicae*, Venice; Aristotle's *Opera*, Venice; his *Problemata*, [Magdeburg, 1483] and his *Secreta secretorum*, [Reutlingen, 1483]; two editions of Bartholomew the Englishman's *De proprietatibus rerum*, Cologne and Nuremberg; Pliny's *Historia naturalis*, Venice; and Theophrastus' *De historia et causis plantarum*, Treviso. In 1484 appeared the *Herbarius latinus*, Mainz, Peter Schoeffer, the first original botanical incunabulum that is not printed from a manuscript preserved from antiquity,[1] and the second printed book with illustrations of plants.

Certain of these publications were very popular and appeared from several presses during the same year. Other important texts were late in finding printers, but most of the major classic scientific writers appeared in print during the second half of the fifteenth century. The following are recorded as important texts, or on account of their value as incunabula: Albertus Magnus' *Liber aggregationis*, London, [1485], printed by Machlinia [S.T.C. 258]; the same author's *Secreta mulierum*, [London, 1485] [S.T.C. 273]; *Herbarius zu Teutsch*, Mainz, Peter Schoeffer, the third great printed herbal and of primary importance;[2] Plato's *Opera*, Florence, [*c.* 1485]; Aniarus' *Compotus manualis cum commento*, [Lyons, 1486]; Pliny's *Historia naturalis*, Venice, 1487; Albertus Magnus' *Meteororum libri IV*, Venice, 1488, and his *Physica sive de physico audito*, Venice, 1488; Albumasar's *Flores astrologiae*, Augsburg, 1488; Johann Engel's *Astrolabium*, Augsburg, 1488; Thomas of Aquinas' *De principiis rerum naturalium*, [Leipzig, 1488]; Pliny's *Historia naturalis*, Venice, 1489; Albert Magnus' *De coelo et mundi*, Venice, 1490; Aristotle's *Chiromantia*, Ulm, 1490; Euclid's *Elementa geometriae*, Vicenza, 1491; Aristotle's *De natura animalium*, [*etc.*], Venice, 1492; Albert von Sachsen's *Quaestiones de coelo et mundo Aristotelis*, Pavia, Venice, 1492; Albertus Magnus' *Metaphysica*, Venice, 1494; the same author's *De generatione et corruptione*, Venice, 1495; his *Meteorologica*, Venice, 1494 [1495]; his *De animalibus*, Venice, 1495; and his *De generatione et corruptione*, Venice, 1495; a Greek version of Aristotle's *Opera*, Venice, 1495–98; Avicenna's *Metaphysica*, Venice, 1495; Puerbach's *Theoricae novae planetarum*, Venice, 1495; Pliny's *Historia naturalis*, Venice, 1496; Aristotle's *Opera*, Venice, 1497; Theophrastus' *De historia plantarum*, Venice, 1497; Aristotle's *De natura animalium*, Venice, 1498;

[1] *VI. International Botanical Congress, Amsterdam, 1935. Catalogue of the exhibition of books.*
[2] Arber Agnes. *Herbals. . . . New edition*, Cambridge, 1938, p. 22.

Albertus Magnus' *De mineralibus*, [Cologne, 1499]; and a Greek version of Dioscorides' *De materia medica*, Venice, 1499. Aristotle's *Problemata*, Paris, [*c.* 1500] is recorded by F. N. L. Poynter[1] as being in the Wellcome Historical Medical Library, the only other known copy being in the Bibliothèque Nationale, while Aristotle's *De physico* of 1500, also in the Wellcome Library, was apparently previously unrecorded.

Following the publication of Klebs' list, George Sarton[2] contributed an interesting study on the same subject, illustrated with sixty facsimiles. Sarton was more interested in the contents of incunabula than in their bibliographical make-up, but has much to say on both these aspects. He deals with bibliographical 'ghosts', the combination of texts in certain incunabula, criticizes the lack of indication of pagination in Klebs' list, and discusses the definition of the term 'scientific' as interpreted by Klebs. Sarton lists the most popular authors of scientific incunabula, 'best-sellers', etc.,[3] and provides a chronological list of authors whose scientific works were published before 1501,[4] together with much additional matter of value to those interested in the subject of scientific incunabula. In a reply to Sarton's analysis of Klebs' work, Zoltán Haraszti[5] stresses the value of the work of the early printers, their scholarship, and the beauty of their productions, concluding that while rare book collectors and bibliographers may tend to over-emphasize the importance of their studies, their work cannot be neglected by historians of science.

Cornelius à Beughem, in his *Incunabula typographiae*, 1688,[6] was the first to use the term 'incunabula', which means 'swaddling-clothes; the cradle; childhood; origin, beginning', and has since been applied to books printed before 1501. This catalogue is also the earliest printed catalogue of books printed in the fifteenth century.

Certain general lists of incunabula are useful for tracing early scientific items, and are here noted as being of value to those interested in this material. Michael Maittaire (1668–1747) was the first important cataloguer of incunabula. He was a Frenchman resident in England, a profound scholar, and he collected an important library which was sold in London the year after his death. Maittaire's *Annales typographici*, 1719–41, includes in chronological order the printed literature issued up to 1664. The work is not exhaustive, the entries are inconsistent, and a supplement was issued by Michael Denis in 1789.

[1] See p. 53.
[2] Sarton, George. The scientific literature transmitted through the incunabula. (An analysis and discussion illustrated with sixty facsimiles.) *Osiris*, 5, 1938, pp. 41–245.
[3] Sarton, George. *Op. cit.*, pp. 182–193.
[4] Sarton, George, *Op. cit.*, pp. 198–226.
[5] Haraszti, Zoltán. Dr. Sarton on scientific incunabula. *Isis*, 32, 1940 (1947), pp. 52–62.
[6] Beughem, Cornelius à. *Incunabula typographiae, sive catalogus librorum scriptorumque proximis ab inventione typographiae annis, usque ad annum Christi M.D. inclusive, in quavis linguâ editorum*, Amsterdam, 1628. This is arranged alphabetically by authors; second edition, 1733.

The *Annalen* of Georg Wolfgang Panzer was published between 1788 and 1805, and covers literature in the German language printed up to 1526. This was followed by the same author's *Annales typographici ad annum 1536*, eleven volumes, Nuremberg, 1793–1803, the entries in which are arranged alphabetically.

The most important catalogue of incunabula is Ludwig F. T. Hain's *Repertorium bibliographicum ad annum 1500*, four volumes, Stuttgart, Tübingen, 1826–38. Hain (1781–1836) based his list upon the collection in the Munich Hofbibliothek, and arranges 16,299 incunabula alphabetically by authors. The entries are full and reliable, and hence still of value. Walter Arthur Copinger (1847–1910) published a supplement to Hain in three volumes, 1895–1902, correcting 7,000 items and describing 6,619 additional incunabula. This also is arranged alphabetically, and the final volume contains as an appendix Konrad Burger's *Printers and publishers of the 15th century, with lists of their works*. This was enlarged and issued separately in 1908. Dietrich Reichling published a further supplement to Hain-Copinger at Munich, 1908–14, listing a further 1,921 incunabula, and correcting previous entries.

Robert George Collier Proctor (1868–1903) issued his *Index to the early printed books in the British Museum . . . to 1500, with notes of those in the Bodleian Library*, five volumes, 1898–1903, which is arranged typographically by countries, towns, presses and dates, which has since been known as "Proctor order". This was expanded in the *Catalogue of books printed in the XVth century now in the British Museum*, Parts 1–8, 1908–49.

Konrad Haebler of the Royal Berlin Library made an exhaustive study of the type, analysing and classifying this material to form an invaluable bibliographical tool. Of his numerous typographical studies the *Typenrepertorium der Wiegendrucke*, six volumes, Halle, Leipzig, 1905–24, is of particular significance.

Catalogues covering the incunabula of special countries, private, public and university libraries have also been published,[1] the most promising having been the catalogue of Mlle. Marie Pellechet (1840–1900), *Catalogue général des incunables des bibliothèques de France*, Paris, 1897–1909, continued after her death by Marie Louis Polain. Unfortunately the list only reached the letter H.

The *Gesamtkatalog der Wiegendrucke* began publication in 1925, and by 1938 Band 7 reached the word 'Eigenschaften'. This authoritative list would, if completed, surpass all previous bibliographies of incunabula, providing complete details of all items, references to other catalogues, and listing important libraries possessing each item.

These general catalogues of incunabula are followed by those more strictly devoted to science, the first of which was Ludwig Choulant's

[1] See Cowley, J. D. *Bibliographical description and cataloguing*, 1939, pp. 214–248.

Graphische Incunabeln für Naturgeschichte und Medizin, [*etc.*], Leipzig, 1858, reprinted at Munich in 1924. This list of illustrated incunabula is not exhaustive, but provides full details of the books and plates, followed by particulars of the various editions, including place of publication, date, size (i.e., folio, etc.), name of printer, and annotation.

Arnold C. Klebs (1870–1943) made an important collection of medical and scientific incunabula, which is now in the Historical Library at the Yale University School of Medicine. He published an important short-title list as *Incunabula scientifica et medica*, Bruges, 1938,[1] which has been mentioned above.[2] References to standard bibliographies are provided, and editions appear in chronological order under authors.

It is remarkable how quickly the art of printing spread between its invention in Europe and the end of the fifteenth century, and how many editions were published during that period. Proctor records 4,157 incunabula from Italy, 3,232 from Germany, and 998 from France, while E. Gordon Duff[3] lists 431 English incunabula. The value of this material to scholarship, as examples of contemporary literature, as guides to the evolution of typography, and as historical material, is incalculable. Incunabula are gradually being absorbed by the larger national and scientific libraries, either by purchase or by donation, but individual collectors can still occasionally acquire these items in the market. It is essential that serious collectors should make themselves acquainted with the history of typography, and with bibliography in general. R. A. Peddie's *Fifteenth century books*[4] contains lists of printed matter on initials, printers' marks, colophons, title-pages, signatures and watermarks, together with a list of catalogues of collections of incunabula arranged by localities, and a list of the Latin names of towns where incunabula were printed. Information on the cataloguing of incunabula is provided by Arnold C. Klebs,[5] and by Henry Guppy,[6] while much useful matter on the advanced treatment of bibliographical material is contained in J. D. Cowley's *Bibliographical description and cataloguing*, 1939. George Sarton[7] has provided interesting details of wrongly dated incunabula, giving examples of incorrect dates in colophons, reversed figures, and information on 'remainders' of incunabula republished with new title-pages in the sixteenth century.

Scientific incunabula are to be found in most of the large national,

[1] In *Osiris*, 4, i, 1938, pp. 1–359. [2] See pp. 46, 50.
[3] Duff, E. Gordon. *Fifteenth-century English books*, [*etc.*], 1917.
[4] Peddie, R. A. *Fifteenth-century books: a guide to their identification. With a list of the Latin names of towns and an extensive bibliography of the subject*, 1913.
[5] Desiderata in the cataloguing of incunabula; with a guide for catalogue entries. *Pap. Bibl. Soc. Amer.*, 10, 1916, pp. 143–163.
[6] *Rules for the cataloguing of incunabula. . . . Second edition, revised*, 1932.
[7] Sarton, George. Incunabula wrongly dated: fifteen examples with eighteen illustrations. *Isis*, 40, 1949, pp. 227–240.

university and scientific libraries, but separate lists of these items are rare. Several medical libraries have printed lists of their incunabula,[1] and many of these items are scientific rather than medical. Individual items must be traced in the appropriate catalogues issued by the larger libraries, or in the reference books noted above.

Erik Waller (1875–1955) collected a remarkable library which he bequeathed to the University of Uppsala, and which contains 150 incunabula. These are described in volume one of the catalogue of the collection compiled by Hans Sallander,[2] among them being works by Albertus Magnus, Aristotle, Pliny and Puerbach.

The Wellcome Historical Medical Library houses one of the largest collections of incunabula contained in a medical library, and F. N. L. Poynter[3] has described these in a finely produced catalogue. This contains a total of 632 items, 117 of which are not in the British Museum, 122 items and four fragments having been printed before 1481. The catalogue lists a number of items devoted to science, including writings by Albertus Magnus, Aristotle, Dioscorides, Pliny and Theophrastus, and contains indexes of names; subjects; signatures, inscriptions and other marks of ownership; printers; countries and towns; and a chronological list.

The National Library of Medicine, Bethesda, U.S.A., contained 232 incunabula in 1916, 449 in 1932, and 493 are listed in the catalogue of the collection compiled by Dorothy M. Schullian and Francis E. Sommer in 1950.[4] Many of these items are scientific, and the authors of the catalogue stress the fact that while the typography, paper, bindings, etc., of incunabula have received considerable attention, little work has so far been done on the actual texts.

The Royal College of Physicians of London houses a collection of over one hundred books printed in the fifteenth century, but only a typescript catalogue of this exists. It is worthy of more permanent record and greater publicity.

Curt F. Bühler,[5] among other information on scientific and medical

[1] See Thornton, pp. 36–38.

[2] *Bibliotheca Walleriana. The books illustrating the history of medicine and science collected by Dr. Erik Waller and bequeathed to the Library of the Royal University of Uppsala. A catalogue compiled by Hans Sallander*, 2 vols., Stockholm, 1955.

[3] Poynter, F. N. L. *A catalogue of incunabula in the Wellcome Historical Medical Library*, London, [*etc.*], 1954. (Publications of the Wellcome Historical Medical Museum, N.S. No. 5.) See also Wellcome Historical Medical Library, *Catalogue of the Wellcome Historical Medical Library. I. Books printed before 1641*, 1962; and Moorat, S. A. J., *ed. Catalogue of Western manuscripts on medicine and science in the Wellcome Historical Medical Library, Vol. I. Manuscripts written before A.D. 1650*, 1962.

[4] Schullian, Dorothy M., and Sommer, Francis E. *A catalogue of incunabula and manuscripts in the Army Medical Library* [*now the National Library of Medicine*], New York, [1950].

[5] Bühler, Curt F. Scientific and medical incunabula in American libraries. *Isis*, 35, 1944, pp. 173–175; see also Bühler, Curt F. *The fifteenth-century book. The scribes: the printers; the decorators*, Philadelphia, 1960.

incunabula in American libraries, lists the holdings of scientific incunabula of ten of these libraries, providing the following figures: Huntington Library, 678; Library of Congress, 536; Boston Medical Library,[1] 398; Army Medical Library,[2] 388; College of Physicians, Philadelphia, 321; Pierpont Morgan Library, 306; Harvard University,[1] 297; Walters Art Gallery, 236; Yale University, 202 (excluding the Klebs Collection); New York Academy of Medicine, 113. The Bibliographical Society of America has published a census of incunabula in American libraries, edited by Margaret Bingham Stillwell.[3] This lists 35,232 copies of 11,132 titles, of which 28,491 are housed in institutions and 6,741 are in private collections.

The study of the contents of these early printed books can generally be accomplished from photographic or facsimile copies, and scholars should be enabled to gain access to these copies in any part of the world. It should be possible for the larger libraries to collect facsimiles, which for many practical purposes are to be preferred to the originals. This would facilitate the comparison of texts and of copies, and would further preserve the continuity of rare items. Not only are these liable to accidental destruction, but many of them are unsuitably housed, and are deteriorating in vaults and ill-ventilated rooms. Large sums of money are represented by certain collections of incunabula of which little practical use is made. The mere fact of ownership, or even an occasional exhibition behind glass, can give little satisfaction compared with that of making the material available to those capable of appreciating not only its monetary value, but its potential value to scholarship.

[1] The Boston Medical Library is now joined with the medical collection at Harvard in the Countway Library.

[2] Now the National Library of Medicine.

[3] *Incunabula in American libraries: second census of fifteenth-century books owned in the United States, Mexico, and Canada. Edited by Margaret Bingham Stillwell*, New York, 1940.

CHAPTER III

Scientific Books of the Fifteenth and Sixteenth Centuries

"The acquisition of any knowledge whatever is always useful to the intellect, because it will be able to banish the useless things and retain those which are good. For nothing can be either loved or hated unless it is first known."

LEONARDO DA VINCI

Without doubt, the most important event of the fifteenth century was the development of printing and its spread throughout Europe. The advancement of learning was hastened by the fact that the writings of earlier scholars were made more generally available, and that contemporary texts were comparatively speedily distributed. There was a consolidation of knowledge, a re-evaluation, and a steady progress forwards which gathered momentum in succeeding centuries. Following the appearance in print of the acknowledged classics, new works began to appear. Philosophy was separated from science, and the latter began to split into a multiplicity of branches, although scholars were still able to cover extensive fields, instead of having to specialize in minute topics. Leonardo da Vinci could contribute usefully to all sub-divisions of science, while lesser men might concentrate on smaller fields of research.

Nicholas Cusanus, or Nicholas of Cusa (1401–64) was born Nicholas Krebs in the village of Cues, and has been called "that almost universal genius", but Lynn Thorndike[1] suggests that his achievements have been greatly overrated, despite the fact that his *De docta ignorantia* embodies "a few brilliant random suggestions". This work was written in 1439–40, but was not published in an English translation until 1954, when it appeared as *Of learned ignorance*, translated by Germain Heron, New Haven, London. A French translation by L. Moulnier was published in Paris in 1930. Nicholas Cusanus attempted to reform the calendar, and recorded careful experiments on growing plants indicating that they absorb material from the air. He wrote on the use of the balance in physical experimentation, and was the author of *De staticis experimentis* which was originally published in a series of papers with the title *Idiotae libri quatuor*, 1476, and translated into English as *The idiot in four books*.

[1] Thorndike, Lynn. *Science and thought in the fifteenth century: studies in the history of medicine and surgery, natural and mathematical science, philosophy and politics*, New York, 1929, Chapter VII, pp. 133–141.

The first and second of wisdome. The third of the minde. The fourth of statick experiments, [*etc.*], London, 1650. An edition of the Latin version was published in Leipzig, 1937.[1] Nicholas Cusanus also wrote *De figura mundi*, and *De quadratura circuli*, which was first printed at Nuremberg in 1533. Two volumes of his *Opuscula varia* were published at Strassburg, 1488–90, with later editions in 1502, 1514 and 1565, and are sometimes erroneously listed as *Opuscula mathematica et theologica*. His *Opera* appeared in three volumes from Basle in 1565, and a modern edition of his philosophical writings began publication in Stuttgart in 1949. A definitive edition of the writings of Cusanus began to appear under the auspices of the Heidelberger Akademie der Wissenschaften in 1932, and six volumes had appeared by 1940. In 1960 it was announced that a further ten volumes would be published.[2,3]

Georg Puerbach (also known as Peuerbach or Purbach) (1423–61) was a professor at Vienna, and became astronomer and astrologer to King Ladislas of Hungary. He was the author of *Theoricae novae planetarum*, and translated six books of Ptolemy's *Almagest*, this translation being completed by Regiomontanus, and printed in 1496. Puerbach also wrote *Epherimedes*, first printed in 1474, and *Tabulae directionum*, which was largely derived from the writings of Giovanni Bianchini. Lynn Thorndike has also suggested that the importance of Puerbach and Regiomontanus has been overestimated.[4] Johannes Müller of Königsberg (1436–76), usually known as Regiomontanus, wrote in 1463 the first systematic treatise on trigonometry, *De triangulis*, which was first printed in 1533 at Nuremberg. The original manuscript of this work was found at Pulkovo Observatory in 1938, and an English translation of the 1533 text, by Barnabus Hughes, was published in 1967. Regiomontanus was born on 6th June, 1436, and died in Rome in June or July, 1476. In 1467 he went to the University of Pressburg (now Bratislava) at the invitation of Mathias Corvinus of Hungary, and there wrote a number of mathematical works. In 1471 he went to Nuremberg, where it was alleged that he wrote the *Commensurator*, a textbook of geometry in three books. A German translation of this with a commentary, and a list of writings issued by the printing works of Regiomontanus have been published by Willhelm Blaschke and Günther Schoppe.[5] However, Marshall Clagett[6] has stated

[1] See Viets, Henry. De staticis experimentis of Nicholas Cusanus. *Ann. med. Hist.*, 4, 1922, pp. 115–135; translation on pp. 126–135; and Hoff, Hebbel F. Nicholas of Cusa, van Helmont and Boyle: the first experiment of the Renaissance in quantitative biology and medicine. *J. Hist. Med.*, 19, 1964 pp. 99–117.

[2] *J. Hist. Med.*, 15, 1960 p. 306.

[3] See also Clemens, F. J. *Giordano Bruno und Nicholas von Cusa*, Bonn, 1847.

[4] Thorndike Lynn. *Op. cit.*, Chapter VIII, pp. 142–150.

[5] Blaschke Willhelm, and Schoppe Günther. Regiomontanus: Commensurator. *Akad. der Wiss. u.d. Lit. Abh. math.-nat. Klasse*, Jhg. 1956 (Nr. 7), pp. 445–529.

[6] Clagett, Marshall. A note on the *Commensurator* falsely attributed to Regiomontanus. *Isis*, 60, 1969, pp. 383–384.

that the *Commensurator* dates from the fourteenth century, and was begun by a person unknown, to be completed shortly after 1343 by the French astronomer and mathematician Johannes de Maris. Clagett proposes to edit the complete work, which cannot now be regarded as an original work by Regiomontanus.[1]

The combination of artist and scientist is by no means uncommon, but Leonardo da Vinci's fame as a painter has tended to obscure his recognition as the most versatile scientist of any century. This was largely due to the fact that while his paintings were recognized as masterpieces during his lifetime, his scientific work remained in manuscript form until long after his death. Leonardo da Vinci (1452–1519) was born probably in Anchiano, a village near Vinci between Florence and Pisa, on 15th April, 1452, and was the first to disregard the teachings of Galen, preferring to experiment for himself. He intended to write a book on human anatomy, which might well have secured for him the niche accorded to Vesalius as founder of scientific anatomy, but nothing of his work was published during his lifetime. He made dissections to improve his artistic skill, and executed some 750 separate anatomical drawings. These he left to Frencesco Melzi, from whom they were obtained by Pompeo Leoni, the sculptor. Prince Charles purchased some of these in 1623, and they were deposited with drawings by Hans Holbein in a secret cupboard in the Royal Library at Windsor. There they were discovered by Richard Dalton, librarian to George III, who showed them to William Hunter. The latter decided that Leonardo "was the best anatomist at that time in the world".[2] Leonardo contributed to anatomy, embryology, and physiology, his notes and drawings illustrating these subjects being recognized as the first scientific illustrations, which have never been surpassed.[3] His

[1] See *Der deutsche Kalender des Johannes Regiomontan, Nuremberg, um* 1474. *Faksimiledruck nach dem Exemplar der Preussischen Staatbibliothek. Mit einer Einleitung von Ernst Zinner.* (*Veröffentlichungen der Gesellschaft für Typenkunde des XV. Jahrhunderts Wiegendruck-gesellschaft. Reihe B. Seltene Frühdrucke in Nachbildungen. Bd.* 1), Leipzig, 1937; Zinner, Ernst. *Leben und Wirken des Johannes Müller von Königsberg genannt Regiomontanus.* (*Schriftenreihe zur bayerischen Landesgeschichte. Herausgegeben v.d. Kommission f. bayerischen Landesgeschichte bei der Bayerischen Akademie der Wissenschaften, Bd. 31*), Munich, 1938; and Zinner, Ernst. Neuer Regiomontanus-Forschungen und ihre Ergebnisse. *Sudhoffs Archiv. Gesch. Med.*, 37, 1953, pp. 104–108.

[2] See Finch, Sir Ernest. The forerunner [Leonardo da Vinci]. Thomas Vicary lecture delivered at the Royal College of Surgeons of England on 29th October, 1953. *Ann. Roy. Coll. Surg. Engl.* 14, 1954, pp. 71–91.

[3] See Belt, Elmer. Leonardo da Vinci, medical illustrator. *Postgrad. Med.*, 16, 1954, pp. 150–157; Belt, Elmer. *Leonardo the anatomist.* (*Logan Clendening Lectures on the History and Philosophy of Medicine. Fourth Series*), Lawrence, Kansas, 1955; Böttger, Herbert. Die Embryologie Leonardos da Vinci. *Centaurus*, 3, 1953–54, pp. 222–235; Heydenreich, Ludwig H. *Leonardo da Vinci*, 2 vols., 1954; and Saunders, J. B. de C. M. Leonardo da Vinci as anatomist and physiologist: a critical evaluation. *Texas Rep. Biol. Med.*, 13, 1955, pp. 1010–1026.

anatomical notebooks have been translated by E. MacCurdy,[1] and J. Playfair McMurrich[2] has contributed a study of Leonardo as an anatomist. His contributions to physiology have also been separately investigated.[3]

Leonardo studied mathematics, astronomy, and flying, constructing a model of a flying machine after carefully investigating the flight of birds,[4] and he suggested the possibility of the helicopter and the parachute, He designed a parabolic compass, made drawings of quick-firing and breech-loading guns as part of his contribution to military science, invented the camera obscura, conducted experiments on hydraulics and on combustion,[5] and was a keen architect. As an enthusiastic botanist he prepared many detailed drawings of plants. The majority of Leonardo's drawings are in the Ambrosian Library at Milan, others being in Paris and in the Royal Library at Windsor. His posthumously published works include the *Tabula anatomica*, [*etc.*], Lüneburg, 1830; *I manoscritti di Leonardo da Vinci della reale biblioteca di Windsor*, [*etc.*], two volumes, Paris, Turin, 1898–1901; *Notes et desseins*, twelve volumes, Paris, 1901]; and *Quaderni d'anatomia. I–VI Fogli della Royal Library di Windsor*, [*etc.*] six volumes, Oslo, 1911–16. It was believed that a number of Leonardo manuscripts existed in the Biblioteca Nacional of Madrid, and after a number of unsuccessful searches they were discovered accidently by Jules Piccus. These deal almost entirely with science and technology, and have been the subject of a paper by Ladislao Reti.[6] A biography by V. P. Zubov[7] provides information on Leonardo's manuscripts and on his published writings, the best bibliography of which is by Ettore Verga,[8] who lists 2,900 items.

Had Francis Bacon (1561–1626), Viscount St. Albans, deserted politics to devote his entire time to scholarship, he might have contributed even more usefully to scientific discovery. Instead, his philosophical writings pointed the way to organized research, and led to the founding of the

[1] *The note books of Leonardo da Vinci. Arranged, rendered into English . . . by E. MacCurdy* 2 vols., 1938.

[2] McMurrich, J. Playfair. *Leonardo da Vinci the anatomist, 1452–1519*, 1930.

[3] See Keele, K. D. *Leonardo da Vinci on movement of the heart and blood*, [*etc.*], 1952; and Baumgartner, Leono. Leonardo da Vinci as a physiologist. *Ann. med. Hist.*, N.S.4, 1932, pp. 155–171.

[4] See Hart, Ivor B. *The world of Leonardo da Vinci, man of science, engineer and dreamer of flight*, 1961; Hart, Ivor B. *The mechanical investigations of Leonardo da Vinci*, 1925, which contains (pp. 200–235) an English translation of Leonardo's MS. *On the flight of birds;* see also the special issue of *Scientia*, 1952–53, containing eight studies of Leonardo as a scientist. These were reprinted in Italian and French editions.

[5] See Reti, Ladislao. Leonardo da Vinci's experiments on combustion. *J. chem. Educ.*, 29, 1952, pp. 590–596.

[6] Reti, Ladislao. The Leonardo da Vinci codices in the Biblioteca Nacional of Madrid, *Technology and Culture*, 8, 1967, pp. 437–455.

[7] Zubov, V. P. *Leonardo da Vinci. Translated from the Russian by David H. Kraus*, Cambridge, Mass., 1968.

[8] Verga, Ettore. *Bibliografia Vinciana, 1493–1930*, 2 parts, Bologna, 1931.

Royal Society. He urged the necessity for experiment, as opposed to taking for granted information derived from books, and his writings remain as evidence of a great mind that might have been more usefully employed in practising the injunctions it urged upon others. Bacon was the subject of a bibliographical study by George Walter Steeves in 1910,[1] and more recently by R. W. Gibson;[2] the following information is largely derived from the latter work, Bacon's *Essayes* was twice printed in London in 1597, one edition being in octavo and the other duodecimo; also, in 1598, 1606, 1612 and 1613, among others, with an Edinburgh edition in 1614. There were eleven Italian editions, including printings in London, 1617 and 1618 (twice); Florence, 1619; Venice, 1619; and Milan, 1620. French translations were published in Paris, 1619, 1621 and 1622 and London, 1619. Latin versions were published at Leyden in 1641, 1644 (three), 1659 and Amsterdam, 1662 and 1685. Dutch translations were published at Leyden in 1646, 1647 and 1649, and one at Rotterdam in 1649. A German translation was published at Nuremberg in 1654, and a Swedish version at Stockholm in 1726. Bacon's *The twoo bookes of . . . the proficience and advancement of learning*, [*etc.*], was published in London, 1605, 1629, and in Oxford, 1633, with a French version printed in Paris in 1624. *De sapientia veterum*, [*etc.*], was published in London, 1609, 1617 and 1634, with editions from Leyden, 1633 (two), 1657; Arnhem, 1680; and Amsterdam, 1680 and 1684. English versions of this were printed in London, 1619 (twice), 1622, 1658, and in Edinburgh 1681. A French translation was printed in Paris in 1619 and 1644, and a German translation at Nuremberg in 1654.

Francis Bacon's *Novum organum* was printed twice in London in 1620; at Leyden in 1645 and 1650; and in Amsterdam in 1660 and 1694. His *Historia ventorum* was published in London, 1622; at Leyden, 1638 and 1648; and at Amsterdam, 1695; with French translations from Paris, 1649 and 1650; and an English translation published in London, 1653. *De dignitate et augmentis scientiarum* was printed on numerous occasions, including editions from London, 1623 and 1623 [1624]; a French translation being printed in Paris in 1632, 1634, 1640 and 1689. An English translation was published under the title *Of the advancement and proficience of learning*, [*etc.*], and was printed at Oxford, 1640 (three issues), and London, 1674. A German translation appeared from Stuttgart, 1665, and Jena, 1719, while a Swedish version was published in Stockholm in 1729 and 1736. The *Historia vitae et mortis* was published in London, 1623; Leyden, 1636 and 1637; Dillingen, 1645; Amsterdam,

[1] Steeves, George Walter. *Francis Bacon: a sketch of his life, works and literary remains, chiefly from a bibliographical point of view*, 1910.

[2] Gibson, R. W. *Francis Bacon: a bibliography of his works and of Baconiana to the year 1750*, Oxford, 1950; contains many facsimiles of title-pages. Privately printed *Supplement*, Oxford, 1959.

1663; and Strassburg, 1712; with an English translation from London, 1638, and a French translation published in Paris, 1647 and 1653.

In 1626 the first edition of Bacon's *Sylva sylvarum: or a natural historie*, [*etc.*], was published in London, and later editions include printings in 1627, 1628, 1631, 1635 and 1651. A French translation was published in Paris in 1631, and Latin versions were printed in Amsterdam, 1648 and 1661, and Leyden, 1648. His *Nova Atlantis* was published at Utrecht in 1643, with an English translation from London, 1659, and a French version printed in Paris in 1702. Bacon's collected writings were published as *Opera omnia* at Frankfort on Main, 1665 (twice); Amsterdam, 1684, 1685, 1696 and 1730; an edition edited by S. J. Arnold was published at Leipzig, 1694, and Copenhagen, 1694. Bacon's *Complete works*, edited by John Blackbourne, were published in London in 1730; his *Philosophical works*, edited by Peter Shaw, appeared in three volumes, London, 1733; and his *Works*, edited by Basil Montague, were published in seventeen volumes, London, 1825–74. A further edition of his *Works* was edited by James Spedding, Robert Leslie Ellis and Douglas Denon Heath in seven volumes, London, 1857–59, and James Spedding's *The letters and the life of Francis Bacon* also appeared in seven volumes, London, 1861–74. The complete fourteen volumes were reprinted, New York, 1968. A selection from his writings was published in 1955.[1] Although Francis Bacon's first work was published in the sixteenth century, his influence was largely felt in the seventeenth and succeeding centuries.[2]

Despite the fact that Copernicus' *De revolutionibus* was printed in 1543, no textbook in common use in Europe during the sixteenth century expounded the Copernican theory, few even mentioning it. In fact, Sacrobosco's *Sphaera* was the most popular text in the sixteenth century, although it had been written over three hundred years earlier.[3] Popular acceptance of the Copernican system did not follow until after the publication of such works as John Wilkins'[4] *Discovery of a world* [*etc.*], 1638, and *A discourse concerning a new world*, 1640, and Fontenelle's[5] *Entretiens sur la pluralité des mondes*, 1686. Astronomy had received attention from earliest times, and it attracted several prominent students

[1] *Selected writings of Francis Bacon. With an introduction and notes by Hugh G. Dick*, New York, [1955].

[2] See Rossi, Paolo. *Francis Bacon, from magic to science. . . . Translated from the Italian by Sacha Rabinovitch*, 1968; Crowther, J. G. *Francis Bacon: the first statesman of science*, 1960; Farrington, Benjamin. *Francis Bacon, philosopher of industrial science*, New York, 1949; Sturt, Mary. *Francis Bacon: a biography*, 1932; and West, Muriel. Notes on the importance of alchemy to modern science in the writings of Francis Bacon and Robert Boyle. *Ambix*, 9, 1961, pp. 102–114.

[3] See Johnson, Francis R. Astronomical text-books in the sixteenth century. In, *Science, medicine and history*, Vol. 1, 1953, pp. 285–302; and Dingle, Herbert. Astronomy in the sixteenth and seventeenth centuries. *Ibid.*, Vol. 1, 1953, pp. 455–468.

[4] See p. 89.

[5] See p. 93.

in the sixteenth century. Ptolemy's *Catalogue of stars* had a sequel in 1437, when Ulugh Beg made public his *Catalogue*. Ulugh Beg (1394–1449) was a Persian astronomer who erected an observatory in Samarkand, and his *Catalogue of stars* appeared in an English edition as recently as 1916.[1]

Nicolaus Copernicus (Coppernicus, or Coppernic) (1473–1543) has been described as the founder of modern astronomy[2] but he also studied mathematics, law and medicine, his work as a medical officer of health being particularly noteworthy. Copernicus was born at Thorn (Torun) in Polish Pomerania, and went to school there before proceeding in 1491 to the University of Cracow. He studied law at Bologna from 1496 to 1500, when he went to Padua and studied medicine. In fact, he appears to have studied at various universities for fifteen years before returning to Poland, where he became successful as a physician. In his absence he had been elected a Canon of Frauenburg Cathedral, and there he made most of his discoveries. After spending many years on his book, Copernicus circulated a résumé in manuscript, before it was printed at Nuremberg in 1543 as *De revolutionibus orbium coelestium libri VI*.[3] Later editions include printings at Basle, 1566; Amsterdam, 1617; and Warsaw, 1754, containing a Polish translation edited by John Baronowski. This is the first edition to contain the author's original preface, the 1548 edition carrying a forged, anonymous preface by the editor, Andreas Ossiander. These were superseded by the edition of Maximilian Curtze, based on a critical examination of original manuscripts and published at Thorn in 1873. A German translation of this text, by C. L. Menzzer, was published at Thorn six years later, and a French translation by Alexandre Koyré of the first eleven chapters was printed at Paris in 1934. In 1616, seventy-three years after initial publication, this first printed book by Copernicus was prohibited by being placed on the *Index*. But Copernicus had written another work before *De revolutionibus*, which was circulated in manuscript, and then lost for three centuries. A copy was found in Vienna, and published by Maximilian Curtze,[4] while a second copy, discovered in the same city, was published by Arvid Lindhagen.[5] Leopold Prowe[6] based his text on these two manuscripts,

[1] Knobel, Edward Ball. *Ulugh Beg's Catalogue of stars. Revised from all Persian manuscripts existing in Great Britain, with a vocabulary of Persian and Arabic words*, Washington, 1917. (Carnegie Institute Publication, No. 250.)

[2] But see Africa, Thomas W. Copernicus' relation to Aristarchus and Pythagoras. *Isis*, 52, 1961, pp. 403–409.

[3] Facsimile edition, Paris, 1927; a further facsimile reprint was announced in 1966.

[4] *Mittheilungen des Coppernicus-Vereins für Wissenschaft und Kunst zu Thurn*, I, 1878, pp. 1–17.

[5] *Bihang till K. Svenska Vet. Akad. Handlingar*, 6, xii, Stockholm, 1881.

[6] Prowe, Leopold. *Nicolaus Coppernicus*, 2 vols., Berlin, 1883–84. Vol. 2, pp. 184–202.

and Edward Rosen has translated Prowe's text into English.[1] Yet another work, *De lateribus et angulis triangulorum*, [*etc.*], Wittenberg, 1542 (Plate 1), was the only other book by Copernicus to be printed during his lifetime. Copernicus was also the author of a résumé of *De revolutionibus* entitled *De hypothesibus motuum coelestium a se constitutis commentariolus*, written about 1509, and of which only a few copies, made after his death, remain; and *Monetae cudendae ratio*, a treatise on currency, written between 1519 and 1525. Copernicus died on 24th May, 1543, and as he lay on his death-bed received the first printed copy of *De revolutionibus*. His collected works, edited by Fritz Kubach, began publication as *Gesamtausgabe*, Munich and Berlin, in 1944.[2] The medical library and other books belonging to Copernicus are preserved in the University of Uppsala, where they were taken by Gustavus Adolphus during the Thirty Years' War. The Copernican system of the heavenly bodies influenced both Kepler and Newton, and it is such germs of thought that have frequently grown to full fruition in the minds of succeeding generations.[3]

The author of several geographical and mathematical books, some of which were printed at his own press, Petrus Apianus (Peter Bienewitz or Bennewitz) (1501–52) is mainly known for his *Astronomicum Caesareum*, Ingolstadt, 1540. Born at Leisnig in Saxony, Apianus eventually became professor of mathematics at the University of Ingolstadt in Bavaria. A facsimile of the *Astronomicum* from the copy owned by Tycho Brahe, and now in Gotha Landesbibliothek, was printed in nine-colour photo-type with some hand-colouring, and bound in full calf, Leipzig, 1967. This contains an introduction by Diedrich Wattenberg in German and English. The meteoroscope used by Apianus was derived from Johannes Werner (1468–1528), whose geographical works Apianus issued in a new edition after the author's death.[4]

Thomas Digges (born *c.* 1545) was an ardent supporter of Copernicus,

[1] Rosen, Edward. The Commentariolis of Copernicus. *Osiris*, 3, 1938, pp. 123–141; see also Rosen, Edward. *Three Copernican treatises.* (*The* Commentariolus *of Copernicus, the* Letter against Werner, *the* Narratio prima *of Rheticus*), New York, 1939. (Records of Civilization No. 30); reprinted by Dover Books, New York, 1959.

[2] Baranowski, Henryk. *Bibliografia kopernikowska, 1509–1955*, Warsaw, 1958.

[3] See Armitage, Angus. *Copernicus: the founder of modern astronomy. Revised edition*, 1957; Buck, R. W. Doctors afield—Nicolaus Coppernic (1473–1543). *New Engl. J. Med.*, 250, 1954, pp. 954–955; Rytel, Alexander. Nicolaus Copernicus, physician and humanitarian. A new approach. *Bull. Pol. Med. Sci. Hist.*, 1, 1956, pp. 3–11; Mizwa, Stephen P. *Nicholas Copernicus. A tribute of nations*, New York, 1945; Rudnicki, Józef. *Nicholas Copernicus (Mikola Kopernik) 1473–1543. . . . Translated from the Polish, by B. W. A. Massey*, [*etc.*], 1943; Rosen, Edward. The authentic title of Copernicus's major work. *J. Hist. Ideas*, 4 1943, pp. 457–474; Dingle, Herbert. Nicolaus Copernicus, 1473–1543. *Endeavour*, 2, 1943 pp. 136–141; and Ravetz, J. R. *Astronomy and cosmology in the achievement of Nicolaus Copernicus*, Warsaw, 1968. (Polish Academy of Sciences, Research Centre for the History of Science and Technics. Monographs on the history of science and technics, Vol. 30.)

[4] See North, J. D. Werner, Apian, Blagrave and the meteoroscope. *Brit. J. Hist. Sci.*, 3, 1966, pp. 57–65.

a pupil of John Dee and a friend of Tycho Brahe. Digges greatly influenced astronomical thought in England, supervised the fortification of Dover Harbour, and became Muster-Master General of English forces in the Netherlands. He was the author of two important astronomical works, and left a third unfinished, entitled *Commentaries on the revolutions of Copernicus*. Thomas Digges' first book was the *Pantometria*, 1571, begun by his father, and which went into a second enlarged edition as *A geometrical practical treatise named Pantometria*, 1591 [S.T.C. 6859]. He wrote a book on a new star that appeared in the constellation of Cassiopeia in 1572. The work was entitled *Alae seu scalae mathematicae*, 1573. Thomas Digges' *A perfit description of the caelestiall orbes according to the most anciente doctrine of the Pythagoreans, lately revived by Copernicus and by geometricall demonstrations approved* was appended to the new edition of his father Leonard Digges' *Prognostication everlastinge*, 1567. The appendix is largely a translation of part of Copernicus' *De revolutionibus*, and is illustrated with diagrams. The work went into many editions, all containing Thomas Digges' additional matter, including printings in 1578, 1583, 1585, 1592, 1596 and 1605. Most of these editions are very rare, and copies are often imperfect. Thomas Digges' work has not been published separately, and has therefore tended to be overlooked.[1]

Tycho Brahe (1546–1601) has been described as the "greatest astronomer of the second half of the sixteenth century", and Herbert Dingle has stated: "certainly without Tycho's work there would have been no Kepler's laws, and possibly Newton would have been unable to formulate his law of gravitation".[2] Born at Knudstrup, Denmark, on 14th December, 1546, Tyge (Latinized to Tycho) Brahe went to Copenhagen University in 1559.[3] He quickly took to astronomy, and after three years went to Leipzig. Returning to Denmark in 1565, Tycho went to Wittenberg, Rostock, Basle and Augsburg, before returning to his native land in 1570. For a time he studied chemistry, but returned to astronomy, travelling extensively in Europe. He was about to settle in Basle, when the King of Denmark became his patron, installing him on the island of Hveen. There Tycho built a house and two observatories, one known as Uraniburgum or Uraniborg, and the other Stellaeburgum. He also installed a printing press and a paper mill, to produce his own writings. His first book, *De nova stella*, had been published at Copenhagen in 1573, but few copies were issued. It was translated into English as *Astronomicall*

[1] Johnson, Francis R. The influence of Thomas Digges on the progress of modern astronomy in sixteenth-century England. *Osiris*, 1, 1936, pp. 390–410; and Johnson, Francis R., and Larkey, Sanford V. Thomas Digges, the Copernican system, and the idea of the infinity of the universe in 1576. *Huntington Lib. Bull.*, No. 5, April, 1934, pp. 69–117. This latter reprints the treatise from the 1576 edition, (pp. 78–95).

[2] Dingle, Herbert. Tycho Brahe, 1546–1601. *Endeavour*, 5, 1946, pp. 137–141.

[3] See Christianson, John. Tycho Brahe at the University of Copenhagen, 1559–62. *Isis*, 58, 1967, pp. 198–203.

coniectur of the new and much admired star which appeared in the year 1572, London, 1632 [S.T.C. 3538]. This was followed by *Epistolarum astronomicarum libri*, [*etc.*], Uraniborg, 1596; *De mundi aetherei recentibus phaenomensis liber secundus*, [*etc.*], Uraniborg, 1588, which was not offered for sale until 1603, although copies were distributed by the author. The first volume was printed at Uraniborg in sections over several years, with the title *Astronomiae instauratae progymnasmata*, and published in 1602, some copies being dated 1603. This was issued with a new title-page, with Volume 2, from Frankfurt in 1610.

Tycho's patron died in 1588, and in 1597 the astronomer left Hveen for Copenhagen, thence to Rostock, and on to the castle at Wandsbeck, near Hamburg. His *Astronomiae instauratae mechanica* was published at Wandsbeck in 1598, a very rare edition, and was reprinted at Nuremberg in 1602. In 1599 Tycho went to Prague, entering the service of Rudolph II, and residing in a castle at Benatky.[1] Early in 1600 he was visited there by Kepler, who worked with him for a period. But the restless Tycho moved back to Prague in 1600, to die there on 24th October, 1601. Kepler received Tycho's manuscripts, which were later sold by Ludwig Kepler to Frederick III of Denmark, and deposited in the Royal Library. A complete edition of Tycho Brahe's *Opera omnia*, edited by J. L. E. Dreyer, was published at Copenhagen in fifteen volumes, 1913–29, and a reprint of this was announced in 1969.[2]

A native of Nola, near Naples, Giordano Bruno (1548–1600) went to London in 1583, and published several books there, some with false imprints. Bruno's philosophy was based on Nicholas of Cusa and on Copernicus, and he asserted that the universe consists of many worlds, that this world is not the centre of the universe, that the universe is infinite in space and time, and is permeated by a common soul. Bruno was burned at the stake in 1600, after being imprisoned by the Inquisition for seven years. Bruno's works include *Philothei J. Bruni recens et completa ars reminiscendi* (*Explicatio triginta sigillorum*), two parts, [London, 1583?]

[1] See Folta, J., *et al. History of exact sciences in Bohemian lands up to the end of the nineteenth century*, Prague, 1961.

[2] See Dreyer, J. L. E. *Tycho Brahe: a picture of scientific life and work in the sixteenth century*, Edinburgh, 1890, (catalogue of Tycho's MSS. in the Royal Library, Copenhagen, pp. 390–392), reprinted New York, 1963 (Dover), and Gloucester, Mass., 1964; Nielsen, Lauritz. *Tycho Brahes Bogtrykkeri. En bibliografiskbokhistorisk Undersøgelse*, Copenhagen, 1946, (a study of his press and writings); Gade, John Allyne. *The life and times of Tycho Brahe*, Princeton, 1947, (not well documented, or free from errors); *Tycho Brahe's Description of his instruments and scientific work as given in* Astronomiae instauratae mechanica (*Wandesburgi 1598*). *Translated and edited by Hans Raeder, Elis Strömgren and Bengt Strömgren* (*Det Kongelige Danske Videnskabernes Selskab*), Copenhagen, 1946; Jones, Sir H. Spencer. Tycho Brahe (1546–1601). *Nature*, 158, 1946, pp. 856–861; Thoren, Victor E. An early instance of deductive discovery: Tycho Brahe's lunar theory. *Isis*, 58, 1967, pp. 19–36; and Hellmann, C. Doris. Was Tycho Brahe as influential as he thought? *Brit. J. Hist. Sci.*, 1, 1963, pp. 295–324.

[S.T.C. 3939]; *La cena de le ceneri, descritta in cinque dialogi*, London, 1584 [S.T.C. 3935]; *De la causa principio et uno*, Venice, [London], 1584 [S.T.C. 3936]; *Del' infinito universo et mondi*, Venice [London], 1584 [S.T.C. 3938]; *Spaccio de la bestia trionfante*, Paris [London], 1584 [S.T.C. 3940]; *De gl' heroici furori*, Paris [London], 1585 [S.T.C. 3937], and *De innumerabilibus immenso et infigurabili; seu de universo et mundis libri octo*, 1591. His *Opera latine conscripta* were published in eight volumes, Florence and Naples, 1879–91, and this edition was reprinted in facsimile, Stuttgart, 1968. John Hayward[1] and A. Nowicki[2] have recorded the locations of early editions of Bruno's writings, and Frances A. Yates[3] has provided an authoritative study of his work.[4]

The work of Johann Kepler (1571–1630) paved the way for Newton's discoveries, and, as noted above, Kepler worked with Tycho Brahe. Johann Kepler was born on 27th December, 1571, at Weil, near Stuttgart, and was educated at Tübingen. He lectured on astronomy at Graz, and in 1596 published his *Prodromus dissertationum cosmographicorum continens mysterium cosmographicum* at Tübingen, with a second edition in 1621. This led to his visiting Tycho, and after the death of the latter, Kepler became Imperial Mathematician to the Emperor Rudolph. In 1612 he went to Linz, and spent several years constructing tables of the planetary motions, based on Tycho's observations. At Ulm, Kepler published his *Tabulae Rudolphinae* in 1627, and later became professor of astronomy at Rostock. He died on 15th November, 1630.

Kepler's investigations into the law of refraction are described in his *Ad vitellionem paralipomena*, Frankfurt, 1604, and the reasons for his formula, which appear to have escaped his commentators, are discussed in a paper by R. A. Houstoun.[5] Kepler also wrote *De stella nova in pede Serpentarii* [*etc.*], Prague and Frankfurt, 1606; and *Astronomia nova*, [*etc.*], Prague, 1609,[6] translated into German by Max Caspar, Munich and Berlin,

[1] Hayward, John. The location of copies of the first editions of Giordano Bruno. *Book Collector*, 5, 1956, pp. 152–157; supplemented by, First editions of Giordano Bruno. Location of additional copies. *Ibid.*, 5, 1956, pp. 381–382.

[2] Nowicki, A. Early editions of Giordano Bruno in Poland. *Book Collector*, 13, 1964, pp. 342–345.

[3] Yates, Frances A. *Giordano Bruno and the Hermetic tradition*, 1964.

[4] See also Singer, Dorothea Waley. *Giordano Bruno, his life and thought. With annotated translation of his work On the infinite universe and worlds*, New York, 1950; Singer, Dorothea Waley. The cosmology of Giordano Bruno (1548–1600). *Isis*, 33, 1941–42, pp. 187–196; Armitage, Angus. The cosmology of Giordano Bruno. *Ann. Sci.*, 6, 1948–50 pp. 24–31; and Salvestrini, Virgilio. *Bibliografia de Giordane Bruno 1582–1950). 2nd ed. postuma, a cura di Luigi Firpo*, Florence, 1958.

[5] Houstoun, R. A. Kepler and the law of refraction. *Institute of Physics Bulletin*, 9, 1958, pp. 3–6; see also Ronchi, Vasco. From seventeenth-century to twentieth-century optics. *International Council of Scientific Unions Review*, 4, 1962, pp. 145–156.

[6] See Wilson, Curtis. Kepler's derivation of the elliptical path. *Isis*, 59, 1958, pp. 5–25; and Russell, L. J. Kepler's laws of planetary motion 1609–1666. *Brit. J. Hist. Sci.*, 2, 1964, pp. 1–24.

1929. This book was the subject of scholarly study by Robert Small in *An account of the astronomical discoveries of Kepler*, London, 1804, which was reprinted with a foreword and index by William D. Stahlman, Madison, 1963. Kepler's *Dissertatio cum nuncio sidereo nuper ad mortales misso a G. Galilaeo*, Prague, 1610, was first completely translated into English by Edward Rosen as *Kepler's Conversation with Galileo's sidereal messenger*, New York, 1965. His *Strena, seu de nive sexangula*, Frankfurt, 1611, is very rare and went into a second edition in 1670, with German translations by R. Klug (1907), F. Rossmann (1943), and by H. Strunz and H. Borm (1958). A modernized text of the Latin edition, with an English translation by Colin Hardie on opposite pages, and essays by L. L. Whyte and B. F. J. Mason, has been published as *The six-cornered snowflake*, Oxford, 1966. This volume also provides bibliographical information on editions, translations and commentaries. Kepler's *Epitome astronomicae copernicanae* was issued in three parts, Linz and Frankfurt, 1618–21, and went into a second edition, Frankfurt, 1635. An English translation of this is contained in *Great books of the western world*, Chicago, 1952, pages 839–1004. Kepler's *Somnium* was published posthumously in 1634, but he had composed the first draft of the manuscript in 1593. This has been translated by Patricia Frueh Kirkwood[1] and by Edward Rosen,[2] who provides a bibliography, many useful appendices, footnotes, and an interesting introduction. Kepler was also the author of *De cometis*, 1618; *Epitome astronomiae Copernicanae*, Linz and Frankfort, 1618–21; and *Harmonice mundi*, Augsburg, 1619. Max Caspar has compiled an invaluable bibliography of Kepler's writings listing 162 items from 1590 to 1930.[3] Kepler's collected works began publication under the editorship of Walther von Dyck before the war in Munich, as *Gesammelte Werke*, and three volumes appeared between 1937 and 1939. Volume eighteen, the last of six volumes of letters, was published in 1959, volume nine, *Mathematische Werke*, appeared in 1960, and volume eight, *Mysterium Cosmographicum*, in 1964.[4]

Mathematical works were very popular with the early printing presses,

[1] See Lear, John. *Kepler's dream. With the full text and notes of* Somnium sive astronomia lunaris, Joannis Kepleri, *translated by Patricia Frueh Kirkwood*, Berkeley, California, 1965; see also a critical examination of this by Rosen, Edward. Kepler's *Dream* in translation. *Isis*, 57, 1966, pp. 392–394.

[2] *Kepler's* Somnium: *The Dream, or posthumous work on lunar astronomy. Translated with a commentary by Edward Rosen*, Madison, London, 1967.

[3] Caspar, Max. *Bibliographia Kepleriana. Ein Führer durch das gedruckte Schrifttum von Johannes Kepler*, Munich, 1936, reprinted 1968; see also Caspar, Max. *Kepler. Translated and edited by C. Doris Hellman*, London, New York, 1959.

[4] See also Armitage, Angus. *John Kepler*, 1966; Baumgardt, Carola. *Johannes Kepler: life and letters*, [*etc.*], 1952; Grant, R. *Johann Kepler. A tercentenary commemoration of his life and work*, Baltimore, 1931; Cajori, Florian, Johannes Kepler, 1571–1630. *Sci. Mon.*, 30, 1930, pp. 385–394; and Rosen, Edward. Galileo and Kepler: their first two contacts. *Isis*, 57, 1966, pp. 262–264.

and it has been estimated that at least sixty-three of these were printed between 1480 and 1490. The first, by an unknown author, is very rare, and was printed at Treviso in 1478.[1] The *Nobel opera de arithmetica*, by Pietro Borgo or Borghi, was first printed in Venice by Ratdolt in 1484. Nothing is known about the author, but sixteen editions of the book appeared from Venetian presses in a century. After the first edition it appeared with the title *Libro de abacho*.[2] The first arithmetic printed in Spain was the *Suma de la art de arismetica*, by Francesch Sanct Climent, published in Barcelona in 1482.[3]

In England, *The mirrour of the world*, printed by Caxton in 1481, contained only one page devoted to arithmetic, and the first book devoted entirely to the subject was an anonymous work published at St. Albans in 1537 by John Herford with the title *An introduction for to lerne to recken with the pen, or with the counters*. This is represented only by a fragment in the British Museum. Later editions were printed in London, 1539 [S.T.C. 14118]; 1546 [S.T.C. 14119]; 1552 [not in S.T.C.]; 1556 [not in S.T.C.]; 1574 [S.T.C. 14120]; 1581 [S.T.C. 14121]; and the eighth and last edition, published with the title *Arithmaticke, or an itroduction* [*sic.*], [*etc.*], 1629 [S.T.C. 14122].[4] This book was actually based on an anonymous Dutch arithmetic, and also included part of a French treatise. The Dutch work was first published at Brussels in 1508 as *Die maniere om te leeren cyffren na die rechte consten algorismi. Int gheheele ende int ghebroken*, and went into a second edition published in Antwerp, [*c.* 1510], with a French translation printed in the same town in 1529. The French book was published at Lyons between 1530 and 1537 as *La vraye maniere, pour apprēdre a chiffrer et cōpter*.[5]

Michael Stifel (1486–1567) has been described as the greatest algebraist of the sixteenth century. He wrote a small treatise on the subject, but his best known mathematical work is the *Arithmetica integra*, published at Nuremberg in 1544. The third part of this is devoted to algebra.

The chief works of the Italian, Niccolo Fontana Tartaglia (1500–57), were *Nova scienza*, 1546, containing his solution of cubic equations; *Quesiti et inventioni deverse*, [*etc.*], 1546, of which there were several translations and editions, including the first complete edition of 1554, of which a facsimile edition was published at Brescia in 1959 by the Atenio di Brescia; and his *Trattato di numeri et misure*, published at Venice in six parts, 1556–60. The first four parts deal with commercial and speculative

[1] Smith, David Eugene. The first printed arithmetic (Treviso, 1478). *Isis*, 6, 1924, pp. 311–331.

[2] Smith, David Eugene. The first great commercial arithmetic. *Isis*, 8, 1926, pp. 41–49.

[3] Karpinski, Louis C. The first printed arithmetic of Spain: Francesch Sanct Climent, *Suma de la art de arismetica*, Barcelona, 1482. *Osiris*, 1, 1936, pp. 411–420.

[4] Richeson, A. W. The first arithmetic printed in English. *Isis*, 37, 1947, pp. 47–56.

[5] Bockstaele, P. Notes on the first arithmetics printed in Dutch and English. *Isis*, 51, 1960, pp. 315–321.

arithmetic, and geometry; the last two parts, on the use of compasses in Euclidean geometry, and on algebra, were edited by his publisher after the author's death. A French edition of Tartaglia's work on arithmetic was published at Paris in 1578, the translation being by Guillaume Gosselin. Tartaglia's collected works were published in Venice in 1606.

Girolana (Giralamo) Cardan (1501–76) formulated "Cardan's rule" for the solution of cubic equations, which he derived from Tartaglia, and published as his own. Cardan was an extensive traveller, and occupied chairs at Milan and Pavia. He was recognized as a distinguished astrologer, and his work on algebra, *Ars magna de rebus algebraicis*, which appeared in 1545, marked the beginning of a new epoch. Cardan's writings were widely read, and greatly influenced later scientists, his *De rerum varietate* containing several items of interest in the history of chemistry. Editions of this were published in Basle, 1557, and Basle, 1581, with several intermediate and later printings and translations which are recorded by David F. Larder.[1] His collected works appeared in ten volumes at Lyons, 1663, and a reprint of these was announced in 1968.[2]

Reinerus Gemma Frisius (1508–55) was born at Dokkum, Friesland, and was professor of medicine at Louvain, where he also lectured on mathematics. He was the author of a number of works, including *Cosmographicus liber Petri Apiani*, Antwerp, 1529; *De principiis astronomia*, Antwerp, 1530; *Libellus de locorum describendorum*, Antwerp, 1533; *Arithmeticae practicae methodus facilis*, Antwerp, 1540; *De radio astronomico et geometrico liber*, Antwerp, 1545; and *De astrolabo catholico liber*, edited by Cornelius Gemma, Antwerp, 1556.

The founder of the British school of mathematics,[3] Robert Recorde, was born about 1510[4] at Tenby, Pembrokeshire, and died in the King's Bench Prison, Southwark, in 1558. He was educated at Oxford, migrated to Cambridge and became doctor of medicine in 1545. He later taught at Oxford, before going to London. Recorde was the author of several textbooks, some of which went into many editions. These include *The grounde of artes, teachyng the perfect worke and practise of arithmetike* London, 1543 [S.T.C. 20795.5], the most popular mathematical book for a hundred years, the last edition being dated 1699.[5] *The pathway to knowledg, containing the first principles of geometrie*, two parts, 1551 [S.T.C.

[1] Larder, David F. The editions of Cardanus' *De rerum varietate. Isis*, 59, 1968, pp. 74–77.
[2] Wykes, Alan. *Doctor Cardano, physician extraordinary*, 1969.
[3] A narrative account of early English mathematical practitioners, with biographical notes and a chronological arrangement of manuscripts and printed works are contained in Taylor, Eva Germaine Rimington. *The mathematical practitioners of Tudor and Stuart England*, Cambridge, 1954.
[4] See Easton, Joy B. On the date of Robert Recorde's birth. *Isis*, 57, 1966, p. 121.
[5] The various editions of Recorde's writings are recorded in Easton, Joy B. The early editions of Robert Recorde's *Ground of artes. Isis*, 58, 1967, pp. 515–532; and Wallis, P. J. Fun with figures. *Accountant's Magazine*, Dec., 1968, pp. 1–4.

20812], with later editions in 1574 [S.T.C. 20813], and 1602 [S.T.C. 20814]; *The castle of knowledge*, 1556 and 1596 [S.T.C. 20796–7]; *The gate of knowledge*, a work on mensuration, is no longer extant; his *Treasure of knowledge*, and other works, were not completed. Recorde's *The whetstone of witte, which is the seconde parte of arithmetike*, 1557, is probably the first English work to contain the signs + and —, and also suggested the sign of equality (=). *Record's arithmeticke*, [*etc.*], first appeared in 1615 [S.T.C. 20815], and another edition was published in 1654.[1]

Dr. John Dee (1527–1608) was a mathematician, alchemist, astrologer and an empiric, who also collected together an extensive library. All his printed works are extremely rare, and many of his books remained in manuscript. Dee contributed to an edition of Recorde's *Ground of arts* and wrote a valuable preface to Henry Billingsley's first English translation of Euclid's *Elements*, published in 1570. He was the author of *General and rare memorials pertayning to the perfecte arte of navigation*, 1577 [S.T.C. 6459];[2] *The cause of floods and ebbs*, and *The philosophical and political occasions and names of the heavenly asterismes*, both written in 1553; *Propadeumata aphoristica*, London, 1558, new edition, 1567; and *Monas hieroglyphica*, Antwerp, 1564. A second edition of this latter was published at Frankfurt in 1591, and this was translated into English by J. W. Hamilton-Jones, London, 1947. A French version of the *Monas hieroglyphica*, translated by E. A. Grillot de Guiry, was published at Paris in 1925, and the English version was based on this French version, both being inaccurate and incomplete. A facsimile of the 1564 text has appeared with a translation by C. H. Josten,[3] who also provides useful biographical and bibliographical material on Dee. The latter also presented a 'supplication' to Queen Mary "for the recovery and preservation of ancient writers and monuments", which, had it been successful, would have brought him honour as the founder of the state National Library.[4,5]

Algebra and algebraic trigonometry were developed at the hands of François Viète or Franciscus Vieta (1540–1603), a French lawyer who devoted much time to the study of mathematics. His *Canon mathematicus*, [*etc.*], Paris, 1579, is extremely rare. He was also the author of an

[1] See Clarke, Frances Marguerite. New light on Robert Recorde. *Isis*, 8, 1926, pp. 50–70; Johnson, Francis R., and Larkey, Sanford V. Robert Recorde's mathematical teaching and the anti-Aristotelian movement. *Huntington Lib. Bull.*, 7, April, 1935, pp. 59–87; and Karpinski, Louis C. The whetstone of witte (1557). *Bibliotheca mathematica*, 13, 1913, pp. 223–228.

[2] See also S.T.C. Nos. 6460–6466.

[3] Josten, C. H. A translation of John Dee's "Monas hieroglyphica" (Antwerp, 1564), with an introduction and annotations. *Ambix*, 12, 1964, pp. 84–221.

[4] See Thornton, John L. Dr. John Dee and his scheme for a national library. *Medical Bookman and Historian*, 2, 1948, pp. 359–362; also in Thornton, John L. *Selected readings in the history of librarianship, 2nd ed.*, 1966, pp. 25–29.

[5] See Deacon, Richard. *John Dee, scientist, geographer, astrologer and secret agent to Elizabeth I* 1968; and Smith, Charlotte Fell. *John Dee (1527–1608)*, [*etc.*], 1909.

important treatise on algebra, *In artem analyticam isagoge*, Tours, 1591, of which an English translation by J. Winfree Smith is printed in Jacob Klein's scholarly *Greek mathematical thought and the origin of algebra*, 1968;[1] *De numerosa potestatum resolutione*, Paris, 1600, containing his researches on the theory of equations, as does *De aequationum recognitione et emendatione*, published posthumously in 1615. Vieta's *Opera mathematica*, edited by F. van Schooten, was published in Leyden, 1646.

Simon Stevin, or Stevinus, of Bruges (1548–1620), a Flemish engineer and inventor, introduced the decimal scheme for representing fractions, and conducted experiments on the relative rate of fall of bodies of different weight. He also occupied himself with the method of the resolution of forces, the law of equilibrium, on inclined planes, and laid the foundations for the science of hydrostatics. Stevin was the author of *Tafelen van interest, midtsgders de constructie der selver*, Antwerp, 1582; *Problematum geometricorum libri V*, Antwerp, 1583; *Dialectike ofte bewysconst. Leerende van allen saecken recht ende constelick oirdeelen*, [*etc.*], Leyden, 1585, and Rotterdam, 1621; *De Thiende*, Leyden, 1585, translated into French by Stevin himself as *La Disme*, Leyden, 1585, and into English with additions by Robert Norton as *Disme: the art of tenths, or dicimall arithmetike*, London, 1608 [S.T.C. 23264]; *L'Arithmétique et la pratique d'arithmétique*, two volumes, Leyden, 1585, with an edition revised by Albert Girard, Leyden, 1625; *De Beghinselen der Weeghconst*, Leyden, 1586; *De Weeghdaet*, Leyden, 1586; *De Beghinselen des Waterwichts*, Leyden, 1586; *Vita politica*, [*etc.*]., Leyden, 1590. Delft, 1611, Middelburg, 1658, and Amsterdam, 1684; *Appendice algébraique contenant règle générale de toutes équations*, Leyden, 1594; *Ideae mathematicae pars prima*, 1593; *De Sterctenbovwing*, on fortification, Leyden, 1599, which went into several editions, including an English translation by Edward Wright (1558?–1615), published as *The haven finding art, by the latitude and variation*, London, 1599 [S.T.C. 23265]; *Wiscontige Ghedachtenissen*, [*etc.*], five volumes, Leyden, 1605–08, with a Latin translation by Willebrord Snel, five volumes, and a French translation by Jean Tuning, four volumes, both published from 1605 to 1608. Stevin's *Oeuvres mathématiques*, [*etc.*] was published in six volumes at Leyden, 1634. A modern edition of the Dutch and Latin texts, with an English translation, edited by the Royal Academy of Sciences, Amsterdam was published as *The principal works of Simon Stevin*, five volumes, Amsterdam, 1955–67.[2]

[1] Klein, Jacob. *Greek mathematical thought and the origin of algebra. . . . Translated by Eva Brann. With an appendix containing Vieta's* Introduction to the analytical art. *Translated by J. Winfree Smith*, Cambridge, Mass., London, 1968.

[2] See also Sarton, George. Simon Stevin of Bruges (1548–1620). *Isis*, 21, 1934, pp. 241–303, (bibliography, iconography, and numerous facsimiles of title-pages); continued as The first explanation of decimal fractions and measures (1585). Together with a history of the decimal idea and a facsimile (no. xvii) of Stevin's Disme. *Isis*, 23, 1935, pp. 153–244.

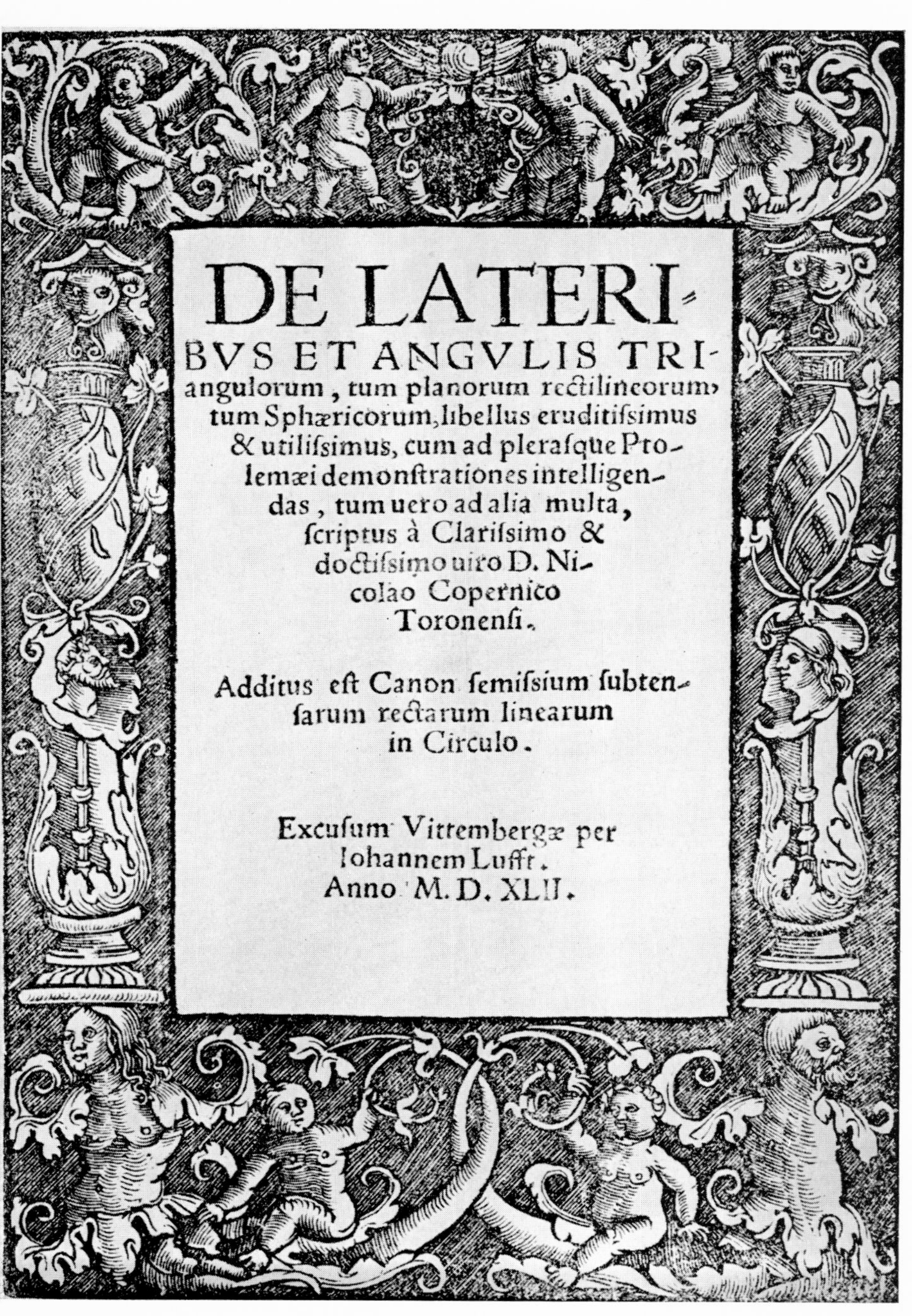

DE LATERI-
BVS ET ANGVLIS TRI-
angulorum, tum planorum rectilineorum,
tum Sphæricorum, libellus eruditiſsimus
& utiliſsimus, cum ad pleraſque Pto-
lemæi demonſtrationes intelligen-
das, tum uero ad alia multa,
ſcriptus à Clariſsimo &
doctiſsimo uiro D. Ni-
colao Copernico
Toronenſi.

Additus eſt Canon ſemiſsium ſubten-
ſarum rectarum linearum
in Circulo.

Excuſum Vittembergæ per
Iohannem Lufft.
Anno M. D. XLII.

PLATE I. Title-page of Nicolaus Copernicus' *De lateribus et angulis triangulorum*, Wittenberg, 1542

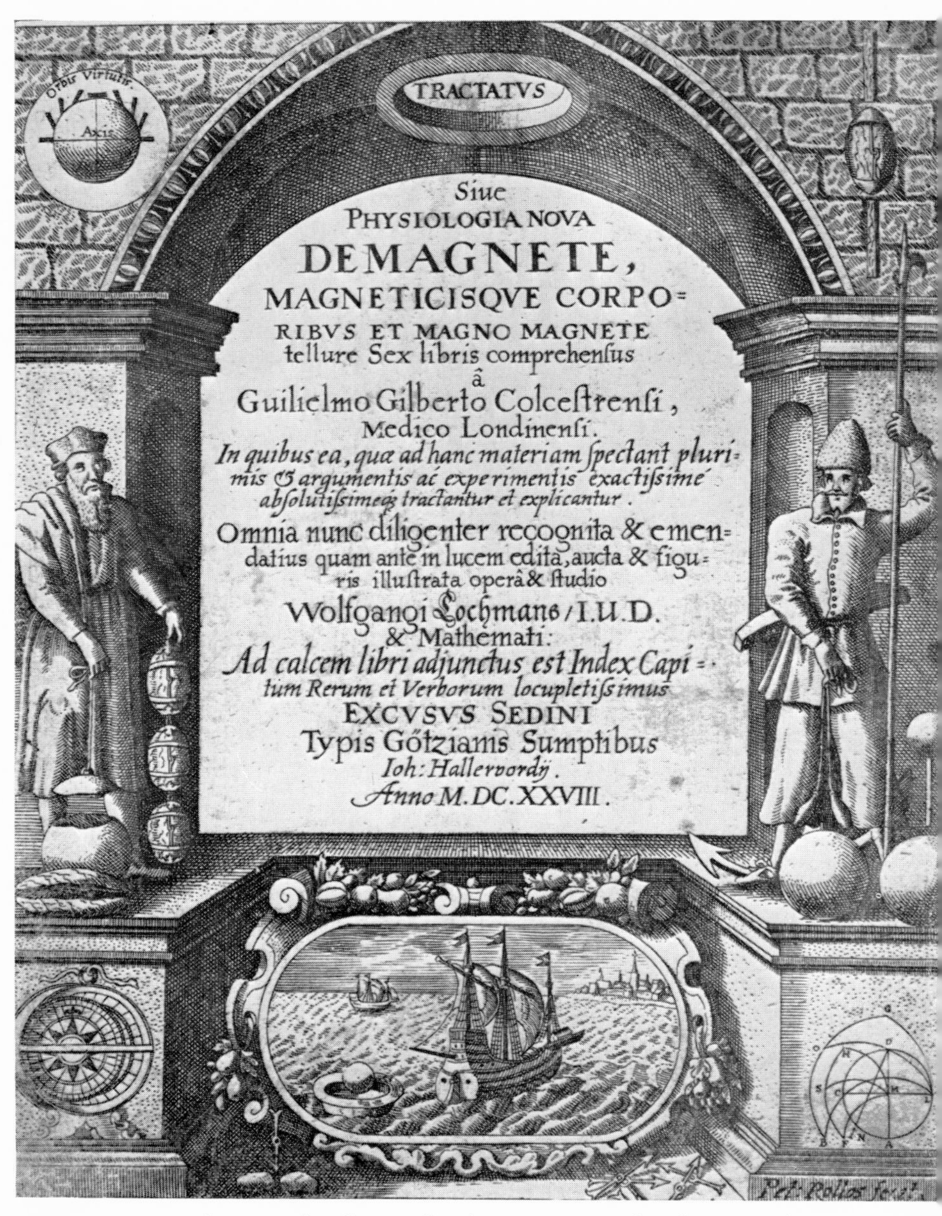
TRACTATVS

Siue

PHYSIOLOGIA NOVA

DE MAGNETE,

MAGNETICISQVE CORPO=
RIBVS ET MAGNO MAGNETE
tellure Sex libris comprehensus
â
Guilielmo Gilberto Colcestrensi,
Medico Londinensi.

*In quibus ea, quæ ad hanc materiam spectant pluri=
mis & argumentis ac experimentis exactissime
absolutissimeq; tractantur et explicantur.*

Omnia nunc diligenter recognita & emen=
datius quam ante in lucem edita, aucta & figu=
ris illustrata operâ & studio

Wolfgangi Lochmans/I.U.D.
& Mathemati:

*Ad calcem libri adjunctus est Index Capi=
tum Rerum et Verborum locupletissimus*

EXCVSVS SEDINI
Typis Gőtzianis Sumptibus
Ioh: Hallervordÿ.
Anno M.DC.XXVIII.

PLATE 2. Title-page of William Gilbert's *Tractatus sive physiologia nova de magnete*, Stettin, 1628.

The *De magnete, magneticisque corporibus, et de magno magnete tellure; physiologia nova*, [*etc.*], London, 1600 [S.T.C. 11883], of William Gilbert (1544–1603) has been described as the "first major original contribution to science that was published in England". A second edition, even more rare than the first, was published in Stettin, 1628 (Plate 2). Gilbert was a native of Colchester, who lived in London from 1573. He was physician to Queen Elizabeth, and a great experimentalist, electricity and magnetism owing much to his pioneer work.[1] His classic work, published at the very end of the sixteenth century, deals primarily with magnetism, but the last section is devoted to the system of the universe, being based chiefly on the views of Bruno. An English version. *On the magnet*, [*etc.*], was translated by Silvanus P. Thompson, London, 1900, following upon a previous translation by P. Fleury Mottelay which was published in New York in 1893 and reprinted in 1958. Thompson included much new material in his study, which the Gilbert Club of London produced as a tercentenary commemorative volume dated 1900, but actually published in 1901. Both these texts were reproduced in cheap editions in 1958. Duane H. D. Roller[2] has also produced a study of Gilbert and his work, in which he provides bibliographical details of the various editions and reprints of *De magnete*, which was also translated into Russian by A. I. Dovatur and printed in Moscow in 1956. Another work by William Gilbert, *De mundo nostro sublunari philosophia nova*, was published at Amsterdam in 1651, many years after his death, having been seen through the press by a surviving brother. A facsimile of this, the only edition to be printed, was published in Amsterdam, 1965, together with a companion study of the work by Sister Suzanne Kelly.[3,4]

Alchemical literature flourished extensively in the fifteenth and sixteenth centuries, and it was not until the following century that the subject became established on a strictly scientific basis. Arnaldo of Brussels compiled an alchemical treatise between 1473 and 1490, containing hundreds of individual recipes, and a manuscript of this at Lehigh University, Bethlehem, Pennsylvania, has been the subject of investigation by W. J. Wilson.[5] Valerius Cordus (1515–44) is generally remembered as the discoverer of ether, although he was probably not the first to

[1] See Singer, Charles. Dr. William Gilbert (1544–1603). *J. Roy. Nav. Med. Serv.*, 2, 1916, pp. 495–510.

[2] Roller, Duane H. D. *The De magnete of William Gilbert*, Amsterdam, 1959. Bibliographical details, pp. 174–182.

[3] Kelly, Suzanne. *The* De mundo *of William Gilbert*, Amsterdam, 1965; see also Kelly, Suzanne. Gilbert's influence on Bacon: a revaluation. *Physis* (*Firenze*), 5, 1963, pp. 249–258.

[4] See also Hesse, Mary B. Gilbert and the historians. *Brit. J. Phil. Sci.*, 11, 1960, pp. 1–10, 130–142; and Dawrant, A. G. William Gilbert of Colchester. *St. Bart's Hosp. J.*, 59, 1955 pp. 47–53.

[5] Wilson, W. J. An alchemical manuscript by Arnaldus de Bruxella, *Osiris*, 2, 1921, pp. 220–405.

prepare it. He conducted botanical studies, and his *Dispensatorium*, printed in 1546,[1] was the first European pharmacopœia, of which French (1578) and Dutch (1662) translations were published. His *Historia plantarum* was written before 1545, but not printed until 1561, and represents one of the most important of the early herbals.[2] His collected works, edited by Conrad Gesner, were published in 1561, and these contain an account of his synthesis of ether. Peter Coudemberg also issued editions of the works of Valerius Cordus in 1568 and 1571.[3]

Basil Valentinus may have been entirely mythical, and his alleged writings may have been written by his editor, Thölde or Thölden. A list of these, together with additional information, is contained in Ferguson,[4] but we must mention his *Triumph-Wagen Antimonii*, first published at Leipzig in 1604, with a rare edition, also from Leipzig, published in 1611, and also one of 1624.

The founder of iatrochemistry, or chemistry applied to medicine, Phillippus Aureolus Theophrastus Bombastes von Hohenheim, who later adopted the name Paracelsus, was born at Einsiedeln, near Zurich, in 1493, or according to Walter Pagel, more probably in 1494. Pagel[5] has produced a thoroughly documented study of Paracelsus, which is a reliable guide to his writings and to biographical studies of him. Paracelsus qualified in medicine, studied metals and alchemy, and spent much of his time travelling throughout Europe. His work in medicine and surgery is mentioned elsewhere,[6] and he was also a pioneer in experimental chemistry. He discovered zinc, introducing this, together with antimony, lead and copper sulphate, into the pharmacopœia. His work is intermingled with mediaeval astrology, and is full of mystical ideas. Karl Sudhoff[7] has published a scientific analysis of the editions and manuscripts of Paracelsus, who died in 1541. He had been a prolific writer, but many of

[1] See Schmitz, Rudolf. Zur Bibliographie der Erstausgabe des Dispensatoriums Valerii Cordi. *Sudhoffs Arch. Gesch. Med.*, 42, 1958, pp. 260–270.

[2] See Sprague, T. A., and Sprague, M. S. The herbal of Valerius Cordus. *J. Linn. Soc.* (*Bot.*), 52, 1939, pp. 1–13.

[3] See Leake, Chaucey D. Valerius Cordus and the discovery of ether. *Isis*, 7, 1925, pp. 14–24; and Tallmadge, Guy K. The third part of the *De extractione* of Valerius Cordus. *Isis*, 7, 1925, pp. 394–395.

[4] Ferguson, pp. 77–82.

[5] Pagel, Walter. *Paracelsus: an introduction to philosophical medicine in the era of the Renaissance*, Basle, New York, 1958; see also Pagel, Walter. Paracelsus and the neo-Platonic and Gnostic tradition. *Ambix*, 8, 1960, pp. 125–166; and Pagel, Walter. The Prime Matter of Paracelsus. *Ambix*, 9, 1961, pp. 117–135.

[6] Thornton, pp. 64–66.

[7] Sudhoff, Karl. *Versuch einer Kritik der Echtheit der paracelsischen Schriften*, 2 vols., Berlin, 1894–99; see also Sudhoff, Karl. *Nachweise zur Paracelsus-Literatur*, Munich, 1932, which is supplemented by Weimann, Karl Heinz. *Paracelsus-Bibliographie 1932–1960. Mit einem Verzeichnis neu entdeckter Paracelsus-Handschriften* (*1900–60*), Wiesbaden, 1963; and *Registerband zu Sudhoffs Paracelsus Gesamtausgabe bearbeitet von Martin Müller. Nova Acta Paracelsica Supplementum 1960*, Basle.

his writings were not published until after his death. Others have been falsely attributed to him, and Sudhoff is the safest guide to the authentic literature. The first collected edition of the writings of Paracelsus was that edited by John Huser and published in ten volumes at Basle, 1589–90, with subsequent printings in 1603 and 1605. This edition is both reliable and useful, and no complete edition followed until the *Sämtliche Werke* edited by Karl Sudhoff in fourteen volumes, 1922–33. A reprint of this was begun in 1954. The chemical contributions of Paracelsus have been investigated by T. P. Sherlock.[1] *Archidocis*, 1570, Paracelsus' handbook of chemistry, went into numerous editions.[2]

The first work on practical metallurgy to be printed was *De la pirotechnia*, Venice, 1540, written by Vannoccio Biringuccio (1480–1538 or 1539), which was used extensively by Agricola in his better-known book. Vannoccio Vincenzio Austino Luca Biringuccio was born in Siena, and travelled extensively, visiting mines and metal works. He became head of the papal foundry and director of the papal artillery. His *Pirotechnia* was very popular, four editions being printed in Venice before 1600, the first in 1540, and there have been six Italian editions, three French, one Latin, one German (1925), and an English translation from the Italian by C. S. Smith and Martha Teach Gnudi (1942).[3] Two English translations of portions of the text appeared in the sixteenth century, and the first volume only of a definitive Italian edition was published in 1914.[4]

Georgius Agricola, or Georg Bauer (1494–1555), was born at Glauchau, in Saxony, and studied at Leipzig, Bologna and Venice before setting up as a physician. Agricola occupied much of his time studying geology and mineralogy, and it is for his pioneer contributions to these subjects that he is remembered. His first printed work was a grammatical treatise entitled *Libellus de prima ac simplici institutione grammatica*, 1520. His first treatise on mining, *Bermannus sive de re metallica*, was published in 1530, appearing in Latin in seven editions between 1530 and 1657; an Italian

[1] Sherlock, T. P. The chemical works of Paracelsus. *Ambix*, 3, 1948, pp. 33–63.

[2] See also *Paracelsus. Selected writings, edited with an introduction by Jolande Jacobi. Translated by Norbert Guterman*, 1951. (Sketch of his life and work; extracts from his writings; with bibliography of these, pp. 339–341. Published in German as *Theophrastus Paracelsus; Lebendiges Erbe*, Zurich, 1942); Strebel, J. Paracelsus als Chemiker und Verfasser des ersten deutschsprachigen Lehrbuches der Chemie. *Praxis*, 38, 1949, pp. 806–814; Stoddart, Anna M. *The Life of Paracelsus Theophrastus von Hohenheim, 1493–1541*, [*etc.*], 1911; Rádl, Em. Paracelsus. Eine Skizze seines Lebens. *Isis*, 1, 1913, pp. 62–94.

[3] *The pirotechnia of Vannoccio Biringuccio. Translated from the Italian, with an introduction and notes, by Cyril Stanley Smith and Martha Teach Gnudi*, New York, 1942, reprinted 1943, 1959 and 1963.

[4] See Zietz, Joseph R. "The Pirotechnia" of Vannoccio Biringuccio. *J. chem. Educ.*, 29, 1952, pp. 507–510; McKie, Douglas. Three historical notes. *Nature*, 163, 1949, p. 628, which mentions references to *The pirotechnia* by Robert Hooke and by Sir Thomas Browne; and Lange, Erwin F. Alchemy and the sixteenth-century metallurgists. *Ambix*, 13, 1966, pp. 92–95.

translation was published in 1550, with German translations in 1778 and 1806. Agricola's *De ortu et causis subterraneorum*, Basle, 1546, was the first book on physical geography, further Latin editions appearing in 1558, 1612 and 1657, with an Italian translation dated 1550, and a German version published between 1806 and 1810. In his *De natura fossilium*, Basle, 1546, he used the word 'fossil' for the first time, and also described many new metals. An Italian translation of this was published in 1612, a German translation appeared from 1809–10, and the first English version was published by the Geological Society of America in 1955.[1] Agricola was also the author of the following: *De peste*, Basle, 1554; *De animantibus subterraneis*, Basle, 1549, with an Italian translation in 1563; and his most important work, *De re metallica*, first published at Basle in 1556, with further printings in 1561, 1621 and 1657. German versions were published at Basle, 1557 and 1621; Frankfurt, 1580; and an Italian translation at Basle, 1563; An English translation of the 1556 edition, with biographical and bibliographical notes by Herbert Clark Hoover and Lou Henry Hoover, was published in London in 1912, and reprinted in New York, 1950. A German edition appeared from Berlin in 1928, and went into a third edition in 1961. Agricola was responsible for the most significant work on mineralogy after Pliny, and his collected writings have been published under the auspices of the Staatlichen Museums für Mineralogie und Geologie zu Dresden[2] in ten volumes, which include a biography, a bibliography listing all editions of his writings from 1520–1955, and a bibliography of writings on Agricola.[3]

Lazarus Ercker (1530–94?) was born at Annaberg, and after holding a number of positions connected with mines and mints, became assayer and chief consultant to the Emperors Maximilian II and Rudolph II, and published his book entitled *Beschreibung allerfürnemisten mineralischen Ertzt und Berckwerks Arten*, Prague, in 1574. Other editions were published in 1580, 1598, and 1629, all from Frankfurt, and the work was then enlarged and issued as *Aula subterranea domina dominantum subdita subditorum* [*etc.*], Frankfurt, 1672, 1684, 1703 and 1736. Meanwhile an English translation had been published in London, 1683, under the title *Fleta minor. The laws of art and nature, in knowing, judging, assaying, fining, refining and inlarging the bodies of confined metals.* The translation of Ercker's book formed the first part of this publication and was made by Sir John Pettus (1613–90), who himself wrote the second part entitled *Essays on*

[1] Agricola, Georgius. De natura fossilium. . . . Translated from the first Latin edition of 1546 by Mark Chance Bandy and Jean A. Bandy for the Mineralogical Society of America. *Geological Society of America, Special paper*, 63, 1955.

[2] Agricola, Georgius. *Ausgewählte Werke*, [*etc.*], Berlin, 1955–65.

[3] See also Boenheim, Felix. Georgius Agricola. *Wissenschaftliche Annalen*, 4, 1955, pp. 657–720; Eyles, Joan M. Georgius Agricola (1494–1555). *Nature*, 176, 1955, pp. 949–950; and *Agricola-Studien*, Berlin, 1957.

metallic words. Sir John Pettus was a colourful character. A Fellow of the Royal Society, he spent his fortune in the Royal cause, and in 1683 was in the Fleet Prison, this being the reason for the book's title. His book was re-issued in 1686 with a different title-page.[1,2] A modern English translation of the 1580 Frankfurt edition appeared in 1951.[3]

Scientific botany had its origin in the study of herbs, and the resultant herbals which described plants with purely utilitarian characteristics. This material has been studied by Eleanour S. Rohde,[4] by Agnes Arber,[5] and also in a more recent volume by Wilfrid Blunt.[6] The oldest Saxon herbal, the *Leech book of Bald*, dating from *circa* A.D. 900–950, is housed in the British Museum, and is complete in 109 leaves. It was previously at Glastonbury, where it was written by Cild, under the direction of Bald. The first printed herbal, stated to be also the oldest existing illustrated herbal, is the translation of the Latin *Herbarium* of Apuleius Barbarus, Apuleius Platonicus or Pseudo-Apuleius. This has been described as "an important compilation of medical recipes", and is mainly compiled from fourth-century Greek material. It was the most influential early Latin herbal, and was first printed at Rome, probably in 1481. A modern edition has been prepared by R. T. Gunther.[7] The *Herbarium* consists of 132 chapters, each dealing with a herb, and four manuscripts of this survive.[8] Charles Singer[9] has written an authoritative article on early Greek herbals, giving the herbal of Diokles of Karystos, written about 350 B.C., as the earliest of its kind. Diokles was followed by Theophrastus of Eresos (380–286 B.C.), whose work was first printed in 1483 at Treviso as *De historia et causis plantarum Latine, Theodoro Gaza interprete;* and by Pedacius Dioscorides, a Greek army surgeon of the first century. His *De materia*

[1] Partington, J. R. *A history of chemistry*, 2, 1961, pp. 104–107, who refers to Beierlein, P. R. *Lazarus Ercker*, Berlin, 1955.

[2] Armstrong, Eva V., and Lukens, Hiram S. Lazarus Ercker and his "Propierbuch". Sir John Pettus and his "Fleta minor". *J. chem. Educ.*, 16, 1939, pp. 553–562.

[3] *Lazarus Ercker's Treatise on ores and assaying, translated from the German edition of 1580, by A. G. Sisco and C. S. Smith*, Chicago, 1951.

[4] Rohde, Eleanour Sinclair. *The old English herbals*, [*etc.*], 1922; contains bibliographies of manuscript herbals, of printed English herbals, and of certain foreign herbals.

[5] Arber, Agnes. *Herbals, their origin and evolution in the history of botany, 1470–1670. . . . A new edition*, [*etc.*], Cambridge, 1938; mainly concerned with the botanical aspect. See also Arber, Agnes. From medieval herbalism to the birth of modern botany. *In*, *Science, medicine and history*, Vol. 1, 1953, pp. 317–336.

[6] Blunt, Wilfrid. *The art of botanical illustration*, [*etc.*], 1950; see also Marcus, Margaret Fairbanks. The herbal as art. *Bull. Med. Libr. Ass.*, 32, 1944, pp. 376–384.

[7] Gunther, R. T. *The Herbal of Apuleuis Barbarus*, Oxford, 1925.

[8] See Cockayne, Oswald. *Leechdoms, wort-cunning and starcraft of early England, being a collection of documents for the most part never before printed, illustrating the history of science in this country before the Norman Conquest*, 3 vols., 1864–66, reprinted 1961; and Payne, Joseph Frank. *The Fitz-Patrick Lectures for 1903. English medicine in the Anglo-Saxon times*, [*etc.*], Oxford, 1904.

[9] Singer, Charles. The herbal in antiquity. *J. Hellenic Stud.*, 47, 1927, pp. 1–52; contains ten coloured plates, many illustrations and references.

medica, translated into Latin by Petrus Paduanensis, was published at Colle in 1478, the Greek text first appearing in 1499, while an English translation, edited by R. T. Gunther, was first published in 1934.[1]

In England, our herbals practically disappeared after the Norman Conquest, but about the middle of the thirteenth century Bartholomaeus Anglicus wrote *De proprietatibus rerum*, the seventeenth book being devoted to herbs. This general encyclopaedia of natural science became extremely popular on the Continent. In 1398 it was translated into English by John de Trevisa, and was first printed at Basle about 1470, at least fourteen editions appearing before 1500. That printed by Wynkyn de Worde in London [1495] is said to be the first book printed on paper made in England.[2] The first printed English book devoted entirely to herbs is that published by Richard Banckes in London, 1525. The author is unknown, but the book is generally known as Banckes' Herbal, and went into numerous editions, many being from other publishers. Printed editions include those issued in 1525, 1526, 1530, 1532–37, 1541, 1546, 1548, 1550 and 1552.[3] Yet another anonymous herbal was the *Grete herball* printed by Peter Treveris in 1526 [S.T.C. 13176–13179], which was actually a translation of the French *Le grand herbier*. The contents of this work are in alphabetical order, and it is poorly illustrated, but was the first English book to contain pictures of plants. Editions are recorded for 1516,[4] 1525,[5] 1526, 1529, 1539, 1550 and 1561.[6]

The title of the father of English botany has been bestowed upon William Turner (1510–68), who studied the subject in a scientific manner and was responsible for almost three hundred first records of British plants.[7] Turner travelled extensively on the Continent, where he did most of his writing, and his first botanical work, *Libellus de re herbarium novus*, was printed by John Byddell in London, 1538 [S.T.C. 24358]. This was privately re-issued in facsimile by Benjamin Daydon Jackson in 1877. Another short work, *The names of herbes in Greek, Latin, Englishe, Duche, and Frenche wyth the commone names that herbaries and apotecaries use, gathered by William Turner*, appeared in print ten years later [S.T.C. 24359], and was reprinted in 1881 as Vol. 34, Series D of the English Dialect Society, edited by J. Britten. Both the *Libellus* and the *Names of*

[1] *The Greek herbal of Dioscorides, illustrated by a Byzantine A.D.* 512, *Englished by John Goodyer A.D.* 1655, *edited and first printed A.D.* 1933 *by Robert T. Gunther*, [*etc.*], Oxford, *for the author*, 1934; contains 396 illustrations.

[2] S.T.C. 1536; also editions of 1535 [S.T.C. 1537] and 1582 [S.T.C. 1538].

[3] See Rohde, Eleanour Sinclair. *Op. cit.*,, pp. 204–206, for complete list.

[4] Mentioned only by Ames.

[5] Mentioned only by Hazlitt.

[6] See Rohde, Eleanour Sinclair. *Op. cit.*, pp. 207–208.

[7] See Nelson, G. A. William Turner's contribution to the first records of British plants. *Proceedings of the Leeds Philosophical and Literary Society, Scientific Section*, 8, 1959, pp. 109–138.

herbes, which is essentially an English-language expansion of the former title, were reprinted together in facsimile, with introductory matter by James Britten, Benjamin Daydon Jackson and William Thomas Stearn, and issued as Publication No. 145 by the Ray Society in 1965. This includes a life of William Turner by Jackson, and several indexes, including Turner's Latin and English plant names with their probable modern identifications. Following the facsimile is a useful transcript of the black-letter text. Turner's major contribution to botany was *A new herball*, the first part being printed by Steven Mierdman at London, 1551 [S.T.C. 24365], and the second and third parts by Arnold Birckman at Cologne, 1561 [S.T.C. 24366]. Birckman also printed the 1568 edition containing all three parts [S.T.C. 24367]. Turner's herbal is particularly notable for the numerous beautiful woodcuts with which it is illustrated, these being mainly derived from Fuchs. Turner also contributed an interesting item to ornithological literature, his *Avium praecipuarum quarum apud Plinium et Aristotelem mentio est brevis & succincta historia*, Cologne, 1544, having been reprinted in 1823, and again in 1903 with notes and an appendix by A. H. Evans. Turner was the author of several other books, including one on wines,[1] and his book on birds followed closely upon a similar work which he had prepared for the press. This was the uncompleted *Dialogus de avibus* by Gybertus Longolius, or Gilbert of Longueil (1507–43), a Dutch doctor, and which was published by Johannes Gymnich at Cologne in 1554 just before Turner's own book on the subject was printed by the same press. This book by Longolius has been overlooked in favour of Turner's well-known work on birds, but it has been recognized as the first book on the subject indicating first-hand observation and scientific speculation. Thomas P. Harrison[2] has investigated this forgotten treatise, showing its relationship to Turner's work, and suggesting that it should be appreciated as an example of the beginnings of scientific investigation into ornithology.

The herbal of John Gerard (1545–1612) was also very popular in its day. Its author was probably born near Wistaston, Cheshire, and lived in Holborn, where he grew over a thousand herbs, of which he published a catalogue entitled *Catalogus arborum, fructicum ac plantarum, tam indigenarum quam exoticarum, in horto Johannis Gerardi civis et chirurgi Londinensis nascentium* in 1596, with a second edition three years later, of which a facsimile reprint was published in a limited edition of two hundred copies, Weinheim, 1962. Gerard's main work, *The herball, or generall historie of plantes*, was printed by John Norton in London, 1597 [S.T.C. 11750], and the second edition was even more successful. This was greatly enlarged and amended by Thomas Johnson, and published in 1633 [S.T.C.

[1] See S.T.C. 24351–24368.

[2] Harrison, Thomas P. Longolius on birds. *Ann. Sci.*, 14, 1958, pp. 257–268.

11751], being illustrated with 2,677 blocks.[1] Robert H. Jeffers[2] has published a useful biography of Gerard, correcting many criticisms of him, and providing useful information on some of his contemporaries. Johnson was a native of Selby, Yorkshire, having been born there about 1597, and he was killed during the defence of Basing House in 1644. He made many botanizing expeditions, of which he published details,[3] and also translated the works of the famous French army surgeon, Ambroise Paré (1634).[4]

A sixteenth-century herbal from the New World was discovered in the Vatican Library by Charles Upson Clark, and is on European-made paper. Known as the Badianus Manuscript, this is a purely Mexican product, having been written in 1552 by an Indian physician, Martinus de la Cruz, in Aztec, and translated into Latin by another Indian, Juannes Badianus. This is the only medical text known to be the work of the Aztec Indians, and it has been beautifully reproduced in colour facsimile, and has been translated twice.[5]

Germany was the home of several eminent sixteenth-century botanists, the first of whom, Otto Brunfels (1489–1534), was the author of the earliest scientific botanical work. A native of Mainz, Brunfels studied theology, later to become a teacher and then a physician. The first part of his *Herbarum vivae eicones* was published by Schott of Strassburg in 1530, the third part following six years later. The text of this work was based on Dioscorides, but the illustrations, by Hans Weiditz, are based on nature, and are far superior to previous efforts. The originals were discovered some years ago at Berne. Jerome Bock (1498–1554), of Heiderback, was the author of *Kreuter Buch*, Strassburg, 1539, the first edition

[1] Another edition of this appeared in 1636, and this has been partly reproduced as *Gerard's Herball; the essence thereof distilled by Marcus Woodward from the edition of Th. Johnson*, 1636, Edinburgh, London, 1927, reprinted London, 1964.

[2] Jeffers, Robert H. *The friends of John Gerard (1545–1612), surgeon and botanist*, Falls Village, Conn., 1967; see also Moore, A. G. N. A herbal found by chance. *Brit. med. J.*, 1962, i pp. 1756–1757.

[3] See *Opuscula omnia botanica Thomae Johnsoni*, 1847; also Kew, H. W., and Powell, H. E. *Thomas Johnson, botanist and royalist*, [*etc.*], 1932, which contains a list of his works; and Power, Sir D'Arcy. Thomas Johnson (1597–1644), botanist and barber surgeon. *Glasg. med. J.*, 133, 1940, pp. 201–203, which corrects certain information contained in Kew-Powell.

[4] See also Barlow, Horace Mallinson. Old English herbals, 1525–1640. *Proc. Roy. Soc. Med.*, 6, 1913, Section of Hist. of Med., pp. 108–149; Payne, J. F. English herbals. [Summary.] *Trans. Bib. Soc.*, 9, 1906–08, pp. 120–123; also, with illustrations, *Trans. Bib. Soc.*, 11, 1909–11, pp. 299–310.

[5] *The Badianus manuscript* (*Codex Barberini, Latin* 241) *Vatican Library; an Aztec herbal of 1552. Introduction, translation and annotations by Emily Walcott Emmart. With a foreword by Henry E. Sigerist*, Baltimore, 1940; also Emmart, Emily Walcott. Concerning the Badianus manuscript. An Aztec herbal, "Codex Barberini. Latin 241" (Vatican Library). *Smithsonian Miscellaneous Collections*, 94, no. 2, 1935; and Gates, William. *The De la Cruz-Badiano Aztec herbal of 1552. Translation and commentary*, Baltimore, 1939 (Maya Society Publication No. 23).

having no illustrations, but that of 1546 carries engravings by David Kandel. Bock's text is a great improvement on that of Brunfels, but the illustrations are inferior.

Leonhardt Fuchs (1501–66), a native of Wemding, in Bavaria, was a physician, and he compiled an important guide to medicinal plants. Once again, the text is inferior, the plants being arranged in alphabetical order without any attempt at classification, but the woodcuts are of outstanding beauty. Fuch's *De historia stirpium*, [*etc.*] was first printed at Basle in 1542, and contains over five hundred full-page woodcuts by Veit Rudolf Speckle (engraver), Heinrich Füllmaurer (who transferred the drawings to the blocks) and Albrecht Mayer (the artist). A German translation entitled *New Kreüterbuch* was published at Basle in 1543, and is probably rarer than the Latin original. The work includes a glossary of botanical terms, and it was reprinted in facsimile in 1964. Fuchs' illustrations were copied in botanical works for many years after his death, and his name is commemorated in the genus "Fuchsia".[1] W. R. LeFanu[2] discovered and described a book in the Royal College of Surgeons of England Library which had previously belonged to Fuchs, and which is inscribed with his name.[3]

Pier Andrea Mattioli (Matthiolus) (1500–77), another physician-botanist, was a native of Siena. His *Dioscoride libri cinque della historia et materia medicinale tradotti in lingua volgare italiana* was first published in Venice in 1544, to be followed by a Latin version ten years later entitled *Commentarii in sex libros Pedacii Dioscordiis*. This work was reprinted numerous times in both Latin and Italian versions, and went into at least forty editions, over 32,000 copies being sold. Although popular, Mattioli's work is considered inferior to that of Fuchs and Brunfels, and is less reliable scientifically.

Rembert Dodoens (Dodonaeus) (1517–85) was a native of Malines, Belgium. He was a physician, becoming professor of medicine at Leyden in 1582, and his herbal, entitled *Cruÿdeboeck*, was published in Flemish at Antwerp in 1554 by van der Loe. Woodblocks from the octavo edition of Fuchs' work were mainly used, but the book is not a translation of the latter. A French translation by Charles de l'Ecluse was published as *Histoire des plantes* in the same year as the original, and an English translation by Henry Lyte from the French version was printed at Antwerp

[1] See Sprague, T. A., and Nelmes, E. The herbal of Leonhardt Fuchs. *J. Linn. Soc.* (*Bot.*), 48, 1931, pp. 545–642; and Ganzinger, Kurt. Ein Krauterbuchmanuskript des Leonhart Fuchs in der Wiener Nationalbibliothek. *Sudhoffs Arch. Gesch. Med.*, 43, 1959, pp. 213–224.

[2] LeFanu, W. R. A volume associated with Leonhart Fuchs. *J. Hist. Med.*, 11, 1956, pp. 344–346.

[3] See also Fichtner, Gerhard. Neues zu Leben und Werk von Leonhart Fuchs aus seinen Briefen an Joachim Camerarius I and II in der Trew-Sammlung. *Gesnerus* (*Aarau*), 25, 1968, pp. 65–82.

in 1578 as *A niewe herball, or histoire of plantes*, [*etc.*]. Dodoens was also the author of *Cosmographica in astronomian et geographiam isagoge*, 1548, of which a facsimile was printed in Niewkoop, 1963, with an introduction by A. Louis. Plantin published several later books by Dodoens, including his collected works, entitled *Stirpium historiae pemptades sex*, 1583.

Not only did l'Ecluse translate several botanical works into French and Latin, but he contributed several important original writings. Jules-Charles de l'Ecluse (Clusius) (1526–1609) was a native of Arras, a notable linguist and a great traveller. He was the author of several works, including *Rariorum aliquot stirpium per Hispanias observatarum historia*, 1576; *Rariorum aliquot stirpium per Pannoniam, Austriam et vicinas historia*, 1583; *Rariorum plantarum historia*, [*etc.*], Antwerp, 1601, consisting of his collected writings to that date; *Exoticorum libri decem; quibus animalium, plantarum*, [*etc.*], Leyden, 1605; and *Curae posteriores . . . aliquot animalium novae descriptiones*, [*etc.*], (Antwerp), 1611. L'Ecluse cultivated numerous foreign plants, and gave remarkably accurate descriptions of those he encountered. The *Fungorum historia* appended to his *Rariorum plantarum historia* is said to be the first published monograph of this nature.

Mathias de l'Obel (Lobelius) (1538–1616) is of particular interest because he spent much of his life in England, dying at Highgate. A native of Lille, de l'Obel came to England in 1559, and after a period in the Low Countries as physician to William the Silent, returned to this country, where he became botanist to James I. De l'Obel made a careful study of British botany, and with Pierre Pena, who had accompanied him to London, was the author of *Stirpium adversaria nova*, London, 1570, of which an enlarged edition was published by Plantin at Antwerp in 1576 as *Plantarum seu stirpium historia.* A Flemish edition was published in 1581, and Plantin also issued the plates separately as *Plantarum . . . icones*, De l'Obel's classification, based on the form of the leaf, is of great interest. and his name is perpetuated in the Lobelia.[1]

Adam Zaluziansky von Zaluzian (1558–1613), a Bohemian, studied in Germany, taught classics at Charles University, where he became professor of Greek literature and Dean of the University, and was also a doctor of medicine. His writings were chiefly medical and theological, but his most significant book was *Methodi herbariae libri tres*, Prague, 1592, which was reprinted at Frankfurt in 1604. Jean Ruel (1479–1537) was also the author of an important botanical treatise, *De natura stirpium libri tres*, Paris, 1536, and this was reprinted at Basle in 1537 and 1543, and at Venice in 1538.

These early botanical works are a fascinating study, and present many interesting features. It will be seen that some woodcuts were printed

[1] See Leclair, E. Matthias de Lobel, médecin et botaniste Lillois (1538–1616). *Monspeliensis Hippocrates*, 11, 1968, No. 41, pp. 11–20; reprinted from *J. Sci. méd. Lille*, No. 36, 1938.

merely in outline, it being intended that the illustrations should be coloured by hand. Some copies were so treated before leaving the publisher, perhaps with colours carefully copied from the artist's original drawings. Others would be coloured by their owners. One can trace the influence of a work upon its successors, noting the frequent use of similar illustrations, indeed of the use of the same blocks, perhaps rearranged, and with new descriptions. Both botanically and bibliographically the early writings on plants are of singular importance, and one is advised to consult the above-mentioned studies of the subject by Eleanour S. Rohde, Agnes Arber, and by Wilfrid Blunt, the latter of which deals in particular with botanical illustration.

Turning from botany to the other branches of biology, we find several important figures whose contributions to science heralded the approach of "the age of scientific endeavour", as the seventeenth century has been termed. It is sometimes difficult to assign them to any particular branch of science, as we find, for example, Albrecht Dürer (1471–1528), who wrote on anatomy, chemistry, mathematics and other subjects. Only two of his 150 works were printed during his lifetime, but his work on perspective, optics, the growth of animal life, and his remarkable flower paintings, stamp Dürer as a mathematician and botanist as well as a remarkable artist. His woodcut of the Indian rhinoceros is probably one of the best-known illustrations in zoological literature, having been reproduced by various authors for over two hundred years. The block was drawn and cut by Dürer in 1515, and the plates bearing this date are very rare, many others being erroneously dated 1513. F. J. Cole[1] has contributed an interesting paper mentioning the various works copying Dürer's rhinoceros, in which Cole states that Gesner was the only author acknowledging the original source.

Andreas Vesalius (1514–64) is regarded as the first scientific anatomist, studying the human body for the preparation of his monumental *De humani corporis libri septem* (1543).[2] He is here mentioned for his work on comparative anatomy, which led to his complete study of the human body. C. D. O'Malley[3] has written a definitive biography of Vesalius, and has also reviewed the writings on Vesalius from 1914 to 1964.[4]

Another pioneer in comparative anatomy was Pierre Belon (Bellonius) (1517–64), a native of Soultière, in France. Belon studied botany under Valerius Cordus, and also studied medicine. He travelled extensively,

[1] Cole, F. J. The history of Albrecht Dürer's rhinoceros in zoological literature. In *Science, medicine and history*, Vol. 1, 1953, pp. 337–356.

[2] See also Thornton, pp. 48–53.

[3] O'Malley, Charles Donald. *Andreas Vesalius of Brussels, 1514–1564*, Berkeley, Los Angeles, 1964.

[4] O'Malley, Charles Donald. A review of Vesalian literature. *History of Science*, 4, 1965, pp. 1–14.

visiting England, Greece, Italy, Egypt and Arabia in his study of the natural history of those countries. Belon's *L'Histoire de la nature des oyseaux*, [*etc.*], Paris, 1555, his most important work, contains a comparison of the skeleton of man and bird. Belon is considered the founder of modern ichthyology, as well as an authority on ornithology, and the following are among his published books: *L'Histoire naturelle des estranges poissons marins, avec le vraie peincture et description du dauphon et de plusieurs autres de son espèce*, Paris, 1551; *Les observationes de plusieurs singularitez et choses mémorables trouvées en Grèce, Asie, Judée, Egypte, Arabie, et autres pays estranges*, [*etc.*], Paris, 1553, with an illustrated edition in 1555; *De aquatibilibus libri duo*, [*etc.*], Paris, 1553, with a French translation, *La nature et diversité des poissons*, [*etc.*], Paris, in 1555. Another medical man, Guillaume Rondelet (1507–66), was a native of Montpellier, where he became professor of medicine and anatomy, and also contributed usefully to ichthyology. His *Libri de piscibus marinis*, [*etc.*], two volumes, Lyons, 1554–55,[1] also appeared in a French translation, [Lyons?], 1558. One of Rondelet's pupils. Volcher Coiter (1534–76), became professor of anatomy at Bologna, and made extensive studies in comparative anatomy. He wrote a book on osteology, *De ossibus et cartilaginibus corporis humanis tabulae*, Bologna, 1566; followed by *Externarum et internarum principalium humani corporis partium tabulae*, [*etc.*], Nuremberg, 1572 and 1573; and *Lectiones Gabrielis Fallopii de partibus similaribus humani corporis*, [*etc.*], Nuremberg, 1575.[2]

Pierre Gilles (1490–1555) was the author of *Ex Aeliani historia per Petrum Gyllum latini factum itemque ex Porphyrio, Heliodoro, Oppiano &c*, Lyons, 1533, which contains a list of fish names in Latin and French, and an index of terrestrial, aquatic and flying animals, with a special index of fishes. Gilles was also the author of a description of the dissection of an elephant, *Elephanti descriptio missa ad R. cardinalem Armaignacum ex urbe Berrhoea Syriaca*, first printed in the third edition of his *Ex Aeliani historia*, [*etc.*], Lyons, 1565.

Carlo Ruini (*c.* 1530–98) trained as a lawyer at Bologna, where he became a senator, and was the author of the first comprehensive monograph devoted to the anatomy of an animal. The first edition of his book on the horse was published as *Dell' anotomia, et dell' infirmita del cavallo*, Bologna, 1598, and is very rare; the first part deals with the anatomy,

[1] Volume 2, which has a separate title-page, and is really a separate work, bears the title *Universae aquatilium historiae pars altera*, [*etc.*].

[2] See Schullian, Dorothy M. New documents on Volcher Coiter, *J. Hist. Med.*, 6, 1951, pp. 176–194; Adelmann, Howard B. The "De ovarum gallinaceorum generationis primo exordio progressuque et pulli gallinacei creationis ordine" of Volcher Coiter. Translated and edited with notes and introduction by Howard B. Adelmann. *Ann. med. Hist.*, 2nd Ser., 5, 1933, pp. 327–341, 444–457; and Schierbeek, A. Volcher Coiter (1534–76). *Biologisch. Jaarbock*, 1957, pp. 148–156.

and the second part with diseases affecting horses. His anatomical descriptions and the plates are remarkably accurate, his description of the pulmonary circulation, and the drawings of the foetus, being particularly noteworthy. This edition was followed by *Anatomia del cavallo, infermità ed i suoi rimedi*, Venice, 1599, and later editions were published in 1602, 1607 and 1618. A German translation by Peter Uffenbach appeared in 1603, and a French translation of the diseases portion was published under the name of Horace de Francini at Paris in 1607.[1]

Conrad Gesner (1516–65) was a native of Zürich who was appointed professor of Greek at Basle. He published a Greek dictionary, followed by *Enchiridio historiae plantarum*, Basle, 1541, and *Catalogus plantarum*, [*etc.*], 1542. Gesner then explored the Mediterranean coast, making a special study of the botany and zoology. He qualified as a physician, and became professor of natural history at the Collegium Carolinum, Zürich. Gesner's *Bibliotheca universalis*, [*etc.*] was published in three volumes at Zürich, 1545–49, an *Appendix* following in 1555. His *Historia animalium*, [*etc.*], four volumes, Zürich, 1555–58, marks the beginnings of modern zoology, and the fifth volume, devoted to snakes, appeared in 1587. There were numerous extracts, shortened editions and versions of Gesner's works, and he left much work uncompleted at his death, including the portion of the *Bibliotheca universalis* devoted to medicine, and a herbal. His other writings include *De rerum fossilium, lapidum, et gemmarum genera, figuris et similitudinibus liber*, [*etc.*], Zürich, 1565, he edited the botanical works of David Kyber (1553), and of Valerius Cordus (1561); Gesner's own botanical works were finally published as *Opera botanica*, [*etc.*], Nuremberg, 1751–71, and one thousand of his water-colours of plants are in Erlangen University Library. Gesner died of the plague in his fiftieth year on 13th December, 1565, and was buried in the cloister of Zürich Grossmünster, but the position of his remains is unidentified. Despite the fact that he left many projects unfinished, what he did accomplish would have occupied any other person several lifetimes.[2]

Another encyclopaedist, Ulisse Aldrovandi (1522–1605), was a native of Bologna, graduating in medicine there, and becoming a prominent naturalist. Aldrovandi was the author of *Ornithologiæ*, Bologna, 1599, which went into several editions, including three-volume editions from

[1] See Bayon, H. P. The authorship of Carlo Ruini's "Anatomia del Cavallo". *J. comp. Path.*, 48, 1935, pp. 138–148; and Eby, Clifford H. "Anatomia del cavallo". *Western Veterinarian*, 7, 1960, pp. 88–91.

[2] See Fischer, Hans. Conrad Gesner (1516–1565) as bibliographer and encyclopedist. *The Library*, 5th Ser., 21, 1966, pp. 269–281; Bay, J. Christian. Conrad Gesner, (1516–1565), the father of bibliography; an appreciation, *Pap. Bib. Soc. Amer.*, 10, 1916, pp. 53–86; Steiger, R. Erschliessung des Conrad-Gesner-Materials der Zentralbibliothek Zürich. *Gesnerus* (*Aarau*), 25, 1968, pp. 29–64; and Théodoridès, J. Conrad Gesner et la zoologie: les invertébrés. *Gesnerus* (*Aarau*), 23, 1966, pp. 230–237.

Bologna, 1603–81, 1610–35 and 1645–46; and Frankfurt, 1610–35. A translation of volume two, book xiv by L. R. Lind was published as *Aldrovandi on chickens. The ornithology of Ulisse Aldrovandi (1600)*, with introduction and notes, Norman, Oklahoma, 1963. This contains an informative biography of Aldrovandi, a bibliography of his writings, and of works devoted to his life and activities. Aldrovandi's anatomical dissections were carefully executed and illustrated, but he sometimes recorded unconfirmed information which later proved erroneous. He was also the author of *De piscibus*, [*etc.*], Bologna, 1613, with nine other editions up to 1644; *Serpentum et draconum historiae libri duo*, [*etc.*], Bologna, 1640; *Quadrupedum omnium bisulcorum historia*, Bologna, 1642; *Monstrorum historia*, [*etc.*], Bologna, 1642–57; and *Opera*, thirteen volumes, Bologna, 1599–1668, 1602–45 and 1646–48.

Ippolito Salviani (1514–72), a native of Città di Castello, Italy, made his name with his work on fishes, but he was also physician, poet and playwright. His *Aquatilium animalum historiae*, [*etc.*], published at Rome, was issued in parts between 1554 and 1558, being printed in the house of the author. It was reprinted in Rome, [1593?], and in Venice, 1600 and 1602.[1]

These were the main writers on comparative anatomy during the sixteenth century, but the following books are also recognized as being of importance in their respective spheres: *De Romanis piscibus libellus*, [*etc.*], 1524, by Paolo Giovio, Bishop of Nocera (1483–1552), of which an Italian translation was published at Venice in 1560; Edward Wotton (1492–1555), *De differentis animalium libri decem*, Paris, 1552, which is not illustrated, but contains decorated initial letters; *Dell' historia naturale, libri XXVIII*, by Ferrante Imperato (1550–1625), Naples, 1509; and Joseph d'Acosta (1539–1600), *Historia natural y moral de la Indias*, [*etc.*], 1590, the first two books of which had originally appeared in Latin as *De natura Noi orbis libri duo*, 1588–89. This work was translated into French (Paris, 1598), Dutch (1598), English (1604), [S.T.C. 94], with another Spanish edition published at Madrid in 1608.[2]

An interesting study of sixteenth-century science has been provided by William Persehouse Delisle Wightman in *Science and the Renaissance*, two volumes, Edinburgh, London, 1962. Volume one consists of an introduction to the study of the emergence of the sciences in the sixteenth century, and deals with subjects such as mathematics, astronomy, medicine and chemistry; and volume two contains 760 numbered, annotated entries consisting of a bibliography of all sixteenth-century books of a scientific nature in Aberdeen University Library.

[1] See also Gudger, E. W. The five great naturalists of the sixteenth century: Belon, Rondelet, Salviani, Gesner and Aldrovandi: a chapter on the history of ichthyology, *Isis*, 22, 1934–35, pp. 21–40.

[2] See also Union Internationale d'Histoire et de Philosophie des Sciences. *La science au seizième siècle. Colloque internationale de Royaumont 1–4 Juillet, 1957*, Paris, 1960.

CHAPTER IV

Seventeenth-Century Scientific Books

"My determination is not to remain stubbornly with my ideas, but I will leave them and go over to others as soon as I am shown no other purpose than to place truth before my eyes so far as it is in my power to embrace it: and to use the little talent I have received to draw the world away from its old heathenish superstitions, and to go over to the truth and to stick to it."

LEEUWENHOEK

In the evolution of most subjects having extensive backgrounds of historical development we can usually pin-point periods of rapid advancement. Figures stand out as eminent contributors to progress; others are prominent as stumbling blocks. There are periods of inactivity, of re-evaluation, and even of reversion, but the seventeenth century stands out as "the age of scientific endeavour", and "the golden age of science". This does not imply that there was a sudden surge forward in all branches of science, for some of the most eminent men of the period clung to theories that we now consider most unscientific. There was, however, a general trend towards the solution of problems by scientific investigation, a movement which gathered momentum as the result of various circumstances. The ground had been prepared by the earlier scientists, whose writings had been made more generally available as the result of the spread of printing, and by the foundation of universities as seats of learning encouraging scholarship. This century was to witness the birth of numerous universities, the development of scientific societies[1] in various parts of the civilized world, and the advent of the scientific periodical.[2] Alchemy progressed towards chemistry by becoming based upon scientific research, which is symbolic of the advance of all the sciences during this period. Men began to experiment for themselves, instead of taking for granted the views of their predecessors. They shared the results of their experiments, debated and discussed them, forming themselves into groups having mutual interests. The age of magic and quackery was slowly passing.[3] The age of scientific experiment and reasoning had begun.

[1] See Chapter VII. [2] See Chapter VIII.

[3] See Thorndike, Lynn. Mediaeval magic and science in the seventeenth century. *Speculum*, 28, 1953, pp. 692–704.

Most of the early scientists contributed to several branches of science, and any classification into groups of biologists, mathematicians, physicists, chemists and astronomers must be arbitrary. Several eminent contributors to these subjects were medical men, and it was possible for a man to hold simultaneously chairs devoted to several sciences. His chief claim to fame may be a debatable point. depending perhaps upon the profession of his most voluble supporters. An anthology of English scientific literature of this century has been edited by Norman Davy,[1] and Douglas McKie[2] has surveyed the same period.

Although Galileo Galilei might be assigned to the sixteenth century with some justification, he is considered here because his works were not printed until early in the seventeenth. Furthermore, he did not die until 8th January, 1642, and his influence was greatest during his latter years. Galileo was born on 15th February, 1564, at Pisa. He studied medicine at the university there, but apparently neglected this subject for the works of Euclid and Archimedes. He went to Siena as lecturer on mathematics, later occupying chairs devoted to this subject at Pisa (1589), and at Padua (1592–1610). In 1582 Galileo devised the pulsimeter (*pulsilogium*), and he also invented a geometrical and military compass and a thermometer. His work was divided roughly into three periods. From 1589 to 1609 he was chiefly engaged in research upon mechanics and other branches of physics;[3] from 1609 to 1632 he studied astronomy;[4] while from 1633 to 1642 he devoted his time mainly to physics. Much of Galileo's work was not published, but many of his manuscripts are preserved. His published writings include *Le operazione del compasso geometrico et militari di Galileo Galilei* [*etc.*], Padua, 1606; *Difesa . . . contro alle columnie e imposture di Baldessa Capra Milanese*, [*etc.*], Venice, 1607; and *Siderius nuncius*, Venice, 1610, which upheld the Copernican doctrine, and represents one of Galileo's most important works.[5] A facsimile of this 1610 edition was announced in 1966. An English translation by E. Stafford Carlos has been published.[6] Other works by Galileo were *Discorso . . . intorno alle cose, che Stanno in su l'acqua o che in quella si muouono*, Florence, 1612;

[1] Davy, Norman, *ed. British scientific literature in the seventeenth century*, London, [*etc.*], 1953 [i.e. 1954]. Contains a chronological table and references, with short extracts from the writings of Bacon, Sprat, Ray, Grew, Boyle, Hooke and others.

[2] McKie, Douglas. Men and books in English science (1600–1700). Part 1. *Sci. Progr.*, 46, 1958, pp. 606–631.

[3] See Hall, A. Rupert. Galileo and the science of motion. *Brit. J. Hist. Sci.*, 2, 1964–65, pp. 185–199.

[4] See Whitrow, G. J. Galileo's significance in the history of astronomy. *Quart. J. Roy. Astron. Soc.*, 5, 1964, pp. 182–195.

[5] See Rosen, Edward. Title of Galileo's *Siderius muncius*. *Isis*, 41, 1950, pp. 287–289.

[6] Galilei, Galileo. *The Siderial messenger . . . with part of preface to Kepler's Dioptrics, containing the original account of Galileo's astronomical discoveries. A translation, introduction and notes by E. Stafford Carlos*, 1960; see also Hall, A. Rupert. Galileo's system of the world. *Quart. J. Roy. Astron. Soc.*, 5, 1964, pp. 304–317.

THE

ORIGINE

OF

FORMES and QUALITIES,

(According to the *Corpuſcular Philoſo-phy*,) Illuſtrated by *Conſiderations* and EXPERIMENTS,

(Written formerly by way of *Notes* upon an *Eſſay* about NITRE)

By the Honourable

ROBERT BOYLE,

Fellow of the *Royal Society*.

Audendum eſt, & Veritas inveſtiganda; quam etiamſi non aſſequamur, omnino tamen propius, quàm nunc ſumus, ad eam perveniemus. Galen.

OXFORD,

Printed by H. HALL Printer to the Univerſity, for RIC: DAVIS. An. Dom. MDCLXVI.

PLATE 3. Title-page of Robert Boyle's *The origine of formes and qualities*, Oxford, 1666.

EXERCITATIONES
DE
Generatione Animalium.

Quibus accedunt quædam
De Partu: de Membranis ac humoribus Uteri: & de Conceptione.

AUTORE
GUILIELMO HARVEO
Anglo, in Collegio Medicorum *Londinensium* Anatomes & Chirurgiæ Professore.

LONDINI,
Typis Du-GARDIANIS; impensis *Octaviani Pulleyn* in Cœmeterio *Paulino*.
M. DC. LI.

PLATE 4. Title-page and frontispiece of William Harvey's *De generatione*, London, 1651.

Istoria e dimostrazione intorno alle macchie Solari e loro accidenti, [*etc.*], Rome, 1613; and *Il saggiatore nel quale bon bilancia esquisita e giusta si ponderano le cose contenute nella libra astronomica et filosophica di Lotario Sarsi Sigensano*, Rome, 1623. This was first printed in its entirety in English in 1960.[1] The *Dialogo . . . sopra id due massime sistemi del mondo, Tolemaico e Copernico*, Florence, 1632 (Frontispiece), has appeared in two recent translations into English, one edited by Giorgio de Santillana[2] and based on the earliest English translation published in 1661 in the first volume of Thomas Salusbury's *Mathematical collections and translations*, and the other newly translated by Stillman Drake[3] from the Italian. Galileo's *Della scienza meccanica* was written about 1599, but only circulated in manuscript until a French translation was published in 1634 with the title *Les méchaniques de Galilée Mathématicien & ingenieur du Duc de Florence*. His *Discorsi e dimostrazione matematiche intorno a due nuove scienze attenenti alla mecanica & i movimenti locali*, was printed at Leyden in 1638;[4] *De motu* was written in Pisa about 1590, and not published by Galileo, but an English translation by I. E. Drabkin was issued with a translation of *Le meccaniche* by Stillman Drake.[5] The latter was written about 1600 and last translated into English by Thomas Salusbury in 1665. Salusbury's *Mathematical collections and translations*, 1661–65 was issued in facsimile with an introduction by Stillman Drake in a limited edition of two hundred copies, two volumes, 1967. Collected editions of Galileo's works appeared in 1655–56, 1718, 1744 and 1808, while more recent editions include *Le opera di Galileo Galilei*, edited by E. Alberi, sixteen volumes, Florence, 1842–56; *Le Opera*, edited by A. Favaro, twenty volumes, Florence, 1890–1909; reprinted with additions at Florence, 1929.[6] Galileo has been credited with being the inventor of the microscope, and he certainly made one as early as 1609. He also used the telescope for his astronomical observations, and is justly honoured as an eminent forerunner of Newton. Galileo's troubles with the Church present a complicated story which has been clarified by

[1] *The controversy on the comets of 1618. Galileo Galilei, Horatio Grassi, Mario Guiducci, Johann Kepler, Translated by Stillman Drake and C. D. O'Malley*, Philadelphia, 1960. This contains the following translations; Grassi: On the three comets of the year 1618; Guiducci: Discourse on the comets; Grassi: The astronomical balance; Guiducci: Letter to Tarquino Galluzzi; Galilei: The assayer; and Kepler: Appendix to the *Hyperaspistes*.

[2] Galilei, Galileo. *Dialogue on the great world systems. Edited by Giorgio de Santillana from the translation by T. Salusbury*, Chicago, 1953.

[3] Galilei, Galileo. *Dialogue concerning the two chief world systems—Ptolemaic & Copernican. Translated by Stillman Drake*, [*etc.*], Berkeley, Los Angeles, 1953.

[4] Galilei, Galileo. *Dialogues concerning two new sciences. Translated by Henry Crew and Alfonse de Salvio*, [*etc.*], New York, 1914, etc.

[5] Galilei, Galileo. *On motion and On mechanics*, Madison, Wisconsin, 1960; see also Drabkin, I. E. A note on Galileo's *De motu*. *Isis*, 51, 1960, pp. 271–277.

[6] Further details are contained in the bibliography (pp. 203–206) in Taylor, F. Sherwood. *Galileo and the freedom of thought*, 1938. A bibliography listing 2,108 items published before 1895 exists in Carli, A., and Favaro, Antonio. *Bibliographia Galileiana (1568–1895)*, Rome, 1896.

Giorgio de Santillana,[1] and is also mentioned by E. N. da C. Andrade[2] in one of many papers published to celebrate the four hundredth anniversary of Galileo's birth.[3]

Two pupils of Galileo, Giovanni Alfonso Borelli and Evangelista Torricelli, applied their master's mathematico-physical methods and ideas to other fields of science. Evangelista Torricelli (1608–47) was born at Faenza, and went to Rome in 1628 to study under Castelli, and in 1643 he measured the height of columns of mercury in the closed tubes of a pump. The space above the mercury became known as the "Torricellian vacuum". Torricelli concerned himself with optical problems, pure mathematics, and the dynamics of liquids. His first publication was *De motu gravium naturaliter descendentium et projectorum*, 1641, which included the enunciation of the theory of the conservation of energy. The *Opera geometrica* was published in Florence in 1644, and on his death three years later he left his works to Cavalieri[4] for publication. Cavalieri, however, died a month later and it was not until 1919 that an incomplete collection of these appeared in five volumes, as *Opere, publiées par G. Loria et G. Vassura*, Faenza.[5]

Giovanni Alfonso Borelli (1608–79) was born in Naples, and showed early mathematical ability which caused him to be sent to Pisa, where he came under the influence of Galileo. From 1649 to 1656 he held a professorial chair at Messina, but then returned to Pisa, and became interested in physiology through the influence of Malpighi. Borelli conducted experiments on air pressure, with particular reference to breathing, and also on the pendulum, on capillary action, and on the velocity of sound. He attempted to provide a complete mathematical mechanical system of the universe in his *Theoricae mediceorum planetarum*, which was published in Florence in 1666. His posthumously published *De motu animalium*, Rome, 1680, was of great significance in the history of physiology, and his mechanical training led him to apply his theories to biology in

[1] Santillana, Giordio de. *The crime of Galileo*, Chicago, 1955.

[2] Andrade, E. N. da C. Galileo. *Notes Rec. Roy. Soc. Lond.*, 19, 1964, pp. 120–130.

[3] See also Brodrick, James. *Galileo: the man, his work, his misfortunes*, 1964; Fahie, J. J. *Galileo his life and work*, 1903; Gebler, Karl von. *Galileo Galilei and the Roman curia. . . . Translated . . . by Mrs. George Sturge*, 1879; Geymonat, L. *Galileo Galilei: a biography and enquiry into his philosophy of science. . . . Texts translated from the Italian with additional notes and appendix by Stillman Drake*, New York, 1965; Golino, Carlo L., *ed. Galileo re-appraised*, Berkeley, Los Angeles, 1966; Haden, Russell L. Galileo and the compound microscope. *Bull. Hist. Med.*, 12, 1942, pp. 242–247; Kaplon, Morton, F., *ed. Homage to Galileo: papers presented at the Galileo quadricentennial, University of Rochester, October 8 and 9, 1964*, Cambridge, Mass., London, 1965, Koyré, A. *Études galileénnes*, 3 parts, Paris, 1939; reprinted together 1966; Olschki, Leonardo. The scientific personality of Galileo. *Bull. Hist. Med.*, 12, 1942, pp. 248–273; and Vaccaro, Leopold. Galileo Galilei. *Ann. med. Hist.*, 2nd Ser., 7, 1935, pp. 372–384.

[4] See p. 95.

[5] See also Middleton, W. E. Knowles. The place of Torricelli in the history of the barometer. *Isis*, 54, 1963, pp. 11–28.

explaining limb movements, and the mechanics of the skeletal and muscular systems. *De motu* was also printed at Leyden in 1685 and 1710, and at Naples in 1734.[1] Other publications of Borelli include *De motionibus naturalibus a gravitate pendentibus*, 1670, and Leyden, 1680, and *De vi percussionis et motionibus naturalibus*, 1670, and Leyden, 1686.

John Wilkins (1614–72), who became Bishop of Chester, was keenly interested in astronomy, and supported the meetings of scientists which led to the foundation of the Royal Society. In 1638 he published anonymously his first book, in which he attempted to prove that the moon was an habitable world. This was entitled *The discovery of a world in the moone, or, a discourse tending to prove that 'tis probable there may be another habitable world in that planet*, London, 1638, and the third edition (1640) contains a *Discourse concerning the possibility of a passage thither*, which forecast the possibility of space travel. A French translation was published in Rouen in 1655. This book was followed by *A discourse concerning a new planet, tending to prove that 'tis probable our earth is one of the planets*, London, 1640; *Mercury, or the secret and swift messenger, showing how a man may with privacy and speed communicate his thoughts to a friend at any distance*, 1641, an ingenious work on cryptography; *Mathematical magick, or the wonders that may be performed by mechanical geometry*, London, 1648, reprinted in 1968; *An essay towards a real character, and a philosophical language*, to which was appended *An alphabetical dictionary wherein all English words according to their various significations are either referred to their places in the Philosophical Tables or explained by such words as are in these Tables*, 1668, in which Wilkins was assisted by John Ray, Francis Willughby and others; and several other sermons and theological works. Some of Wilkins' writings were collected together and published with a short life of the author as *The mathematical and philosophical works of the Right Reverend John Wilkins*, London, 1708, and reprinted in two volumes in 1802. His house, together with his books and manuscripts were destroyed by the Great Fire of 1666.[2]

Christian Huygens made valuable contributions to astronomy, mathematics, mechanics and optics. He was born at The Hague on 14th April, 1629, and studied at Leyden and Breda. He travelled extensively, visiting England on several occasions, where he became a member of the Royal Society. Huygens enjoyed correspondence with most of the eminent scientists of his period, and Newton was profoundly impressed by his

[1] See also *Aeronautical Classics*, No. 6, 1911, published by the Aeronautical Society of Great Britain, pp. 30–32, which contains a translation by T. O'B. Hubbard and J. H. Ledeboer of Borelli's section on the analysis of the flight of birds; and Koyré, A. La mécanique céleste de J. A. Borelli. *Rev. Hist. Sci., Paris*, 1952, pp. 101–138; and *Discovery*, 19, 1958, p. 314.

[2] See Bowen, E. J., and Hartley, Sir Harold. The Right Reverend John Wilkins, F.R.S. (1614–72). *Notes Rec. Roy. Soc. Lond.*, 15, 1960, pp. 47–56.

genius. From 1666 to 1681 Huygens worked in Paris, and in the latter year returned to Holland, where he devoted himself to optics. Huygens made observations on Saturn, the results being published in his *Systema Saturnium*, published in 1659. His earlier works include *Theoremata de quadratura hyperboles, ellipsis et circuli*, Leyden, 1651; and *De circuli magnitudine inventa*, Leyden, 1654, both printed by Elzevir: and *Horologium*, 1658. His most famous work, the *Horologium oscillatorium, sive De motu pendulorum ad horologia aptato demonstrationes geometricae* was printed in Paris, 1673, and a reprint was announced in 1966. His *Discours de la cause de la pesanteur* was published in 1690 as a supplement to his *Traité de la lumière*, Leyden, 1690, and a facsimile of this was published in 1966. His *Tractatus de vi centrifuga* was published posthumously in 1703. Huygen's *Kosmotheros sive de terris coelestibus*, [*etc.*], was published at The Hague in 1698 and 1699, while an English translation was printed in London in 1698. The *Oeuvres complètes de Christiaan Huygens, publiées par la Société hollandaise des Sciences* was published in twenty-two volumes between 1888 and 1950. Huygens has been described as "one of the greatest scientific geniuses of all time". Not only did he discover Saturn's ring and the satellite Titan, but he was the founder of the wave theory in light, invented the pendulum clock, discovered the principle of the spiral-spring regulator, studied the collision of elastic bodies, and greatly developed the use of the telescope. He died in 1690, and his manuscripts are preserved in Leyden University Library.[1] These contain much unpublished material, including Huygens' notes on his observations through the microscope, which reveal him as a pioneer investigator to be considered with Leeuwenhoek, Hooke, and similar observers in this field of research.[2]

A clergyman of promise as an astronomer, but who died in his twenty-second year, predicted and observed the transit of Venus of 1639. This was Jeremiah Horrocks (died 1641), who taught himself astronomy and wrote a book in 1640, which was published as *Venus in sole pariter visa*, Danzig, in 1662.[3]

The contributions of the Paris Observatory and of the Greenwich Observatory are of particular significance in the history of astronomy. At the former, the activities of Cassini and Römer are of special interest. Giovanni Domenico Cassini (1625–1712) was an Italian civil engineer, before going to Paris. He measured the periodic rotations of Mars and

[1] Bell, A. E. *Christian Huygens and the development of science in the seventeenth century*, 1947; see also Crommelin, C. A. The clocks of Christiaan Huygens. *Endeavour*, 9, 1950, pp. 64–69.

[2] See Rooseboom, Maria. Christiaan Huygens et la microscopie. *Arch. néerl. Zool.*, 13, 1958, Suppl., pp. 59–73.

[3] See Dingle, Herbert. Astronomy in the sixteenth and seventeenth centuries. In, *Science, medicine and history*, Vol. 1, 1953, pp. 455–468.

Jupiter, and discovered four satellites of Saturn. He made a table of Jupiter's satellites, and also made observations on Mars and on the sun's parallax. His works were published as *Divers ouvrages d'astronomie*, The Hague, 1731. Ole Rømer, or Olaus Römer (1644–1710), after working for a while in Paris, made his chief observations at the Round Tower, Copenhagen, and later in his own house. He was the first to show that light has a definite velocity.[1] Although Rømer's manuscripts and instruments were destroyed in a fire in 1728, his pupil Peder Horrebow provided details of these in his *Basis astronomiae*, Copenhagen, 1735.

Johannes Hevelius was born at Danzig on 28th January, 1611, and died on his birthday in 1687. He was educated at Danzig and in Poland, being particularly interested in mathematics during his early years. In 1630 he went to Leyden University, and the following year found him in London. Hevelius returned to Danzig in 1634 and commenced to study astronomy. He was mainly interested in eclipses of the sun, sun-spots, the planets, and particularly the moon, while he measured the positions of over 1,500 fixed stars. The first important work of Hevelius was the *Selenographie, sive lunae descriptio*, 1647, which was followed by *Prodromus cometicus*, [*etc.*], 1665; *Cometographie*, [*etc.*], 1668; *Machina coelestis pars prior*, 1673, and *pars posterior*, 1679. In 1676 Edmond Halley visited Hevelius, and the two made joint observations. The latter's town house and observatory were completely destroyed in 1679, but these were rebuilt. In 1685 appeared his *Annus climactericus*, and after the death of Hevelius, his widow published his *Catalogus stellarum fixarum*, 1687; *Firmamentum sobiesciarum, sive uranographia*, [*etc.*]; and finally his *Prodromus astronomiae*, 1690, containing his catalogue of 1,564 stars. *Selenographie* was reprinted in 1969 with a foreword by H. Lambrecht, and a reprint of *Machina coelestis* was announced in the same year.[2]

The Greenwich Observatory owed its early fame to John Flamsteed, who was born at Denby, Derbyshire, on 19th August, 1646. He was educated at a free school at Derby, leaving at the age of sixteen. He showed an early interest in astronomy and mathematics, and in 1670 came to London, before passing on to Cambridge, where he entered Jesus College. Flamsteed published many papers in the *Philosophical Transactions*, and in 1674–75 he was appointed King's Astronomer, the observatory at Greenwich being erected for him. He erected instruments at his own expense. Flamsteed was elected F.R.S. in 1676–77, but later his subscription lapsed and his name was removed. In 1680 he published anonymously his tract *The doctrine of the sphere grounded on the motion of*

[1] See Cohen, I. Bernard. Roemer and the determination of the velocity of light. *Isis*, 31, 1939–40, pp. 327–370.

[2] See MacPike, Eugene Fairfield. *Hevelius, Flamsteed and Halley: three contemporary astronomers and their mutual relations*, 1937.

earth. After numerous delays in printing, Flamsteed's observations were published in 1712, without the author's permission, by Halley, but they were from an incomplete copy. A corrected version was published in the *Historia coelestis Britannica*, three volumes, 1725, which was issued posthumously, as he died on 31st December, 1719.[1]

Edmond Halley succeeded Flamsteed as Astronomer Royal in 1720, and with Newton had been involved in disputes with Flamsteed. Halley was born on 8th November, 1656, and was educated at St. Paul's School, before becoming a student at Queen's College, Oxford. He became interested in mathematics and astronomy at an early age, and his first paper was printed in the *Philosophical Transactions* in October, 1676, to be followed by many others. In 1676 Halley went to St. Helena to make observations on the southern stars. On his return in 1678 he published his *Catalogus stellarum Australium, sive supplementum catalogi Tychonici*, 1679, which is now rare, and of which a French edition was published in Paris in the same year. Also in 1678, Halley received an M.A. from Oxford, and was elected F.R.S., while in the following year he visited Hevelius. After travelling extensively in France and Italy between 1680 and 1682, he became a firm friend of Newton. Halley stimulated the latter to complete his *Principia*, saw it through the press, and paid for its publication by the Royal Society, of which Halley had become assistant secretary. In 1696 he became Deputy Comptroller of the Mint at Chester, but in 1698 he sailed to the West Indies as captain of a small vessel, repeating the trip the following year. Halley conducted important work on tides and navigation, and in 1703 was elected to the Savilian chair of geometry at Oxford. His *Synopsis of the astronomy of comets* was published in 1705, and he published Flamsteed's astronomical observations seven years later. In 1715 Flamsteed obtained 300 of the 400 copies printed by Halley, extracted ninety-seven sheets containing his authentic observations, and burned the remainder. After Flamsteed's death his widow published these sheets as part of the *Historia coeslestis*.[2] Halley became Astronomer Royal in 1720, and his astronomical tables were published posthumously in Latin in 1749, and reprinted several times in English and French. He died on 14th January, 1741–42, and is remembered eponymously by "Halley's comet", which he observed in 1682, and the return of which he correctly predicted would occur in 1758. Halley edited *Miscellanea curiosa. A collection of some of the principal phenomena in nature*

[1] See MacPike, Eugene Fairfield. *Op. cit.;* and *An account of the Revd. John Flamsteed, first Astronomer Royal. Compiled from his own manuscripts and other authentic documents never before published*, 1835; *Supplement to the account of the Revd. John Flamsteed, . . . by Francis Bailey*, 1837. These were reprinted in facsimile with the addition of an author index, 1966.

[2] See Bullard, Sir Edward. Edmond Halley (1656–1741). *Endeavour*, 15, 1956, pp. 189–199; and Dingle, Herbert. Edmond Halley: his times and ours. *The Observatory*, 76, 1956, pp 117–131.

accounted for by the greatest philosophers of this age, [*etc.*], London, 1705–07, the third edition of which was published in three volumes, London, 1723–27. A re-issue of this was announced in 1969.[1]

Although primarily regarded as a philosopher, Bernard Le Bovier de Fontenelle should be noted for his part in popularizing the Copernican system of astronomy by his publication *Entretiens sur la pluralité des mondes*, 1686. Born at Rouen in 1657, Fontenelle, a weak child, was not expected to live long, but survived to within a month of his centenary. A nephew of Corneille, with whom he collaborated in two operas, he was the author of many works, including *Nouvelles dialogues des morts*, 1683; *Doutes sur le système physique des causes occasionnelles*, 1686; *Relation sur l'ile de Borneo*, 1686; and *Histoire des oracles*, 1686. The *Entretiens sur la pluralité des mondes* combines a popular defence of Copernican astronomy and Cartesian physics. R. Shackleton[2] has edited this, and records that at least twenty-eight editions were published during the author's life, while there was a total of ten editions in four different translations into English, and Italian and Russian translations. Fontenelle became permanent secretary to the Académie des Sciences in 1697 and was also a member of the Arcadian Academy of Rome, the Royal Society of London, and the Academies of Berlin, Nancy and Rouen. Fontenelle was thus in an ideal position to keep abreast of the progress in astronomy, and this is reflected in the many revisions made to the various editions of his book. Fontenelle planned and was part author of the *Histoire de l'Académie des Sciences depuis son établissement en 1666 jusqu'à 1686*, Paris, 1733. His *Eloges* of deceased members of the Académie are of particular value. Originally contributed to the volumes of *Mémoires* (*Histoire de l'Académie Royal des Sciences . . . avec les Mémoires, etc.*) these sixty-nine *Eloges* were issued as a collection in 1744, Paris, two volumes, and a selection was re-issued in Paris in 1883.[3]

Mathematics and geometry prospered at the hands of several important figures during the seventeenth century.[4] A number of mathematicians who are remembered chiefly for their writings are considered here. The French mathematician Gérard Desargues (1593–1663) was a native of Lyons, and became an architect and military engineer. He lectured in

[1] See Armitage, Angus. *Edmond Halley*, London, [*etc.*], 1966; MacPike, Eugene Fairfield. *Dr. Edmond Halley (1656–1742). A bibliographical guide to his life and work arranged chronologically, preceded by a list of sources, including references to the history of the Halley family*, 1939; MacPike, Eugene Fairfield. *Hevelius, Flamsteed and Halley: three contemporary astronomers and their mutual relations*, 1937; and MacPike, Eugene Fairfield. *Correspondence and papers of Edmond Halley*, Oxford, 1932.

[2] *Entretiens sur la pluralité des mondes. Digression sur les anciens et modernes. Edited by Robert Shackleton*, Oxford, 1955.

[3] See also McKie, Douglas. Bernard le Bovier de Fontenelle, F.R.S., 1657–1757. *Notes Rec. Roy. Soc. Lond.*, 12, 1957, pp. 193–200.

[4] See Whiteside, Derek Thomas. Patterns of mathematical thought in the later seventeenth century. *Archive for the History of Exact Sciences*, 1, 1961, pp. 179–388.

Paris, and was esteemed by both Descartes and Pascal, who made use of his theorems. He wrote on stone-cutting, and a work on perspective (1636), but his most important contribution, on conics, was his *Brouillon project d'une atteinte aux evénémens des rencontres d'un cône avec un plan,* [*etc.*], published at Paris in 1639. His collected works were published as *Oeuvres . . . réunies et analysées par M. Poudra*, two volumes, Paris, 1854.

René Descartes (1596–1650) was born at La Haye, near Tours, and on leaving school in 1612 he went to Paris, where he settled down to study mathematics. He joined the Army in 1618, but four years later resigned his commission. He spent the next five years in travel, but, after staying in Paris between 1626 and 1628, moved to Holland, where he remained for twenty years. Descartes spent several years writing *Le monde*, attempting a physical theory of the universe, but finding religious hostility to its publication he left it uncompleted. It was later published in this form in 1664. He then wrote his treatise on universal science, *Discours de la méthode pour bien conduire sa raison et chercher la vérité dans les sciences*, Leyden, 1637. This has three appendices, which were possibly issued in 1638, entitled *La dioptrique*, *Les météores* and *La géometrie*, the last being his most important contribution. The first Latin edition of the latter was printed at Leyden in 1649, with a second, in two volumes, from Amsterdam, 1659–61, and a third from Amsterdam, 1683. The first separately printed French edition was published at Paris in 1664. An English translation by David Eugene Smith and Marcia L. Latham was published in 1925.[1] A very rare English translation of the *Discours* was printed in London, 1649 and this was reprinted in 1966. Descartes published his views on philosophy as *Meditationes*, 1641, and three years later appeared his *Principia philosophiae*, Amsterdam, 1644, devoted to physical science.[2] Although written in 1637, his book on physiological psychology, *De homine*, was not published until 1662, when it was printed at Leyden. Descartes' *Regulae ad directionem ingenii* was first published at Amsterdam in 1701 in a volume entitled *Opuscula posthuma physica et mathematica*.[3] The complete works of Descartes, edited by Victor Cousin, were published in eleven volumes, Paris, 1824–26, but are considered inferior to the *Oeuvres, publiées par C. Adam et P. Tannery, sous les auspices du Ministère de l'Instruction Publique*, twelve volumes, Paris, 1897–1910, with an index volume published in 1913. Volume twelve consists of a biography of Descartes by Ch. Adam. Descartes has been called the father of modern

[1] Descartes, René. *The Geometry of René Descartes. Translated from the French and Latin by David Eugene Smith and Marcia L. Latham. With a facsimile of the first edition, 1637*, Chicago, 1925; reprinted New York, 1954.

[2] Blackwell, Richard J. Descartes' laws of motion. *Isis*, 57, 1966, pp. 220–234.

[3] See Beck, L. J. *The method of Descartes: a study of the Regulae*, Oxford, 1952; and Scott, J. F. *The scientific work of René Descartes (1596–1650). . . . With a foreword by H. W. Turnbull*, 1952.

philosophy, and his work on the analytical method of solving geometrical problems is also of outstanding importance. He is remembered eponymously by the terms 'Cartesian ovals', certain curves having important optical properties, and 'Descartes' rule of signs'. Gregor Sebba[1] has compiled a critical bibliography of literature relating to Descartes, providing summaries, references to reviews and adequate indexes.[2]

One of the most important mathematicians of his period, Bonaventura Cavalieri (1598–1647), was professor of mathematics at Bologna from 1629 until his death. He is mainly remembered for his use of the principle of indivisibles, formulated by him in 1629, but not published until it appeared in his *Geometriae indivisibilibus continuorum nova quadam ratione promota*, 1635. This was expanded in his *Exercitationes geometricae*, 1647, and re-issued with corrections in 1653.

Blaise Pascal (1623–62) contributed to physical science and mechanics, in addition to mathematics, and, of course, religious thought. At the age of sixteen he wrote an essay on conic sections, published as a broadside in 1640 as *Essai pour les coniques*, Paris.[3] Together with Pierre de Fermat (1601–65), a magistrate of Toulouse, Pascal created the calculus of probabilities, but much of the work of the former did not appear until it was published as *Varia opera mathematica*, [*etc.*], Toulouse, 1670–79 (reprinted Brussels, 1969), with a modern edition by Paul Tannery and Charles Henry, three volumes, Paris, 1891–96. This contains Fermat's correspondence with various mathematicians, including Pascal. The latter used his arithmetical triangle in 1653, but no account of this appeared until 1665, when it was published as *Traité du triangle arithmétique*. His *Traités de l'equilibre des liqueurs et de la pesanteur de la masse de l'air* was published posthumously in 1663, was reprinted in 1819, and a facsimile of this reprint was published in Paris in 1956; he also wrote on the cycloid (1658). Collected editions of his works include L. Brunschwig, P. Boutroux and F. Gazier's *Oeuvres de Blaise Pascal*, [*etc.*], fourteen volumes, Paris, 1904–14, and the better edition by F. Strowski entitled *Oeuvres complètes de Pascal*, three volumes, Paris, 1921–31.[4]

Several British mathematicians of note published important con-

[1] Sebba, Gregor. *Bibliographia Cartesiana. A critical guide to the Descartes literature 1800–1960*, The Hague, 1964.

[2] See also Crombie, A. C. Descartes. *Sci. Amer.*, 201, 1959, pp. 160–173; Armitage, Angus. René Descartes (1596–1650) and the early Royal Society. *Notes Rec. Roy. Soc. Lond.*, 8, 1950–51, pp. 1–19; and Aiton, E. J. The Cartesian theory of gravity. *Ann. Sci.*, 15, 1959, pp. 27–49.

[3] See Smith, David Eugene. "Essay pour les coniques" of Blaise Pascal. Translated by Frances Marguerite Clark. *Isis*, 10, 1928. Facsimile and English translation, pp. 16–20.

[4] See also Boutroux, E. *Pascal*, seventh edition, Paris, 1919; Strowski, F. *Pascal et son temps*, 3 vols., Paris, 1931; Maire, Albert. *Bibliographie générale des oeuvres de Blaise Pascal*, 2 vols., Paris, 1925–26; and *The physical treatises of Pascal. The equilibrium of liquids and The weight of the mass of the air. Translated by I. H. B. and A. G. H. Spiers, with introduction and notes by Frederick Barry*, New York, 1937. Bibliographical note, pp. 171–173.

tributions during the seventeenth century, the first of whom, John Napier (1550–1617), was a native of Merchiston in Scotland. He invented the modern notation of fractions, and also logarithms. As early as 1594 Napier had privately communicated to Tycho Brahe his results, and his invention was publicly announced in his *Mirifici logarithmorum canonis descriptio*, Edinburgh, 1614, of which an English translation was published in the following year. The method by which the logarithms were calculated was not made available until after Napier's death, when it was explained in *Mirifici logarithmorum canonis constructio*, published in 1619,[1] although written before the *Descriptio*. Napier was also the author of *Randologia*, published in 1617, containing a description of "Napier's bones", which were designed to simplify multiplication and division.[2]

William Oughtred (1575–1660) made important contributions to mathematics, originated the use of the symbol X for multiplication, and was the inventor of the slide rule. His *Arithmeticæ in numeris et speciebus institutio . . . atque adeo totius mathematicae quasi clavis est*, 1631, went into several editions, including a rare printing in 1647 which has been described by P. J. Wallis.[3] The work was translated into English as *The key of the mathematicks new forged and filed*, 1647. This used a number of mathematical symbols including that for multiplication. His pupil William Forster published the first description of the slide rule in his translation of Oughtred's *The circle of proportion and The horizontall instrument*, 1632, and other dates. Another important work was *Trigonometria: hoc est, modus computandi triangularum latera & angulos*, [*etc.*], edited by Richard Stokes, London, 1657, which made use of symbols that were unfortunately neglected until reintroduced by Euler. An English translation of this was published in the same year.[4]

John Wallis (1616–1703) was born at Ashford, Kent, and in 1649 was appointed to the Savilian chair of geometry at Oxford. He published a treatise on conic sections in 1655, but his most important mathematical work was *Arithmetica infinitorum*, 1655, which contains the germ of the differential calculus, and extended and systematized the methods of analysis of Descartes and Cavalieri. Wallis's *De algebra tractatus*, 1685, contains the first systematic use of formulae. The work also includes a number of tracts, including *A treatise of angular sections*, 1684, which has

[1] A translation of the *Constructio*, by W. R. Macdonald was published at Edinburgh in 1889. This contains a bibliography of Napier's writings.

[2] See Hobson, E. W. *John Napier and the invention of logarithms, 1614*. Cambridge, 1914; and *Napier Tercentenary Memorial Volume*, Edinburgh, 1915.

[3] Wallis, P. J. William Oughtred's "Arithmeticae in numeris" or "Clavis mathematicae" 1647; an unrecorded (unique?) edition in Aberdeen University Library. *The Bibliotheck*, 5, 1968, pp. 147–148.

[4] See Wallis, P. J. William Oughtred's "Circles of proportion" and "Trigonometries". *Trans. Cambridge Bib. Soc.*, 4, 1968, pp. 372–382.

been the subject of a paper by Christoph J. Scriba,[1] who has also indexed the correspondence of Wallis.[2] This includes 818 letters written between 1641 and 1703, most of which are preserved in the Royal Society and the Bodleian Library, Oxford. They include letters to and from Henry Oldenburg, Edmond Halley, Richard Waller, Hans Sloane, Christiaan Huygens, John Flamsteed and others. The collected writings of John Wallis were printed as *Opera mathematica*, three volumes, Oxford, 1693–98.[3]

In addition to his theological writings, Isaac Barrow (1630–1677) published several works on mathematics and optics. He was Gresham Professor of Geometry at Cambridge (1662–63) and Lucasian Professor of Mathematics (1663–1669), resigning this post in favour of his pupil, Isaac Newton. Barrow's writings include *Euclidis elementa*, 1655; *Euclidis data*, 1657; *Mathematicae lectiones*, 1664–66; *Lectiones opticae*, 1669, 1670 and 1674; and *Archimedis Opera*. An edition of his mathematical works was published in 1860.

Another Scotsman, James Gregory (1638–75), was born at the manse of Drumoak, near Aberdeen. His description of the earliest reflecting telescope is contained in *Optica promota*, [*etc.*], London, 1663, and in that year he came to London, whence he proceeded to Italy for three years. While in Padua he wrote *Vera circuli et hyperbolae quadratura*, [*etc.*], Padova, 1667, another edition appearing in 1668; and *Geometriae pars universalis*, [*etc.*], Padova, 1668. His *Exercitationes geometricae* was also published in London in 1668, and in the same year he was elected a Fellow of the Royal Society. When Charles II established a chair of mathematics at St. Andrews, Gregory became its first occupant, but in 1674 he moved to a similar position at Edinburgh.[4]

Thomas Harriot (1560–1621) was born at Oxford, and was sent on an expedition to Virginia by Sir Walter Raleigh. On his return to London Harriot lived at Sion House, the seat of Henry Percy, Earl of Northumberland, and wrote several volumes of manuscripts, some in a secret script.[5]

[1] Scriba, Christoph J. John Wallis' *Treatise of angular sections* and Thâbit ibn Quarra's generalization of the Pythagorean theorem. *Isis*, 57, 1966, pp. 56–66.

[2] Scriba, Christoph J. A tentative index of the correspondence of John Wallis, F.R.S. *Notes Rec. Roy. Soc.*, 22, 1967, pp. 58–93.

[3] See Scott, J. F. *The mathematical work of John Wallis, D.D., F.R.S., (1616–1703)* [*etc.*], 1938. Contains lists of his papers in the *Philosophical Transactions*, and in his *Opera;* also Scott, J. F. John Wallis as a historian of mathematics. *Ann. Sci.*, 1, 1936, pp. 335–357.

[4] See Turnbull, Herbert Westren, *ed. James Gregory tercentenary memorial volume. Containing his correspondence with John Collins and his hitherto unpublished mathematical manuscripts, together with addresses and essays communicated to the Royal Society of Edinburgh, July, 1938. Edited by Herbert Westren Turnbull*, 1939; and Turnbull, Herbert Westren, and Bushnell, George Herbert. *University of St. Andrew's. James Gregory Tercentenary. Record of the celebrations held in the University Library July fifth MCMXXXVIII*, St. Andrew's, 1939.

[5] See Seaton, E. Thomas Hariot's secret script. *Ambix*, 5, 1956, pp. 111–114.

These are preserved in the British Museum[1] and at Petworth House. Baron von Zach (1754–1832), an astronomer, found the Harriot papers in 1784, and for almost fifty years attempted to have them published by Oxford University. S. P. Rigaud examined the papers in the British Museum, and attacked von Zach, who has been vindicated by Johs Lohne[2] for his appreciation of Harriot as an outstanding mathematician and astronomer. Many scientists had access to his papers, and were probably indebted to his observations, none of which was published during his lifetime. "De reflexione corporum rotundorum", on the collision of elastic bodies, was written in 1618, and was ready for press; "De numeris triangularibus" was almost completed; several other works were lost. Of his scientific writings only *Artis analyticae praxis*, London, 1631 was printed, and that posthumously. It is not his most important work, but developed the results of Vieta, and contributed to the development of algebra.[3] The circumstances of the publication of this book have been recorded by Rosalind C. H. Tanner,[4] and several other papers devoted to Harriot's work have been published.[5] There is still a need for a thorough investigation of his manuscripts, with eventual publication of an edited version of the significant items. The only book by Harriot to be printed during his lifetime was his *Briefe and true report of the new found land of Virginia*, 1588. A well-documented study of atomism from Harriot to Newton has been provided by Robert Hugh Kargon in *Atomism in England from Hariot to Newton*, Oxford, 1966, which stresses the potential value of the unexplored scientific material to be found in Harriot's manuscripts, and also devotes chapters to Francis Bacon, Thomas Hobbes, Descartes, Charleton and Boyle.

André Tacquet (died 1660) wrote *Cylindricorum et annularium libri iv*, [*etc.*], Antwerp, 1651; *Eléments de géométrie plane et solide*, 1654; *Arith-*

[1] British Museum, Harleian MSS., 6001–6002, 6083; Add. MSS. 6782–6789.

[2] Lohne, Johs. Thomas Harriott (1560–1621). The Tycho Brahe of optics; preliminary notice. *Centaurus*, 6, 1959, pp. 113–121; Lohne, Johs. The fair fame of Thomas Harriott. Rigaud *versus* Baron von Zach. *Centaurus*, 8, 1963, pp. 69–84; Lohne, Johs. Dokumente zur Revalidierung von Thomas Harriott als Algebraiker. *Archive for the History of Exact Sciences*, 3, 1966–67, pp. 185–205.

[3] Cajori, Florian. A revaluation of Harriott's Artis analyticae praxis. *Isis*, 11, 1928, pp 316–324.

[4] Tanner, Rosalind C. H. On the role of equality and inequality in the history of mathematics. *Brit. J. Hist. Sci.*, 1, 1962–63, pp. 159–169.

[5] See Morley, F. V. Thomas Hariot—1560–1621. *Sci. Monthly*, 14, 1922, pp. 60–66; Jaquot, Jean. Thomas Harriott's reputation for impiety. *Notes Rec. Roy. Soc.*, 9, 1952, pp. 164–187; Kargon, Robert H. Thomas Hariot, the Northumberland circle and early atomism in England. *J. Hist. Ideas*, 27, 1966, pp. 128–136; Tanner, Rosalind C. H. The study of Thomas Harriot's manuscripts. I. Harriot's will. *History of Science*, 6, 1957, pp. 1–16; Pepper, Jon V. The study of Thomas Harriot's manuscripts. II. Harriot's unpublished papers. *History of Science*, 6, 1957, pp. 17–40; Tanner, Rosalind C. H. Thomas Harriot as mathematician—a legacy of hearsay; part 1. *Physis*, 9, 1967, pp. 235–247; see also Kargon, Robert H. *Atomism in England from Hariot to Newton*, Oxford, 1966.

métique théorique et pratique, 1656, of which later editions appeared at Antwerp, 1665 and 1682, and Brussels, 1683; and *Opera mathematica*, [*etc.*], Antwerp, 1669.[1] Yet another Frenchman of whom little is known, François Le Gendre (died between 1672 and 1678), was the author of three extant works, *Traicté arithmétique contenant les règles necessaires tant pour les finances que pour les marchandise . . . ensemble un abrégé de l'arithmétique aux jettons . . . nouvellement augmenté d'un traicté d'arpentage*, Paris, 1646; *L'Arithmétique en sa perfection*, [*etc.*], Paris, 1657, 1668, 1672, 1682, and many other editions up to 1812; and *La vraye maniere de tenir livres de comptes ou de raison par parties doubles*, [*etc.*], Paris, 1658.[2]

It is of interest to note the earliest book devoted to arithmetic published in America, which was by Pedro de Paz, and entitled *Arithmetica*, City of Mexico, 1623. This was an elementary textbook on the subject, and is very rare, no complete copy being known. Books on other mathematical subjects published in Mexico before 1623 were Juan Diez Freile's *Sumario*, 1556; Garcia de Palacio's *Instruccion nauthica*, 1587; and Fray Alejo Garcia's *Kalendario perpetuo*.

Physics and its subdivisions were greatly advanced during this century, and in this section we consider several eminent figures who contributed equally to other branches of science. Boyle, Hooke and Newton were our greatest scientists of the period, and their work contributed enormously towards giving the seventeenth century its historical significance. Pierre Gassend (Gassendi) (1592–1655) was a French philosopher, naturalist and physicist, who expressed noteworthy ideas on the atomic structure of matter, on light and optics, and who was the first to attempt to measure the velocity of sound. His most important book, *De proportione qua gravia decidentia accelerantur epistolae tres*, Paris, 1646, proved the law of acceleration of falling bodies by experiment, and *De motu impresso a motore translato*, a series of letters published in 1642, contains his thoughts on the principle of inertia.[3] Gassendi's *Institutio astronomica*, 1647, went into at least five more editions up to 1702, and his influence on early American writers was widespread.[4] His *Opera omnia* were first published in six volumes, Lyons, 1658, and this edition was reprinted in Stuttgart-Bad Connstatt, 1968.[5]

Henry Power (1623–68) was a pioneer in microscopical research in

[1] Bosmans, H. André Tacquet (S.J.) et son traite d'"Arithmetique théorique et pratique". *Isis*, 9, 1927, pp. 66–82.

[2] See Sanford, Vera. François Le Gendre, arithmeticien. *Osiris*, 1, 1936, pp. 510–518.

[3] Pav, Peter Anton. Gassendi's statement of the principle of inertia. *Isis*, 57, 1966, pp. 24–34.

[4] Gorman, Mel. Gassendi in America. *Isis*, 55, 1964, pp. 409–417.

[5] See also Clark, Joseph T. Pierre Gassendi and the physics of Galileo. *Isis*, 54, 1963, pp. 352–370; Gregory, Tullio. *Scetticismo ed empirismo, studio su Gassendi*, Bari, 1961; Centre International de Synthèse. *Pierre Gassendi, 1592–1652, sa vie et son oeuvre*, Paris, 1955; Centre National de la Recherche Scientifique. *Tricentenaire de Pierre Gassendi, 1655–1955*, Paris, 1957.

this country, and was a disciple of Sir Thomas Browne, Power's writings reminding one of the style of the author of *Religio medici*. Power, who was a doctor of medicine and an original Fellow of the Royal Society, published only one work, which was entitled *Experimental philosophy, in three books; containing new experiments, microscopical, mercurial, magnetical,* [*etc.*], London, 1664 [1663], but his poem on the microscope has been published in recent times[1]. *Experimental philosophy* was reprinted with an introduction by Marie Boas Hall, New York, London, 1966. Power's microscopical observations are of great interest, and his work anticipated that of Hooke. The book contains the first general essay on microscopy published in England, and Power also left a number of unpublished manuscripts, which are now in the British Museum. One of these (Sloane MS. 1343) is dated 1652 and contains the results of numerous experiments on the circulation. This manuscript has been published in a paper by F. J. Cole,[2] and a physiological treatise, "Historia physico-anatomica", which had been prepared for publication in 1666, has been discussed in a paper by C. Webster.[3]

The air-pump and its development connects the work of three scientists, Guericke, Schott and Papin. Otto von Guericke (1602–86) made a great contribution to experimental science by his invention of the air-pump. The date of the invention is uncertain, but he gave a public demonstration in Magdeburg in 1654. The first published account of Guericke's air-pump was given by Kaspar Schott (1608–66) in his *Mechnica. Hydraulico pneumatica*, Wurzburg, 1657. Guericke was born in Magdeburg in 1602, and studied jurisprudence, mathematics and mechanics. His work encountered considerable opposition and this impelled him to publish his own account in *Experimenta nova (ut vocantur) Magdeburgica de vacuo spatio*, Amsterdam, 1672. The most important section of this work is Book III, *De propriis experimentalis*,[4] but Book IV contains the description of Guericke's machine for generating electrical charges.[5] Guericke died in Hamburg in 1686. Schott, a pupil of Kircher, had repeated Guericke's experiments at the latter's request, and his book *Mechanica. Hydraulico pneumatica* already mentioned was responsible for Boyle's interest in the construction of an air-pump. This work was translated into English by Richard Waller, in 1684. Schott was also the author of a number of works similar in type to those of Kircher, e.g. *Thaumaturgus physicus. Sive magiae universalis naturae et artis*, Wurzburg, 1657–59; *Physica curiosa sive*

[1] See Cowles, Thomas. Dr. Henry Power's poem on the microscope. *Isis*, 1, 1936, pp. 510–518.

[2] Cole, F. J. Henry Power on the circulation of the blood. *J. Hist. Med.*, 12, 1957, pp. 291–324.

[3] Webster, C. Henry Power's experimental philosophy. *Ambix*, 14, 1967, pp. 150–178.

[4] German translation by F. Dannemann, Ostwald's *Klassiker*, 59.

[5] Heathcote, N. H. de V. Guericke's sulphur globe. *Ann. Sci.*, 6, 1950, pp. 293–305.

mirabilia naturae et artis libri XII, Wurzburg, 1662, 1664, 1667, 1697; and *Technica curiosa sive mirabilia artis*, Nurnberg, 1664. Denis Papin (1647–*c.* 1722) was born at Coudrais.[1] After acting as an assistant to Huygens, he acted in a similar capacity to Boyle, becoming a Fellow of the Royal Society in 1680. In 1687 he was appointed to the chair of mathematics at the University of Marburg, moving to Cassel in 1696, and returning to London in 1707. His later movements are obscure. Among his publications were *Nouvelles expériences du vide avec la description des machines qui servent à les faire*, Paris, 1674. Translated into Latin by (or for) Boyle as *Experimentorum novorum physico-mechanicorum continuatio secunda*, [*etc.*], London, 1680, and into English, 1682, which describes his air-pump; *Fasciculus dissertationum*, Marburg, 1695, a collection of his writings also published in French as *Recueil des divers pièces touchant quelques nouvelles machines*, Cassel, 1695; *A new digester, or engine for softening bones*, London, 1681; *A continuation of the new digester . . . together with some improvements and new uses of the air-pump, tryed both in England and Italy*, London, 1687; and *Ars nova ad aquam ignis adminiculo efficacissime elivandem*, Cassel, 1707. His correspondence with Liebniz and Huygens has been published.[2,3]

The Honourable Robert Boyle was born at Lismore, County Cork, on 25th January, 1626, the son of Richard Boyle, Earl of Cork. He was the fourteenth child and seventh son of his father. Robert Boyle was educated at Eton, and then went to Geneva, where he resided for almost five years. He suffered from renal calculus for the greater part of his life, which affected his activities. When Boyle set up a laboratory at Oxford, Robert Hooke became his assistant, and the two invented a "pneumatical engine", which resulted in the enunciation of Boyle's law. Boyle was the friend of John Evelyn, and of Henry Oldenburg, to whom he sent his contributions from Oxford for the Royal Society. In 1669 Boyle left Oxford to live with his sister, Lady Ranelagh, in Pall Mall. He died on 30th December, 1691, and was buried in St. Martin-in-the-Fields. Robert Boyle was the author of forty-two books and numerous scientific papers, his writings having been the subject of bibliographical study by Sir Geoffrey Keynes[4] and more exhaustively by John F. Fulton.[5] The following details are mainly derived from the latter's monumental work, which mentions that many 'ghosts' exist, that many of Boyle's works are very

[1] See Dickinson, H. W. Tercentenary of Denis Papin. *Nature*, 160, 1947, pp. 422–423.

[2] Gerland, Ernst. *Leibnizens und Huygens' Briefwechsel mit Papin, nebst der Biographie Papins*, Berlin, 1881.

[3] Andrade, E. N. da C. The early history of the vacuum pump. *Endeavour*, 16, 1957, pp. 29–35.

[4] *The Honourable Robert Boyle. A handlist of his works. G. L. K*[*eynes*] *for and from J. F. F*[*ulton*], 1932.

[5] Fulton, John Farquhar. *A bibliography of the Honourable Robert Boyle. Fellow of the Royal Society. . . . Second edition*, Oxford, 1961. The section on biography and criticism (pp. 155–196) contains a wide selection of references up to 1960.

obscure bibliographically, and that there are many 'issues' with slight differences, such as cancelled title-pages. Boyle's first published book,[1] generally known from its running title as *Seraphick love*, was *Some motives and incentives to the love of God*, [*etc.*], London, 1659, of which there were numerous editions and translations. His first scientific work, *New experiments physico-mechanicall, touching the spring of the air, and its effects*, (*made for the most part, in a new pneumatical engine*), [*etc.*], Oxford, 1660, contained the results of experiments on the air pump, the first of which had been made by Hooke. Boyle's law was not enunciated in this first edition, but in the *Defence* against Linus contained in the second, also published at Oxford in 1662. A continuation was published at Oxford, 1669, a second continuation in Latin, Boyle's only work to appear first in Latin, London, 1680, and Geneva, 1682, with a translation into English, London, 1682. Boyle's law has been described as the result of a co-operative enterprise, with Henry Power, Richard Towneley, Robert Hooke and others making contributions, and C. Webster[2] and others[3] have contributed papers on the subject. Latin editions of the *New experiments* were published at Oxford, 1661; The Hague, 1661; London, 1663; Rotterdam, 1669; Geneva, 1677; "Coloniae Allobrogum", 1680; and another issue with the imprint, Geneva, 1680. *Certain physiological essays, written at distant time, and on several occasions*, was first published in London, 1661, with a second edition in 1669. Latin editions were printed in London, 1661 and 1667; Amsterdam, 1667 (two editions); and there were others.

The Sceptical chymist: or chymico-physical doubts & paradoxes, [*etc.*], London, 1661,[4] is rare, and there was no Oxford edition printed in this year as sometimes stated, but a facsimile reprint of the London printing was issued in 1965. An early version of this work was copied by Henry Oldenburg into his commonplace book late in 1660 under the title *Reflexions on the experiments vulgarly alledged to evince the 4 peripatetique elements, or ye 3 chymicall principles of mixt bodies*. It was written in 1658[5] and the substance of the *Reflexions* occurs in the *Sceptical chymist*. A tran-

[1] His first published writing was signed "Philaretus", and appeared in Samuel Hartlib's *Chymical, medicinal and chyrurgical addresses*, [*etc.*], 1655. See Rowbottom, Margaret E. The earliest published writing of Robert Boyle. *Ann. Sci.*, 6, 1948–50, pp. 376–389, where Boyle's article is reprinted; and Maddison, R. E. W. The earliest published writing of Robert Boyle, *Ann. Sci.*, 17, 1961 [1963], pp. 165–173.

[2] Webster, C. Richard Towneley and Boyle's law. *Nature*, 197, 1963, pp. 226–228; and Webster, C. The discovery of Boyle's law, and the concept of the elasticity of air in the seventeenth century. *Archive Hist. exact. Sci.*, 2, 1965, pp. 441–502.

[3] See Cohen, I. Bernard. Newton, Hooke and "Boyle's law" (discovered by Power and Towneley). *Nature*, 204, 1964, pp. 618–621; and Neville, Roy G. The discovery of Boyle's law, 1661–62. *J. chem. Educ.*, 39, 1962, pp. 356–359.

[4] See Davis, Tenney L. The first edition of the Sceptical chymist. *Isis*, 8, 1926, pp. 71–76.

[5] See Webster, C. Water as the ultimate principle of nature: the background to Boyle's Sceptical chymist. *Ambix*, 13, 1966, pp. 96–107.

script of Oldenburg's version has been published by Marie Boas.[1] The second English version of the *Sceptical chymist* was published with the imprint Oxford, 1680, but with an advertisement stating that this date should be 1679.[2] Latin editions were printed in London, 1662; Rotterdam, 1662 and 1668; Geneva, 1677 and 1680, and later editions and translations also were printed. Correspondence between Samuel Hartlib (died 1662) and Boyle reveals close scientific relationship between the two scientists, and the Hartlib papers at Sheffield University are worthy of thorough investigation, particularly Hartlib's manuscript daybook kept between 1634 and 1660.[3]

Boyle's *Some considerations touching the usefulness of experimental naturall philosophy*, [*etc.*], was published at Oxford, 1663 and 1664 (two issues), with the first edition of Volume 2 printed at Oxford, 1671. Latin editions were printed at Lindau, 1662 (two issues), and Geneva, 1694. It is believed that Boyle wrote a tract entitled *The history of colours begun*, dated 1663, but Fulton was unable to trace a copy. A volume entitled *Experiments and considerations touching colours*, [*etc.*] went into numerous editions, including impressions dated London, 1664 and 1670, with Latin versions published in London, 1665, and Amsterdam, 1667. A facsimile of the 1664 edition with an introduction by Marie Boas Hall was published in 1964. H. A. M. Snelders[4] suggests that Boyle was the first to use 'spot-tests' for analytical purposes, and described the technique in this work, although it had been mentioned by Pliny.

Boyle's *Occasional reflections upon several subjects*, [*etc.*], was first published in London in 1665, and his *New experiments and observations touching cold, or an experimental history of cold, begun*, [*etc.*], was printed in London, 1665 and 1683. Fulton has suggested that this was never translated into Latin, but Marie Boas Hall[5] states that a translation was made and printing begun, but that booksellers in Holland offered to take the entire impression. No copy has been traced. *Hydrostatical paradoxes, made out by new experiments*, [*etc.*], was printed in Oxford, 1666, with Latin editions from Oxford, 1669; Rotterdam, 1670; Geneva, 1677; and "Colonia Allobrogum", 1680, together with later editions. *The origine of formes and*

[1] Boas, Marie. An early version of Boyle's *Sceptical chymist*. *Isis*, 45, 1954, pp. 153–168; see also Neville, Roy G. "The sceptical chymist", 1661. A tercentenary tribute. *J. chem. Educ.*, 38, 1961, pp. 106–109.

[2] Fulton, 34.

[3] See Webster, Charles, *ed. Samuel Hartlib and the advancement of learning*, Cambridge, 1970; O'Brien, J. J. Samuel Hartlib's influence on Robert Boyle's scientific development. Pt. 1. The Stalbridge period. *Ann. Sci.*, 21, 1965, pp. 1–14; Pt. 2. Boyle in Oxford. *Ann. Sci.*, 21, 1965, pp. 257–276; and Wilkinson, Ronald Sterne. The Hartlib papers and seventeenth-century chemistry. *Ambix*, 15, 1968, pp. 54–59.

[4] Snelders, H. A. M. Reply to query "Was Boyle the first to use spot-test analysis?" *Isis*, 56, 1965, pp. 210–211.

[5] Hall, Marie Boas. What happened to the Latin edition of Boyle's *History of cold. Notes Rec. Roy. Soc. Lond.*, 17, 1962, pp. 32–35.

qualities, (according to the corpuscular philosophy), [*etc.*], appeared from Oxford, 1666[1] (Plate 3) and 1667, with another issue of the latter with a cancel title-page lacking the imprimatur on the reverse. Latin editions of this work were published at Oxford, 1669; Amsterdam, 1671; Geneva, 1688; and "Coloniae Allobrogum", 1688. Two issues of *Tracts . . . about the cosmicall qualities of things*, [*etc.*], bear the imprint Oxford, 1671, while Latin editions were issued from Amsterdam, 1671 (three issues); London, 1672; Geneva, 1677; "Coloniae Allobrogum", 1680, and another issue of this with the imprint Geneva, 1680. Boyle's observations on the effect of temperature and pressure on the size of air bubbles in water are contained in *Tracts . . . of a discovery of the admirable rarefaction of the air*, [*etc.*], of which a very rare Latin edition was published in London, 1670, with an English translation, London, 1671, and of which there was possibly a reissue with errata in the same year. Latin versions also appeared from London, 1671; Geneva, 1677; and "Coloniae Allobrogum", 1680, with the inevitable accompanying issue bearing the imprint Geneva, 1680. The first scientific contribution in the history of crystallography is represented by *An essay about the origine & virtues of gems*, [*etc.*], London, 1672, with Latin editions from London, and also Hamburg, 1673; Geneva, 1677; and "Coloniae Allobrogum", and also Geneva, 1680. *Tracts . . . containing new experiments, touching the relation between flame and air*, [*etc.*], was first published in London, 1672, with a reissue containing a cancel title-page dated 1673; a Latin version appeared from Geneva in 1696. One of Boyle's significant but less widely known works is *Essays of . . . effluviums*, [*etc.*], London, 1673[2] (three issues), with a Latin version also published in London in the same year, and several other editions. His *Tracts consisting of observations about the saltness of the sea*, [*etc.*], contains six tracts, and was published in London, 1674, with Latin editions, London, 1676; and Geneva, and also "Coloniae Allobrogum", 1680. *Experiments, notes &c. about the mechanical origine or production of divers particular qualities*, [*etc.*], contains eleven tracts on electricity, magnetism, taste and smell, among other subjects, and was printed in London, 1675, of which another issue with a cancel title-page is dated 1676. Latin versions appeared from London, 1692, and Geneva, 1694. Several of these tracts were also published separately. *Of a degradation of gold made by an anti-elixir: a strange chymical narrative*, was published in London, 1678, and reprinted, London, 1739. Two tracts on phosphorescence, first published separately, and then

[1] See Kargon, R. Walter Charleton, Robert Boyle, and the acceptance of Epicurean atomism in England. *Isis*, 55, 1964, pp. 184–192.

[2] The full title reads *Essays of the strange subtilty, determinate nature, great efficacy of effluviums. To which are annext new experiments to make fire and flame ponderable: together with a discovery of the previousness of glass*, five tracts in one volume which has been the subject of an interesting study by, McKie, Douglas. The Hon. Robert Boyle's *Essays of effluviums* (1673). *Sci. Progr.*, 29, 1934–35, pp. 253–265.

together, are *The aerial noctiluca*, [*etc.*], London, 1680, with a German edition from Hamburg, 1682; and *New experiments and observations made upon the icy noctiluca*, [*etc.*], London, 1681–82. These were published together in Latin from London, 1682; and Geneva, 1692, with another issue dated 1693. The most important of Boyle's medical writings was his *Memoirs for a natural history of humane blood*, [*etc.*], London, 1683–84, which has a cancel title-page, and of which several copies of the original issue have recently been discovered bearing the date 1684. Latin editions appeared from London, 1684: Geneva, 1685 and 1686, with an edition bearing the imprint "Coloniae Allobrogum" and the latter date; *Experiments and considerations about the porosity of bodies*, [*etc.*], London, 1684, with Latin editions from London, 1684; and Geneva, and also "Coloniae Allobrogum", 1686; *Short memoirs for the natural experimental history of mineral waters*, [*etc.*], London, 1684–85; *An essay of the great effects of even languid and unheeded motion*, [*etc.*], London, 1685, with two other issues, 1685 and 1690; *Of the reconcileableness of specifick medicines to the corpuscular philosophy*, [*etc.*], London, 1685, with Latin editions from London, 1686; and Geneva, 1687, and a French edition from Lyons, 1688, with another issue having the date altered to 1689; *A free enquiry into the vulgarly receiv'd notion of nature*, [*etc.*], London, 1685–86, with Latin editions bearing the imprint London, 1688; and Geneva, 1688; *Some receipts of medicines, for the most part parable and simple. Sent to a friend in America*, London, 1688, which Fulton failed to trace, but of which pamphlet there is a copy in the British Museum,[1] recorded in Wing [4043], and which was later contained in *Medical experiments; or, a collection of choice remedies*, [*etc.*], London, 1692; Volume 2, London, 1693; and Volume 3, London, 1694, there being several further editions of these; *A disquisition about the final causes of natural things*, [*etc.*], London, 1688 (two issues); with a Dutch translation, Amsterdam, 1688. This contains as an appendix a separate tract with the title *Some uncommon observations about vitiated sight*, which includes classical accounts of exophthalmic ophthalmoplegia and adaptation to dark, together with other contributions to ophthalmology, and has been the subject of a paper by Richard A. Hunter and F. Clifford Rose.[2]

Medicina hydrostatica: or hydrostaticks applyed to the materia medica, [*etc.*], was printed in London, 1690, with a Latin edition from Geneva, 1693; *Experimenta & observationes physicae*, [*etc.*], London, 1691; *The general history of the air*, [*etc.*], London, 1692, in which Boyle was assisted by John Locke (1632–1704), who saw it through the press after Boyle's death;[3]

[1] See Maddison, R. E. W. The first edition of Robert Boyle's *Medicinal experiments. Ann. Sci.*, 18, 1962 [1964], pp. 43–47.

[2] Hunter, Richard A., and Rose, F. Clifford. Robert Boyle's "Uncommon observations about vitiated sight". (London, 1688). *Brit. J. Ophthal.*, 42, 1958, pp. 726–731.

[3] See Dewhurst, Kenneth. Locke's contribution to Boyle's researches on the air and on human blood. *Notes Rec. Roy. Soc. Lond.*, 17, 1962, pp. 198–206.

General heads for the natural history of a country, great or small, [*etc.*], London, 1692; and a Latin translation from Geneva, 1696. Robert Boyle contributed several papers to the *Philosophical Transactions*, and was the author of several non-scientific works. Other spurious writings have been attributed to him. Among numerous collections and epitomes of Boyle's *Works*, the following are noteworthy: the five-volume edition, London, 1744; a six-volume edition, London, 1772; and a Latin edition, three volumes, Venice, 1696–97. Many of Boyle's letters and manuscripts are preserved in the Royal Society, the British Museum, and elsewhere. R. E. W. Maddison had contributed an index of his correspondence,[1] and also written an invaluable series of papers devoted to Boyle and his contemporaries,[2] culminating in a full-scale biography.[3] This includes Boyle's autobiographical sketch, a geneological chart, and incorporates material from Boyle's letters, papers and notebooks in the Royal Society, and manuscript material in the British Museum and at Chatsworth. Very well-documented and illustrated, this biography contains Boyle's last will and testament, and provides information on the fate of his collection of books.[4]

Benedict de Spinoza (1632–77) was born in Amsterdam, his parents being Jewish refugees from Portugal. Spinoza studied mathematics and philosophy, and was a skilful maker of lenses, but he is remembered mainly as a philosopher. His *Tractatus theologico-politicus*, Hamburg [Amsterdam], 1670, was published anonymously, even the name of the printer, "Henricum Kunraht", being pseudonymous. There were several issues of this, all of which are rare. This was the only book by Spinoza to be published during his lifetime, but he was the author of a work entitled *Ethica ordini geometrico demonstrata*, 1677, and also of *Algebraic calculation of the rainbow*, 1687, a facsimile of the original Dutch text of which, with an introduction by G. Ten Doesschate, was published in Niewkoop, 1963. Spinoza's *Opera posthuma* were published in Latin and

[1] Maddison, R. E. W. A tentative index of the correspondence of the Honourable Robert Boyle, F.R.S. *Notes Rec. Roy.Soc. Lond.*, 13, 1958, pp. 128–201.

[2] Maddison, R. E. W. Studies in the life of Robert Boyle, F.R.S. *Notes Rec. Roy. Soc. Lond.*, 9, 1951–52, pp. 1–35, 196–219; 10, 1952, pp. 15–27; 11, 1954, pp. 38–53, pp. 159–188; 18, 1963, pp. 104–124; 20, 1965, pp. 51–77; see also Maddison, R. E. W. The portraiture of the Honourable Robert Boyle, F.R.S. *Ann. Sci.*, 15, 1959, pp. 141–214.

[3] Maddison, R. E. W. *The life of the Honourable Robert Boyle, F.R.S.*, 1969.

[4] See also Boas, Marie. *Robert Boyle and seventeenth century-chemistry*, Cambridge, 1958; Hall, Marie Boas. *Robert Boyle on natural philosophy. An essay with selections from his writings*, Bloomington, 1965; Hall, Marie Boas. Robert Boyle. *Scientific American*, 217, 1967, pp. 97–102; More, Louis Trenchard. *The life and works of the Honourable Robert Boyle*, London, [*etc.*], 1944; [Keynes, Sir Geoffrey L.] Robert Boyle. *St. Bart's Hosp. J.*, 39, 1931–32, pp. 184–189; Fulton, John F. Robert Boyle and his influence on thought in the seventeenth century. *Isis*, 18, 1932, pp. 77–102; McKie, Douglas, Boyle's law. *Endeavour*, 7, 1948, pp. 148–151; Fulton, John Farquhar. The Honourable Robert Boyle, F.R.S. (1627–1692). *Notes Rec. Roy. Soc. Lond.*, 15, 1960, pp. 119–135; and Maddison, R. E. W. A summary of former accounts of the life and work of Robert Boyle. *Ann. Sci.*, 13, 1957, pp. 90–108.

in Dutch in 1677. Spinoza was acquainted with many of the most prominent scientists of his period, and his influence extended through the centuries following his death.[1]

Described as "probably the most inventive man who ever lived", Robert Hooke (1635–1703) contributed to many branches of knowledge, including physics, meteorology, astronomy, geology, biology, combustion and respiration. He was born on 18th July, 1635, at Freshwater, Isle of Wight, and after a brief apprenticeship to a portrait painter, became a pupil at Westminster School, before proceeding to Christ Church, Oxford in 1653. Hooke took an early interest in geometry, mathematics and astronomy, and became assistant to Thomas Willis and then to Robert Boyle. Hooke invented and constructed in 1658–59 the air-pump with which Boyle carried out his experiments on combustion and respiration, and Mayow's later works on these subjects owed much to the work of Boyle and Hooke.[2] Among Hooke's numerous inventions was that of the balance spring for watches,[3] of the anchor-escapement, a wheel barometer,[4] a marine barometer, a thermometer, a wind-gauge, a rain-gauge, and the iris diaphragm. In 1662 Hooke became curator of experiments for the entertainment of members of the Royal Society. Unfortunately, many of his researches were inconclusive, and he made himself unpopular with both Henry Oldenburg and Sir Isaac Newton.[5] Hooke was elected F.R.S. in 1663, and later became professor of geometry at Gresham College, where he delivered the Cutlerian Lectures. He produced a model for rebuilding London after the Great Fire, and was responsible for the building of the Monument, Bedlam Hospital, Montague House, the College of Physicians in Warwick Lane, Aske's Hospital at Hoxton, and assisted Wren, by whom he was employed as his paid deputy, in designing numerous other buildings. In 1677, on the death of Oldenburg, Hooke became secretary of the Royal Society, and from 1679 to 1682 edited the *Philosophical Collections*, which temporarily took the place of the *Philosophical Transactions*. He acquired a Lambeth M.D. in 1691, and died on 3rd March, 1702–3. Robert Hooke has been neglected for many years, but recent studies of the man and his work have resulted in a re-evaluation of his numerous contributions to science, and Richard

[1] See Sarton, George. Spinoza. *Isis*, 10, 1928, pp. 11–15.

[2] See McKie, Douglas. Fire and the flamma vitalis: Boyle, Hooke and Mayow. In, *Science, medicine and history*, Vol. 1, 1953, pp. 469–488; and Turner, H. D. Robert Hooke and Boyle's air pump. *Nature*, 184, 1959, pp. 395–397.

[3] See Hesse, Mary. Hooke's vibration theory and the isochromy of springs. *Isis*, 57, 1966, pp. 433–441; and Hesse, Mary B. Hooke's philosophical algebra. *Isis*, 57, 1966, pp. 67–83.

[4] Middleton, W. E. K. A footnote to the history of the barometer: an unpublished note by Robert Hooke, F.R.S. *Notes Rec. Roy. Soc. Lond.*, 20, 1965, pp. 145–151.

[5] See Westfall, Richard S. Newton's reply to Hooke and the Theory of colors. *Isis*, 54, 1963, pp. 82–96.

S. Westfall[1] rightfully suggests that modern historians have gone too far in supporting certain of Hooke's claims to prior discovery. Sir Geoffrey Keynes[2] has published a characteristic bibliography of Hooke, from which the information provided below is largely derived, and Margaret 'Espinasse[3] has contributed a useful biography which places Hooke in his rightful position as a great scientist and architect. Hooke discovered cells in plants, and corresponded with Leeuwenhoek, verifying his experiments. Many of his papers appeared in the *Philosophical Transactions*, and among his publications, in chronological order, the following are particularly notable. It has been suggested that Robert Hooke's initials are incorporated in the title of *New Atlantis. Begun by the Lord Verulam, Viscount St. Albans and continued by R. H. Esquire,*[*etc.*], London, 1660, although there is no direct evidence of this. The book is very rare, and Sir Geoffrey Keynes considers it to be by Hooke. *An attempt for the explication of the phaenomena, observable in an experiment published by the Honourable Robert Boyle*, [*etc.*], London, 1661 (Keynes 2], also by 'R.H.', which is Hooke's first scientific publication, and deals with the problem of capillary attraction. This was translated into Latin and printed at Amsterdam in 1662, but the only copy of this edition traced by Keynes is in Leyden University Library. A pamphlet of which no copy has been identified bore the title *A discourse of a new instrument to make more observations in astronomy*, London, 1661 [Keynes 4]. Hooke's monumental *Micrographia: or some physiological descriptions of minute bodies made by magnifying glasses. With observations and inquiries thereupon* was published in London in April, 1665,[4] [Keynes 6], the *imprimatur* of the Royal Society being dated Novem. 23, 1664. There were slight alterations in the second issue published in 1667, and various extracts were later issued. A. Rupert Hall[5] published a lecture delivered in commemoration of the tercentenary of the publication of *Micrographia*, which he suggests to be of vital significance in the founding of modern biology.

An attempt to prove the motion of the earth, 1664 [Keynes 16], was the first of Hooke's Cutlerian Lectures to be published. His *Animadversions on the first part of the machina coelestis*, [*etc.*], [Keynes 18], was published in London, 1674, and *A description of helioscopes*, [*etc.*], [Keynes 19], in 1676. This was

[1] Westfall, Richard S. Hooke and the law of universal gravitation. A reappraisal of a reappraisal. *Brit. J. Hist. Sci.*, 3, 1967, pp. 245–261.

[2] Keynes, Sir Geoffrey Langdon. *A bibliography of Dr. Robert Hooke*, Oxford, 1960. This contains lists of MSS., which are mostly in the Royal Society, the British Museum, the Bodleian Library, and Trinity College, Cambridge (pp. 75–84); letters from printed sources (pp. 85–86); and a section on biography and criticism (pp. 87–96).

[3] 'Espinasse, Margaret. *Robert Hooke*, London, [*etc.*], 1956. (Contemporary Science Series.)

[4] Reprinted in facsimile by Dover Books, New York, 1961.

[5] Hall, A. Rupert. *Hooke's Micrographia 1665–1965. A lecture in commemoration of the tercentenary of the publication of Micrographia: or some physiological descriptions of minute bodies by Robert Hooke, delivered at the Middlesex Hospital Medical School on 25th November, 1965*, 1966.

followed by *Lampas: or, descriptions of some mechanical improvements of lamps and waterpoises,* [*etc.*], London, 1677 [Keynes 20], and *Lectures de potentia restitutiva, or of spring, explaining the power of springing bodies,* [*etc.*], London, 1678 [Keynes 21], which contains Hooke's law of the spring, one of his most significant statements. These tracts were collected together as *Lectures and collections,* [*etc.*], London, 1678, and *Lectiones Cutlerinanae, or a collection of lectures,* [*etc.*], London, 1679 [Keynes 22–23]. Collections of Hooke's writings include *Philosophical collections,* seven parts, London, 1679–82 [Keynes 24]; *Posthumous works . . . containing his Cutlerian lectures and other discourses,* [*etc.*], edited by Richard Waller, London, 1705 [Keynes 25] (reprinted 1968), the first section of which was translated into Russian (Moscow and Leningrad, 1948); and *Philosophical experiments and observations,* London, 1726 [Keynes 36] (reprinted 1967), which was to have served as volume two of *Posthumous works,* but Waller died before it was ready for publication, and William Derham (1657–1735) edited this volume. Hooke contributed extensively to the first twenty-two volumes of the *Philosophical Transactions,* these being listed by Keynes (pp. 56–58), as are Hooke's contributions to other books. Hooke's *Diary* in the Guildhall Library has been edited by Henry W. Robinson and Walter Adams.[1] Hooke's life and writings have been the subject of several studies by R. T. Gunther.[2] Robert Hooke contributed usefully to most branches of science, but he was irritable, probably because of ill health, and differed with some of his fellow scientists, often over trivial matters. Had he concentrated more upon specific objectives, he might have provided more valuable results, but his would have been a less interesting personality.[3]

The immortal Sir Isaac Newton stands out among his fellow scientists of the seventeenth century, for he made important progress in mathematics, optics, gravitation, astrophysics, dynamics and chemistry, while his work on gravitation is known superficially to every schoolboy. Newton was born on 25th December, 1642, at Woolsthorpe, near

[1] Hooke, Robert. *The Diary of Robert Hooke, M.A., M.D., F.R.S., 1672–1680. Transcribed from the original in the possession of the Corporation of the City of London* (*Guildhall Library*). *Edited by Henry W. Robinson and Walter Adams,* [*etc.*], 1935.

[2] Gunther, R. T. *Early science in Oxford,* Vols. 6–7. *The life and work of Robert Hooke, Parts I–II,* Oxford, *for the author,* 1930. Volume 8, 1931, consists of a facsimile of Hooke's *Cutler Lectures,* and Volume 13, 1938, of his *Micrographia.*

[3] See Andrade, E. N. da C. Wilkins Lecture. Robert Hooke. *Proc. Roy. Soc.,* A., 201, 1950, pp. 439–473; Andrade, E. N. da C. Robert Hooke, F.R.S. (1635–1703). *Notes Rec. Roy. Soc. Lond.,* 15, 1960, pp. 137–145; Patterson, Louise Diehl. Robert Hooke and the conservation of energy. *Isis,* 38, 1947–48, pp. 151–156; Patterson, Louise Diehl. Hooke's gravitation theory and its influence on Newton. I. [–II]. *Isis,* 40, 1949, pp. 327–341; and *Isis,* 41, 1950, pp. 32–45; Robinson, Henry W. Robert Hooke, M.D., F.R.S., with special reference to his work in biology. *Proc. Roy. Soc. Med.,* 38, 1944–45, pp. 485–489; and, Singer, Charles. The first English microscopist: Robert Hooke (1635–1703). *Endeavour,* 14, 1955, pp. 12–18.

Grantham in Lincolnshire. He was admitted as subsizar at Trinity College, Cambridge, in 1661, and three years later was elected a scholar. Even before graduation he made observations with a prism, and discovered the binomial theorem. Derek T. Whiteside[1] has published two papers dealing with Newton's development as a mathematician and the events leading to the writing of the *Principia*, based on a careful study of his unpublished mathematical papers. J. A. Lohne[2] has undertaken a similar task with particular emphasis on Newton's optics, indicating a gradual development rather than the 1apid burst of inspiration suggested by Whiteside for Newton the mathematician. Newton wrote on fluxions, discovered the refractive properties of coloured light, constructed a reflecting telescope, and in 1669 was elected Lucasian professor of mathematics at Cambridge. His later appointments include that of Warden (1696) and Master (1699) of the Mint, and he was elected president of the Royal Society in 1703. Two years later Newton was knighted by Queen Anne at Trinity College. He died on 20th March, 1727. Bibliographies of Newton's writings were published by George J. Gray,[3] and more recently by H. Zeitlinger.[4] The following information is based largely on their lists. Newton's early printed work consists of sixteen papers on optics published in the *Philosophical Transactions* between 1672[5] and 1676. His *Philosophiae naturalis principia mathematica*, generally known as the *Principia*, was published in London, 1687[6] there being two variants of the title-page of this. The imprint of one reads, "Prostat apud plures Bibliopolas"; while in the other, a cancel, it reads, "Prostant Venales apud Sam Smith". It is possible that the copies bearing the latter were intended for the Continental market, but this explanation has been doubted.[7] A second edition of the *Principia*, edited by Roger Cotes, was published at Cambridge, 1713, and the extensive correspondence between Cotes

[1] Whiteside, Derek T. Isaaac Newton: birth of a mathematician. *Notes Rec. Roy. Soc.*, 19, 1964, pp. 53–62; and Whiteside, Derek T. Newton's early thoughts on planetary motion: a fresh look. *Brit. J. Hist. Sci.*, 2, 1964, pp. 117–137.

[2] Lohne, J. A. Isaac Newton: the rise of a scientist 1661–1671. *Notes Rec. Roy. Soc.*, 20, 1965, pp. 125–139.

[3] Gray, George J. *A bibliography of the works of Sir Isaac Newton. Together with a list of books illustrating his works. . . . Second edition*, [*etc.*], Cambridge, 1907; facsimile reprint 1966. First edition 1888.

[4] Zeitlinger, H. A Newton bibliography. *In*, Greenstreet, W. J., *ed.*, *Isaac Newton, 1642–1727. A memorial volume edited for the Mathematical Association by W. J. Greenstreet*, 1927.

[5] See Cohen, I. Bernard. Versions of Isaac Newton's first published paper. With remarks on the question of whether Newton planned to publish an edition of his early papers on light and color. *Arch. int. Hist. Sci.*, 11, 1958, pp. 357–375; and four papers by Richard S. Westfall: The development of Newton's theory of colour. *Isis*, 53, 1962, pp. 339–358; Newton and his critics on the nature of colors. *Arch. int. Hist. Sci.*, 15, 1962, pp. 47–58; Newton's optics: the present state of research. *Isis*, 57, 1966, pp. 102–107; Newton defends his first publication: the Newton-Lucas correspondence. *Isis*, 57, 1966, pp. 299–314.

[6] A facsimile reprint was published in London, 1953.

[7] See Munby, A. N. L. The two title-pages of the "Principia". *Times Literary Supplement*, 21st December, 1951, p. 828, and 28th March, 1952, p. 228.

and Newton was edited by James Edleston and published as *Correspondence of Sir Isaac Newton and Professor Cotes, including letters of other eminent men*, 1850. A third and final edition by Henry Pemberton was published in London, 1726. Twenty-five copies of this were printed on large paper.[1] The 1726 edition was frequently reprinted, and there have been several abridgements. The second edition was pirated at Amsterdam in 1714 and 1723, and an edition of the 1726 edition was reprinted for Sir William Thomson, later Lord Kelvin, at Glasgow, 1871. Other editions include printings in Geneva, three volumes, 1739–42; "Coloniae Allobrogum", three volumes, 1760; Prague, two volumes 1780–85; Glasgow, [*etc.*], four volumes, 1822; and London, Glasgow, two volumes, 1833. Pemberton's edition was prepared under the supervision of Newton, and the latter's ideas were interpreted by Pemberton in *A view of Sir Isaac Newton's philosophy*, London, 1728. Pemberton also advertised a translation of the *Principia*, but he was anticipated by Andrew Motte, and Pemberton's translation, although completed, has never been traced.[2]

The English translation by Andrew Motte was published in two volumes, London, 1729, and in three volumes, London, 1803 and 1819. This translation was published in New York in 1848 and 1850. A facsimile reprint of the first printing of Motte's translation, with a critical, historical and bibliographical introduction by I. Bernard Cohen, was published in two volumes, 1968. Another edition was published as *Sir Isaac Newton's Mathematical principles of natural philosophy, and his System of the world. Translated into English by Andrew Motte in 1729. The translations revised: and supplied with an historical and explanatory appendix by Florian Cajori*, Cambridge, 1934, of which a paper-bound edition appeared in 1962. An English translation by Robert Thorpe, of which only Volume 1 was published, appeared in London, 1777. A translation into French by Gabrielle Emilie de Tonnelier de Breteuil, Marquise de Chastelet, was published in 1759, while a German edition, translated by Jacob Philipp Wolfers, appeared in 1872.[3] John Herivel[4] has published a study of Newton's researches on dynamics, based on documents in the Portsmouth Collection in the University Library Cambridge. Part two of his book comprises a reproduction of the original manuscripts, including the tract "De motu corporum", of which there are five known versions.

Newton's *Opticks: or, a treatise of the reflextions, refractions, inflexions, and colours of light*, [*etc.*], was first published in London, 1704, without his

[1] See Brasch, Frederick E. A survey of the number of copies of Newton's *Principia* in the United States, Canada, and Mexico. *Scripta Mathematica*, 18, 1952, pp. 53–67.
[2] See Cohen, I. Bernard. Pemberton's translation of Newton's *Principai*, with notes on Motte's translation. *Isis*, 54, 1963, pp. 319–351.
[3] See also Ball, W. W. Rouse. *An essay on Newton's "Principia"*, London, New York, 1893.
[4] Herivel, John. *The background to Newton's* Principia. *A study of Newton's dynamical researches in the years 1664–84*, Oxford, 1965 [1966].

name on the title-page, although his initials appeared at the end of the advertisement. This first edition was a quarto, and was followed by a second edition, London, 1717, which was reissued with additions in 1718. Among the manuscripts in the Portsmouth Collection is a draft of revisions which Newton contemplated for this second edition. It is headed "The Third Book of Opticks. Part II. Observations concerning the medium through which light passes, & the agent which emits it", and was written between 1715 and 1717. Henry Guerlac[1] suggests that this explains why the Third book has Part I but no Part II. The third edition appeared in 1721, with a fourth in 1730, the latter having been reprinted in 1931 and in 1952, with a foreword by Albert Einstein and an introduction by Sir Edward Whittaker. A Latin translation by Samuel Clarke, also a quarto, was printed in London, 1706,[2] with a second edition, containing the mathematical treatises, London, 1719. Other editions appeared from Continental publishing houses, and there were French translations by Pierre Coste, two volumes, Amsterdam, 1720, and Paris, 1722, and by Jean Paul Marat, two volumes, Paris, 1787. A German translation by Wilhelm Abendroth was published in Ostwald's *Klassiker*, 1898. Newton's *Optical lectures read in the publick schools of the University of Cambridge, Anno Domini, 1669*, [*etc.*], London, 1728, consisted of the first part only, the complete Latin edition appearing in the following year. There were several editions of *The method of fluxions and infinite series*, [*etc.*], London, 1736, and of his *Arithmetica universalis*, [*etc.*], first published at Cambridge, 1707. Other editions appeared from London, 1722; Leyden, 1732; Milan, 1732 (three volumes); Amsterdam, 1761 (two volumes); and Leyden, 1761. An English translation by Joseph Raphson was published in London, 1720, 1728 and 1769; while a French translation by Noël Beaudeux was printed in two volumes in 1802. The collected works of Newton were published as *Opera mathematica, philosophica et philologica*, [*etc.*], Lausanne and Geneva, three volumes, 1744; and *Opera quae exstant omnia*, [*etc.*], five volumes, London, 1779–85, of which a facsimile reprint was published, Stuttgart-Bad Constatt, 1968. It is also of interest to note that Isaac Newton's first book was his edition of Bernhardus Varenius' *Geographia generalis*, [*etc.*], Cambridge, 1672. A

[1] Guerlac, Henry. Newton's optical aether. His draft of a proposed addition to his *Opticks*. *Notes Rec. Roy. Soc. Lond.*, 22, 1967, pp. 45–57.

[2] A discussion of the Leibniz-Newton controversy, with facsimiles of the original page 315 and the cancel, in this edition, are contained in Koyré, Alexandre, and Cohen, I. Bernard. The case of the missing *Tanquam:* Leibniz, Newton & Clarke. *Isis*, 52, 1961, pp. 555–566. See also Koyré, Alexandre, and Cohen, I. Bernard. Newton & the Leibniz-Clarke correspondence, with notes on Newton, Conti & Des Maizeaux. *Arch. int. Hist. Sci.*, 15, 1962, pp. 64–126; Kubrin, David. Newton and the cyclical cosmos: providence and the mechanical philosophy. *J. Hist. Ideas*, 28, 1967, pp. 325–346; Thackray, Arnold. "Matter in a nut-shell": Newton's *Opticks* and eighteenth-century chemistry. *Ambix*, 15, 1968, pp. 29–53; and Whiteside, Derek T. Newton's marvellous year: 1666 and all that. *Notes Rec. Roy. Soc. Lond.*, 21, 1966, pp. 32–41.

selection of papers from the Portsmouth Collection in the University Library, Cambridge, has been edited and translated by A. Rupert Hall and Marie Boas Hall as *Unpublished scientific papers of Isaac Newton*, Cambridge, 1962. Derek T. Whiteside has issued two volumes of *The mathematical works of Isaac Newton* in the Sources of Science series, 1964–67. Whiteside and others are also editing *The mathematical papers of Isaac Newton*, to appear in eight volumes, of which volumes one to three have been published, 1967–69. Recent research on Newton's work has been based on MSS. and letters in Cambridge University Library, Trinity College and King's College, Cambridge. I. Bernard Cohen[1] reveals the inadequacy of the biographies of Newton, and reminds us that there is no adequate collected edition of his works, and no catalogue of Newton manuscripts. J. W. Herivel[2] traces the development of Newton's thought, as revealed in unpublished documents, in his discovery of the law of centrifugal force, one of the most momentous discoveries in the history of science. Although Newton published no writings devoted to alchemy, an investigation of his manuscripts reveals his interest in the subject, and King's College MS. 38 in Newton's hand, entitled *Sententiae notabiles*, consists of extracts from his favourite writers on alchemy. This is reprinted in a paper by F. Sherwood Taylor,[3] and this aspect of Newton's work has also been the subject of other articles.[4] Facsimiles of Newton's physical, chemical and other papers, with translations where necessary, have been edited by I. Bernard Cohen and Robert E. Schofield.[5] The Royal Society is publishing *The correspondence of Isaac Newton*, edited by H. W. Turnbull and J. F. Scott, which will contain over 1,500 letters, many of which are published for the first time, translations being provided of those originally in Latin. This work is planned to cover seven volumes and the first four volumes have appeared, Cambridge, 1959–67. Alexandre Koyré's scholarly papers on Newton have been edited by I. Bernard Cohen and published as *Newtonian studies*, 1965, and Derek T. Whiteside has reviewed research on Newton,[6] and also evaluated current facsimile reprints of Newton's writings.[7] Frank E. Manuel[8] has written a book

[1] Cohen, I. Bernard. Newton in the light of recent scholarship. *Isis*, 51, 1960, pp. 489–514.

[2] Herivel, J. W. Newton's discovery of the law of centrifugal force. *Isis*, 51, 1960, pp. 546–553.

[3] Taylor, F. Sherwood. An alchemical work of Sir Isaac Newton. *Ambix*, 5, 1956, pp. 59–84.

[4] See Forbes, R. J. Was Newton an alchemist? *Chymia*, 2, 1949, pp. 27–36; Geoghegan, D. Some indications of Newton's attitude towards alchemy. *Ambix*, 6, 1957, pp. 102–106; and McGuire, J. E. Transmutation and immutability: Newton's doctrine of physical qualities. *Ambix*, 14, 1967, pp. 69–95.

[5] *Isaac Newton's Papers and letters on natural philosophy* [etc.], Cambridge, Mass, Cambridge, 1958.

[6] Whiteside, Derek T. The expanding world of Newtonian research. *History of Science*, 1, 1962, pp. 16–29.

[7] Whiteside, Derek T. A face-lift for Newton: current facsimile reprints. *History of Science*, 6, 1967, pp. 59–68.

[8] Manuel, Frank E. *A portrait of Isaac Newton*, Cambridge, Mass., 1968.

stressing Isaac Newton's paranoid controversies with fellow scientists and former friends. Augustus De Morgan (1806–71) prepared a now rare volume entitled *Newton: his friend: and his niece*, which was published in 1885, and reprinted in 1968. J. D. North[1] has contributed a very brief, but comprehensive illustrated biography of Newton, and there are numerous other studies of his life and works.[2,3]

Gottfried Wilhelm Leibniz (1646–1716), a native of Leipzig, propounded the principle of the conservation of energy in physics, but is also renowned as a philosopher and mathematician. He invented an intricate calculating machine, and was largely instrumental in founding the Deutsche Akademie der Wissenschaften zu Berlin. He invented the differential and integral calculus in 1674, which resulted in a controversy with Newton over priority. A paper by Leibniz entitled "Tentamen de motuum coelestium causis" (*Acta Eruditorum*, 1689) has been misinterpreted by scholars of the past and present, and E. J. Aiton[4] in a series of papers has proved that the work of Leibniz on the development of celestial mechanics was original, and deserves recognition. Several of his most important papers appeared in *Miscellanea Berolinensia*, [*etc.*], the proceedings of the Berlin Academy, and his other writings, mainly philosophical, have been the subject of a bibliography.[5] A collection of about

[1] North, J. D. *Isaac Newton*, 1967.

[2] See Brewster, Sir David. *Memoirs of the life, writings and discoveries of Sir Isaac Newton*, two volumes, Edinburgh, 1855, printed in facsimile with a new introduction by R. S. Westfall, 1966; More, Louis Trenchard. *Isaac Newton: a biography . . . 1642–1727*, New York, London, 1934, reprinted as a Dover paperback, 1963; Anthony, H. D. *Sir Isaac Newton*, London, [*etc.*], 1960; Turnbull, H. W. *The mathematical discoveries of Newton*, London, Glasgow, 1947; Greenstreet, W. J., *ed. Isaac Newton, 1642–1727. A memorial volume edited for the Mathematical Association by W. J. Greenstreet*, 1927; History of Science Society. *Sir Isaac Newton, 1727–1927. A bicentenary evaluation of his work. A series of papers prepared under the auspices of the History of Science Society*, [*etc.*], 1928; Royal Society. *Newton tercentenary celebrations, 15–19th July, 1946*, Cambridge, 1947; Brodetsky, Selig. *Sir Isaac Newton: a brief account of his life and work*, 1927; Andrade, E. N. da C. *Isaac Newton*, [*etc.*], 1950, (Personal Portraits Series); Sullivan, J. W. N. *Isaac Newton, 1642–1727*, [*etc.*], 1938; De Villamil, R. *Newton: the man*, [*etc.*], [1931]; McKie, Douglas. Newton and chemistry. *Endeavour*, 1, 1942, pp. 141–144; and Huxley, G. L. Two Newtonian studies. (I. Newton's boyhood interests. II. Newton and Greek geometry.) *Harvard Lib. Bull.*, 13, 1959, pp. 348–361.

[3] See also *A descriptive catalogue of the Grace K. Babson Collection of the works of Sir Isaac Newton, and the material relating to him in the Babson Institute Library, Babson Park, Mass., with an introduction by Roger Babson Webber*, New York, 1950, with a supplement by Henry P. Macomber published by the Babson Institute, 1955; and Andrade, E. N. da C. A Newton collection. *Endeavour*, 12, 1953, pp. 68–75. For information on Newton's library and MSS. see p. 344.

[4] Aiton, E. J. The celestial mechanics of Leibniz. *Ann. Sci.*, 16, 1960 [1962], pp. 65–82; The celestial mechanics of Leibniz in the light of Newtonian criticism. *Ann. Sci.*, 18, 1962 [1964], pp. 31–41; The celestial mechanics of Leibniz: a new interpretation. *Ann. Sci.*, 20, 1964 [1965], pp. 111–123; and An imaginary error in the celestial mechanics of Leibniz. *Ann. Sci.*, 21, 1965 [1966], pp. 169–173.

[5] See Ravier, Emile. *Bibliographie des oeuvres de Leibniz*, Paris, 1937. Errata and addenda to this are provided by Schrecker, Paul. Une bibliographie de Leibniz. *Revue philosophique*, 63, 1938, pp. 324–346.

seventy items by Leibniz was published as *Philosophical papers and letters*, two volumes, Chicago, 1956, edited and translated by Leroy E. Loemker, but the critical edition of his writings begun early this century, and interrupted by two wars, has not reached completion.[1] His mathematical writings were published as *Leibniz' mathematische Schriften. Herausgegeben von G. I. Gerhardt*, seven volumes, Berlin, and Halle, 1849–63, and were reprinted in Hildesheim, 1961. Many of his manuscripts are preserved in the State Library of Lower Saxony, formerly the Royal Library of Hanover, and the University of Pennsylvania acquired a microfilm of this collection. A copy of this microfilm is also available in the University of London Library.[2] A copy of a text of Leibniz was discovered in 1956 in the Archives of L'Académie Royale des Sciences de Paris. It includes his *L'Essay de dynamique* and *Règle générale de la composition des mouvements*, written in 1692, and the texts are reproduced, with commentaries, by Pierre Costabel.[3]

Chemistry of the seventeenth century was in a transitory state, with the illusions of alchemy being displaced by the results of scientific experiment. This was but a gradual process, and alchemical theories persisted for a considerable period. Robert Fludd (1574–1637) was born at Bearsted in Kent, and studied medicine and chemistry on the Continent for six years after leaving Oxford with a master's degree. He later became a Fellow of the College of Physicians. His *Utriusque cosmi maioris scilicit et minoris metaphysica, physica atque technica*, 1617 formed the basis of his cosmological system, and was followed by *Anatomiae amphitheatrum effigie triplici, more et conditione varia designatum*, 1623, describing the scientific and mystical anatomy of the body. *Philosophia moysaica*, 1638, was translated into English in 1659, and deals with medicine and nature. Fludd's work "A philosophicall key" was not published, but is preserved at Trinity College, Cambridge (Western MSS. 1150 [0.2.46]). It was written about 1618.[4] Johann Baptista van Helmont (1579–1644) was a native of Brussels, who, after commencing to study the arts, turned to medicine, graduating M.D. in 1609. He then became interested in

[1] *Sämtliche Schriften und Briefe. Herausgegeben von der Preussischen Akademie der Wissenschaften*, Darmstadt (and Leipzig), 1923–57. This has not yet reached his scientific writings.

[2] See Leibniz material for London. *Times Literary Supplement*, 25th October, 1957, p. 78.

[3] Costabel, Pierre. *Leibniz et la dynamique. Les textes de 1692*, Paris, 1960. (Histoire de la pensee, 1.) See also Cajori, Florian. Leibniz, the master builder of mathematical notations. *Isis*, 7, 1925, pp. 412–429; and Cohen, I. Bernard. Leibniz on elliptical orbits: as seen in his correspondence with the Académie Royale des Sciences in 1700. *J. Hist. Med.*, 17, 1962, pp. 72–82.

[4] See Josten, C. H. Robert Fludd's "Philosophicall key" and his alchemical experiments on wheat. *Ambix*, 11, 1963, pp. 1–23; Debus, Allen G. Renaissance chemistry and the work of Robert Fludd. *Ambix*, 14, 1967, pp. 42–59; and the earlier version of this in Debus, Allen G., and Multhauf, Robert P. *Alchemy and chemistry in the seventeenth century. Papers read . . . at the Clark Library Seminar, 12th March, 1966*, Los Angeles, 1966. Also Debus, Allen G. *The English Paracelsians*, 1965.

chemistry, basing his ideas largely on the writings of Paracelsus. Van Helmont was the founder of the Iatrochemical School, and recognized the physiologic importance of ferments. He conducted experiments on growing plants, and on gases, being the first to use the term 'gas'.[1] He rejected the theory of four elements, believing there to be two only, air and water. His work on urinary calculi, *De lithiasi*, contains descriptions of many chemical experiments, and Boyle was influenced by his work. Van Helmont conducted experiments on metabolism, and on the physiology of digestion, discovering acid to be the agent of gastric secretion, and producing hydrochloric acid by the distillation of salt and dried potter's clay.[2] Apart from his medical writings, van Helmont published little, but after his death his son collected together his manuscripts, publishing them as *Ortus medicinae, id est initia physicae inaudida*, 1648, the best edition of these collected works being that published at Amsterdam in 1652. An English translation was published from London in 1662, this being reissued with a new title-page two years later. Van Helmont's *De tempore*, first published at Amsterdam in 1648, has been the subject of several studies,[3] but his work had little influence upon science until his writings were translated, over a hundred years after his death.[4] His complete works were issued as *Opera omnia*, [*etc.*], Frankfurt, 1682 and 1707.

A native of Lorraine, Jean Beguin, flourished during the later half of the sixteenth and early seventeenth centuries. Little is known about his life, but James Ferguson (1837–1916) left behind two unpublished papers devoted to Beguin, upon which T. S. Patterson[5] based an authoritative study of his life and writings. Beguin studied chemistry and pharmacy, and taught these subjects in Paris. He edited the *Novum lumen chymicum* of Michael Sendivogius, published at Paris in 1608, and in connection with his teaching prepared a manual of chemistry for his students. This was first printed as *Tyrocinium chymicum* in 1610, but was not regularly published. A comparison of this with the 1597 edition of Libavius'

[1] See Pagel, Walter. The "Wild spirit" (gas) of John Baptist van Helmont (1579–1644) and Paracelsus. *Ambix*, 10, 1962, pp. 1–13.

[2] See Pagel, Walter. Van Helmont's ideas on gastric digestion and the gastric acid. *Bull. Hist. Med.*, 30, 1956, pp. 524–536.

[3] See Pagel, Walter. J. B. van Helmont, De tempore, and biological time. *Osiris*, 8, 1948, pp. 346–417; a summary of this appeared as John Baptist van Helmont: *De tempore* and the history of the biological concept of time. *Isis*, 33, 1941–42, pp. 621–623; Pagel, Walter. *The religious and philosophical aspects of van Helmont's science and medicine*, Baltimore, 1944. (Supplement No. 2 to *Bull. Hist. Med.*); Weiss, Helene. Notes on the Greek ideas referred to in Van Helmont, De tempore. *Osiris*, 8, 1948, pp. 418–449; also Patterson, T. S. Van Helmont's ice and water experiments. *Ann. Sci.*, 1, 1936, pp. 462–467.

[4] See also Nève de Mévergnies, P. *Jean Baptiste van Helmont, philosophe*, 1935; and Partington, J. R. Jean Baptista van Helmont. *Ann. Sci.*, 1, 1936, pp. 359–384.

[5] Patterson, T. S. Jean Beguin and his "Tyrocinium chymicum". *Ann. Sci.*, 2, 1937, pp. 243–298; see also Garman, Mel. Some copies of Jean Beguin's textbook of chemistry. *J. chem. Educ.*, 35, 1958, pp. 575–577.

Alchemia suggests that lengthy passages were copied by Beguin.[1] A pirated edition of *Tyrocinium* appeared from Cologne in 1611, causing Beguin to issue a revised edition from Paris in 1612. The first French edition, under the title *Les élémens de chymie*, appeared in Paris in 1615, and an English translation by Richard Russell in 1669. There were also two translations into German. The various editions of these works are recorded in the above-mentioned article by T. S. Patterson, who illustrates his paper with numerous facsimiles of title-pages.[2]

William Davidson [Davidsoune; Davisson; Davissone; D'Avissone; or D'Avissonus] was born in Aberdeenshire about 1593, graduated M.A. at Marischal College in 1617, after apparently qualifying in medicine and pharmacy, and migrated to Paris. He was the first occupant of any chair of chemistry in France, and in 1647 became the first professor of chemistry at the Jardin du Roi. He left Paris in 1651 and gave up chemistry for medicine in the service of the King of Poland. Among other writings, Davidson was the author of *Philosophia pyrotechnica*, an early textbook on chemistry of which the first and second parts were published in Paris, 1635, together with the original versions of the third and fourth parts, which had been printed in 1633. There were also editions in 1640, 1641 and 1642, with a French translation by Jean Hellot published in Paris, 1651 and 1657. Davidson also published *Les élémens de la philosophie de l'art du feu ou chemie*, 1651, which went into a second edition, with a new title-page and a minor alteration at the beginning, in 1657.[3]

Antonio Neri (died 1614) was a priest, who during his travels throughout Italy collected a great deal of information on the manufacture of glass. Neri also worked with glass and made several discoveries, incorporating these in his *De arte vetraria*, Florence, 1612. This work, embellished by other hands, formed the basis of a work on glass which was the most authoritative on the subject until almost the end of the eighteenth century. Reissued in Florence, 1661, and Venice, 1681, this was translated into English by Christopher Merrett (1614–95), who made considerable additions to the work and added illustrations. This translation, *The art of glass*, London, 1662 (reprinted by Sir T. Phillips, Middle Hills, Worcs., 1826), was translated into Latin as *De arte vitraria libri septem*, Amsterdam, 1668, and a German translation of Neri by F. Geissler, *Sieben Bücher handlend von der künstlichen Glass-und Crystallen-*

[1] See Kent, Andrew, and Hannaway, Owen. Some new considerations on Beguin and Libavius. *Ann. Sci.*, 16, 1960 [1963], pp. 241–250.

[2] See also *Bibliotheca Chemica*, Vol. 1, 1906, pp. 93–94.

[3] See Hamy, E. T. William Davisson, Intendant au Jardin du Roi, et Professeur de Chimie (1645–51). *Nouvelles Archives du Museum d'Histoire Naturelle*, 3me Serie, 10, i, 1898, to which is appended a detailed bibliography of Davidson's writings; and Read, John. William Davidson of Aberdeen. The first British professor of chemistry. *Ambix*, 9, 1961, pp. 70–101; also *Aberdeen University Studies*, No. 129, 1951.

Arbeit, Frankfurt and Leipzig, appeared in 1678. Christopher Merrett[1] was a physician, and became the first Harveian librarian of the College of Physicians. He was also among the earliest to be elected a Fellow of the Royal Society. He was the author of *Pinax rerum naturalium Britannicorum*, [*etc.*], London, 1666 (most copies dated 1667), which is devoted chiefly to botany and medicine, but also contains the earliest known list of British birds. The joint work of Neri and Merrett appeared once more with notes by J. Kunckel, as the first part of his *Ars vitraria experimentalis, oder Volkommene Glassmacher-Kunst*, [*etc.*], Frankfurt and Leipzig, 1679, the second part being Kunckel's own work. Johann Kunckel (1630 or 1638–1703), born at Hutten, near Rendsburg, was the son of an alchemist. He had no scholastic training but carried out many experiments on metals and glass. Kunckel wrote his books in German, often giving them Latin titles. In addition to the *Ars vitraria experimentalis* he wrote *Nützliche Observations oder Anmerckungen von den Fixen und fluchtigen Salzen*, [*etc.*], Hamburg, 1676, with a Latin translation, London and Rotterdam, 1678; *Chymische Anmerckungen*, [*etc.*], Wittenberg, 1677, with a Latin translation, Amsterdam, 1694; *Oeffentliche Zuschrift von der Phosphoro Mirabili*, [*etc.*], Leipzig, 1678; *Collegium physico-chymicum experimentale*, Hamburg and Leipzig, 1716, 1722 and 1738, and Berlin, 1767, of which a translation into French is given by F. Hoefer,[2] and which was published posthumously.[3] Kunckel died, probably in Stockholm, on 20th March, 1703.[4]

Another iatrochemist, Andreas Libavius [Libau or Liebau] (1540?–1616), was born at Halle, Germany, studied medicine, history and philosophy, and became the director of the Gymnasium at Coburg. He was the first to prepare ammonium sulphate, and discovered stannis chloride, known as *Spiritus fumans Libavii*, and made considerable progress in analytical technique. He published a number of works, including *Rerum chymicarum epistolica*, Frankfurt, 1595–99, and *Alchemia*, Frankfurt, 1597, from which Jean Beguin apparently copied passages which were printed in the anonymous first edition of his *Tyrocinium chymicum*, 1610.[5] A German version of *Alchemia* was published as *Alchymistische Practic*, [*etc.*], Frankfurt, 1603, which appeared in Latin, Frankfurt, 1604, 1605 and 1607. The *D.O.M.A. Alchemia Andreae Libavia* is his most important work, and was probably the first real textbook of chemistry, including a plan of an ideal chemical laboratory and excellent descriptions of apparatus.

[1] Dodds, Sir Charles. Christopher Merrett, F.R.C.P. (1614–1695), first Harveian Librarian. *Proc. Roy. Soc. Med.*, 47, 1954, pp. 1053–1055.

[2] Hoefer, H. *Histoire de la chimie*, 2nd ed., 1869, Vol. 2.

[3] Partington, J. R. The early history of phosphorus. *Sci. Progr.*, 30, 1936, pp. 402–412.

[4] Partington, J. R. *A history of chemistry*. Vol. 2, 1961, pp. 361–377.

[5] See Kent, Andrew, and Hannaway, Owen. Some new considerations on Beguin and Libavius. *Ann. Sci.*, 16, 1960 [1963], pp. 241–250.

This work had a commentary which was issued separately, and in consequence is missing from some copies. A second edition appeared in 1606, including the commentary, which was also issued in three folio parts, Frankfurt, 1611–15.[1]

A controversial figure, Athanasius Kircher (1602–80), is mentioned briefly for the interest of his publications. Born at Geisa, near Fulda, 2nd May, 1602, Kircher, a Jesuit father, became professor at Würzburg, and in 1631 moved to the Jesuits' College in Rome, where his collection of fossils, antiques and apparatus, the "Museum Kircheriane", still exists.[2] His publications were extensive—thirty-five volumes ranging in subject matter from archaeology and china to medicine and magic.[3] The following, although full of superstition and fantasy, are worthy of note: *Ars magna lucis et umbrae*, Rome, 1646, second edition 1671, the latter describing the magic lantern.[4] *Oedipus Aegyptiacus. Hoc est universalis hieroglyphicae veterum doctrinae temporum iniuria abolitae Instauratio*, three volumes, Rome, 1652–53, is most valuable for its criticism of alchemy. *Scrutinium physico-medicum contagiosae luis, quae pestis dicitur*, Rome, 1658, Leipzig, 1659, and Frankfurt, 1663, on the cause of the plague; and *Mundus subterraneus in XII libros digestus; quo divinum subterrestris mundi opificium . . . exponuntur*, two volumes, Amsterdam, 1664, reprinted 1665; second edition, 1668, and third edition, 1678, Kircher's best-known book, based on his experience of an earthquake in Calabria, and visits to the volcanoes of Italy and Sicily, but containing much invention and extravagant theory.[5]

Johann Rudolf Glauber (1604–68) was born at Karlstadt, but finally settled in Holland. He made numerous contributions to our knowledge of chemistry, and his name is perpetuated in Glauber's salts, which he discovered. Glauber investigated the problems of decomposition and dyeing, and conducted research on other salts. He was also engaged upon the search for the philosopher's stone, and his numerous printed works include *De suri tincture, sive auto potabili vero*, [*etc.*], 1646, 1650 and 1652; *Furni novi philosophici*, [*etc.*], three parts, Amsterdam, 1648–50, of which an English edition was published in 1651; *Opus minerale*, [*etc.*], Amsterdam, 1651, with a Latin translation published 1651–52, and a French version

[1] Partington, J. R. *A history of chemistry*. Vol. 2, 1961, pp. 244–267.

[2] A description of the contents of the museum was published in Amsterdam, 1678: *Romani Collegii Societates Jesu Museum celeberrinum*, pp. 61–66, which comprises a bibliographical survey of Kircher's writings, including books not yet published.

[3] See Chapman, J. S. The strange books of Athanasius Kircher. *Proc. Roy. Inst. G.B.*, 38, 2, no. 171, pp. 259–268; and Fletcher, John. Athanasius Kircher and the distribution of his books. *The Library*, 5th Ser., 23, 1968, pp. 108–117.

[4] See Houstoun, R. A. Athanasius Kircher and the magic lantern. *Sci. Progr.*, 45, 1957, pp. 462–464.

[5] Partington, J. R. Lignum nephriticum. *Ann. Sci.*, 11, 1955, pp. 1–26; and Partington, J. R. *A history of chemistry*, Vol. 2, 1961, pp. 328–333.

printed in 1659; *Miraculum mundi*, [*etc.*], Rotenburg, 1653, with editions from Hanau and Amsterdam in the same year, and one from Prague in 1704; *Des Teutsch-Landes Wohlfarth*, six parts, Amsterdam, 1656–61, and Prague, 1704; and *Novum lumen chimicum*, [*etc.*], Amsterdam, 1664. Glauber's collected writings were published as *Opera omnia*, four volumes, 1651–56, in seven volumes, Amsterdam, 1661; and as *Opera chymica*, two volumes, Frankfurt, 1658–59.

The author of a best-seller in the seventeenth century, Christophle Glaser (*c.* 1615–78) was born in Basle, where he qualified in medicine and pharmacy. He travelled extensively in Europe before settling in Paris about 1658, where he became Apothecary in Ordinary to the King. Glaser delivered public lectures in chemistry and pharmacy, and maintained a private laboratory. His *Traité de la chymie*, Paris, 1663, went into a second edition in 1667, and at least six more editions in French, five in German, and an English version translated from the fourth French edition as *The compleat chymist*, [*etc.*], 1677. The book is eminently practical, and Lémery's *Cours de chymie*, 1675, was based on it, to replace it in popularity.[1]

Only recently has the work of John Mayow (1641–79) received recognition, and a critical study of his writings is badly needed. The evaluation of his chemical work by T. S. Patterson[2] has been criticized by J. R. Partington,[3] who suggests that even modern writers have failed to give Mayow sufficient credit, but W. Boehm[4] has published an interesting study of him and his contemporaries. Mayow was born at Morval in Cornwall, and graduated in law and medicine, but devoted much of his time to chemical experiments. He was a pupil of Thomas Willis (1621–75) and a friend of Richard Lower (1631–91), making several important contributions to chemistry and physiology. Mayow made some ingenious experiments on combustion and respiration. His *Tractatus quinqui medico-physici*, [*etc.*], Oxford, 1674, deals with the source of body heat, among other matters, and was printed at The Hague, 1681; Geneva, 1685 and 1699; and there were eight other editions, including translations into Dutch, German, French and English.[5] In addition to a work on rickets, Mayow also wrote *Tractatus duo*, Oxford, 1668, which was reprinted with a new title-page in 1669 (both of which are rare); and Leyden, 1671 and

[1] See Neville, Roy G. Christophle Glaser and the *Traité de la chymie*, 1663. *Chymia*, 10, 1965, pp. 25–52.

[2] Patterson, T. S. John Mayow in contemporary setting. A contribution to the history of respiration and combustion. *Isis*, 15, 1931, pp. 47–96, 504–546.

[3] Partington, J. R. The life and work of John Mayow (1641–1679). Part I (–II). *Isis*, 47, 1956, pp. 217–230, 405–417; and Partington, J. R. Some early appraisals of the work of John Mayow. *Isis*, 50, 1959, pp. 211–226, which details authors of the seventeenth and eighteenth century mentioning Mayow.

[4] Boehm, W. John Mayow and his contemporaries. *Ambix*, 11, 1963, pp. 105–120.

[5] An English translation was published in 1907 in the series Alembic Club Reprints No. 17.

1708. John F. Fulton[1] has published an exhaustive bibliography of Mayow's printed works, together with those of Lower. Richard Lower (1631–91) has been dealt with elsewhere,[2] but his work on the circulation is particularly noteworthy, and his *Tractatus de corde*, [*etc.*], is probably the most significant of his writings. It was printed in London, 1669 (two issues); Amsterdam, 1669; London, 1680; Geneva, 1685 and 1699; Leyden, 1708, 1722, 1728, 1740 and 1749; Paris, 1679; and Oxford, 1932. This last contains a translation into English by K. J. Franklin,[3] who suggests that *De corde* first appeared in 1668, there being a copy of the 1669 edition in the British Museum which originally belonged to Walter Charleton (1619–1707), and which is annotated and dated 1668 by him.

Johann Joachim Becher (1635–82) was born in Speyer, Germany, becoming professor of medicine at Mainz in 1663, leaving a year later for Munich, where as physician to the Elector he had the use of the best laboratory in Europe. He visited England in 1679, where he studied mining in Cornwall, and it is believed he died in London. He wrote on many subjects, but he is of most interest here for his books on chemistry; brief details of some of these follow. J. R. Partington[4] gives good description of these, and refers to a full bibliography in Hassinger.[5] His most important chemical work, usually called *Physica subterranea*, is *Actorum laboratii chymici monacensis seu physicae subterraneae libri duo*, [*etc.*], Frankfurt, 1667, which in fact consists of one book only. This was republished by Stahl with a long introduction, with the title *Physica subterranea*, [*etc.*], Leipzig, 1703, with the three supplements which had appeared in 1671, 1675 and 1680. There were other editions from Frankfurt in 1681, 1702[?] and 1738. Becher published a German translation in 1680, reissued with a new title-page dated 1690. Another edition of the third supplement was published in London in the same year (1680) as the original. *The Naturkundigung der Metalle*, [*etc.*], Frankfurt, 1660 (on frontispiece), 1661 (on title) appeared also with the same frontispiece but with the titles dated 1679 and 1705. *Parnassi illustrati pars tertia, mineralogia*, [*etc.*], Ulm, 1662, a part of which was issued as: *Parnassus medicinalis illustratus, oder eine neues . . . Thier-Kräuter-und Berg-Buch*, [*etc.*], Ulm, 1663; *Institutiones chimicae prodromae*, [*etc.*], Frankfurt, 1664, and Amsterdam in the same

[1] Fulton, John Farquhar. A bibliography of two Oxford physiologists, Richard Lower, 1631–1691, John Mayow, 1643–1679. *Oxford Bib. Soc. Proc. Pap.*, i, 1934, 62 pp., Oxford, 1935.

[2] See Thornton, pp. 85–86.

[3] Gunther, R. T. *Early science in Oxford. Vol. IX. De corde, by Richard Lower, London, 1669. With introduction and translation by K. J. Franklin*, Oxford, 1932. A list of Lower's writings is given on pp. xxxi–xxxii. See also Franklin, Kenneth James. Some textual changes in successive editions of Richard Lower's *Tractatus de corde item de motu & colore sanguinis et chyli in eum transitu. Ann. Sci.*, 4, 1939, pp. 283–294.

[4] Partington, J. R. *A history of chemistry*, 2, 1961, pp. 637–652.

[5] Hassinger, H. J. J. Becher. Ein Beitrag zur Geschichte der Merkantilismus, *Veröffi. d. Kommission f. neuere Geschichte Österreichs*, 37, 1951. (Bibliography on pp. 254–272.)

year, and 1665, Frankfurt, 1705 and 1716 with a supplement. Possibly also Mainz, 1662. German translations appeared in 1680 and 1690, and *Tripus Hermeticus Fastidicus*, [*etc.*], Frankfurt, 1689 and 1690, which contains information gained on Becher's visit to Cornwall. With G. E. Stahl, Becher may be regarded as the joint founder of the phlogiston theory,[1] and his proposal to the government of Holland for the manufacture of gold is of interest to the historian of alchemy.[2]

Little is known of Sébastien Matte La Faveur beyond the fact that he was succeeded as Professor of Chemistry at Paris University by Lémery, and was author of *Pratique de chymie*, 1671, a link between Glaser's *Traité de la chymie*, 1663, and Lémery's *Cours de chymie*, which is probably based on the two earlier books.[3] The *Cours de chymie*, Paris, 1675, of Nicolas Lémery (1645–1715) is notable for the clarity of its language, and its rejection of mystification. Describing the practical chemical knowledge of the period, the book exerted considerable influence, and was translated into English, Latin, German, Italian and Spanish, reaching a total of eleven editions in France. The English translation by Walter Harris was published as *A course of chymistry*, London, 1677. Lémery was also the author of *Dictionnaire des drogues simples*, Paris, 1698.

A return to Aristotelian views was advanced by the phlogiston theory of Georg Ernst Stahl (1660–1734), of Ansbach, Bavaria, an eminent medical man who became physician to Frederick William I of Prussia. The theory originated with the Greeks, and its origin and rise has been described by Douglas McKie.[4] Stahl was also one of the foremost chemists of his period, contributing extensively to the literature of that subject, and also of medicine. His *Observationum chymico-physico-medicarum*, [*etc.*], Frankfurt and Leipzig, 1697–98, consists of a collection of essays published at monthly intervals, and his *Oeuvres medico-philosophiques et pratiques* was published in six volumes in Paris, 1859–64.[5]

The literature of biology is vast, and it is difficult to isolate the most significant contributions. Authorities differ regarding the value of certain writings, and it is impossible adequately to divide biology at this early

[1] McKie, Douglas. The phlogiston theory. *Endeavour*, 18, 1959, pp. 144–147.
[2] Holmyard, E. J. *Alchemy*, Harmondsworth, 1957.
[3] See Neville, Roy G. The "Pratique de Chymie" of Sébastien Matte de Faveur. *Ambix*, 10, 1962, pp. 14–28; and Neville, Roy G. Christophle Glaser and the *Traité de la chymie*, 1663. *Chymia*, 10, 1965, pp. 25–52.
[4] McKie, Douglas. The phlogiston theory. *Endeavour*, 18, 1959, pp. 144–147.
[5] See Beck, Curt W. Georg Ernst Stahl, 1660–1734. *J. chem. Educ.*, 37, 1960, pp. 506–509; King, Lester S. Stahl and Hoffmann: a study in eighteenth-century animism. *J. Hist. Med.*, 19, 1964, pp. 118–130; Rappaport, R. Rouelle and Stahl – the phlogistic revolution in France. *Chymia*, 7, 1961, pp. 73–102; Strube, Irene. Die Phlogistonlehre Georg Ernst Stahls (1659–1734) in ihrer historischen Bedeutung. *Z. Gesch. Naturw. Techn. Med.*, 1, 1961, pp. 27–51; and Strube, Wilhelm. Die Ausbreitung der Naturanschauung G. E. Stahls unter den deutschen Chimikern des 18. Jahrhunderts. *Z. Gesch. Naturw. Techn. Med.*, 1, 1961, pp. 52–61.

stage into its composite subdivisions. Writers on natural history in general contributed to botany and comparative anatomy, while many qualified medical men feature prominently in the history of biology. F. J. Cole and Nellie B. Eales[1] have compiled a statistical analysis of the literature of comparative anatomy, providing graphs plotted to indicate writings on comparative anatomy from the sixteenth century to 1860. This shows the work accomplished in various European countries, the writings on various groups of animals, and the influence of events, scientific societies and individual authors on the history of anatomical thought. Incidentally, F. J. Cole's[2] study of the history of comparative anatomy presents an authoritative picture of the subject, particularly during the seventeenth century, and is a model of an authentic, well-documented and illustrated history of a specialized subject. These two works emphasize the wealth of material on comparative anatomy, and in this chapter we unite the other branches forming biology, the 'science of life'.

Hieronymus Fabricius ab Aquapendente (Girolamo Fabrizio) (1536–1619) was born at Aquapendente, between Rome and Sienna, and was a pupil of Fallopius at Padua, where Fabricius became professor of surgery in 1565. He built the anatomical theatre there, which still exists, and his fame as a teacher attracted to Padua students from all over Europe. Fabricius retired in 1604, and then settled down to the composition of his works on comparative anatomy, embryology and physiology. As early as 1574 he had seen the valves in the veins, but was not their discoverer; nor did he appreciate their function. This latter explanation was left to his pupil, William Harvey, and it is mainly as the teacher of Harvey that Fabricius is remembered. In addition to his strictly medical writings,[3] Fabricius was the author of *De formato foetu*, Venice, 1600;[4] *De brutorum loquela*, Padua, 1603; *De venarum ostiolis*, Padua, 1603. This first edition is usually found bound with other of his works, with a general title-page bearing the date 1625, but the tract itself is dated

[1] Cole, F. J., and Eales, Nellie B. The history of comparative anatomy. Part 1. A statistical analysis of the literature. *Sci. Progr.*, 11, 1916–17, pp. 578–596.

[2] Cole, F. J. *A history of comparative anatomy from Aristotle to the eighteenth century*, 1944; see also Pagel, Walter. The reaction to Aristotle in seventeenth-century biological thought: Campanella, van Helmont, Glanvill, Charleton, Harvey, Glisson, Descartes. In, *Science, medicine and history*, Vol. 1, 1953, pp. 489–509; and Schierbeek, A. The main trends of zoology in the 17th century. *Janus*, 50, 1963, pp. 159–175.

[3] These are recorded in Thornton, pp. 79–80.

[4] This is the imprint on the title-page, but the colophon gives Padua, 1604, while the author's dedication is dated 1606. Apparently the title-page belonged to his *De visione*, Venice, 1600, and the imprint was allowed to stand without alteration. See *The embryological treatises of Hieronymus Fabricius of Aquapendente. The formation of the egg and of the chick.* [*De formatione ovi et pulli.*] *The formed fetus.* [*De formato foetu.*] *A fascimile edition, with an introduction, a translation, and a commentary by Howard B. Adelmann*, Ithaca, New York, 1942. This is based on the texts of the first editions, contains a life of Fabricius (pp. 6–35), and a bibliographical note (pp. 122–134).

1603;[1] *De musculi artificio, de ossium dearticulationibus*, Vicentiae, 1614; *De respiratione, et ejus instrumentis*, Padua, 1615; *De totius animalis integumentis*, Padua, 1618; *De gula, ventriculo intestinus tractatus*, Padua, 1618; *De motu locali animalium*, Padua, 1618; and *De formatione ovi et pulli tractatus accuratissimus*, Padua, 1621 and 1628. The anatomical and physiological writings of Fabricius have been collected together as *Opera anatomica*, [*etc.*], Padua, 1625; and *Opera omnia anatomica et physiologica*, Leipzig, 1687.

Although William Harvey (1578–1657) was a physician, and has been considered elsewhere in greater detail,[2] he must be mentioned here for his researches on comparative anatomy. Having seen the valves in the veins while studying under Fabricius at Padua, Harvey investigated their function, which he demonstrated, and then explained in his *Exercitatio anatomica de motu cordis et sanguinis in animalibus*, first published at Frankfort in 1628.[3] He collected together a vast amount of material on the development of insects and on comparative anatomy, which was destroyed during the Civil War in 1642 when Harvey's lodgings in London were raided by a mob. Harvey collected together the remnants of his work on embryology, and towards the end of Harvey's life Sir George Ent (1604–89) persuaded him to publish it, Ent seeing the book through the press. It was first published as *Exercitationes de generatione animalium*, London, 1651 (Plate 4). This has been hailed as a masterpiece by several specialists, and it undoubtedly contains a wealth of information on generation, much of which is original, but it probably suffered severely by the early loss of much potential material. A. W. Meyer[4] states that it is erroneous in many sections, and a failure in most respects. C. Webster[5] has produced evidence suggesting that the book was virtually completed by 1638, and the value of *De generatione* should be reassessed in view of the fact that dates mentioned in the text fall within the period 1625 and 1637. However, it is valuable as a mirror of contemporary embryology, and records many interesting experiments.[6] The tercentenary of Harvey's death brought forth a spate of literature concerning

[1] See Fabricius, Hieronymus. *De venarum ostiolis*, 1603. . . . *Facsimile edition with introduction, translation and notes by K. J. Franklin*, Springfield, Baltimore, 1933.

[2] See Thornton, pp. 80–84.

[3] Other editions of this and Harvey's other writings are recorded in Keynes, Sir Geoffrey L. *A bibliography of the writings of Dr. William Harvey, 1578–1657. Second edition revised*, Cambridge, 1953; see also Keynes, Sir Geoffrey L. Harvey and his books. . . . Being the Harveian Lecture delivered to the Harveian Society in London, in March, 1953. *St Bart's Hosp. J.*, 57, 1953, pp. 177–182, 212–216.

[4] Meyer, Arthur William. *An analysis of the De generatione animalium of William Harvey*, Stanford, London, 1936.

[5] Webster, C. Harvey's *De generatione:* its origins and relevance to the theory of circulation. *Brit. J. Hist. Sci.*, 3, 1967, pp. 262–274.

[6] See also Bayon, H. P. William Harvey (1578–1657), his application of biological experiment, clinical observation, and comparative anatomy to the problems of generation. *J. Hist. Med.*, 2, 1947, pp. 51–96; and Wendell-Smith, C. P. William Harvey, man-midwife. *St Bart's Hosp. J.*, 53, 1949, pp. 212–214.

his life and writings, of which the following were the most significant. The existence of several Harvey manuscripts in the British Museum has been known for many years, and among them his *De motu locali animalium* (Sloane MS. 486, ff. 69–118v) which had never previously been published. Gweneth Whitteridge,[1] on behalf of the Royal College of Physicians of London, transcribed and translated, with valuable annotations, this interesting document, producing a notable addition to Harvey's writings, and later occupied herself with a similar task on the *Prelectiones*, also existing in manuscript in the British Museum, but previously published in facismile, with an inadequate and inaccurate transcription in 1886.[2] Another notable publication was a new translation of *De motu cordis* by Kenneth J. Franklin,[3] which he followed up by a translation of Harvey's essays addressed to Riolan and nine letters to others.[4] Harvey's work on embryology and on animals in general was studied by F. J. Cole[5] and by C. P. Wendell-Smith.[6] The Harveian Oration of 1958 by Sir Geoffrey Keynes[7] presents an interesting re-evaluation of John Aubrey's views on Harvey, concluding that Aubrey was "sometimes inaccurate, but he was never untruthful". The mention of Harvey in Boyle's writings was the subject of an interesting paper by Richard A. Hunter and Ida Macalpine,[8] revealing several new references. E. B. Krumbhaar[9] made a general survey of the bibliographies of Harvey, with notes on his writings, including mention of supposed lost ones, and Frederick G. Kilgour has contributed a checklist of Harvey's manuscripts.[10] A scholarly appreciation of Harvey's biological ideas has been

[1] Harvey, William. *De motu locali animalium, 1627. Edited, translated and introduced by Gweneth Whitteridge*, Cambridge, 1959; see also Whitteridge, Gweneth. Harvey's Galen. *St Bart's Hosp. J.*, 61, 1957, pp. 174–175, describing copy of Galen's *Opuscula varia*, 1640, now in British Museum, with marginal comments by Harvey.

[2] *The anatomical lectures of William Harvey. Prelectiones anatomie universalis. De musculis. Edited, with an introduction, translation and notes by Gweneth Whitteridge*, Edinburgh, London, 1964.

[3] *Movement of the heart and blood in animals. An anatomical essay by William Harvey. Translated from the original Latin by Kenneth J. Franklin, and now published for the Royal College of Physicians of London*, Oxford, 1957.

[4] *The circulation of the blood. Two anatomical essays by William Harvey, together with nine letters written by him. The whole translated from the Latin and slightly annotated by Kenneth J. Franklin*, Oxford, 1958.

[5] Cole, F. J. Harvey's animals. *J. Hist. Med.*, 12, 1957, pp. 106–113.

[6] Wendell-Smith, C. P. Harvey and embryology. *St Bart's Hosp. J.*, 61, 1957, pp. 180–183.

[7] Keynes, Sir Geoffrey Langdon. Harvey through John Aubrey's eyes. *Lancet*, 1958, II, pp. 859–865; also separately issued in an expanded version.

[8] Hunter, Richard A., and Macalpine, Ida. William Harvey and Robert Boyle. *Notes Rec. Roy. Soc. Lond.*, 13, 1958, pp. 115–127.

[9] Krumbhaar, E. B. Thoughts on bibliographies and Harvey's writings. *J. Hist. Med.*, 12, 1957, pp. 235–240; see also McMichael, John, *ed. Circulation. Proceedings of the Harvey Tercentenary Congress held on June 8th 1957 at the Royal College of Surgeons of England, London*, Oxford, 1958. This contains several interesting articles on Harvey.

[10] Kilgour, Frederick G. Harvey manuscripts. *Pap. Bibl. Soc. Amer.*, 54, 1960, pp. 177–179; see also Kilgour, Frederick G. William Harvey and his contributions. *Circulation*, 23, 1961, pp. 286–296.

written by Walter Pagel,[1] who appreciates the true value of Harvey's embryological investigations, and evaluates him in the context of seventeenth-century scientific thought. K. D. Keele[2] has written a biographical study of Harvey, and Sir Geoffrey Keynes[3] has contributed a definitive biography of exceptional value to all students of Harvey and his period.

Claude Perrault (1613–88), a native of Paris, has been described as "a naturalist who did more to promote the study of comparative anatomy in the seventeenth century than any other worker". Although a student of anatomy, mathematics and medicine, he abandoned medicine to take up architecture, at which he was eminently successful. However, his contributions to comparative anatomy are many and varied. Perrault was particularly interested in the mechanics of the muscles, and his observations were published as *Observations qui ont este faites sur an grand poisson dissequé dans la Bibliothèque du Roy . . . sur un lion,* [*etc.*], Paris, 1667, published anonymously, as were the two following: *Description anatomique d'un cameleon, d'un castor, d'un dromadaire, d'un ours, et d'une gazelle*, Paris, 1669; *Mémoires pour servir à l'histoire naturelle des animaux*, [*etc.*], 1671, 1676, with a translation into English by A. Pitfield, London, 1688, the engraved title-page of which is dated 1687; and *Essais de physique*, four volumes, Paris, 1680–1688.[4]

Yet another early contributor to the literature of comparative anatomy was also a medical man, Samuel Collins (1618–1710).[5] He became Censor and President of the College of Physicians, and was also physician to Charles I. Collins' *A systeme of anatomy, treating of the body of man, beasts, birds, fish, insects and plants*, two volumes, London, 1685, has been severely criticized, but has also been described as an "important early contribution to comparative anatomy". Much of this work was executed during his retirement, and he drew freely upon the writings of others, including his friend Edward Tyson, so that Collins' book may be regarded rather as a compilation than as an original work.

Walter Charleton (1619–1707), another medical man who also held office as President of the College of Physicians, was physician to both Charles I and Charles II. His *Physiologia Epicuro-Gassendo-Charltoniana: or, a fabrick of science natural, upon the hypothesis of atoms,* [*etc.*], London, 1654, is regarded as his most important work, and deals with general natural philosophy, based on the atomic theory of Epicurus and Gassendi. A reprint with an introduction by Robert Hugh Kargon was published

[1] Pagel, Walter. *William Harvey's biological ideas: selected aspects and historical background*, Basle, New York, 1967.

[2] Keele, Kenneth David. *William Harvey: the man, the physician and the scientist*, London, [*etc.*], 1965.

[3] Keynes, Sir Geoffrey Langdon. *The life of William Harvey*, Oxford, 1966.

[4] See Cole, pp. 424–434.

[5] See Cole, pp. 156–174.

in New York, 1966. This was the first work in English on atomism, and presents a critical, historical review of the subject.[1] A folio volume of almost five hundred pages was issued as "the first part", but apparently Charleton did not follow it up. He was also the author of *Exercitationes physico-anatomicae de oeconomia animali*, Amsterdam, 1659, the second edition of which was published in English as *Natural history of nutrition, life, and voluntary motion. Containing all the new discoveries of anatomists and most probable opinions of physicians concerning the oeconomie of human nature*, London, 1659, which has been called the first textbook of physiology in English. The third edition was published as *Oeconomia superstructa et mechanice explicata*, [*etc.*], London, 1659–60, and the fourth as *Oeconomia animalis novis in medicina hypothesibus superstructa et mechanice explicata*, 1659–60. This went into at least six editions, and a rewritten version appeared in English with the title *Enquiries into human nature in VI anatomic praelectiones*, London, 1680.[2] Charleton's *Onomasticon zoicon*, [*etc.*], first published in London in 1668, has been regarded as a classic treatise on zoology, despite the fact that it is largely a compilation. A reissue of the book, with a new title-page, and with the first sheet reset, was printed in London, 1671, while a second, enlarged edition was published with the title *Exercitationes de differentiis et nominibus animalium*, [*etc.*], three volumes in one, Oxford, 1677. Charleton's writings covered an extensive field, including physics, physiology, zoology, philosophy, psychology and religion, and he was the author of the anonymously published *Natural history of the passions*, London, 1674. Several authorities have mistakenly regarded this as a translation of a book by J. F. Senault (1601–72) entitled *De l'usage des passions*, Paris, 1641, which was also translated into English.[3]

Although chiefly remembered as a diarist, and as a country gentleman rather than as a scientist, John Evelyn (1620–1706) has several claims for consideration here. He was an early member of the Royal Society, served as its secretary, and was keenly interested in experimental science, horticulture and navigation. He has been the subject of a full-scale bibliography by Sir Geoffrey Keynes,[4] from which the following information

[1] See Kargon, Robert Hugh. Walter Charleton, Robert Boyle, and the acceptance of Epicurean atomism in England. *Isis*, 55, 1964, pp. 184–192; and McKie, Douglas. English writers on atomism before Dalton. *Endeavour*, 25, 1966, pp. 13–15.

[2] See Rolleston, Sir Humphry Davy. Walter Charleton, D.M., F.R.C.P., F.R.S. *Bull. Hist. Med.*, 8, 1940, pp. 403–416; this contains a list of other books and translations by Charleton (pp. 414–416).

[3] See Hunter, Richard A., and Cuttler, Emily. Walter Charleton's "Natural history of the passions" (1674) and J. F. Senault's "The use of passions" (1649). A case of mistaken identity. *J. Hist. Med.*, 13, 1958, pp. 87–92.

[4] Keynes, Sir Geoffrey Langdon. *John Evelyn: a study in bibliophily with a bibliography of his writings.* (*Second edition*), Oxford, 1968; first edition limited to 300 copies, Cambridge, 1937; a preliminary list was issued in a limited edition of twenty-five copies as *A handlist of the works of John Evelyn . . . and of books connected with him*, [by Geoffrey L. Keynes and Augustus Theodore Bartholomew], Cambridge, 1916.

is derived. Many of Evelyn's writings were of a political nature, his first book being a translation from the French, *Of liberty and servitude*, [*etc.*], 1649, published 1648–49, but the following are of more scientific interest. He translated *Le jardinier françois* by Nicolas de Bonnefons, as *The French gardiner: instructing how to cultivate all sorts of fruit-trees, and herbs for the garden*, [*etc.*], London, 1658, of which there were three issues, with a second edition in 1669, and a fifth in 1691. Evelyn's *Fumifugium: or the inconvenience of the aer and smoak of London dissipated*, [*etc.*], London, 1661 (two issues); London, 1772; and which was also published in 1930 (twice), 1933, and 1961. His *Sylva: or a discourse of forest-trees, and the propagation of timber in His Majesties Dominions*, [*etc.*], London, 1664, was the first book published by the Royal Society. Editions followed in 1670, 1679, 1706 and 1729 (fifth edition), with an abridged edition published in 1827, and a two-volume edition [1908], a reprint of the fourth. An extensively annotated version edited by Alexander Hunter was published in 1776, 1786, 1801, 1812 and 1825, all except the first of these being published in two volumes. Evelyn's other publications include *Kalendarium hortense: or, the gard'ners almanac*, [*etc.*], London, 1664, with second and third editions in 1666 and 1669 respectively, and of which several other editions appeared; *Navigation and commerce*, [*etc.*], London, 1674; *Philosophical discourse of earth, relating to the culture and improvement of its vegetation, and the propagation of plants*, [*etc.*], London, 1676, of which a second edition was published in 1678, and a sixth in 1787; *The compleat gard'ner*, [*etc.*], London, 1693; and *Acetaria, A discourse of sallets*, London, 1699, with a second issue in 1706, and a second edition also published in that year. Evelyn was the author of several papers published in the *Philosophical Transactions*, and his "Diary" first appeared as *Memoirs, illustrative of the life and writings of John Evelyn, Esq., F.R.S.*, [*etc.*], edited by William Bray, two volumes, London, 1818. Many editions and versions of this have since appeared, and *John Evelyn's manuscript on bees from Elysium britannicum* edited by D. A. Smith, has been published by the Bee Research Association, Gerrards Cross, 1966.[1]

Gerard Blaes, or Blasius (*c.* 1625–92), was a native of Holland, who graduated M.D., and became professor of medicine at Amsterdam. He was a keen student of comparative anatomy, and was specially interested in embryology. His descriptive anatomy of the dog is particularly notable,[2] and Blasius was the author of the following separately published works, the last of which is mainly a compilation: *Miscellanea anatomica, hominis brutorumque variorum, fabricam diversam magna parte exhibentia*, Amsterdam, 1673; *Observata anatomica in homine, simia, equo, vitulo, ove,*

[1] See also Ponsonby, Arthur. Lord Ponsonby of Shulbrede. *John Evelyn, Fellow of the Royal Society, author of Sylva*, 1933.

[2] See Cole, pp. 150–155.

testudine, echino, glire, serpente, ardea, variisque animalibus aliis, Leyden and Amsterdam, 1674; and *Anatome animalium*, Amsterdam, 1681.

The physician, naturalist and poet Francesco Redi (1626–97) was a native of Arezzo, Tuscany, and was educated at the University of Pisa. He spent several years in travel, and became deeply interested in natural science, becoming an active member of the Accademia del Cimento, which had been organized in 1657. Redi's first scientific paper, *Osservazione intorno alle vipere*, Florence, 1664, was devoted to poisonous snakes and their venom, but his masterpiece was the book on the generation of insects, *Esperienze intorno alla generazione degli insetti*, Florence, 1668. This went into five editions in twenty years, and an English translation by Mab Bigelow was published in 1909. Redi was also the author of *Esperienze intorno a diverse cose naturali*, [*etc.*], Florence, 1671; *Osservazione . . . intorno agli animali viventi che si trovani negli animali viventi*, Florence, 1684, devoted to animal parasites; his works were published as *Opere*, seven volumes, Venice, 1712–30, with a new edition, also in seven volumes, Naples, 1748. A volume containing extracts from Redi's works was published in 1925.[1]

John Ray, or Wray[2] (1627–1705), was probably the greatest naturalist of the seventeenth century, and with him must be associated Francis Willughby (1635–72). Ray was born the son of a blacksmith at Black Notley, Essex, on 29th November, 1627. He was educated at Cambridge, where he studied the local plants, of which he published a catalogue. He met Francis Willughby, who financed an expedition to the Continent, where the two studied natural history in the field. Willughby died young, leaving his friend an annuity, and Ray faithfully edited and published Willughby's writings, to which much of his own work was added. Ray was elected F.R.S. in 1667, and his services to botany, and to zoology in general, are outstanding. He was a founder of systematic biology, and a pioneer of the natural system of classification. The writings of John Ray have been the subject of a comprehensive bibliography by Sir Geoffrey Keynes,[3] from which the following details of Ray's scientific writings are derived. His *Catalogus plantarum circa Cantabrigiam nascentium*, [*etc.*], Cambridge, 1660 (Plate 5), was again issued with a cancel title-page bearing the imprint London, 1660. An appendix to this was published at Cambridge, 1663, of which a second edition appeared in 1685. Ray's

[1] See Cole, Rufus. Francesco Redi (1626–1697), physician, naturalist and poet. *Ann. med. Hist.*, 8, 1926, pp. 347–359; Belloni, Luigi. Francesco Redi als Vertereter der italienischen Biologie des XVII, Jahrhunderts. *Münch. med. Wschr.*, 101, 1959, pp. 1617–1624; and Corti, Alfredo. Omero sapeva quel che il Redi dimostro. *Riv. Storia Sci. med. nat.*, 45, 1954, pp. 114–135.

[2] Ray changed the spelling of his name from Wray in 1670.

[3] Keynes, Sir Geoffrey Langdon. *John Ray: a bibliography*, 1951; illustrations and facsimiles; a preliminary handlist was privately printed in a few copies as K[eynes], G[eoffrey] L. *John Ray F.R.S. A handlist of his works*, Cambridge, 1944.

Catalogus plantarum Angliae, et insularum adjacentium, [*etc.*], London 1670, went into a second edition in 1677. An appendix to this appeared as *Fasciculus stirpium Britanicarum, post editum Plantarum Angliae catalogum observatarum a Joanne Raio et ab amicis,* [*etc.*], London, 1688. His tour on the Continent was described in *Observations topographical, moral, and physiological made in a journal through part of the Low-Countries, Germany, Italy, and France,* [*etc.*], London, 1673, which went into a second edition in 1738. Francis Willughby's *Ornithologiae libri tres,* [*etc.*], was published in London in 1676, Ray's work as editor being very significant. The engraved figures of birds were paid for by Willughby's widow and others. The work represents a monumental contribution to science, and is a handsome production, some copies being on large paper. An English translation was published in London in 1678. Ray's system of classification appeared in his *Methodus plantarum nova,* [*etc.*], London, 1682, of which a second issue appeared from Amsterdam in the same year, with a cancel title-page and five additional plates. A second edition was published bearing the imprint London, 1703, but despite the title-page it was printed in Holland. Further issues were printed in Amsterdam, 1710 and 1711, while a third edition was published in London, 1733. Willughby's *De historia piscium libri quatuor,* [*etc.*], was published at Oxford, 1686.[1] Samuel Pepys, then President of the Royal Society,[2] contributed fifty pounds towards the cost of the plates, the book being dedicated to him. Five hundred copies of the work were printed, but it did not sell, and several officers of the Royal Society were paid their salaries in copies, Ray receiving twenty copies as his total payment. A second issue with a cancel title-page appeared with the imprint London, 1743. Ray's most important work, the *Historiae plantarum,* [*etc.*], was published in London, the three volumes being dated 1686, 1688 and 1704. This was not illustrated, but after Ray's death James Petwer published *A catalogue of Mr. Ray's English herbal illustrated with figures in folio copper plates,* [*etc.*], London, *c.* 1715–64. John Ray's synopsis of British plants, summarizing his previous writings on the subject, appeared as *Synopsis methodica stirpium Britannicarum,* [*etc.*], London, 1690. This went into two further editions in 1696 and 1724, a second issue of the third edition also appearing in the latter year. Ray's "most popular and influential achievement", *The wisdom of God manifested in the works of the creation,* [*etc.*], London, 1691, went into numerous English editions, the last being published by the Wernerian Club, London, 1844–46, while a French translation was published at Utrecht in 1714. This work was followed by *Miscellaneous*

[1] The plates of this had been published as *Ichthyographia,* [*etc.*], London, 1685, consisting of 184 full-page copper engravings by Paul van Somers.

[2] See Andrade, E. N. da C. Samuel Pepys and the Royal Society. *Notes Rec. Roy. Soc. Lond.,* 18, 1963, pp. 82–93.

discourses concerning the dissolution and changes of the world, [*etc.*], London, 1692, 1693 and 1713, with a Dutch edition from Rotterdam, 1694 (of which no copy is known), and 1783; and German translations [1698] (no copy now known), and Leipzig, 1732. Ray's *Synopsis methodica animalium quadrupedum et serpentini generis*, [*etc.*], was published in London in 1693, and he also edited *A collection of curious travels & voyages*, [*etc.*], two volumes in one, London, 1693, of which there were three issues, with a second edition from London, 1705, and another 'second' [third], London, 1738. Ray was also the author of *Stirpium Europaearum extra Britannias nascentium sylloge*, [*etc.*], London, 1694; *De variis plantarum methodis dissertatio brevis*, [*etc.*], London, 1696, and *Methodus insectorum*, [*etc.*], London, 1705. This latter consists of only ten leaves, and was preliminary to his *Historia insectorum*, [*etc.*], published in London, 1710. Ray's *Synopsis methodica avium & piscium*, [*etc.*], was published in London in 1713, the *Synopsis avium* portion being translated into French and published at Paris in 1767. John Ray made several communications to the Royal Society, and was the author of several non-scientific writings, all of which are recorded by Sir Geoffrey Keynes, while Ray's life has been the subject of an exhaustive biography by Charles E. Raven.[1]

Marcello Malpighi (1628–94), who has been called the founder of histology, was born at Crevalcuore, near Bologna, and in 1645 went to Bologna to study philosophy. He graduated in medicine and philosophy in 1653, and three years later became professor of medicine at Bologna University. The same year, however, he went to Pisa, and later Messina, but after three years was back at his *alma mater*. Two of his letters announcing the discovery of the structure of the lung, and his observation of the capillaries, written in 1660, were published as *De pulmonibus observationes anatomicae*, Bologna, 1661. In his short tract *De omento piguedine et adiposis ductibus*, published in 1665, Malpighi demonstrated the red corpuscles of the blood, which he failed adequately to describe. His embryological writings, *De ovo incubato observationes*, London, 1673, and *De formatione pulli in ovi*, London, 1673, however, are of great significance. He investigated the papillae of the tongue, discovered the Malpighian layer of the skin, and his book on the viscera, *De viscerum structura exercitatio anatomica*, Bologna, 1666, and Frankfurt, 1683, contains physiological investigations on the liver, spleen and kidneys.[2] A French edition was printed in Paris,

[1] Raven, Charles E. *John Ray, naturalist; his life and works.* (*Second edition*), Cambridge, 1950; see also Arber, Agnes. A seventeenth-century naturalist: John Ray. *Isis*, 34, 1942–43, pp. 319–324; Stevenson, I. P. John Ray and his contributions to plant and animal classification. *J. Hist. Med.*, 2, 1947, pp. 250–261; Raven, Charles E. *English naturalists from Neckam to Ray; a study of the making of the modern world*, Cambridge, 1947; and Sawyer, F. C. John Ray (1627–1705)—a portrait in oils. *J. Soc. Bib. Nat. Hist.*, 4, 1963, pp. 97–99.

[2] See Hayman, J. M. Malpighi's "Concerning the structure of the kidneys"; a translation and introduction. *Ann. med. Hist.*, 7, 1925, pp. 242–263.

1683, and an Italian translation was published in *Celebrazione Malpighiane. Discorsi e scritti*, Bologna, 1966 (pp. 97–169). This book also contains an account of the cortex of the brain, stating that grey matter is composed of minute glands, and white matter of tubes and ducts. Edwin Clarke and J. G. Bearn[1] investigated this by examining brain tissue by means of a seventeenth-century type of microscope, and by modern techniques, contending that Malpighi described artefacts and not neurones, the 'glands' being areas of tissue outlined by capillaries. Many of Malpighi's writings were short papers communicated to the Royal Society, and were published in the *Philosophical Transactions*. His monumental work on the silkworm, the first monograph on an invertebrate, was also published by the Royal Society as *Dissertatio epistolica de bombyce*, London, 1669 (Plate 6). Malpighi described accurately what he saw through his primitive microscope, and his histological investigations are outstanding. In 1684 his house was burned, all his microscopes and manuscripts being destroyed. He became personal physician to Pope Innocent XII in 1691, but died on 29th November, 1694.[2] Malpighi's contribution to botany is also worthy of note. His *Anatome plantarum* was published in London, in two parts, 1675–79. This work illustrates the seedling structure in numerous species, and stresses the developmental aspect. A translation of this into German from the original Latin, edited by H. Möbius, was published at Leipzig in 1901. The collected works of Malpighi, of which there were several editions, were published as *Opera omnia*, two volumes in one, London, 1686; *Opera posthuma*, [*etc.*], London, 1697; and *Opera medica et anatomica*, [*etc.*], Venice, 1698. Howard B. Adelmann[3] has contributed a monumental study of Malpighi, the first volume being devoted to his life and works, the second volume covering currents of embryological thought before Malpighi, and Malpighi's contribution to embryology, and the final three volumes being devoted to his writings. Carlo Frati[4] has published a bibliography of Malpighi's writings.[5]

Nehemiah Grew (1641–1712) worked at the same time as Malpighi on a similar subject. Grew was a native of Warwickshire, and studied medicine at Leyden, before setting up in practice in London. He was one

[1] Clarke, Edwin, and Bearn, J. G. The brain 'glands' of Malpighi elucidated by practical history. *J. Hist. Med.*, 23, 1968, pp. 309–330; see also Meyer, Alfred. Marcello Malpighi and the dawn of neurohistology. *J. neurol. Sci.*, 4, 1967, pp. 185–193.

[2] See Young, James. Malpighi. *N.Z. med. J.*, 20, 1921, pp. 1–19.

[3] Adelmann, Howard B. *Marcello Malpighi and the evolution of embryology*, 5 vols., Ithaca, New York; London, 1966.

[4] Frati, Carlo. *Bibliografia Malpighiana. Catalogo descrittivo delle opere a stampa di Marcello Malpighi e degli scritti che lo riguardano*, Milan, 1897; reprinted London, 1960.

[5] See also Barbensi, Gustavo. Marcello Malpighi (1628–1694). *Sci. med. ital.*, 2nd ser., 3, 1954, pp. 3–13; Lerner, R. Marcello Malpighi: ein ikonographischer und biobibliographischer Überblick. *Mitteilungen des Instituts für die Geschichte der Medizin, Würzburg*, 6, 1960, pp. 1–4; Scott, Flora Murray. Marcello Malpighi, Doctor of Medicine. *Sci. Mon.*, 25, 1927, pp. 546–553; and Thornton, pp. 100–101.

of the earliest Fellows of the Royal Society, and served as its secretary in 1677. Both Malpighi and Grew presented the results of their study of plant anatomy to the Royal Society in 1671, and although Grew incorporated some of Malpighi's observations in his own writings he gave due credit to their originator. In fact the two scientists were friendly, although there is one significant difference between them. The Italian's work, although published in London, was in Latin, while Grew's book is in English. Grew wrote no medical books, but was keenly interested in psychological medicine, herbals and plant anatomy. His printed works include *The anatomy of vegetables begun*, London, 1672; *An idea of a phytological history . . . with a continuation of the anatomy of vegetables*, [*etc.*], London, 1673; *The comparative anatomy of trunks*, London, 1675; *Experiments in consort of the luctation arising from the affusion of several menstruums upon all sorts of bodies, exhibited at the Royal Society, 13th April and 1st June, 1676*, London, 1678, of which a facsimile reprint has been published, Cambridge, 1962; *Musaeum Regalis Societatis. Or a catalogue and description of . . . rarities belonging to the Royal Society. Whereunto is subjoyned the comparative anatomy of the stomach and guts*, London, 1681; *The Anatomy of plants*, [*etc.*], London, 1682 (Plate 7), reprinted with a new introduction by Conway Zirkle, New York, 1965; and *Cosmologia sacra*, [*etc.*], London, 1701. Grew was also the author of a tract on sea water (1684), and an important work on the atomic theory, *A discourse made before the Royal Society Decemb. 10, 1674, concerning the nature, causes, and power of mixture*, London, 1675, of which a French translation was published in Paris, 1679.[1] Robert Sharrock (1630–84), who became Archdeacon of Winchester, was a forerunner of Grew in botanical studies, and was the author of *The history of the propagation and improvement of vegetables*, Oxford, 1660, which went into a second edition in 1672. This is a practical handbook containing some purely botanical descriptions, but indicates that Sharrock understood the construction of seeds, and is an early study of the comparative anatomy of plants. Sharrock collected together a library, which, together with that of his son, was sold in a collection of over 2,000 volumes as the *Bibliotheca Sharrockiana*.[2]

Jan Swammerdam (1637–80) was a native of Amsterdam, and studied medicine at Leyden, Saumur and Paris, to graduate M.D. in 1667 at Leyden. He was particularly interested in anatomy, but died at an early age after suffering from religious melancholia. He was the author of

[1] See Arber, Agnes. Nehemiah Grew (1641–1712) and Marcello Malpighi (1628–94): An essay in comparison. *Isis*, 34, 1942–43, pp. 7–16; Arber, Agnes. Nehemiah Grew and Marcello Malpighi. *Proc. Linn. Soc.*, 153rd Session (1940–41), Pt. 2, pp. 218–238; and Arber, Agnes. Tercentenary of Nehemiah Grew (1641–1712). *Nature*, 147, 1941, pp. 630–632.

[2] See Arber, Agnes. Robert Sharrock (1630–84): a precursor of Nehemiah Grew (1641–1712) and an exponent of "Natural law" in the plant world. *Isis*, 51, 1960, pp. 3–8.

Tractatus physico-anatomico-medicus de respiratione usuque pulmonum, Leyden, 1667, his M.D. thesis; *Historia insectorum generalis, ofte Algemeene Verhandeling von bloedloose diertjens*, Utrecht, 1669, on bloodless animalculae, of which French (1682) and Latin (1685) versions were published; *Miraculum naturae, sive uteri muliebris fabrica*, Leyden, 1672; and *Ephemeri vita*, Amsterdam, 1675, on the may-fly, of which an English version was published in London, 1681. Swammerdam destroyed many of his manuscripts before his death, but others were published posthumously in two magnificent volumes as *Bybel der Naturae*, Leyden, 1737–38.[1] This contains his more important microscopical studies, and was translated into English by Thomas Flloyd as *The book of nature*, [*etc.*], London, 1758. This contains a biography of the author by Hermann Boerhaave, but a comprehensive biographical study of Swammerdam has since been published by Abraham Schierbeek[2], who provides bibliographical details of further editions of the above-mentioned books, and other writings by Swammerdam, many of which are very rare.

The contributions to microscopy of Antony van Leeuwenhoek (1632–1723) are of special interest because he had no scientific training. Leeuwenhoek was born at Delft, in Holland, on 24th October, 1632, and later occupied a civic post in his native town. He spent his spare time grinding lenses, and recording his observations. Regnier de Graaf heard of Leeuwenhoek's discoveries, and suggested that the Royal Society should invite the latter to communicate to them his observations. His first letter was received in London in 1673, and Leeuwenhoek sent about two hundred during the next fifty years, many of them being printed in the *Philosophical Transactions*. Leeuwenhoek observed protozoa in 1674, and bacteria in 1675.[3] He conducted research on spermatozoa, the optic nerve, on muscle, the skin, on insects; in fact on anything that attracted his attention. Leeuwenhoek contributed twenty-six notes to the French Académie des Sciences, and was visited at his home by many eminent personages. He died, aged ninety, on 26th August, 1723, and bequeathed twenty-six of his microscopes to the Royal Society, none of which now survives. Leeuwenhoek had been elected F.R.S. in 1680. Several collections of his writings have been printed, including *Ontledengen en Ontdekkingen*, six volumes, Leyden and Delft, 1693–1718; *Arcana naturae*, four volumes, Delft, 1695–1719; and *Opera omnia*, four volumes, Leyden,

[1] See Cole, F. J. The *Biblia naturae* of Swammerdam. *Nature*, 165, 1950, p. 511, for notes on some of the original drawings from this, which are preserved in the University of Leyden.

[2] Schierbeek, Abraham. *Jan Swammerdam (12th February, 1637–17th February, 1680). His life and works*, Amsterdam, 1967; Dutch edition published 1947.

[3] See *The Leeuwenhoek letter: a photographic copy of the letter of the 9th of October, 1676, sent by Antony van Leeuwenhoek to Henry Oldenburg, Secretary of the Royal Society of London; and a translation into English by Barnett Cohen*, Baltimore, 1937. The original letter is still preserved by the Royal Society, and contains the first recognizable description of bacteria.

CATALOGUS
PLANTARUM
CIRCA
CANTABRIGIAM
nascentium:

In quo exhibentur
Quotquot hactenus inventæ sunt, quæ
vel sponte proveniunt, vel in
agris seruntur;

Unà cum
Synonymis selectioribus, locis natalibus
& observationibus quibusdam
oppidò raris.

Adjiciuntur in gratiam tyronum,
Index Anglico-latinus, Index locorum,
Etymologia nominum, & Explicatio
quorundam terminorum.

CANTABRIGIÆ:
Excudebat *Joann. Field*, celeberrimæ
Academiæ Typographus.
Impensis Gulielmi Nealand, *Bibliopolæ.*
Ann. Dom. 1660.

PLATE 5. Title-page of John Ray's *Catalogus plantarum circa Cantabrigiam nascentium*, Cambridge, 1660.

MARCELLI MALPIGHII

PHILOSOPHI

&

MEDICI

BONONIENSIS

DISSERTATIO EPISTOLICA

DE

BOMBYCE,

SOCIETATI REGIÆ,

Londini ad Scientiam Naturalem promovendam institutæ, dicata.

LONDINI,

Apud Joannem Martyn & Jacobum Alleſtry

Regiæ Societatis Typographos, 1669.

PLATE 6. Title-page of Marcello Malpighi's *De bombyce*, London, 1669.

1719–22, which were reprinted in 1968. A complete edition of Leeuwenhoek's collected letters began publication in 1939,[1] and Clifford Dobell[2] has published an authoritative volume on Leeuwenhoek's bacteriological work, from which a more recent study of his life and works by A. Schierbeek[3] is largely derived. Samuel Hoole translated certain of Leeuwenhoek's writings as *Select works*. These were published by subscription in two volumes, and printed for the translator. Volume I was in two parts, the first dated 1798, but it actually appeared in 1797, this being very rare. Part I again appeared in 1798 with the word 'microscopical' replacing the misprint 'miscroscopical' on the title-page of the first issue. Part two is dated 1799, while Volume 2 did not appear until 1807, and contains an index to the two volumes (three parts). Unfortunately Hoole omitted the passages dealing with spermatozoa and reproduction, and used only the Dutch and Latin editions. Thus he did not incorporate material published solely in the *Philosophical Transactions*. There were several editions and issues of Hoole's translation, the last appearing in 1816.[4] Leeuwenhoek's biological researches have been the subject of papers by A. W. Meyer,[5] F. J. Cole,[6] which is of particular significance for its bibliographical details, and Frank N. Egerton.[7]

The first to study British conchology in a scientific manner, Martin Lister (1638–1712), was a physician in addition to being a naturalist. Born at Radclive, near Buckingham, he graduated in the arts from Cambridge and in medicine from Oxford, having been elected F.R.S. in 1671. Lister made several important contributions to comparative anatomy, particularly on the mollusca, and presented a thousand specimens of shells to the Ashmolean Museum at Oxford. He was the author of numerous works, including over sixty papers in the *Philosophical Transactions*. His first book was *Historiae animalium Angliae tres tractatus*, London, 1678, to which an appendix was published in 1681. *Historia conchyliorum* was published in five parts from 1685 to 1689, and was

[1] *The collected letters of Antoni van Leeuwenhoek. Edited, illustrated and annotated by a Committee of Dutch scientists*, [*etc.*], Parts 1–8, Amsterdam, 1939–67. Title-page, text, etc., in Dutch and English on opposite pages. All are from Leeuwenhoek's original letters and drawings, as often the printed translations are incorrect.

[2] Dobell, Clifford. *Anthony van Leeuwenhoek and his "little animals": being some account of the father of prozoology and bacteriology, and his multifarious discoveries in these disciplines*, [*etc.*], 1932. Reprinted New York, 1958 and 1960.

[3] Schierbeek, A. *Measuring the invisible world: the life and works of Antoni van Leeuwenhoek. With a biographical chapter by Maria Rooseboom*, London, New York, 1959.

[4] See Dobell, Clifford. Samuel Hoole, translator of Leeuwenhoek's *Select works:* with notes on that publication. *Isis*, 41, 1950, pp. 171–180.

[5] Meyer, A. W. Leeuwenhoek as an experimental biologist. *Osiris*, 3, 1938, pp. 103–122.

[6] Cole, F. J. Leeuwenhoek's zoological researches. Part 1[–2]. *Ann. Sci.*, 2, 1937, pp. 1–46, 185–235.

[4] Egerton, Frank N. Leeuwenhoek as a founder of animal demography. *J. Hist. Biol.*, 1 1968, pp. 1–22.

illustrated by two of Lister's daughters. Further editions of this followed in 1699, 1770 and 1823. Lister also wrote *Appendix ad historiae conchyliorum librum IV*, London, 1692; *Exercitatio anatomica*, [*etc.*], London, 1694; *Exercitatio anatomica altera*, [*etc.*], London, 1695; and *Conchyliorum bivalvium utriusque aquae exercitatio anatomica tertia*, London, 1696. He was the author of an early work on English mineral waters, *Novae ac curiosae exercitationes et descriptiones thermarum ac fontium medicatorum Angliae*, [*etc.*], London, 1682, and 1686 and also of *A journey to Paris in the year 1698*, London, 1699, with a French translation, Paris, 1873. Raymond Phineas Stearns[1] has edited a modern printing of this book, adding a biography and bibliography of Lister. Certain of Lister's manuscripts are in the Bodleian Library, including three volumes of letters from Pepys, Grew, Hooke and Sloane. Others are with the Sloane manuscripts in the British Museum.[2]

Edward Tyson (1650–1708) was born at Bristol, and was educated at Magdalen Hall, Oxford, coming to London about 1677. Three years later he took his doctorate of medicine at Cambridge, and practised as physician to Bridewell and Bethlem Hospitals. Tyson wrote several important papers for the *Philosophical Transactions*, including the anatomy of a porpoise (1680); on the rattlesnake, tapeworm, roundworm and collared peccary (1683); on an opossum (1698), and another on the same subject, followed by a paper by William Cowper, also on the opossum (1704). His major work, however, was *Orang-outang, sive homo-sylvestris; or, the anatomy of a pygmie compared with that of a monkey, an ape, and a man. To which is added a philological essay concerning the pygmies . . . of the ancients*, [*etc.*], London, 1699 (Plate 8). A facsimile reprint of this was published in 1966 with an introduction by M. F. Ashley Montagu. William Cowper executed the drawings, and also wrote the section on the muscles. The ape described was actually a chimpanzee, and its skeleton is still preserved in the British Museum (Natural History). A second edition of this work, entitled *The anatomy of a pygmy*, [*etc.*], was published in London, 1751, but this appears to be a reissue of the 1699 edition with a new title-page and the addition of Tyson's papers on the rattlesnake, the musk-hog and the round-worm, etc., from the *Philosophical Transactions*.[3] The *Philological essay* was edited by Bertram C. A. Windle and issued in a limited edition of 550 copies from London in 1894. Tyson's *Opera omnia*, 1751, consists of copies of the 1699 edition of the work on the 'pygmie', bound

[1] Lister, Martin. *A journey to Paris in the year 1698. Edited, with annotations, a life of Lister, and a Lister bibliography, by Raymond Phineas Stearns*, Urbana, [*etc.*], 1967.

[2] See Cole, pp. 231–245; Wood, S. Martin Lister, zoologist and physician. *Ann. med. Hist.*, N.S.1, 1929, pp. 87–104; and Wilkins, G. L. Notes on the "Historia conchyliorum" of Martin Lister (1683–1712). *J. Soc. Bib. Nat. Hist.*, 3, 1957, pp. 196–205.

[3] See Thornton, John L. In our Library. X. Tyson's Pygmies, 1699. *St. Bart's Hosp. J., War Edition*, 3, 1941–42, p. 78.

up with reprints of his other works.[1] Edward Tyson has been the subject of a lengthy study by M. F. Ashley Montagu,[2] who contributes later information in an additional paper.[3]

Returning to botany, we must mention Rudolf Jacob Camerarius (1665–1721), whose theory of sexuality in plants is sometimes overlooked. Camerarius was born at Tübingen, where he studied philosophy and medicine, and after travelling in England and on the Continent, returned as director of the Botanic Garden in his native town. In the following year (1689) he became professor of natural philosophy there. Camerarius studied the writings of the ancients on the pollination of dates, and made many experiments before publishing his *De Sexu plantarum epistola* in 1695. This represents an important step in botanical literature, being based on a careful study of previous writings, and on the results of numerous experiments, Johann Christian Mikan collected together the scattered writings of Camerarius, and published them as *R. J. Camerarii Opuscula botanici argumenti*, [*etc.*], Prague, 1797.

Several other contributors to biology in the seventeenth century deserve mention in brief. Edward Topsell (died 1638) was the author of *The historie of foure-footed beastes*, [*etc.*], London, 1607, which went into several later editions. It is mainly a compilation, and was accompanied by *The historie of serpents; or, the second booke of living creatures*, [*etc.*], London, 1608, which is generally found bound with the above. A later edition of these two entitled *The history of four-footed beasts and serpents and insects*, [*etc.*], was published in two volumes, London, 1658, and a facsimile edition with an introduction by Willy Ley was published in three volumes, New York, 1967. The 1658 edition contains *The theater of insects* by Thomas Moffet [Moffett, Moufet; Mouffet; Muffet] (1553–1604). This had been published posthumously in Latin as *Insectorum sive minimorum animalium theatrum*, London, 1634,[4] having been written about 1590, when it was ready for press. It was partly compiled from the writings of Edward Wotton, Conrad Gesner and papers left to Moffet by Thomas Penny. After Moffet's death the manuscript was sold to Sir Theodore Mayerne, who published it in 1634. The English version of 1658 was translated by John Rowland, and Moffet's book was the earliest entomological work to contain scientifically accurate illustrations.[5] Daniel Sennert (1572–1637), a professor of medicine at Wittenberg, attempted

[1] See Cole, pp. 198–221.

[2] Montagu, M. F. Ashley. *Edward Tyson, M.D., F.R.S., 1650–1708, and the rise of human and comparative anatomy in England; a study in the history of science*, [*etc.*], Philadelphia, 1943. (Memoirs of the American Philosophical Society, Vol. XX.) List of Tyson's writings, pp. 462–465. See also review of this by F. J. Cole, *Nature*, 152, 1943, pp. 611–612.

[3] Montagu, M. F. Ashley. Tysoniana. *Isis*, 36, 1945–46, pp. 105–108.

[4] Manuscript in British Museum (Sloane 4014).

[5] See Friedman, Reuben. Thomas Moffet. The tercentenary of his contribution to scabies. *Medical Life*, 41, 1934, pp. 620–637.

a classification of animal life in his *Epitome naturalis scientiae*, [*etc.*], [Venice?], 1618, and Venice, 1651. Nicholas Tulp (1593–1674),[1] a Dutch physician and naturalist, first used the name 'orang-outang', but applied it to the chimpanzee, in his *Observationum medicarum libri tres*, Amsterdam, 1641, of which there were several printings, including an edition in 1762.

Joannes Jonstonus, or John Johnston (1603–1675), was born abroad, but was of Scottish descent. He studied at St. Andrews, London and Cambridge, and after travelling on the Continent, settled in Silesia. Johnston wrote extensively on natural history, including *Thaumatographia naturalis in decem classes distincta*, [*etc.*], Amsterdam, 1632, 1633, 1641 and 1665. This was translated into Dutch, and into English (London, 1657), while the ornithology section was translated into French, two volumes, Paris, 1773–74. His *Historia naturalis*, six parts in one volume, Amsterdam, 1657, was also published in six separate parts at Frankfurt, 1657–58.

Sir Thomas Browne (1605–82), physician and author of the immortal *Religio medici*, has been considered elsewhere,[2] but he conducted numerous experiments on embryology, which are recorded in his *Pseudodoxia epidemica*, [*etc.*], first published in 1646. A. G. Debus[3] has discussed Browne's study of colours in this work, and Stephen Merton[4] has written on his physiological views. Sir Thomas Browne also wrote on the birds of Norfolk,[5] on fishes, and his *Garden of Cyrus* contains much botanical lore.[6] Browne corresponded with Henry Power, and was interested in sap and in the reproduction of plants. He grew herbs, flowers and fruit, and appears to have been on the verge of several discoveries made by his contemporaries. Sir Geoffrey Keynes has compiled a bibliography of Browne's writings,[7] and also edited his *Works* in six volumes, 1928–31, of which a revised edition was published in four volumes, London, 1964. An earlier edition of Browne's *Works*, edited by Simon Wilkin was also issued in four volumes, London, Norwich, 1835–36. Frank Livingstone Huntley[8] and Leonard Nathanson[9] have published biographical studies of Browne.[10]

[1] See Thornton, pp. 105–106. [2] See Thornton, pp. 92–94.

[3] Debus, A. G. Sir Thomas Browne and the study of colour indicators. *Ambix*, 10, 1962, pp. 29–36.

[4] Merton, Stephen. Old and new physiology in Sir Thomas Browne: digestion and some other functions. *Isis*, 57, 1966, pp. 249–259.

[5] *Notes and letters on the natural history of Norfolk, more especially on the birds and fishes. From the MSS. of Sir Thomas Browne, M.D. (1605–82), in the Sloane Collection in the Library of the British Museum and in the Bodleian Library, Oxford. With notes by Thomas Southwell*, 1902.

[6] See Merton, E. S. The botany of Sir Thomas Browne. *Isis*, 47, 1956, pp. 161–171.

[7] Keynes, Sir Geoffrey Langdon. *A bibliography of Sir Thomas Browne, Kt., M.D. . . . Second edition revised and augmented*, Oxford, 1968; first edition Cambridge, 1924. See also Keynes, Sir Geoffrey Langdon. Sir Thomas Browne. *Brit. med. J.*, 1965, II, pp. 1505–1510.

[8] Huntley, Frank Livingstone. *Sir Thomas Browne: a biographical and critical study*, Ann Arbor, 1962. [9] Nathanson, Leonard. *The strategy of truth. A study of Sir Thomas Browne*, Chicago, London, 1967. [10] See also Chalmers, Gordon Keith. Sir Thomas Browne, true scientist. *Osiris*, 2, 1936, pp. 28–79.

Jan Goedart (1620–68), a native of Middelburg, was the author of the earliest work on insect metamorphosis, *Metamorphosis et historia naturalis insectorum*, Middelburg, 1662. Two later volumes were added to this, Volume 2 being edited by Paul Veezaerdet, 1667, while Jan de Mey edited Volume 3, 1668. This was translated into French, three volumes, Amsterdam, 1700; an English work entitled *Of insects* was translated by Martin Lister, and published at York, 1682.

Although Thomas Willis (1621–75) is mainly remembered for his contributions to medicine,[1] his work on comparative neurology is of vital significance. Willis was assisted by Richard Lower in his experiments, while (Sir) Christopher Wren made drawings of his preparations, including some of the original drawings for the *Cerebri anatome*, [*etc.*], London, 1664. This contains a most accurate description of the nervous system, and described the "nerve of Willis", and the "circle of Willis". His *De anima brutorum*, [*etc.*], Oxford, 1672, was published in London in the same year, and includes a description of the nervous system. William Feindel has edited a commemorative work to mark the tercentenary of the publication of *Cerebri anatome*, containing biographical and bibliographical information, together with a facsimile of the English translation by Samuel Pordage. This was published by McGill University Press as *The anatomy of the brain and nerves*, 2 vols., 1965. Hansruedi Isler[2] has written a biographical study of Willis which contains a short list of his writings.[3]

Sir Christopher Wren (1632–1723) has been noted above as an artist, but his work as a scientist has received little recognition. His solution of a geometrical problem sent from France in 1658, and surviving as a printed broadside in the archives of the Royal Society, has been printed in facsimile with a translation from the Latin by A. Rupert Hall.[4] In 1662 Wren showed an automatic rain-gauge to the Royal Society, but details were not published. A Frenchman, B. D. Monconys, gave an account of it after visiting England in 1663, but ascribed it to "M. Renes" instead of to Wren.[5]

Walter Needham (*c.* 1631–69) studied at Oxford under Willis, and was the author of an important work on comparative anatomy, his *De formato*

[1] See Thornton, pp. 95–97; also Cole, pp. 222–231.

[2] Isler, Hansruedi. *Thomas Willis, 1621–75, doctor and scientist*, New York, London, 1968.

[3] See also Dewhurst Kenneth. Willis in Oxford: some new MSS. *Proc. Roy. Soc. Med.*, 57, 1964, pp. 682–687; Hierons, Raymond and Mayer, Alfred. Willis' place in the history of muscle physiology. *Proc. Roy. Soc. Med.*, 57, 1964, pp. 687–692; Meyer, Alfred, and Hierons, Raymond. A note on Thomas Willis' views on the corpus striatum and the internal capsule. *J. neurol. Sci.*, 1, 1964, pp. 547–554; and Meyer, Alfred. Karl Friedrich Burdach on Thomas Willis. *J. neurol. Sci.*, 3, 1966, pp. 109–116.

[4] Hall, A. Rupert. Wren's problem. *Notes Rec. Roy. Soc. Lond.*, 20, 1965, pp. 140–144.

[5] See Biswas, Asit K. The automatic rain-gauge of Sir Christopher Wren, F.R.S. *Notes Rec. Roy. Soc. Lond.*, 22, 1967, pp. 94–104.

foetu, first published in London, 1667, was reprinted at Amsterdam in the same year. The work was dedicated to Robert Boyle, and is the only item by Needham still worthy of note.[1]

Frederik Ruysch (1638–1731), a native of The Hague, became professor of anatomy at Amsterdam, and is mainly remembered for his remarkable museum, the specimens for which were injected by means of a method invented by Ruysch. He was the author of numerous writings, including *Dilucidatio valvularum in vasis lymphaticis et lacteis*, The Hague, 1665, of which a facsimile reprint was published in 1964, *Museum anatomicum Ruyschianum*, Amsterdam, 1691; *Thesaurus anatomicus*, i–x, Amsterdam, 1701–16; *Thesaurus animalium primus*, Amsterdam, 1710; and his writings are collected together in *Opera omnia anatomica-medico-chirurgico*, [*etc.*], four volumes, Amsterdam, 1737.[2]

Nicolaus Steno (Niels Stensen) (1638–86), a native of Copenhagen, contributed extensively to anatomy, geology and philosophy. He corresponded with William Croone (1633–84), who communicated his letters to the Royal Society.[3] Croone was the author of *De ratione motus musculorum*, London, 1664, which was published anonymously at first, and the second edition of 1667 was revised in the light of Steno's work on the same subject in which he was influenced by his friend Swammerdam.[4] Steno made pioneer investigations into the structure of crystals,[5] and was the author of *De solido intra solidum naturaliter contento dissertationis prodromus*, Florence, 1669, of which a facsimile edition, edited by W. Junk, was published in Berlin in 1904. This deals with fossils contained in rocks, and Henry Oldenburg edited the English translation published in 1671. A modern translation by John Garrett Winter appeared in 1916.[6] Steno's numerous writings include *Observationes anatomicae*, [*etc.*], Leyden, 1662 and 1680; *De musculis et glandulis observationum specimen*, [*etc.*], Copenhagen, 1664; *Elementorum myologiae specimen, seu musculi descriptio geometrica*, [*etc.*], Florence, 1667, and Amsterdam, 1669, which included his first geological woik, *Canis carchariae dissectum caput*, translated into

[1] See Thornton, John L. In our Library. I. Needham's *De formato foetu*, 1668. *St. Bart's Hosp. J.*, 50, 1946–47, p. 46.

[2] See Cole, pp. 305–311.

[3] Rome, Remacle. Nicolas Sténon et la Royal Society of London. *Osiris*, 12, 1956, pp. 244–268; see also article by F. N. L. Poynter, note 3, opposite.

[4] See Wilson, Leonard G. William Croone's theory of muscular contraction. *Notes Rec. Roy. Soc. Lond.*, 16, 1961, pp. 158–178; and Payne, Leonard M., Wilson, Leonard G., and Hartley, Sir Harold. William Croone, F.R.S. (1633–84). *Notes Rec. Roy. Soc. Lond.*, 15, 1960, pp. 211–219.

[5] See Seifert, Hans. Nicolaus Steno als Bahnbrecher der modernen Krystallographie. *Sudhoffs Arch. Gesch. Med.*, 38, 1954, pp. 29–47.

[6] *The Prodromus of Nicolaus Steno's Dissertation concerning a solid body enclosed by process of Nature within a solid. An English version with an introduction and explanatory notes by John Garrett Winter*, New York, 1916; reprinted New York, 1968.

English by Axel Garboe,[1] and *Discours sur l'anatomie du cerveau*, [*etc.*], a paper which had been read in 1665, Paris, 1669, of which a facsimile edition was published in Copenhagen in 1950, a Latin translation having been printed in Leyden in 1671. His *Opera philosophica*, edited by Vilhelm Maar, was published in two volumes, Copenhagen, 1910, as was his *Opera theologica*, edited by K. Larsen and G. Scherz, Copenhagen, 1941–47.[2] A series of outstanding papers on "Steno and brain research in the seventeenth century" was published in 1968.[3]

[1] Steno, Nicolaus. *The earliest geological treatise (1667). Translated from Canis carchariae dissectum caput. With introduction and notes by Axel Garboe*, London, New York, 1958.

[2] See also Eyles, V. A. Nicolaus Steno, seventeenth-century anatomist, geologist and ecclesiastic. *Nature*, 174, 1954, pp. 8–10; Schlichting, Th. H. Das Tagebuch von Niels Stensen. *Centaurus*, 3, 1953–54, pp. 305–310; and Scherz, Gustav. Niels Stensens Smaragdreise. *Centaurus*, 4, 1955–56, pp. 51–57.

[3] Steno and brain research in the seventeenth century. Proceedings of the International Historical Symposium . . . held in Copenhagen 18th–20th August, 1965. Edited by Gustav Scherz. *Analecta Medico-Historica*, 3, 1968. Papers include the following: Clarke, Edwin S. Brain anatomy before Steno, pp. 27–34; Schulte, B. P. M. Swammerdam and Steno, pp. 35–41; Dewhurst, K. Willis and Steno, pp. 43–48; Rothschuh, K. E. Descarte, Stensen und der *Discours sur l'anatomie du cerveau* (1665), pp. 49–57; Djørup, Frans. Stenos' ideas on brain research, pp. 111–114; Faller, A. Die Hirnschnitt-Zeichnungen in Stensens Discours sur l'anatomie du cerveau, pp. 115–145; Bastholm, E. Niels Stensen's myology, pp. 147–153; Belloni, L. Stensen-Andenken in Italien, pp. 171–180; Herrlinger, R. Niels Stensens Discours sur l'anatomie du cerveau im deutschen medizinischen Schrifttum, pp. 181–184; Keele, Kenneth D. Niels Stensen and the neurophysiology of pain, pp. 225–231; Poynter, F. N. L. Nicolaus Steno and the Royal Society of London, pp. 273–280; and Scherz, G. Nicolaus Steno the humanist, pp. 295–302.

CHAPTER V

Scientific Books from 1701-1800

"The seventeenth century bequeathed a great heritage to its successor; and the eighteenth century proved itself a worthy heir of the age of genius."

A. WOLF

The previous century has been described as "the age of scientific endeavour", and saw the foundations of many new universities and the initiation of numerous scientific societies. These fostered the development of scientific literature, and led to the founding of periodicals, many of these being published as official organs of societies. The pioneer work of the outstanding scientists of that century opened up fresh fields to enquiring minds, and there was no lack of these in the eighteenth century. Their efforts added impetus to the work initiated by Malpighi, Leeuwenhoek and Hooke, as examples from one branch of science, and rapid progress in most of the sciences was to be expected, and was achieved.

Astronomy was studied by numerous eminent scientists during the eighteenth century, but few prominent books were devoted to the subject. Apparatus and theories were more properly described in articles in journals, and several prominent astronomers are omitted because they were not the authors of books.[1] James Bradley (1693–1762), a friend of both Newton and Halley, succeeded the latter as Astronomer Royal (1742–62), and had been elected a Fellow of the Royal Society in 1718. A graduate of Oxford, Bradley became Savilian professor of astronomy there in 1721. He wrote an account of the comet of 1723,[2] and Bradley's discovery in collaboration with Samuel Molyneux (1689–1728) of the aberration of light is also described in the *Philosophical Transactions*.[3]

[1] See Spencer-Jones, Sir Harold. Astronomy through the eighteenth century. *Phil. Mag.*, Commemoration issue, July, 1948, pp. 10–27.

[2] *Phil. Trans.*, 33, 1724, p. 41.

[3] *Phil. Trans.*, 35, 1728, pp. 637–651. This volume is antedated, as the paper was not communicated until 1729. See also Sarton, George. Discovery of the aberration of light. With facsimile reproduction (no. xii) of James Bradley's letter to Edmond Halley giving an account of his discovery. (*Philosophical Transactions*, 35, no. 406, 637–661, 1729.) *Isis*, 16, 1931, pp. 233–265; and Forbes, Eric Gay. Dr. Bradley's astronomical observations. *Quart. J. Roy. Astron. Soc.*, 6, 1965, pp. 321–328.

Bradley also discovered and explained aberration in the stars, and the mutation of the earth's axis. His writings were collected together and edited by S. P. Rigaud as *Miscellaneous works and correspondence of the Rev. James Bradley*, Oxford, 1832.

Tobias Mayer (1723–62) was born in Marbach, near Stuttgart, and became Professor of Mathematics at Göttingen in 1751. Many of his unpublished manuscripts, some correspondence and drawings are preserved in the University of Göttingen. Mayer published many papers, and the following books: *Neue und allgemeine Art, alle Aufgaben aus der Geometrie*, [*etc.*], Esslingen, 1741; *Mathematischer Atlas*, [*etc.*], Augsburg, 1745; and *Germaniae atque in ea locorum principaliorum mappa critica*, [*etc.*], Nuremberg, 1750. Some of his miscellaneous works were published posthumously in *Opera inedita*, *I*, Göttingen, 1775, but no further volumes appeared.[1]

Nevil Maskelyne (1732–1811) became Astronomer Royal in 1765, two years after the appearance of his *British mariner's guide*, 1763. He was responsible for the *Nautical Almanac* for 1767 and for the next forty-four annual issues. The *Almanac* is still current, and the *Nautical Almanac and Astronomical Ephemeris for 1967* contains a special introduction reproducing Maskelyne's original preface to the first volume, with his description of the method of longitude determination.[2] He was elected F.R.S. in 1758, and described night myopia in a paper published in the *Philosophical Transactions*.[3]

Thomas Wright (1711–86), a native of Durham and a follower of Newton, established himself as a tutor in astronomy and mathematics to the aristocracy. Certain of Wright's ideas influenced Immanual Kant in his explanation of the theory of the universe, and Wright was the author of several books, some of which are very rare. His explanation of the Milky Way is contained in *An original theory or new hypothesis of the universe*, [*etc.*], London, 1750, of which a second edition was published in America in 1837, but his theory was abandoned in a sequel, probably written after 1771. This was discovered among papers auctioned in 1966, and was published as *Second or singular thoughts upon the theory of the universe*, London, 1968, edited by M. A. Hoskin. Wright's work on navigation, *Pannauticon*, was published in 1734, and no copy of this is now known; *The universal vicissitude of seasons*, 1737, contains diagrams drawn

[1] See Forbes, Eric Gay. The life and work of Tobias Mayer (1723–62). *Quart. J. Roy. Astron. Soc.*, 8, 1967, pp. 227–251.

[2] See Forbes, Eric Gay. The bicentenary of the *Nautical Almanac* (1767). *Brit. J. Hist. Sci.*, 3, 1967, pp. 393–394; and Sadler, D. H. The bicentenary of the Nautical Almanac. *Quart. J. Roy. Astron. Soc.*, 8, 1967, pp. 161–171.

[3] An attempt to explain a difficulty in the theory of vision, depending on the different refrangibility of light. *Phil. Trans.*, 79, 1789, pp. 256–264; see also Levene, John R. Nevil Maskelyn, F.R.S., and the discovery of night myopia. *Notes Rec. Roy. Soc. Lond.*, 20, 1965, pp. 100–108.

and coloured by hand, but the only known copy of this is in Durham University Library: *The use of the globes* was printed in 1740; *Clavis coelestis*, [*etc.*], 1742, is a popular introduction to astronomy; *Louthiana*, 1749, deals with the antiquities of the County of Louth; and Wright's final book, *The longitude discover'd without use of graduated instruments*, was published in 1773.[1] Immanual Kant (1724–1804) is better known as a philosopher, but his early writings were scientific, and include *Allgemeine Naturgeschichte und Theorie des Himmels* [*etc.*], Konigsberg and Leipzig, 1755, with later editions 1798, 1890 (*Ostwald's Klassiker*, Nr. 12), and 1908, with an English translation by W. Hastie which was published as *Kant's Cosmogony, as in his essay on the theory of the heavens. With introduction, appendices and a portrait of Thomas Wright, of Durham*, [*etc.*], Glasgow, 1900.[2]

A native of Rothiemay, Banffshire, James Ferguson (1710–76) lectured in London on astronomy. He made numerous orreries and other instruments, and was the author of several pamphlets and books. *The astronomical rotula* was printed in 1741 and 1753 and was very popular; it was followed by his *Astronomical card dial; select mechanical exercises*, London, 1773; *Description and use of the astronomical rotula*, London, 1775; and *Astronomy explained upon Sir Isaac Newton's Principles*, London, 1756, 1757, 1764, 1770, with several other editions and translations up to 1821.[3]

Friederich Wilhelm Herschel, generally known as Sir William Herschel (1738–1822), was a native of Hanover who spent most of his life in England. He discovered the planet Uranus in 1781, constructed many telescopes, and received numerous honours for his brilliant researches. He discovered a large number of double stars, two new satellites of Saturn, and a large number of nebulae. With his sister Caroline (1750–1848), who ably assisted him in his astronomical observations, Herschel worked at Slough for forty years, having constructed an enormous telescope there. His results are recorded in about seventy papers printed in the *Philosophical Transactions* between 1780 and 1821, and were gathered together as *Collected papers*, two volumes, 1912. In 1800 Herschel contributed a series of papers on thermal radiation to the *Philosophical Transactions*,[4] describing investigations to prove that light is essentially different from radiant heat,[5] and his astronomical work has been the

[1] See Paneth, F. A. Thomas Wright of Durham. *Endeavour*, 9, 1950, pp. 117–215.
[2] See Whitrow, G. J. Kant and the extragalactic nebulae. *Quart. J. Roy. Astron. Soc.*, 8, 1967, pp. 48–56.
[3] See Henderson, E. *Life of James Ferguson, F.R.S., in a brief autobiographical account, and further extended memoir*, [*etc.*], Edinburgh, [*etc.*], 1867.
[4] *Phil. Trans.*, 1800, pp. 253–283, 284–292, 293–326, 437–538.
[5] See Lovell, D. J. Herschel's dilemma in the interpretation of thermal radiation. *Isis*, 59, 1968, pp. 46–60.

subject of studies by Angus Armitage[1] and Michael A. Hoskin.[2,3]

Several French astronomers of the eighteenth century assumed particular significance on account of their discoveries and writings. Nicholas Louis de La Caille (1713–62) was an assistant at the Paris Observatory, and later professor of mathematics at the Mazarin College. He led an expedition to the Cape of Good Hope sent by the Académie des Sciences in 1750, the results of his journey being recorded in *Journal historique du voyage fait au Cap de Bonne-Espérance*, Paris, 1763. His *Stellarum australium catalogus* (*coelium australe stelliferum*) was also published at Paris in the same year. La Caille surveyed many new nebulae and star clusters, measured the acceleration of gravity at the Cape and at Mauritius, and made observations of Mars and Venus. His *Astronomiae Fundamenta*, Paris, 1757, a catalogue of about four hundred of the brightest stars, long retained its value, and was made more generally accessible when it was reprinted in 1833, with an introduction and notes by Francis Baily.[4] The *Tabulae solares*, 1757–58, records La Caille's extensive observations on the sun, and his astronomical work has been the subject of an interesting study by Angus Armitage.[5]

Joseph Jérôme Le Français de La Lande (1732–1807) was sent to Berlin to make observations simultaneously with La Caille at the Cape. He was the author of a very popular book, *Traité d'astronomie*, first published in two volumes, 1764, which went into several editions. His masterpiece, *Histoire céleste française*, was published in 1802, and contains a catalogue of over 47,000 stars.

Pierre Simon, Marquis de Laplace (1749–1827), was the son of a Normandy farmer, and became professor of mathematics at the Ecole Militaire in Paris. His greatest book was the *Mécanique céleste*, five volumes, 1799–1825, his main work on the stability of the solar system. Laplace's nebular hypothesis, which was popular for almost a century, was contained in his *Exposition du système du monde*, 1796. Laplace was probably the greatest astronomer France has produced, and his contributions to mathematics and physics are not inconsiderable. His *Oeuvres complètes* appeared in fourteen volumes between 1878 and 1912.

A survey of eighteenth-century mathematics is provided by J. F. Scott[6] in a paper mentioning 110 mathematicians. Chief among these is the Swiss Bernoulli family, which produced a number of eminent

[1] Armitage, Angus. *William Herschel*, London, 1962.

[2] Hoskin, Michael A. *William Herschel and the construction of the heavens*, London, 1963.

[3] See also Lubbock, *Lady* Constance A. *The Herschel chronicle: the life-story of William Herschel and his sister Caroline Herschel. Edited by his granddaughter Constance A. Lubbock*, Cambridge, 1933; and Sidgwick, J. B. *William Herschel, explorer of the heavens*, 1953.

[4] *Mem. Roy. Astron. Soc.*, 5, 1833, p. 93.

[5] Armitage, Angus. The astronomical work of Nicholas Louis de Lacaille. *Ann. Sci.*, 21, 1956, pp. 163–191.

[6] Scott, J. F. Mathematics through the eighteenth century. *Phil. Mag.*, Commemoration issue, July, 1948, pp. 67–91.

mathematicians. Jacob Bernoulli (1654–1705) was a native of Basle, and became professor of mathematics there in 1687. He established the calculus of probabilities, and also systematized the calculus of Leibniz. Bernoulli published five tracts at Basle between 1689 and 1704,[1] and his chief work was *Ars conjectandi*, Basle, 1713.[2] His collected works were published in two volumes in 1744, and the first volume of *Die gesammelten Werke von Jakob Bernouli*, to be issued in four volumes, was announced from Basle in 1969. His brother Johann Bernoulli (1667–1748) succeeded him in the chair at Basle, and was also interested in chemistry and medicine, but his particular study was problems involving maxima and minima, and analytical trigonometry. Johann Bernoulli's *Lectiones mathematicae de methodo integralium*, which was written in 1691 and 1692 and published in 1742,[3] contains some of his lectures on the methods of the integral calculus. Another work by him was discovered in manuscript in Basle University Library,[4] and his *Opera omnia* appeared in four volumes in 1743. Johann's second son, Daniel Bernoulli (1700–82) became professor of mathematics at St. Petersburg, later returning to Basle. He wrote a book on the motion of fluids entitled *Hydrodynamica*, Strassburg, 1738, and was the author of several other works.[5] A new edition of the collected works of the mathematicians Bernoulli began publication in 1955,[6] and will contain letters, diaries, manuscripts and printed works of the three major members of the family, mentioned above, and of Nicholas Bernoulli (1687–1759), Nicholas Bernoulli (1695–1726), Johann Bernoulli (1710–90), Johann Bernoulli (1744–1807), and Jacob Bernoulli (1759–89), over 7,500 letters being included.[7]

Abraham De Moivre (1667–1754) elaborated the theory of permutations and combinations, and his *The doctrine of chances, or a method of calculating the probability of events in play*, first appeared in 1718. Editions followed in 1738 and 1756. De Moivre was also the author of *Miscellanea analytica de seriebus et quadraturis*, 1730. Some copies of this have a supplement, and a very few others contain a second supplement which was added to copies still in stock three years after publication. This second supplement also appeared in the second and third editions of his *Doctrine of chances*,[8] the second edition (1738) of which was reprinted in facsimile in 1967.

Leonhard Euler (1707–83), also of Basle, and a pupil of Johann Bernoulli, became professor of mathematics at St. Petersburg Academy, and in

[1] Contained in Ostwald's *Klassiker*, No. 171. [2] Ostwald's *Klassiker*, Nos. 107–108.
[3] German version in Ostwald's *Klassiker*, No. 194.
[4] Published in Ostwald's *Klassiker*, No. 211.
[5] See Huber, Friedrich. *Daniel Bernoulli (1700–82) als Physiologe und Statistiker*, 1958. (*Basler Veröffentlichungen zur Geschichte der Medizin und der Biologie*, Fasc. VIII.)
[6] Spies, Otto. *Der Briefwechsel von Johann Bernoulli*, [*etc.*], Bd. 1, Basle, 1955.
[7] See Truesdale, C. The new Bernoulli edition. *Isis*, 49, 1958, pp. 54–62. [8] See Archibald, R. C. A rare pamphlet of Moivre and some of his discoveries. *Isis*, 8, 1926 pp. 671–684.

1741 went to the Prussian Academy of Sciences at Berlin, returning to St. Petersburg in 1766. He published numerous papers in the transactions of the Academy, and is believed to have published about seven hundred papers and forty-five volumes. Euler was particularly interested in analysis, but contributed to mathematics in general, and to astrology, "Euler's equations" being of special significance in dynamics. His major publications include *Methodus inveniendi lineas curvas maximi minimive proprietate gaudentes sive solutio problematis isoperimetrici latissimo sensu accepti*, Lausanne, Geneva, 1774;[1] *Introductio in analysin infinitorum*, Lausanne, 1748, a reprint of which was announced in 1967; *Institutiones calculi differentialis*, 1755; *Institutiones calculi integralis*, 1768–70; and a work on algebra entitled *Vollständige Anleitung zur Algebra*, 1770.[2] His collected works began publication in Leipzig in 1911 as *Opera omnia*, and will be completed in seventy-four volumes. Euler's popular lectures on natural philosophy were published as *Letters of Euler on different subjects in physics and philosophy*, London, 1802, a reprint of which was announced in 1966.

Jean le Rond d'Alembert (1717–83) was a native of Paris, and was elected a member of the French Academy in 1741, following mathematical essays written in 1738 and 1740. Most of his mathematical writings were produced between 1743 and 1754, the first, *Traité de dynamique*, Paris, 1743, being of importance as a landmark in the development of mechanics. This enunciates the principle known by his name, that the "internal forces of inertia must be equal and opposite to the forces which produce the acceleration". This principle was applied to fluids in his *Traité de l'équilibre et du mouvement des fluides*, Paris, 1744, a theory of the motion of fluids. Among other writings this was followed by *Réflexions sur la cause générale des vents*, 1747; *Essai d'une nouvelle théorie sur la résistance des fluides*, 1752; and *Opuscules mathématiques*, 1761–80. D'Alembert's writings on physical astronomy were collected together in his *Recherches sur différents points importants du système du monde*, three volumes, Paris, 1754. He collaborated with Denis Diderot in writing the mathematical and philosophical portions of the *Encyclopédie, ou dictionnaire raisonné des sciences, des arts et des métiers*, 17 volumes, of which the first seven were printed in Paris, (1751–57), eight to seventeen, 'Neuchâtel', 1765, with subsequent plates, supplements and tables printed up to 1780. Bibliographical descriptions of this and subsequent editions are provided by John Lough,[3] who also lists the contributions of

[1] Ostwald's *Klassiker*, No. 46.

[2] See also Oldfather, W. A., Ellis, C. A., and Brown, D. M. Leonhard Euler's elastic curves. (De curvis elasticis, Additamentum I to his Methodus inveniendi lineas curvas maximi minimive proprietate gaudentes, Lausanne and Geneva, 1744.) *Isis*, 20, 1933, pp. 72–160.

[3] Lough, John. *Essays on the* Encyclopédie *of Diderot and D'Alembert*, London, [*etc.*], 1968.

d'Alembert. A good biography of D'Alembert has been published by Ronald Grimsley.[1]

A historian of mathematics, Jean Etienne Montucla (1725–99) was born at Lyons, and was educated at the Jesuit College there. He studied law at Toulouse, and then went to Paris, where he was clerk in superintendence of the royal buildings at Versailles. His first book was *Histoire de recherches sur la quadrature du cercle*, [*etc.*], Paris, 1754, of which a second edition edited by S. F. Lacroix was published in 1831. This was followed by his *Histoire des mathématiques*, two volumes, Paris, 1758, with a new edition in four volumes, 1799 to 1805.[2]

Euler was succeeded at the Berlin Academy by Joseph Louis Lagrange (1736–1813), who has been described as "the greatest mathematician of his period". Lagrange was born in Turin of French extraction, and in 1786 moved to Paris, where he lectured at the Ecole Polytechnique. He further developed the calculus of variations, and was the author of numerous papers, some being devoted to astronomy. His major publications include *Mécanique analytique*, Paris, 1788, Lagrange's greatest work, of which a German edition appeared at Göttingen, 1797; *Théorie des fonctions analytiques*, [*etc.*], Paris, 1797, of which a second edition appeared in 1813, a Portuguese translation from Lisbon, 1798, and a German translation from Berlin in the same year; *De la résolution des équations numériques de tous les degrés*, Paris, 1798; and *Leçons sur le calcul des fonctions*, 1801, 1804 and 1806. A German edition of his mathematical writings, edited by C. A. Crelle,was published as *Mathematische Werke*, Berlin, 1823–24, and his collected works appeared as *Oeuvres*, fourteen volumes, Paris, 1867–92.[3]

A native of Toulouse, Adrian Marie Legendre (1752–1833) became professor of mathematics at the Ecole Militaire, Paris, and wrote extensively on astronomy, geometry, mechanics and physics, in addition to mathematics, his main work being on the attractions of ellipsoids and on geodesy. Legendre's chief publications include *Essai sur la théorie des nombres*, Paris, 1798, with a second edition in 1808, and appendices in 1816 and 1825; a third edition entitled *Théorie des nombres* was published in Paris, two volumes, 1830; *Exercises de calcul intégral*, published in three volumes, 1811–26, the bulk of this being ultimately published in his *Traité des fonctions élliptiques*, two volumes, Paris, 1825–26, with three supplements, published 1828–32, forming a third volume; *Nouvelle théorie des parallèles*, 1803; and *Nouvelles méthodes pour la détermination des orbites des comètes*, Paris, 1805. His *Eléments de géométrie* was first published

[1] Grimsley, Ronald. *Jean D'Alembert (1717–83)*, Oxford, 1963.

[2] See Sarton, George. Montucla (1725–1799), his life and works. *Osiris*, 1, 1936, pp. 519–567; this contains a bibliography of sources and facsimiles of title-pages.

[3] See Loria, Gino. Saggio di una bibliografia Lagrangiana *Isis*, 40, 1949, pp. 112–117.

in 1794, and appeared in many editions and translations. An English translation by Sir David Brewster was published in 1823.

A survey of English mathematicians from 1714 to 1840 has been provided by Eva G. R. Taylor,[1] who provides short bibliographies of 2,282 practitioners, with narrative surveys of the main branches of the subject. George Berkeley, Bishop of Cloyne (1685–1753), who is chiefly remembered as a philosopher, published an attack on the logical basis of the method of fluxions in *The Analyst: or, a discourse addressed to an infidel mathematician*, 1734, which was answered by other writers. His criticism of Newton on space and matter has been examined by W. A. Suchting.[2]

Brook Taylor (1685–1731), the founder of the calculus of finite difference, was the author of *Methodus incrementorum directa et universa*, 1715, on fluxions, which contains "Taylor's theorem". Thomas Simpson (1710–61) applied fluxions to problems of physics and astronomy with marked success in his *New treatise on fluxions*, 1737, and was also the author of *The Nature and laws of chance*, 1740; and *Doctrine of annuities and reversions*, 1742.

Colin Maclaurin (1698–1746) was appointed to the chair of mathematics at Marischal College, Aberdeen, at the age of nineteen, and later occupied a similar position at Edinburgh. He was the author of *Geometria organica*, 1720; *Treatise on fluxions*, two volumes, Edinburgh, 1742; *Treatise on algebra*, 1748; and *An account of Sir Isaac Newton's Philosophy*, 1748 (facsimile reprint 1968), the two latter appearing posthumously.[3]

An early worker on elliptic integrals, John Landen (1719–90), practised in Peterborough as a surveyor, and presented several mathematical papers to the Royal Society. Several of these were published in the *Philosophical Transactions*, and he was the author of several books including *Mathematical lucubrations*, 1755; *Mathematical memoirs*, [*etc.*], two volumes, 1780–89; *Discourse concerning the residual analysis*, [*etc.*], 1758; and *The residual analysis*, 1764, only Book I of which was published.[4]

Descriptive geometry advanced at the hands of Gaspard Monge (1746–1818), who also contributed to the differential geometry of surfaces. Monge was a native of Beaune in Burgundy, and became professor at Mézières, and later at the Ecole Polytechnique, Paris. He first published

[1] Taylor, Eva G. R. *The mathematical practitioners of Hanoverian England, 1714–1840*, Cambridge, 1966.

[2] Suchting, W. A. Berkeley's criticism of Newton on space and motion. *Isis*, 58, 1967, pp. 186–197; see also Pastore, N. Samuel Bailey's critique of Berkeley's theory of vision. *J. Hist. behav. Sci.*, 1, 1965, pp. 321–337.

[3] See Turnbull, Herbert Western. *Bi-centenary of the death of Colin Maclaurin (1698–1746)*, [*etc.*], Aberdeen, 1951. (Aberdeen University Studies, No. 127.)

[4] See Green, H. Gwynedd, and Winter, H. J. J. John Landen, F.R.S. (1719–90)—mathematician. *Isis*, 35, 1944, pp. 6–10.

his account of descriptive geometry in a periodical,[1] and then as *Géometrie descriptive*, 1798.[2]

Several prominent figures dominated eighteenth-century physics,[3] and this period saw the initiation of important researches in electricity. Francis Hauksbee the Elder (died 1713) was the inventor of the glass electrical machine, and conducted experiments on frictional electricity. Hauksbee was a Fellow and Curator of the Royal Society. His *Physico-mechanical experiments on various subjects*, [*etc.*], London, 1709, which was printed for the author, contains a list of his early papers, more of which were included in the second, posthumous edition published in 1719. Several of Hauksbee's papers were published in the *Philosophical Transactions*. He described an improved air-pump, and his early work on electricity is significant.[4]

Stephen Gray (1666/67–1736) has been called the father of electrical communication, and his experimental work, especially that on conductors and non-conductors, is of great significance in the history of electricity. He communicated twenty-one papers to the *Philosophical Transactions*, and was elected a Fellow of the Royal Society in 1733. He was also the author of numerous letters[5] which are preserved at the Royal Greenwich Observatory, the British Museum and the Royal Society, one of his letters to Sir Hans Sloane (B.M. Sloane MSS. 4041, folio 83), dated 3rd January, 1707/8, containing significant information on his electrical experiments which is superior to that published in his article in the *Philosophical Transactions* in 1720.[6] A list of his papers published in that periodical between 1696 and 1736 appears in a study of Gray contributed by I. Bernard Cohen.[7] John Freke (1688–1756), a surgeon at St. Bartholomew's Hospital, and a Fellow of the Royal Society, was also interested in electricity, and wrote *An essay to show the cause of electricity and why some things are non-electricable*, [*etc.*], 1748 (third edition, 1752). This was republished with two other items as *A treatise on the nature and property of fire*, 1752.[8]

Benjamin Franklin (1706–90) was a native of Boston, and in his youth

[1] *Journal des Ecoles Normales*, 1795 [2] Ostwald's *Klassiker*, No 117

[3] See Dingle, Herbert Physics in the eighteenth century. *Phil Mag* , Commemoration issue, July, 1948, pp 28–46.

[4] See Home, Roderick W. Francis Hauksbee's theory of electricity. *Archive for History of Exact Sciences*, 4, 1967, pp. 203–217.

[5] See Chipman, Robert A. The manuscript letters of Stephen Gray, F.R.S. (1666/7–1736). *Isis*, 49, 1959, pp 414–433 Contains chronological list of his known letters and published papers.

[6] See Chipman, Robert A. An unpublished letter of Stephen Gray on electrical experiments. 1707–08. *Isis*, 45, 1954, pp. 33–40.

[7] Cohen, I. Bernard. Neglected sources for the life of Stephen Gray (1666 or 1667–1736). *Isis*, 45, 1954, pp. 41–50. List of his published papers, pp. 48–50.

[8] See Chalstrey, John. The life and works of John Freke (1688–1756). *St. Bart's Hosp. J.*, 61, 1957, pp. 85–89, 108–112.

THE

ANATOMY

OF

PLANTS.

WITH AN

IDEA

OF A

Philoſophical Hiſtory of Plants.

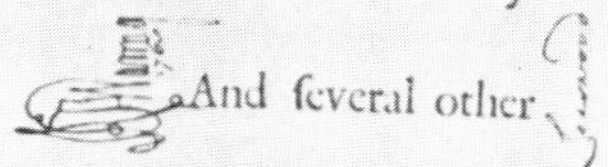

And ſeveral other

LECTURES,

Read before the

ROYAL SOCIETY.

By *NEHEMJAH GREW* M.D. Fellow of the *ROYAL SOCIETY*, and of the *COLLEGE* of *PHYSICIANS*.

Printed by *W. Rawlins*, for the Author, 1682.

PLATE 7. Title-page of Nehemiah Grew's *The anatomy of plants*, 1682.

Orang-Outang, sive Homo Sylvestris:

OR, THE

ANATOMY

OF A

PYGMIE

Compared with that of a

Monkey, an *Ape*, and a *Man*.

To which is added, A

PHILOLOGICAL ESSAY

Concerning the

Pygmies, the *Cynocephali*, the *Satyrs*, and *Sphinges* of the ANCIENTS.

Wherein it will appear that they are all either *APES* or *MONKEYS*, and not *MEN*, as formerly pretended.

By *EDWARD TYSON* M. D.

Fellow of the Colledge of Physicians, and the Royal Society: Physician to the Hospital of *Bethlem*, and Reader of Anatomy at *Chirurgeons-Hall*.

LONDON:

Printed for *Thomas Bennet* at the *Half-Moon* in St. *Paul's* Church-yard; and *Daniel Brown* at the *Black Swan* and *Bible* without *Temple-Bar* and are to be had of Mr. *Hunt* at the Repository in *Gresham-Colledge*. M DC XCIX.

PLATE 8. Title-page of Edward Tyson's *Orang-outang, sive homo sylvestris: or, the anatomy of a pygmie*, London, 1699.

was apprenticed to the printing trade. Keenly interested in science, he was constantly experimenting, and is mainly remembered for his demonstration by means of a kite of the electrical nature of lightning. After 1757 Franklin became involved in politics, and was American Ambassador in Paris, but his scientific interest never waned. Several of his papers were published in the *Philosophical Transactions*.[1] His work on electricity was published as *Experiments and observations on electricity*, [*etc.*], London, 1751, followed by *Supplemental experiments, . . . Part II*, [*etc.*], London, 1753, and *New experiments and observations, . . . Part III*, [*etc.*], London, 1754, The second English edition was published in two parts only as *New experiments and observations on electricity*, [*etc.*], 1754, and the third edition, in three parts, appeared between 1760 and 1765. On the title-page of Part 3 of this the edition is incorrectly given as the fourth. The fourth and fifth English editions appeared with the title *Experiments and observations on electricity*, [*etc.*], in 1769 and 1774 respectively. French editions were published at Paris in 1752, 1756 (two volumes), and 1773 (two volumes), the last bearing the title *Oeuvres*. An Italian edition translated by Carlo Giuseppe Campi appeared from Milan in 1774, and a German translation by Johan Carl Wilcke was published at Leipzig in 1758. The fifth English edition has been reprinted in recent years, with a critical and historical introduction by I. Bernard Cohen.[2] But Franklin was interested in many other aspects of science and in medicine, and a series of papers by Denis I. Duveen and Herbert S. Klickstein investigate his connection and collaboration with Lavoisier,[3] mentioning his acquaintance and correspondence with numerous other eminent chemists. Franklin and Lavoisier collaborated in three particular studies, explosives, the aerostatic balloon, and animal magnetism, and the above-mentioned fully documented papers contain full information concerning the relationship between these two remarkable men.[4] Yale University Press is pub-

[1] See Andrade, E. N. da C. Benjamin Franklin in London. *Journal of the Royal Society of Arts*, 104, 1956, pp. 216–234; and, Panorama of progress. Honoring Benjamin Franklin, 1706–1956. *Journal of the Franklin Institute*, 261, 1956, pp. 1–188. This contains nineteen papers covering various aspects of Franklin's activities.

[2] *Benjamin Franklin's experiments. A new edition of Franklin's Experiments and observations on electricity. Edited: with a critical and historical introduction by I. Bernard Cohen*, Cambridge, Mass., 1941.

[3] Duveen, Denis I., and Klickstein, Herbert S. Benjamin Franklin (1706–90) and Antoine Laurent Lavoisier (1743–94). Part I. Franklin and the new chemistry. [Part II. Joint investigations. Part III. Documentation.] *Ann. Sci.*, 11, 1955, pp. 103–128, 271–308; 13, 1957, pp. 30–46; see also Lopez, Claude A. Saltpetre, tin and gunpowder: addenda to the correspondence of Lavoisier & Franklin. *Ann. Sci.*, 16, 1960 [1962], pp. 83–94.

[4] See also Franklin, Benjamin. *Benjamin Franklin's autobiographical writings. Selected and edited by Carl van Doren*, 1946; Andrade, E. N. da C. The scientific work of Benjamin Franklin. *Nature*, 177, 1956, pp. 60–61; Cohen, I. Bernard. A note concerning Diderot and Franklin. *Isis*, 46, 1955, pp. 268–272; Cohen, I. Bernard. *Franklin and Newton: an inquiry into speculative Newtonian experimental science and Franklin's work in electricity as an example thereof*, Boston, 1966; and Heathcote, N. H. de V. Franklin's introduction to electricity. *Isis*, 46, 1955, pp. 29–35.

lishing an extensive collection of *The papers of Benjamin Franklin*, edited by Leonard W. Labaree and others, sponsored by the American Philosophical Society and Yale University, and expected to occupy forty volumes. Volume one appeared in 1959. The same sponsors were responsible for *The autobiography of Benjamin Franklin*, New Haven, 1964, prepared from Franklin's original manuscript now in the Henry E. Huntington Library, and the best available edition of the *Autobiography*.

One of the leading French scientists interested in electricity during this period was the Abbé Jean Antoine Nollet (1700–70). He used, and incidentally named, the Leyden jar, investigated the effect of electricity on germination and plant growth, and also constructed an electroscope. His main publications consist of *Lettres sur l'électricité dans lesquelles on examine les découverts qui ont été faites sur cette matière depuis l'année 1752 et les conséquences que l'on en peut tirer*, Paris, 1764, which first appeared in Italian from Venice in 1755; *Recherches sur les causes particulières des phénomènes électriques*, [*etc.*], Paris, 1764; *Essai sur l'électricité des corps*, Paris, 1750 and 1765; and *L'Art des expériences, ou avis aux amateurs de la physiques, sur la choix, la construction et l'usage des instruments*, [*etc.*], three volumes, Paris, 1770.

Roger Joseph Boscovich (1711–87) was born Rudjer Josip Bošković in Ragusa (Dubrovnik), and went to Rome in 1725. In recent years great interest has been taken in his scientific work, a large collection of papers having been acquired by the University of California.[1] This includes over 180 manuscripts devoted to astronomy, philosophy, geodesy, hydrography, mathematics, mechanics, optics, etc.; about 2,000 letters; and numerous diplomas, certificates, official documents and fragments of a diary kept in 1750. The major work of Boscovich was his *Philosophiae naturalis theoria redacta ad unicam legem vivium in natura existentium*, 1758, which was translated into English by J. M. Child, Chicago and London, 1922. A chronological list of the published writings of Boscovich, numbering over one hundred, together with information on his life and works are contained in a volume edited by Lancelot Law Whyte.[2]

Although described as shy, retiring and eccentric, the Honourable Henry Cavendish (1731–1810) must be considered as one of the most eminent scientists of his period. He was very wealthy, and opened his extensive library for the use of the scientific public. Cavendish contributed usefully to both chemistry and physics, and joined the Royal Society in 1760, contributing several papers to its *Philosophical Transactions*. In 1766 appeared therein his first paper, "On factitious airs",

[1] See Hahn, Roger. The Boscovitch archives at Berkeley. *Isis*, 56, 1965, pp. 70–78.

[2] Whyte, Lancelot Law, *ed. Roger Joseph Boscovich, S.J., F.R.S., 1711–1787. Studies of his life and work on the 250th anniversary of his birth*, 1961; see also Siegfried, Robert. Boscovich and Davy: some cautionary remarks. *Isis*, 58, 1967, pp. 236–238.

but the results of most of his experiments were left in manuscript. His "Experiments on air" was published in the *Philosophical Transactions* in 1784, and was separately published in a French translation as *Expériences sur l'air*, [*etc.*], London, in 1785. His important paper on electricity also first appeared in the above periodical, with the title "An attempt to explain some of the principal phenomena of electricity, by means of an elastic fluid" (1771), but his major writings have been collected together in the two following works: *The electrical researches of the Honourable Henry Cavendish, F.R.S., written between 1771 and 1781, edited from the original MSS. in the possession of the Duke of Devonshire, by J. Clerk Maxwell*, Cambridge, 1879, many of the items appearing there in print for the first time; and *The Scientific papers of the Honourable Henry Cavendish, F.R.S. Volume I. The electrical researches. Edited from the published papers, and the Cavendish Manuscripts in the possession of His Grace the Duke of Devonshire, by James Clerk Maxwell. Revised by Sir Joseph Larmor. Volume II. Chemical and dynamical. Edited . . . by Sir Edward Thorpe*, [*etc.*], two volumes, Cambridge, 1921.[1] Although he was not the author of a book, we must mention an assistant to Cavendish (1782 or 73 to 1789), who carried out important research on the freezing points of solutions. Sir Charles Blagden (1748–1820) was at one time a surgeon in the British Army, but became secretary to the Royal Society. His papers were mainly printed in the *Philosophical Transactions.*[2]

Charles Augustin Coulomb (1736–1806), born at Angoulême and educated in Paris, became a member of the Académie des Sciences in 1782. He invented the torsion balance, and investigated the theory of magnetic attraction and repulsion, proving that "the force exerted between two charged bodies depends on the charges and the distance between them". Coulomb established the inverse square law of magnetic force, and his name is perpetuated in the practical unit of quantity in measuring electricity, the "coulomb". He was the author of numerous papers published in the *Mémoires de l'Académie Royale des Sciences*, and his *Théorie des machines simples* was presented to the Académie des Sciences in 1784. A new edition was published at Paris in 1820.

The founder of electrophysiology, Luigi Galvani (1737–98), a native of Bologna, is remembered for his experiments on frogs, and his *De viribus electricitatis in motu musculari* was first published in the proceedings of the Bologna Academy in 1791, and then appeared as a separate book, Bologna, 1791, being reprinted twice in that year. In the following year two editions were printed at Modena, 1792, one having the plates

[1] See Berry, Arthur John. *Henry Cavendish: his life and scientific work*, 1960; and Wilson, George. *The life of the Honble. Henry Cavendish, including abstracts of his more important scientific papers, and a critical inquiry into the claims of all the alleged discoverers of the composition of water*, 1851, (Works of the Cavendish Society).

[2] See Getman, Frederick H. Sir Charles Blagden, F.R.S. *Osiris*, 3, 1938, pp. 69–87.

printed in red ink. There were several other printings of *De viribus electricitatis*, including a facsimile edition (Berlin, 1925); German translations by Johann Mayer (Prague, 1793), and by Arthur John von Oettingen (Leipzig, 1894); an Italian translation (Bologna, 1937); a Russian translation (Moscow, Leningrad, 1937); a French translation (Paris, 1939); and two English translations both published in the same year, one by Robert Montraville Green (Cambridge, Mass., 1953), and the other by Margaret Glover Foley (Norwalk, Conn., 1953).[1] This work together with his other writings, were published as *Opere edite ed inedite*, [*etc.*], two volumes, Bologna, 1841–42, and in 1937 appeared *Memorie ed experimenti inedite*, Bologna, which contains the first Italian translation of *De viribis*, with the Latin text in footnotes, a description of Galvani's existing manuscripts, and a contribution to the bibliography of his writings.[2] Galvani's theory of "animal electricity" created world-wide interest, but was overthrown by Alessandro Volta (1745–1827), who invented the electric pile. This was the prototype of modern batteries and accumulators. Many of Volta's papers were published in the *Philosophical Transactions*, and he substituted the term "metallic electricity" for "animal electricity". A complete edition of his scientific works was published as *Le opere di Alessandro Volta*, seven volumes, Milan, 1918–29, and his letters appeared as *Epistolario di Alessandro Volta*, five volumes, Bologna, 1949–55.[3]

A dissertation by William Cleghorn (1754–83) contains what is believed to be the first detailed exposition of the material theory of heat, later known as the caloric theory. Cleghorn was born at Granton on 30th October, 1754, graduated B.A. at Trinity College, Dublin, in 1777, followed by an M.D. from Edinburgh in 1779 for his dissertation *De igne*. Douglas McKie[4] has written a paper evaluating this work, reproducing the Latin text with an English translation, and giving the location of copies of the original. Cleghorn became joint lecturer in anatomy at Dublin in 1781, but unfortunately died on 20th April, 1783.

Although he was not the author of a book, the eponymous connection

[1] Galvani, Luigi. *Commentary on the effects of electricity on muscular motion. Translated into English by Margaret Glover Foley. With notes and a critical introduction by I. Bernard Cohen. Together with a facsimile of Galvani's De viribus electricitatis in motu musculari (1791), and a bibliography of the editions and translations of Galvani's book prepared by John Farquhar Fulton and Madeline E. Stanton*, Norwalk, Conn., 1955. This bibliography was used for the information provided above.

[2] See also Pupilli, Guilio Cesare. Luigi Galvani. *Sci. med. ital.*, 2, 1955, pp. 5–24; and Hoff, Hebbel F. Galvani and the pre-Galvanian electrophysiologists. *Ann. Sci.*, 1, 1936, pp. 157–172.

[3] See Osman, W. A. Alessandro Volta and the inflammable-air eudiometer. *Ann. Sci.*, 14, 1958, pp. 215–242; see also Ostwald's *Klassiker*, Nos. 114, 118; and Scolari, Felici. *Guide bibliografiche. Alessandro Volta per incario delle R. Commissione per l'Edizione Nationale delle Opere Voltiane. Appendice: Indici dei volumi dell' Edizione Nazionale delle Opere del Volta e relativi cenni illustrativi a cura del Francesco Massardi*, Rome, 1927.

[4] McKie, Douglas. William Cleghorn's *De igne* (1779). *Ann. Sci.*, 14, 1958, pp. 1–82.

with thermometry of Daniel Gabriel Fahrenheit (1686–1736) demands mention of his name. Fahrenheit substituted mercury for alcohol in his thermometers in 1721, and he also constructed a hypsobarometer, a thermometer that also indicated the atmospheric pressure under which the liquid was boiling. His only publications were five articles in Latin printed in the *Philosophical Transactions*.[1,2] The Swedish astronomer Anders Celsius (1701–44) constructed a mercury thermometer, which later led to the adoption of the centigrade scale.

Pierre Bouguer (1698–1758) succeeded his father as professor of hydrography at Croisic in lower Brittany, and was later professor of hydrography at Le Havre and associate geometer to the Académie des Sciences. In 1729 appeared his *Essai d'optique sur la gradation de la lumière*, defining the quantity of light lost by passing through a given extent of atmosphere. Bouguer decided that the light of the sun was three hundred thousand times more intense than that of the moon, and was responsible for some of the earliest measurements in photometry. He invented a heliometer, experimented with the pendulum, and in 1735 visited Peru to measure a degree of the meridian near the equator. This is described in his *Figure de la terre détérminée*, Paris, 1749, and Bouguer's publications on navigation include *Nouveau traité de navigation et du pilotage*, Paris, 1755. His *Éssai d'optique* was augmented and published posthumously as *Traité d'optique sur la gradation de la lumière* in 1760, and has been translated into English by W. E. Knowles Middleton.[3]

Significant publications on light were compiled by Johann Heinrich Lambert (1728–77) as the result of his researches on photometry, and entitled *Photometria, sive de mensura et gradibus luminis, colorum et umbrae*, Augsburg, 1760;[4] by Robert Smith (1689–1768) in his *Compleat system of opticks in four books*, Cambridge, 1738, a comprehensive textbook on light which was translated into French and German. Smith, who was the Master of Trinity College, Cambridge, was also the author of a valuable textbook on *Harmonics*, 1748.

Although primarily an engineer, James Watt (1736–1819) was also a scientist. He was born on 19th January, 1736, at Greenock, Renfrewshire, and spent 1756–57 in London learning the trade of instrument-maker. He later proceeded to Glasgow, and set up shop in the precincts of the College as "Mathematical Instrument Maker to the University", becoming very friendly with Joseph Black. He set up in the city with a partner as Watt and Craig, and later in Birmingham as Boulton and

[1] *Phil. Trans.*, 33, 1924–25, pp. 1–3, 78–84, 114–118, 140–141, 179–180. Republished in Ostwald's *Klassiker*, No. 57, 1894.
[2] See also Cajori, Florian. Notes on the Fahrenheit scale. *Isis*, 4, 1921–22, pp. 17–22.
[3] *Pierre Bouguer's Optical treatise on the gradation of light. Translated, with introduction and notes by W. E. Knowles Middleton*, Toronto, 1961.
[4] German translation in Ostwald's *Klassiker*, Nos. 31–33.

Watt. Watt took part in the meetings of the Lunar Society at Birmingham, and among the many honours that came to him was that of F.R.S. in 1785. His invention of a steam engine was of vital importance in industrial development, and although James Watt did not record his researches in print, his name must be written large in the history of steam power.[1] The full text of the surviving correspondence between Watt and Black, together with Watt's notebook of his experiments on heat have been published in a book edited by Eric Robinson and Douglas McKie.[2]

Dictionaries are invaluable guides to the historical development of a subject, and are themselves milestones of history. David Layton[3] has examined the treatment of science in general English dictionaries from the seventeenth century, through general scientific dictionaries to specialized dictionaries dealing with mathematics, etc. The earliest chemical dictionary was William Johnson's *Lexicon chymicum*, two volumes, London, 1652–53, which went into several editions. The second, printed in 1704, has been published in a facsimile edition, including the 1710 supplement, two volumes, New York, London, 1966. The first technical dictionary was compiled by John Harris (*c.* 1666–1719), a clergyman who was keenly interested in science, and was elected F.R.S. in 1696. Harris taught mathematics privately, and lectured on the subject in the Marine Coffee House, Birchin Lane, London. He delivered the Boyle Lectures in 1698, and was the author of books on astronomy, geography and geology, as well as collections of sermons. His *Lexicon technicum: or, an universal English dictionary of arts and sciences, explaining not only the terms of art, but the arts themselves*, London, 1704, was a folio, and was published by subscription, a supplementary volume appearing in 1710. There were several editions of the *Lexicon*, the fifth and last appearing in two folio volumes, 1736, with a supplementary volume dated 1744.[4] The treatment of science in encyclopaedias is also an interesting study, and Arthur Hughes[5] has contributed a series of papers on the subject.

Several eminent chemists made important contributions to their subject during the eighteenth century, and a brief survey of developments during that period has been provided by J. R. Partington.[6] Herman Boerhaave (1668–1738) has been described as the "greatest clinical teacher

[1] See Dickinson, H. W. *James Watt, craftsman and engineer*, Cambridge, 1936; and Hart, Ivor B. *James Watt and the history of steam power*, New York, 1949.

[2] Robinson, Eric, and McKie, Douglas, *eds. Partners in science: James Watt and Joseph Black*, 1970.

[3] Layton, David. Diction and dictionaries in the diffusion of scientific knowledge: an aspect of the history of the popularization of science in Great Britain. *Brit. J. Hist. Sci.*, 2, 1965, pp. 221–234.

[4] See McKie, Douglas. John Harris and his *Lexicon technicum. Endeavour*, 4, 1945, pp. 53–57.

[5] Hughes, Arthur. Science in English encyclopaedias. *Ann. Sci.*, 7, 1951, pp. 340–370; 8, 1952, pp. 323–367; 9, 1953, pp. 233–264; 11, 1955, pp. 74–92.

[6] Partington, J. R. Chemistry through the eighteenth century. *Phil. Mag.*, Commemoration

of the eighteenth century", but he also contributed in no uncertain manner to many branches of science. He was appointed to the chair of botany and medicine at Leyden University in 1709, and his contributions to medicine are mentioned elsewhere,[1] but he also held the chair of chemistry there (1718), his treatise on chemistry being one of his most important writings.[2] Students collected his lecture notes and published a book under his name containing numerous errors, which decided Boerhaave to issue an authorized edition. The spurious publication appeared as *Institutiones et experimenta chemiae*, two volumes, Paris, 1724, and did not bear the name of the printer. It has been suggested that this was actually published at Leyden. Another edition in two volumes bears the imprint Venice, 1726, and an English translation by Peter Shaw[3] (1694–1760) and E. Chambers, with the title *A new method of chemistry*, was printed in London, 1727. A second edition of the latter was published in 1741 (two volumes) and 1753, but meanwhile Boerhaave had produced a genuine edition with the title *Elementa chemiae*, two volumes, Leyden, 1732,[4] the author putting his signature on the reverse of the title-page of every copy of this edition. The book became very popular, and there were many other editions in English, French, German and Latin, including Latin ones from Leyden, 1732 (the second in that year), in two volumes; London, 1732, also in two volumes; Paris, 1732; and Paris, 1733. French editions, some of which were abridged, appeared from Paris, 1741; The Hague, 1748; Leyden, 1752; and Paris, 1754, in six volumes. German translations were printed in Halberstadt, 1732–34 (in nine volumes), Berlin, 1762 and 1782. English abridged translations were printed in London in 1733 and 1737. The *Elements of chemistry* appeared from London, in 1732, with a translation by Timothy Dallowe published in two volumes, in London, 1735.[5,6] Boerhaave was the author

issue, July, 1948, pp. 47–66; and Trengove, Leonard. Chemistry at the Royal Society of London in the eighteenth century. I[–III]. *Ann. Sci.*, 19, 1963 [1965], pp. 183–237; 20, 1964 [1965], pp. 1–57; 21, 1965 [1966], pp. 81–130.

[1] Thornton, pp. 122–123.

[2] See Lindeboom, Gerrit Arie. *Bibliographia Boerhaaviana: list of publications written by or provided by H. Boerhaave, or based on his works and teaching*, Leiden, 1959. (Analecta Boerhaaviana, Edidit G. A. Lindeboom, Vol. 1); also Hertzberger, Menno. *Short-title catalogue of books written and edited by Hermann Boerhaave*, Amsterdam, 1927.

[3] See Gibbs, F. W. Peter Shaw and the revival of chemistry. *Ann. Sci.*, 7, 1951, pp. 211–237; and Oldham, Graham. Peter Shaw. *J. chem. Educ.*, 37, 1960, pp. 417–419.

[4] Certain copies of this edition lack pp. 423–424, and others have pp. 187–188 in their place. See Neville, Roy G. Boerhaave's "Elementa chemiae", 1732. (Bibliographical note.) *Book Collector*, 8, 1959, pp. 428–429; and Poynter, F. N. L. "Elementa chemiae", 1732. (Bibliographical note.) *Book Collector*, 9, 1960, p. 64.

[5] See Davis, Tenney L. The vicissitudes of Boerhaave's Textbook of chemistry. *Isis*, 10, 1928, pp. 33–46.

[6] See also Gibbs, F. W. Boerhaave's chemical writings. *Ambix*, 6, 1958, pp. 117–135; Jevons, F. R. Boerhaave's biochemistry. *Med. Hist.*, 6, 1962, pp. 343–362; and Kerker, Milton. Herman Boerhaave and the development of pneumatic chemistry. *Isis*, 46, 1955, pp. 36–49.

of *Some experiments concerning mercury. . . . Translated from the Latin, communicated by the author to the Royal Society*, London, 1734, and it has been suggested that Samuel Johnson might have been responsible for this translation. Latin editions were published at Utrecht in 1736 and Venice in 1737. It has also been suggested that Johnson might have been responsible for the translation of a section of Boerhaave's *Elementa chemiae*, the first eight sheets of this having been published as *Elements of chemistry. Being the annual lectures of Hermann Boerhaave. . . . Englished by a Gentleman of the University of Oxford*, number 1, London, price one shilling and sixpence. This is not dated, but was published on 10th January, 1732, and copies have only been traced in the Medical Society of London and in the British Museum. Johnson was keenly interested in chemistry and in Boerhaave, and F. W. Gibbs[1] advances evidence in favour of Johnson having translated both these items, which are very scarce. This suggestion has been questioned by Arthur Sherbo,[2] but the evidence is inconclusive. G. A. Lindeboom[3] has published a monumental biography of Boerhaave, and also edited his correspondence with scientists throughout Europe.

William Lewis (1708–81) has been neglected by historians, but he was the author of an important work on the application of physics and chemistry to technological problems. Born in Richmond, he taught chemistry, and was a pharmacologist. Elected F.R.S. in 1745, he edited George Wilson's *Complete course of chemistry*, 1746, and with Alexander Chisholm issued the *New dispensatory*, 1753, which went into several editions. Lewis's most important publication was *Philosophical commerce of arts*, 1763, which is mainly a description of the laboratory of an experimental chemist, and a comprehensive account of physics and chemistry. F. W. Gibbs[4] provides a list of his writings, and many of his notebooks are in the British Museum, Cardiff Central Library and the Wedgwood Museum, Barlaston.[5]

The first major western study of Russian science is that written by Alexander Vucinich[6], but one of the most important Russian-born scientists was undoubtedly Mikhail Vasilevich Lomonosov (1711–65). He achieved prominence as a chemist and physicist, and a collection of his

[1] Gibbs, F. W. Dr. Johnson's first published work? *Ambix*, 8, 1960, pp. 24–34.
[2] Sherbo, Arthur. The translation of Boerhaave's *Elementa chemiae*. *Ambix*, 13, 1966, pp. 108–117.
[3] Lindeboom, Gerrit Arie. *Herman Boerhaave. the man and his work*, 1968; *Boerhaave's Correspondence. Edited by G. A. Lindeboom*, 2 vols., Leyden, 1962–64; see also Lindeboom, Gerrit Arie, *ed. Iconographia Boerhaavii*, Leyden, 1963; Lindeboom, Gerrit Arie. Herman Boerhaave (1668–1738), teacher of all Europe. *J. Amer. Med. Assoc.*, 206, 1968, pp. 2297–2301; and Underwood, E. Ashworth. Boerhaave after three hundred years. *Brit. med. J.*, 4, 1968, pp. 820–825.
[4] Gibbs, F. W. A notebook of William Lewis and Alexander Chisholm. *Ann. Sci.*, 8, 1952, pp. 202–220.
[5] See also Swin, Nathan. William Lewis (1708–1781) as a chemist. *Chymia*, 8, 1962, pp. 63–88.
[6] Vucinich, Alexander. *Science in Russian culture.* (*Volume I.*) *A history to 1860*, London, 1965.

scientific papers was published in 1904 by B. N. Menshutkin as *Lomonosov as a physical-chemist*. Much of this was translated by Max Speter and published as No. 178 of Ostwald's *Klassiker* in 1910, and B. N. Menshutkin[1] has provided a further study of Lomonosov.[2] In addition to numerous published papers, Lomonosov translated several textbooks into Russian, and was the author of *Oratio de meteoris vi electrica ortis*, St. Petersburg, 1754, and *De origine lucis*, St. Petersburg, 1757. He was the first to observe the atmosphere of Venus (1761), and to record the freezing of mercury. Lomonosov has been acclaimed as the founder of Russian science.[3]

Pierre Joseph Macquer (1718–84) graduated in medicine, but confined his researches to chemistry, and became professor of chemistry at the Jardin du Roi. His researches on chemistry and chemical technology were numerous, as were his writings, but of the latter the most significant was his dictionary, the first truly scientific work of this nature. Entitled *Dictionnaire de chymie*, [*etc.*], this was first published anonymously in two volumes, Paris, 1766, of which there appears to be two distinct impressions, but without textual differences.[4] Later editions include printings from Paris, 1777 (three volumes); 1778 (in two formats, four volumes in 8vo and two volumes in 4to); and Neuchâtel, 1789, in five volumes edited by H. Struve. This was translated into Danish and published at Copenhagen in 1771 and 1772; into German, Leipzig, 1768 (three parts); Leipzig, 1781–83 (six volumes), with a second edition of this translation by J. G. Leonhardi, in seven volumes, Leipzig, 1788–91; into English by James Keir (1735–1820), two volumes, London, 1771, with a second edition in three volumes, London, 1777; into Italian, Pavia, 1783–84, and Venice, 1784–85. Macquer was also the author of the following works devoted to chemistry: *Elémens de chymie théorique*, Paris, 1749, 1753 and 1756, with a German translation from Leipzig, 1752; and *Elémens de chymie pratique*, two volumes, Paris, 1751 and also 1756. These two works were translated into English by Andrew Reid as *Elements of the theory and practice of chemistry*, London, two volumes, 1758, with a second edition in 1764, a third in three volumes, Edinburgh, 1768, probably a fourth from London in 1775, and a fifth published in Edinburgh in 1777; into German, 1768; into Dutch, 1773, 1775; into Russian, 1774–75. Macquer was joint author with Antoine Baumé of *Plan d'un cours de chymie expéri-*

[1] Menshutkin, B. N. *Russia's Lomonosov*, Princeton, N.J., 1952.

[2] See also Pomper, Philip. Lomonosov and the discovery of the law of the conservation of matter in chemical transformations. *Ambix*, 10, 1962, pp. 119–127; and Tchenpkal, V. L. M. V. Lomonosov and his astronomical works, in commemoration of the 250th anniversary of his birth. *The Observatory*, 81, 1961, pp. 183–189.

[3] See also Leicester, Henry M. Boyle. Lomonosov, Lavoisier, and the corpuscular theory of matter. *Isis*, 58, 1967, pp. 240–244.

[4] See Neville, Roy G. Macquer's "Dictionnaire de chymie", 1766. *Book Collector*, 15, 1966, pp. 484–485; and Neville, Roy G. Macquer and the first chemical dictionary, 1766. A bicentennial tribute. *J. chem. Educ.*, 43, 1966, pp. 486–490.

mentale et raisonnée avec un discours historique sur la chymie, Paris, 1757. Many of Macquer's manuscripts are in the Bibliothèque Nationale, Paris, and among them is a sixteen-page essay in his handwriting entitled "Sur la pierre philosophale", undated and apparently unpublished, but probably written between 1753 and 1766. This has been described, but not reproduced, by W. A. Smeaton.[1] Another unpublished manuscript in the possession of Roy G. Neville[2] has been translated into English by the owner.[3]

Born at Bordeaux with an Irish father and Scottish mother, Joseph Black (1728–99) was educated at Glasgow, where he assisted William Cullen (1712–90) in his chemical researches. Unfortunately, Cullen published nothing on this subject but he was keenly interested in chemistry, and inspired others to pursue this interest.[4] Black went on to Edinburgh to complete his medical education. In 1754 he was awarded an M.D., and his inaugural dissertation was entitled *De humore acido a cibis orto et magnesia alba*. It was also communicated to the Philosophical Society of Edinburgh, and published in *Edinburgh Physical and Literary Essays*, 1756,[5] but this version differs from the original dissertation in that the study of digestion is omitted, and a section on the constitution of chalk and lime is added. An enlarged version of this thesis was later published as *Experiments upon magnesia alba, quick-lime and other alkaline substances*, [*etc.*], Edinburgh, 1777 and 1782. Black was also the author of *Directions for preparing aerated medicinal waters, by means of the improved glass machines at Leith Glass-Works*, Edinburgh, 1787; and *Lectures on the elements of chemistry, delivered in the University of Edinburgh, . . . now published from his manuscripts by John Robison*, two volumes, Edinburgh, London, 1803.[6] The *Lectures* were prepared after Black's death from his manuscript notes to forestall the publication of students' notes of his lectures. An American edition in three volumes was published at Philadelphia, volume one of which is dated 1807, and volumes two and three, 1806. It has been suggested that the printing of the lectures was begun in 1806,

[1] Smeaton, W. A. Macquer on the composition of metals and the artificial production of gold and silver. *Chymia*, 11, 1966, pp. 81–88.

[2] Neville, Roy G. "Observations sur la mine de fer de Bagory" (1767), an unpublished manuscript by P.-J. Macquer. *Chymia*, 8, 1962, pp. 89–96.

[3] See also McKie, Douglas. Macquer, the first lexicographer of chemistry. *Endeavour*, 16, 1957, pp. 133–136; and McKie, Douglas. The descent of Pierre Joseph Macquer. *Nature*, 163, 1948, p. 628.

[4] See Wightman, William P. D. William Cullen and the teaching of chemistry. [–2]. *Ann. Sci.*, 11, 1955, pp. 154–165; 12, 1956, pp. 192–205. Also, Dobbin, Leonard. A Cullen chemical manuscript of 1753. *Ann. Sci.*, 1, 1936, pp. 138–156; and Crosland, M. P. The use of diagrams as chemical 'equations' in the lecture notes of William Cullen and Joseph Black. *Ann. Sci.*, 15, 1959, pp. 75–90.

[5] Also Alembic Club Reprints, No. 1.

[6] See Neave, E. W. J. Joseph Black's Lectures on the elements of chemistry. *Isis*, 25, 1936, pp. 372–390.

but that a cancel title-page bearing the date 1807 was inserted in volume one to make it appear up to date upon publication.[1] Unfortunately the text was already obsolete, and the sale of copies relied mainly upon the reputation of Black in America. Douglas McKie[2] has published a series of papers on manuscripts of Black's lectures, notably those of George Cayley, which are in the possession of York Medical Society, and Thomas Cochrane,[3] preserved in the Andersonian Library of the University of Strathclyde. McKie compares the various manuscript versions with the printed texts, and has also published some of Black's correspondence, including letters between him and James Watt, covering 1768 to 1799.[4] Black was the pioneer of quantitative analysis and pneumatic chemistry, investigated latent heat, and was the originator of the theory of "specific heat".[5]

Joseph Priestley (1733–1804) was a Unitarian pastor, a teacher, and then companion and librarian to the Lord Shelburne, Marquis of Lansdowne, in which occupation he had ample leisure for scientific research. In 1791, owing to his sympathy with the French Revolution, his home was raided by a mob,[6] and in 1794 he followed his three sons to America, dying there in 1804. Priestley first became interested in electricity, and made early experiments with carbon dioxide. He investigated the action of plants on the air, isolated nitric oxide and hydrogen chloride (1772), collecting the latter over mercury. He discovered ammonia (1773), sulphur dioxide (1774), and in the following year in a letter to Sir John Pringle announced his isolation of what we now call oxygen.[7] Priestley had been elected F.R.S. for his electrical experiments,[8] and in 1773 he

[1] See Miles, Wyndham D. Joseph Black, Benjamin Rush and the teaching of chemistry at the University of Pennsylvania. *Lib. Chron.*, 22, 1956, pp. 9–18.

[2] McKie, Douglas. On some MS. copies of Black's chemical lectures. II[–VI]. *Ann. Sci.*, 15, 1959, pp. 65–73; 16, 1960, pp. 1–9; 18, 1962 [1964], pp. 87–97; 21, 1965, pp. 209–255; 23, 1967, pp. 1–33.

[3] Cochrane, Thomas. *Notes from Dr. Black's lectures on chemistry 1867–8. Edited with an introduction by Douglas McKie*, Wilmslow, 1966; see also McKie, Douglas. On Thom. Cochrane's MS. notes of Black's chemical lectures, 1767–68. *Ann. Sci.*, 1, 1936, pp. 101–110.

[4] McKie, Douglas, and Kennedy, David. On some letters of Joseph Black and others. *Ann. Sci.*, 16, 1960 [1962], pp. 129–170; and Robinson, Eric, and McKie, Douglas. *eds. Partners in science. Letters of James Watt and Joseph Black, Edited with introductions and notes by Eric Robinson and Douglas McKie*, 1970.

[5] See Ramsay, Sir William. *The life and letters of Joseph Black*, [*etc.*], 1918; Guerlac, Henry. Joseph Black and fixed air: a bicentenary retrospective, with some new or little known material. *Isis*, 48, 1957, pp. 124–151, 433–456; and Joseph Black. *Endeavour*, 14, 1955, pp. 115–116.

[6] See Maddison, R. E. W., and Maddison, Francis R. Joseph Priestley and the Birmingham Riots. *Notes Rec Roy. Soc. Lond.*, 12, 1956, pp. 98–113; and McKie, Douglas. Priestley's laboratory and library and other of his effects. *Notes Rec. Roy. Soc. Lond.*, 12, 1956, pp. 114–136.

[7] *Phil. Trans.*, 1775, p. 387. See also Partington, J. R. The discovery of oxygen. *J. chem. Educ.*, 39, 1962, pp. 123–125.

[8] See Schofield, Robert E. Electrical researches of Joseph Priestley. *Arch. int. Hist. Sci.*, 16, 1963, pp. 277–286.

received the Copley Medal.[1] Priestley was interested in anatomy, botany and astronomy, and his writings include several theological items. His published works have been the subject of a study by John F. Fulton and Charlotte H. Peters,[2] and an appraisal of Priestley's writings by John F. Fulton,[3] from which sources the following details of Priestley's scientific writings are derived. A bibliography of his writings has been compiled by Ronald E. Crook,[4] which includes works relating to Priestley and a select alphabetical list of his articles in periodicals. Fuller bibliographical descriptions and a more convenient arrangement of entries would enhance the value of a revised edition. The Birmingham Library was founded in 1779 and was reorganized by Priestley, who went to Birmingham in 1780. He was a member of the Lunar Society, several members of whom were among the proprietors of the Library.[5] Priestley's first scientific work was *The history and present state of electricity, with original experiments*, London, 1767, which went into a second edition in 1769, in which year *Additions* to the first edition were published, *Additions* to the second edition appearing in 1772, in which year a third edition was published in two volumes. The fourth edition, London, 1775, was followed by a fifth in 1794. There were a French translation, three volumes, Paris, 1771; a German translation, Berlin, 1772; and a Dutch translation, Amsterdam, 1772–73. This was followed by *A familiar introduction to the study of electricity*, two volumes, London, 1768, which went into further editions in 1769, 1777 and 1786, with a translation into German published in 1778. His *Directions for impregnating water with fixed air; in order to communicate to it the peculiar spirit and virtues of Pyrmont water, and other mineral waters of a similar nature* was also published in London in 1772. The most important record of Priestley's experiments is *Experiments and observations on different kinds of air*, three volumes, 1774–77, the first two volumes of which went into a second edition from 1775–76, with a third edition of volume one in 1781. A three-volume abridgement by the author was published at Birmingham in 1790, and a three-volume French edition was published in 1777. Following *Experiments and observations* Priestley published a further series of three volumes, which he treated as a continuation of the series, with the title *Experiments and observations relating to various branches of natural philosophy; with a continuation of the observations on air*, London, one volume, 1779, and Birmingham, two volumes, 1781 and 1786. His

[1] See McKie, Douglas. Joseph Priestley and the Copley Medal. *Ambix*, 9, 1961, pp. 1–22.

[2] Fulton, John Farquhar, and Peters, Charlotte H. An introduction to a bibliography of the educational and scientific works of Joseph Priestley. *Pap. Bib. Soc. Amer.*, 30, 1936, pp. 150–167. Issued as a short-title list in 1937.

[3] Fulton, John Farquhar. The scientific writings of Joseph Priestley (1733–1804). *Atti del XIV Congresso Internazionale di Storia della Medicina*, 1954, pp. 1–8.

[4] Crook, Ronald E. *A bibliography of Joseph Priestley, 1733–1804*, 1966.

[5] See Parish, Charles. *History of the Birmingham Library*, [*etc.*], 1966. Chapter III. Priestley's association with the Library, pp. 12–31.

other writings include several papers in the *Philosophical Transactions*, and also *Experiments and observations relating to the analysis of atmospherical air*, 1796; *Considerations on the doctrine of phlogiston*, 1796; and *The doctrine of phlogiston established*, 1800, with a second edition in 1803. A very rare pre-print of one of Priestley's important papers has been described by Henry Guerlac.[1] This was *Observations on different kinds of air. Printed for the Philosophical Transactions, 1773, Vol. LXII*, London, 1772, which was presented to the Royal Society in March, 1772, but which did not actually appear in the *Philosophical Transactions* until 1773. The very rare pre-print is identical in substance with the later paper, but does not bear the same pagination, and appeared after 29th October, 1772. Fulton[2] also records an important paper by Priestley, apparently not previously noticed by his biographers, which was read to the American Philosophical Society in Philadelphia three months before his death, and was published as "Observations and experiments relating to equivocal or spontaneous generation" (*Trans. Amer. Phil. Soc.*, 6, 1804, pp. 119–129). This was an attack on Erasmus Darwin's attempt to revive the theory of spontaneous generation, Priestley's experiments having decisively disproved the possibility of this.[3] Priestley greatly advanced our knowledge of gases, but some historians, with the advantage of hindsight, consider that his researches were hampered by his rigid adherence to the phlogiston theory, which caused him to call oxygen "dephlogisticated air". Ira V. Brown[4] has edited selections from Priestley's writings, Robert E. Schofield[5] has published an edition of his autobiography, among other papers, and F. W. Gibbs[6] was the author of the first complete biography of Priestley, although this was not intended to be definitive. Nevertheless, it is a scholarly production and is the best available study of Priestley's life and works.[7]

Richard Kirwan (1733–1812) was a native of Co. Galway, and was educated at Poitiers and Paris. On his return to Ireland he first studied

[1] Guerlac, Henry. Joseph Priestley's first papers on gases and their reception in France. *J. Hist. Med.*, 12, 1957, pp. 1–12.

[2] Fulton, John Farquhar. *Op. cit.*

[3] See Abrahams, Harold J. Priestley answers the proponents of abiogenesis. *Ambix*, 12, 1964, pp. 44–71.

[4] *Joseph Priestley. Selections from his writings. Edited by Ira V. Brown*, University Park, Pa., 1962.

[5] Schofield, Robert E., *ed. A scientific autobiography of Joseph Priestley (1773–1804); selected scientific correspondence*, Cambridge, Mass., 1966; see also the following papers by Robert E. Schofield: The scientific background of Joseph Priestley. *Ann. Sci.*, 13, 1957, pp. 148–163; Joseph Priestley, the theory of oxidation and the nature of matter. *J. Hist. Ideas*, 25, 1964, pp. 285–294; Joseph Priestley, natural philosopher. *Ambix*, 14, 1967, pp. 1–15.

[6] Gibbs, F. W. *Joseph Priestley. Adventurer in science and champion of truth*, 1965.

[7] See also *Memoirs of Dr. Joseph Priestley, to the year 1795, written by himself. With a continuation . . . by his son Joseph Priestley*, 2 vols., 1806–07; Holt, Anne. *A life of Joseph Priestley. . . . With an introduction by Francis W. Hirst*, 1931; McKie, Douglas. Joseph Priestley (1733–1804), chemist. *Sci. Progr.*, 28, 1933, pp. 17–35; Walker, W. Cameron. The beginnings of the scientific career of Joseph Priestley. *Isis*, 21, 1934, pp. 81–97; Hartog, Sir Philip J. The newer views of Priestley and Lavoisier. *Ann. Sci.*, 5, 1941–47, pp. 1–56; *J. chem. Educ.*, Vol. 4, February, 1927, is a special Priestley Number.

law, but came to London, where he pursued science from 1777 to 1787. He contributed several papers to the Royal Society, and was elected F.R.S. in 1780, being awarded the Copley Medal two years later. Kirwan was a friend of Cavendish, Priestley and Sir Joseph Banks. In 1787 he returned to Ireland, and published much of his work in the *Transactions* of the Royal Irish Academy. He was an authority on the mineralogy of Ireland, his *Elements of mineralogy*, 1784, becoming a standard textbook that was translated into French. German and Russian, and Kirwan was also the author of *Geological essays*, 1799, and *Logick*, 1807. He introduced several new chemical terms, and investigated combining proportions of acids and bases. His two important chemical books were *An essay on phlogiston and the constitution of acids*, London, 1787, which was translated into French by Madame Lavoisier, with notes by her husband attacking the phlogiston theory; and *An essay on the analysis of mineral waters*, London, 1799.[1]

Torbern Olof Bergman (1735–84) was intimately connected with Uppsala University, where he contributed to astronomy, mathematics, physics, natural history and, from 1767, to chemistry. All his scientific papers were collected together as *Opuscula physica et chemica*, [*etc.*], six volumes, Stockholm, Uppsala and Aboe (Volumes 4–6, Leipzig), 1779–90. The first three volumes were edited by Bergman, the remaining three being edited after the author's death by E. B. G. Hebensteit. Several of his writings were translated, including *Outlines of mineralogy*, [*etc.*], translated by William Withering, Birmingham and London, 1783; *Physical and chemical essays*, translated by Edmund Cullen, two volumes, London, 1784, and in three volumes, London, 1788–91; *An essay on the usefulness of chemistry*, [*etc.*], a rare translation from the German by Jeremy Bentham and Franz Xaver Schwediauer (1748–1824), London, 1783; and *A dissertation on elective attractions*, translated by Thomas Beddoes, London, 1785, of which a facsimile reprint was announced in 1968. This was a translation of Bergman's "Disquisitio de attractionibus electivis", first published in Latin in a periodical in 1775, a French extract of which likewise appeared in a journal in 1778. A revised and enlarged Latin version was published by Bergman in 1783 in volume three of his *Opuscula physica et chemica*. François Joseph Bonjour (1754–1811) translated the "Disquisitio" into French, consulting Beddoes' English translation, and publishing the work as *Traité des affinités chymiques, ou attractions électives; traduit du Latin, sur la dernière édition de Bergman*, [*etc.*], Paris, 1788.[2]

[1] See Reilly, J., and O'Flynn, N. Richard Kirwan, an Irish chemist of the eighteenth century. *Isis*, 13, 1929–30, pp. 298–319.

[2] See Linder, Bertel, and Smeaton, W. A. Schwediaur, Bentham and Beddoes: translators of Bergman and Scheele. *Ann. Sci.*, 24, 1968 [1969], pp. 259–273; Smeaton, W. A. F.-J. Bonjour and his translation of Bergman's "Disquisitio de attractionibus electivis". *Ambix*, 7, 1959, pp. 47–50; see also Moström, Birgitta. *Tobern Bergman. A bibliography of his works*, Stockholm, 1957.

Although not himself a major figure, Louis Bernard Guyton de Morveau (1737–1816) took part in the reformation of chemical nomenclature, and championed Lavoisier's new chemistry.[1] He was joint editor of the *Annales de Chimie*, and was mainly responsible for the adoption of chlorine hydrochloric acid as a disinfectant. He was the author of *Discours sur l'état actuel de la jurisprudence*, [*etc.*], Paris, 1768, which was published anonymously; *Digressions académiques*, Dijon, 1772; *Elémens de chymie théorique et pratique*, [*etc.*], three volumes, Dijon, 1777;[2] *Mémoire sur les denominations chimiques*, 1782; and *Traité des moyens de désinfecter l'air de prévenir la contagion, et d'en arrêter les progrès*, [*etc.*], Paris, 1801, which describes fumigation with chlorine.

Carl Friedrich Wenzel (1740–93) went to Leipzig in 1766 to study mathematics, physics and chemistry, which latter subject later occupied most of his time. He experimented on the composition of salts and the rate of solution of metals in acids. His writings on chemistry and metallurgy include *Einleitung zur höheren Chymie*, Leipzig, 1774 [1773]; *Chymische Untersuchung des Flufsspaths*, Dresden, 1783; and *Lehre von der Verwandschaft der Körper*, Dresden, 1777 [1776], 1779 and 1782, which contains the results of much of his experimental work.[3]

Described as "an experimental genius", Carl Wilhelm Scheele (1742–86) was born at Stralsund, Pomerania, then Swedish. He was apprenticed to an apothecary, and later opened a pharmacy. He conducted experiments on every branch of chemistry, proving that air consisted of oxygen and nitrogen, discovering hydrogen persulphide, chlorine, manganese, prussic acid, baryta, various inorganic acids and numerous other substances. Delay in publication prevented his recognition as discoverer of oxygen before Priestley. Scheele was a founder of organic chemistry. Many of his papers appeared in the *Transactions* of the Swedish Academy and other periodicals. His *Chemische Abhandlung von der Luft und dem Feuer*, Uppsala and Leipzig, 1777, second edition, Leipzig, 1782, was translated into French as *Traité chymique de l'air et du feu* by Philippe-Frédéric, Baron de Dietrich, Paris, 1781, and into English by J. R. Forster as *Treatise on air and fire*, London, 1780. His *Opuscula chemica et*

[1] See Duveen, Denis L., and Klickstein, Herbert S. A letter from Guyton de Morveau to Macquart [*sic* Macquer] relating to Lavoisier's attack against the *phlogiston* theory (1778); with an account of de Morveau's conversion to Lavoisier's doctrines in 1787. *Osiris*, 12 1956, pp. 342–367; Duveen, Denis I., and Klickstein, Herbert S. A case of mistaken identity: Macquer and not Macquart. *Isis*, 49, 1958, pp. 73–74; and Smeaton, W. A. The contributions of P.-J. Macquer, T. O. Bergmann and L. B. Guyton de Morveau to the reform of chemical nomenclature. *Ann. Sci.*, 10, 1954, pp. 87–106.

[2] See Smeaton, W. A. Guyton de Morveau's course of chemistry in the Dijon Academy. *Ambix*, 9, 1961, pp. 53–69; Smeaton, W. A. Guyton de Morveau and chemical affinity. *Ambix*, 11, 1963, pp. 55–64; and Smeaton, W. A. Louis Bernard Guyton de Morveau, F.R.S. (1737–1816) and his relations with British scientists. *Notes Rec. Roy. Soc. Lond.*, 22, 1967, pp. 113–130.

[3] See Ferguson, Vol. 2, pp. 544–545.

physica, two volumes, Leipzig, 1788–89, was translated into German, Berlin, 1793, with a facsimile reprint in two volumes, 1891; into English, London, 1786, of which a facsimile reprint was issued in 1966; and into French, two volumes, Paris, 1785–88.[1]

"Only a moment to cut off his head, and perhaps a hundred years before we see such another." So wrote Lagrange of Antoine Laurent Lavoisier (1743–94), a native of Paris, who was educated at the Collège Mazarin. He first studied law, and then became interested in mathematics, astronomy, mineralogy, geology,[2] anatomy, meteorology and chemistry. In 1766 Lavoisier was awarded a gold medal for an essay on the lighting of streets, and in 1768 became a member of the Académie des Sciences at the age of twenty-five. In 1775 he was appointed inspector of gunpowder, and ten years later became a member of the Committee of Agriculture in Paris. His chemical experiments included investigations into calcination of metals; he studied the changes occurring during breathing and combustion, made experiments on the composition of water, and on respiration and transpiration. In 1771 he married Marie Anne Pierrette Paulze, who greatly aided him in his work. On 8th May, 1794, Lavoisier and many others were executed during the French Revolution.[3] Denis I. Duveen and Herbert S. Klickstein have contributed numerous important papers on Lavoisier and his work, but their major project from our view-point is their monumental compilation *A bibliography of the works of Antoine Laurent Lavoisier, 1743–1794*,[4] a guide to the bibliographical maze of his writings, which contains many facsimiles of title-pages. Section A is devoted to his contributions to periodicals arranged in chronological order; section B deals with Lavoisier's four major works; section C lists his minor separate works and contributions to separate works; section D includes the miscellaneous works containing material written by Lavoisier; section E deals with reports by Lavoisier to the Académie on other works; and section F lists the collected works of Lavoisier, or works containing more than one of his writings. Denis I. Duveen[5] has compiled a supplement to this *Bibliography* incorporating corrigenda, 225 new entries, an index of selected topics in the two

[1] See Ferguson, Vol. 2, pp. 330–332.

[2] See Rappaport, Rhoda. Lavoisier's geologic activities, 1763–1792. *Isis*, 58, 1967, pp. 375–384; and Gough, J. B. Lavoisier's early career in science: an examination of some new evidence. *Brit. J. Hist. Sci.*, 4, 1968, pp. 52–57.

[3] See Duveen, Denis I. Antoine Laurent Lavoisier and the French Revolution. (i–iv.] *J. chem. Educ.*, 31, 1954, pp. 60–65; 34, 1957, pp. 502–503; 35, 1958, pp. 233–234, 470–471; and Duveen, Denis I., and Klickstein, Herbert S. Some new facts relating to the arrest of Antoine Laurent Lavoisier. *Isis*, 49, 1958, pp. 347–348.

[4] Duveen, Denis I., and Klickstein, Herbert S. *A bibliography of the works of Antoine Laurent Lavoisier, 1743–1794*, [*etc.*], 1954; see also review of this by W. A. Smeaton in *The Library*, 5th ser., 11, 1956, pp. 130–133.

[5] Duveen, Denis I. *Supplement to a bibliography of the works of Antoine Laurent Lavoisier 1743–1794*, 1964.

volumes, and a short-title check-list of the more important books and articles dealing with Lavoisier published between 1787 and 1963. Over one hundred of Lavoisier's papers were published in scientific periodicals, but many of his early memoirs were held back even after acceptance for publication, but were later included in his collected works. Most of his important articles were published in *Mémoires de l'Académie Royale de Sciences*, the Abbé Rozier's *Observations sur la Physique, sur l'Histoire Naturelle, et sur les Arts*, and in *Annales de Chimie*. The first volume was the only one published of Lavoisier's *Opuscules physiques et chimiques*, Paris, 1774, there being two issues of both this edition and of the second published in Paris in 1801. An English translation by Thomas Henry was printed as *Essays physical and chemical*, London, in 1776, and a German translation by Christian Ehrenfried Weigel was published at Greifswald in 1783. Of the *Méthode de nomenclature chimique, proposée par MM. de Morveau, Lavoisier, Bertholet* [*sic*] *et de Fourcroy*, [*etc.*], Paris, 1787, there were two issues and a further separate printing. This went into several editions, including an English translation by James St. John, London, 1788; a partial German translation, Berlin, 1791; and a complete German translation by Karl Freyherrn von Meidinger, Vienna, 1793. Lavoisier's *Traiteé élémentaire de chemie*, [*etc.*], was published in two volumes, Paris, 1789, but there were some author's pre-publication copies, in one volume, which have been described as the first issue of the first edition. Douglas McKie[1] mentions the discovery of seven of these copies, but rightly concludes that these were not regularly published, and cannot therefore be termed first issues of this edition.[2] The second edition also appeared in 1789, but in three volumes, the third being a new issue of *Méthode de nomenclature chimique*. A pirated "second edition" was published in Paris in 1793, of which there were three printings. There were many further editions and translations of the *Traité*, including the third edition, two volumes, Paris, 1801; others in three volumes from Avignon, 1804, and Paris, 1805; and an abridgement was published in Paris in 1937. An English translation by Robert Kerr, Edinburgh, 1790, was printed five times in three editions, the last in 1802, and there was a German translation by Sigismund Friedrich Hermbstädt in two volumes, Berlin and Stettin, 1792. Two Italian translations were printed, one translated from the 1789 French edition by Count Vincenzo Dandolo (1758–1819), four volumes, Venice, 1791, with a second edition in 1792; the other was

[1] See McKie, Douglas. On some pre-publication copies of Lavoisier's *Traité* (1789). *Ambix*, 9, 1961, pp. 37–46.

[2] See Duveen, Denis I., Antoine Lavoisier's Traité de chimie: a bibliographical note. *Isis* 41, 1950, pp. 168–171; and Duveen, Denis I., and Klickstein, Herbert S. A bibliographical study of the introduction of Lavoisier's *Traité éléméntaire de chimie* into Great Britain and America. *Ann. Sci.*, 10, 1954, pp. 321–338.

prepared concurrently by Luigi Parisi and Gaetano La Pira, Naples, 1791–92.[1]

Lavoisier had planned to publish an edition of his memoirs in eight volumes, but left only volume two, most of volume one, and some of volume four as partially completed proofs. Mme. Lavoisier published these as *Mémoires de chimie*, two volumes, in 1805, but some copies may have been distributed in 1803. These copies were all presented, and the work was not sold until after the death of Mme. Lavoisier in 1836. These volumes have no title-pages, only half-titles, with no indication of printer or publisher, no index, table of contents or illustrations, but contain numerous printing errors. The first collected edition of Lavoisier's writings appeared as *Physikalisch-chemische Schriften*, five volumes, Greifswald, 1783–94, the first three volumes having been edited by Christian Ehrenfried Weigel, and the last two volumes by H. F. Link. The French Government published Lavoisier's collected work as *Oeuvres de Lavoisier*, six volumes, 1862–93, these having been reprinted in the U.S.A. in 1965. A seventh volume, containing Lavoisier's correspondence, edited by René Fric, began publication in 1955 with fascicule one, covering 1763 to 1769; fascicule two covers 1770 to 1775, Paris, 1957; and fascicule three covers 1776 to 1783. It has been suggested that fuller annotations would improve later fascicules.

There have been numerous studies of Lavoisier in recent years dealing with various aspects of his career, such as his membership of, and activities within various organizations and societies,[2] of his publications in various periodicals,[3] many of his contributions having appeared anonymously, and being therefore elusive. Lavoisier initiated the birth of modern

[1] See Debus, A. G. A forgotten chapter in the introduction of the new chemistry in Italy. *Ambix*, 11, 1963, pp. 153–157.

[2] See Smeaton, W. A. Lavoisier's membership of the Assembly of Representatives of the Commune of Paris, 1789–90. *Ann. Sci.*, 13, 1957, pp. 235–248; Smeaton, W. A. Lavoisier's membership of the Société Royale d'Agriculture and the Comité d'Agriculture. *Ann. Sci.*, 12, 1956, pp. 267–277; Smeaton, W. A. Lavoisier's membership of the Société Royale de Médicine, *Ann. Sci.*, 12, 1956, pp. 228–244; and Smeaton, W. A. The early years of the Lycée and the Lycée des Arts. A chapter in the lives of A. L. Lavoisier and A. F. de Fourcroy, 1. The Lycée of the Rue de Valois. [2. The Lycée des Arts.] *Ann. Sci.*, 11, 1956, pp. 257–267, 309–319. Scheler, L. *Lavoisier et Révolution française*, 2 vols., Paris, 1961. Vol. 1. reproduces Lavoisier's contributions to the journal of the Lycée des Arts, and Vol. 2, the previously unpublished journal kept by Auguste-Denis Fougeroux de Bondaroy of the Académie Royale des Sciences from 12th July to 29th August, 1789.

[3] See Duveen, Denis I., and Hahn, Roger. A note on some Lavoisiereana in the "Journal de Paris". *Isis*, 51, 1960, pp. 64–66; Smeaton, W. A. "L'Avant-Coureur." The journal in which some of Lavoisier's earliest research was reported. *Ann. Sci.*, 13, 1957, pp. 219–234; and Smeaton, W. A. Two unrecorded publications of the "Régie des Poudres et Saltpêtres" probably written by Lavoisier. *Ann. Sci.*, 12, 1956, pp. 157–159. See also Duveen, Denis I., and Klickstein, Herbert S. The American edition of Lavoisier's *L'Art de fabriquer le salin et la potasse. William & Mary Quart.*, 13, 1956, pp. 493–498; and Storrs, F. C. Lavoisier's technical reports: 1768–94. Part I [–II]. *Ann. Sci.*, 22, 1966 [1967], pp. 251–275; 24, 1968, pp. 179–197.

chemistry,[1] and his premature tragic death cut short a career that fully justified the quotation by Lagrange mentioned above, which was not an exaggeration.[2]

The first professor of chemistry at Berlin University, Martin Heinrich Klaproth (1743–1817), improved mineral analysis and made several important discoveries, including zirconium oxide and uranium in 1789, titanium compounds in 1795 and mellitic acid in honeystone in 1799. His writings were published as *Beiträge zur chemischen Kenntniss der Mineralkörper*, 1795–1810, translated into English as *Analytical essays towards promoting the chemical knowledge of mineral substances*, two volumes, London, 1801; and *Chemisches Wörterbuch*, 1807–10, translated into French with notes by E. J. B. Bouillon Lagrange and H. A. Vogel, as *Dictionnaire de chimie*, [*etc.*], four volumes, Paris, 1810–11.

Claude Louis Berthollet (1748–1822) was born in Savoy of French parents, and studied medicine. In 1780 he was elected a member of the Académie des Sciences, and was technical director of a dye-works. Berthollet introduced the use of chlorine for bleaching, and in 1792 became Director of the Mint, and two years later was Professor of Chemistry at the École Polytechnique, Paris. He went to Egypt with Napoleon and in 1799 read a paper in Cairo on the action of mass. This was published as *Recherches sur les lois de l'affinité*, Paris, An IX [1801], and further enlarged as *Essai de statique chimique*, two volumes, Paris, 1803. Berthollet also investigated the chemical nature of ammonia, discovered hydrocyanic acid and sulphuretted hydrogen, and conducted research on chlorine, hypochlorites and chlorates.[3]

Another French chemist, Antoine François de Fourcroy (1755–1809), was a native of Paris, and after graduating in medicine, he turned to chemistry, occupying the chairs devoted to this subject at the Jardin des Plantes, the École Polytechnique and the Faculty of Medicine, in addition

[1] See Guerlac, Henry. *Lavoisier—the crucial year. The background and origin of his first experiment on combustion in 1772*, Ithaca, New York, 1961.

[2] See also Smeaton, W. A. New light on Lavoisier: the research of the last ten years. *History of Science*, 2, 1963, pp. 51–69; McKie, Douglas. *Antoine Lavoisier, the father of modern chemistry*, [*etc.*], 1935; McKie, Douglas. *Antoine Lavoisier, scientist, economist, social reformer*, 1952; McKie, Douglas. Antoine Laurent Lavoisier, F.R.S., 1743–94. *Notes Rec. Roy. Soc. Lond.*, 7, 1949–50, pp. 1–41; Meldrum, A. N. Lavoisier's early work in science, 1763–71. *Isis*, 19, 1933, pp. 330–363; 20, 1933, pp. 396–425; Berthelot, M. *La révolution chimique. Lavoisier. Ouvrage suivi de notices et extraits des registres inédits de laboratoire de Lavoisier*, Paris, 1890, (Bibliothèque scientifique internationale, LXIX); Hartley, Sir Harold. Antoine Laurent Lavoisier, 26th August, 1743–48, May, 1794. *Proc. Roy. Soc.*, A, 189, 1947, pp. 427–456; Fric, René. Contribution a l'étude de l'evolution des idées de Lavoisier sur la nature de l'air et sur la calcination des métaux. *Arch. int. Hist. Sci.*, 12, 1959, pp. 137–168; Underwood, E. Ashworth. Lavoisier and the history of respiration. *Proc. Roy. Soc. Med.*, 37, 1943–44, pp. 247–262; and Guerlac, Henry. Lavoisier and his biographers. *Isis*, 45, 1954, pp. 51–62.

[3] See Holmes, Frederic L. From elective affinities to chemical equilibria: Berthollet's law of mass action. *Chymia*, 8, 1962, pp. 105–145; and Kapoor, S. C. Berthollet, Proust, and proportions. *Chymia*, 10, 1965, pp. 53–110.

to several public appointments. Fourcroy was interested in animal chemistry, and also popularized Lavoisier's antiphlogiston theory, but he contributed to the entire field of chemistry. A brilliant lecturer and writer, he was the author of the following books: *Leçons élémentaires d'histoire naturelle et de chimie*, [*etc.*], two volumes, Paris, 1782, a second edition in four volumes appearing in 1786, while the fifth edition in five volumes was published in 1794. An Italian translation was published in Milan, 1785. *Mémoires et observations de chimie*, [*etc.*], Paris, 1784; *Principes de chimie*, two volumes in one, Paris, 1787, which was reprinted in the same year, and reissued in 1788 with a new title-page and preface. *Analyse chimique de l'eau sulfureuse d'Enghien*, Paris, 1788, written in collaboration with Delaporte; *Philosophie chimique*, [*etc.*], Paris, 1792, reprinted Leyden, 1794, with further editions in 1795, 1797 and 1806, and several translations, including a German one from Leipzig in 1796;[1] *Tableaux synoptiques de chimie*, [*etc.*], Paris, (1800), which is scarce, and a second edition appeared in 1805, with translations into English and German; and Froucroy's chief work, *Système des connoissances chimiques*, [*etc.*], the first edition of which was published as six quarto volumes in five, Paris, (1801–02), with an octavo edition in eleven volumes dated 1801. An English edition of this, also in eleven volumes, appeared from London in 1804. G. Kersaint[2] has published an extensive study of the life and work of Fourcroy, and an analysis of his contributions to science is contained in a biography by W. A. Smeaton.[3] This also provides a good bibliography of Fourcroy's writings, including books, translations, contributions to periodicals, speeches and reports, and manuscripts.[4]

Jean Antoine Chaptal, Comte de Chanteloup (1756–1832), became professor of chemistry at Montpellier, proposed the name 'nitrogen', and is mainly remembered for his efforts to promote agriculture. He established a factory for the manufacture of mineral acids, and in 1800 became Minister of the Interior. Chaptal's writings include *Tableau analytique du cours de chymie*, Montpellier, 1783; *Élémens de chimie*, three volumes, Montpellier, 1790, which was reissued in the same year with a Paris imprint, and was also translated into English; *L'Art de faire les eaux-de-vie*, [*etc.*], Paris, 1805; and *Chimie appliquée aux arts*, four volumes, Paris, 1807.

[1] See Smeaton, W. A. Some unrecorded editions of Fourcroy's *Philosophie chimique*. *Ann. Sci.*, 23, 1967 [1968], pp. 295–298.

[2] Kersaint, G. Antoine François Fourcroy (1755–1809). Sa vie et son oeuvre. *Mémoires du Muséum national d'Histoire naturelle*, N.S., série D, Sciences physico-chimiques, Tome II, 1966; an abbreviated version appeared in *Rev. d'Hist. Pharmacie*, 15, 1967, pp. 589–596.

[3] Smeaton, W. A. *Fourcroy, chemist and revolutionary, 1755–1809*, Cambridge, *for the author*, 1962.

[4] See also Smeaton, W. A. Antoine-François de Fourcroy (1755–1809). *Nature*, 175, 1955, pp. 1017–1018; and Smeaton, W. A. Fourcroy and the anti-phlogistic theory. *Endeavour*, 18, 1959, pp. 70–74.

A native of Silesia, Jeremias Benjamin Richter (1762–1807) was convinced that chemistry is a branch of mathematics, and although he conducted several important pieces of research his discoveries were not credited to their originator. He used the name 'stoichiometry' for the combining proportions of substances, and his chief publications are *Anfängsgründe der Stochyometrie oder Messkunst chemischer Elemente*, three volumes, Breslau and Hirschberg 1792–94; and *Ueber die neueren Gegenstände der Chemie*, 1791–1802.

John Dalton (1766–1844) was born at Eaglesfield, near Cockermouth, Cumberland, of Quaker parents, and he was largely self-taught. At the age of fifteen he went to Kendal as assistant in a boarding school, later becoming principal there. In 1793 Dalton went to New College, Manchester, as tutor in mathematics and natural philosophy. Later he privately taught mathematics and chemistry in Manchester and elsewhere. Dalton became a member of the Literary and Philosophical Society of Manchester in 1794, and many of his papers were contributed to the Society's *Memoirs*, and to other periodicals. Dalton was colour-blind,[1] and published the first scientific paper on the subject,[2] but he is mainly remembered for his atomic theory of matter.[3],[4] He invented a system of chemical symbols, discovered the law of expansion of gases by heat, and the law of partial pressures. Dalton's first published work was *Meteorological observations and essays*, London, 1793, which is scarce, there being a second issue in the same year and a second edition in 1834. His chief independent publication, however, was *A new system of chemical philosophy*, Parts 1 and 2, Manchester and London, 1808–10, and Volume 2, Part 1, dated 1827, which is rare. The work was reprinted in facsimile in 1953. Dalton distributed a prospectus of the first part of this from his house in Manchester in May, 1808, and the prospectus is rare. A copy was sent to John Bostock (1773–1846), a Liverpool physician, who ordered four copies of the book. Two letters from Dalton to Bostock, together with a copy of the

[1] See Duveen, Denis I., and Klickstein, Herbert S. John Dalton's autopsy. *J. Hist. Med.*, 9, 1954, pp. 360–362; and Snyder, C. Our ophthalmic heritage. The eyes of John Dalton. *Arch. Ophthal.*, 67, 1962, pp. 671–673.

[2] Extraordinary facts relating to the vision of colours: with observations. Read 31st October 1794. *Mem. Lit. Phil. Soc. Manchester*, 5, Part 1, 1798, pp. 28–45.

[3] See Rose, J. John Dalton the atomist. *Nature*, 211, 1966, pp. 1015–1016; Thackray, Arnold W. Documents relating the origins of Dalton's chemical atomic theory. *Manchester Lit. Phil. Soc. Mem. Proc.*, 108, 1965–66, pp. 21–42; Thackray, Arnold W. The emergence of Dalton's chemical atomic theory. *Brit. J. Hist. Sci.*, 3, 1966, pp. 1–23; Thackray, Arnold W. The origin of Dalton's chemical atomic theory: Daltonian doubts resolved. *Isis*, 57, 1966, pp. 35–55; Thackray, Arnold W. John Dalton—accidental atomist. *Discovery*, 27, ix, 1966, pp. 28–33; Nash, Leonard K. The origin of Dalton's chemical atomic theory. *Isis*, 47, 1956, pp. 101–116; and Partington, J. R. Seventeenth-century chemistry, the phlogiston theory and Dalton's theory. *Nature*, 174, 1954, pp. 291–293.

[4] An examination of the possible influence of J. B. Richter on Dalton is provided by Guerlac, Henry. Some Daltonian doubts. *Isis*, 52, 1961, pp. 544–554; see also Siegfried, Robert. Further Daltonian doubts. *Isis*, 54, 1963, pp. 480–481.

prospectus, were among the Bostock family papers acquired by Roy G. Neville.[1] Dalton was elected F.R.S. in 1822. The bicentenary of John Dalton's birth was celebrated by the publication of numerous papers relating to his work,[2] including the proceedings of a conference, edited by D. S. L. Cardwell.[3] A collection of Dalton's letters has been published by Arnold W. Thackray,[4] and Frank Greenaway[5] has written a bicentenary celebration biography for the general reader. A. L. Smyth[6] has published a bibliography of works by and about Dalton, which includes separately published works, translations and edited texts; articles in journals and encyclopaedias, manuscripts, papers read and lectures; and works about him, letters, portraits and sculptures. Much Dalton material was destroyed in 1940 when the home of the Manchester Literary and Philosophical Society was burnt, but the remainder has been salvaged and added to.[7]

Although the atomic theory of William Higgins (1763–1825) preceded that of Dalton, it is inferior. Higgins was probably born in 1763 in Callooney, County Sligo, and as a youth worked in London with his uncle, Bryan Higgins, a physician and chemist. From 1786 to 1788 he was a lecturer-assistant at Oxford, in 1792, he was chemist at Apothecaries' Hall, Dublin, and in 1795 became lecturer to the Royal Dublin Society, where he eventually held a professorship. From 1795 to 1822 Higgins was also part-time chemist to the Irish Linen Board. While still at Oxford he wrote *A comparative view of the phlogistic and antiphlogistic theories*, [*etc.*], London, 1789, which went into a second edition in 1791, and in this he anticipated the law of multiple proportions and valency bonds. It has been suggested that he did nothing of distinction during

[1] Neville, Roy G. Unrecorded Daltoniana: two letters to John Bostock and a prospectus to the "New system", 1808. *Ambix*, 8, 1960, pp. 42–45.

[2] See Thackray, Arnold W. "In praise of famous men"—the John Dalton bicentenary celebrations, 1966. *Notes Rec. Roy. Soc. Lond.*, 22, 1967, pp. 40–44; Greenaway, Frank. John Dalton. *Endeavour*, 25, 1966, pp. 73–78; Greenaway, Frank. John Dalton as a historical figure. *Nature*, 211, 1966, pp. 1013–1014; Greenaway, Frank. John Dalton in London. *Proc. Roy. Instn.*, 41, 1966, pp. 162–177.

[3] Cardwell, D. S. L., *ed. John Dalton & the progress of science. Papers presented to a conference of historians of science held in Manchester, September 19–24, 1966, to mark the bicentenary of Dalton's birth*, Manchester, New York, 1968.

[4] Thackray, Arnold W. Fragmentary remains of John Dalton. Pt. I. Letters. *Ann. Sci.*, 22, 1966, pp. 145–174.

[5] Greenaway, Frank. *John Dalton and the atom*, 1966; and Greenaway, Frank. The biographical approach to John Dalton. *Mem. Proc. Manchester Lit. Phil. Soc.*, 100, 1958–59, pp. 1–98.

[6] Smyth, A. L. *John Dalton, 1766–1844: a bibliography of works by and about him*, Manchester, 1966.

[7] See also Henry, William Charles. *Memoirs of the life and scientific researches of John Dalton*, 1854, (Works of the Cavendish Society); Brockbank, E. M. *John Dalton: some unpublished letters of personal and scientific interest, with additional information about his colour-vision & atomic theories*, Manchester, 1944. (Publications of the University of Manchester, No. CCLXXXVII); and Brockbank, E. M. *John Dalton as experimental physiologist*, Manchester, 1929.

the remainder of his life. In 1814, and subsequently, Higgins claimed to have anticipated Dalton as originator of the atomic theory as then current, and his *Experiments and observations on the atomic theory and electrical phenomena*, London, 1814, contains an attack on Dalton. Higgins was also the author of *An essay on the theory and practice of bleaching*, [*etc.*], London, printed for the author in 1799. His life and writings have been studied by T. S. Wheeler and J. R. Partington[1] in a book which includes a reprint of the second edition of Higgins' first book, and also of his *Experiments and observations*.

Geology and related subjects came into greater prominence in the eighteenth century, the foundations having been laid in the previous century by descriptions of rocks and fossils, and by the study of the results of earthquakes. Extracts from some of the writings on these subjects are given in a series of papers by J. Challinor.[2] There was much literature on geological subjects published during the eighteenth century, but critical examination of it is sparse, and much of it needs investigation by experts.[3] The Italian, Anton Lazzaro Moro (1687–1740), explained his theory of the creation of the earth in *De' Crostacei e degli altri marini corpi che si truovano su monti*, Venice, 1740, as did Benoit de Maillet (1656–1738), a French diplomat, in his *Telliamed ou entretiens d'un philosophe indien avec une missionaire français*, Amsterdam, 1748, published some years after his death. Johann Jacob Scheuchzer (1672–1733), a Swiss, wrote extensively on fossils, and was the author of *Piscium querelae et vindiciae*, 1708, and *Herbarium diluvianum*, 1709, his most important work, which describes a large number of fossil plants. He also wrote a natural history of Switzerland in Latin and German with the title *Helvetiae historia naturalis*, three parts, Zürich, 1716–18.

Jean Etienne Guettard (1715–86) was born at Étampes, near Paris, and studied botany, geology and medicine. He made a geological survey of France, and with Monnet published an *Atlas et description minéralogiques de la France*, 1780. Guettard recognized the Auvergne as a volcanic area, but there have been attempts to deny him this priority.[4] Guettard's work

[1] Wheeler, T. S., and Partington, J. R. *The life and work of William Higgins, chemist (1763–1825). Including reprints of "A comparative view of the phlogistic and antiphlogistic theories", and "Observations on the atomic theory and electrical phenomena", by William Higgins*, Oxford, [*etc.*], 1960; see also Partington, J. R. William Higgins, chemist (1763–1825). *Nature*, 176 1955, pp. 8–9; and Spronsen, J. W. van. William Higgins. *Arch. internat. d'Hist. des Sci.* 19, 1966, pp. 263–270.

[2] Challinor, J. The early progress of British geology. I. From Leland to Woodward, 1538–1728. [II. From Strachey to Michell, 1719–1788. III. From Hutton to Playfair, 1788–1802.] *Ann. Sci.*, 9, 1953, pp. 124–153; 10, 1954, pp. 1–19; 10, 1954, pp. 107–148; see also Schneer, Cecil. The rise of historical geology in the seventeenth century. *Isis*, 45, 1954, pp. 256–268.

[3] See Rappaport, Rhoda. Problems and sources in the history of geology, 1749–1810. *History of Science*, 3, 1964, pp. 60–77; and Eyles, V. A. The history of geology: suggestions for further research. *History of Science*, 5, 1966, pp. 77–86.

[4] See de Beer, Gavin. The volcanoes of Auvergne. *Ann. Sci.*, 18, 1962 [1964], pp. 49–61.

was continued by Nicholas Desmarest (1725–1815), who published many articles in journals, and a work on *Physical geography*, four volumes, 1794–1811. Horace Benedict de Saussure (1740–99), of Geneva, was an early experimental geologist, and he attempted to prove Desmarest's theory that basalt was the result of the fusion of granite by volcanic fire. De Saussure introduced the term 'geology' in 1779, and was the author of *Voyages dans les Alpes*, three volumes, 1779–96.

Several British geologists of this period are of special significance because of their writings. John Woodward (1665–1728) was elected Professor of Physic at Gresham College in 1692, and in the next year was elected F.R.S. A founder of experimental plant physiology, he recognized the existence of strata in the earth's crust, and was the author of the following publications: *An essay towards a natural history of the earth*, 1695, reprinted 1702 and 1723; *Brief instructions for making observations in all parts of the world and sending over natural things*, [etc.], 1696; *Naturalis historia telluris illustrata et aucta*, three parts, 1714; *An attempt towards a natural history of the fossils of England*, [etc.], two volumes, 1728–29; *Fossils of all kinds digested into a method*, 1728.[1] John Strachey (1671–1743) studied the types of geological formation in the south-west of England, and was the author of *Observations on the different strata of earths and minerals*, [etc.], 1727. William Smith (1769–1839) was a geologist and engineer. A founder of stratigraphical geology, he projected a map of English strata, and his collection of fossils was acquired by the British Museum. His publications include *Strata identified by organised fossils*, 1816; *A stratigraphical system of organised fossils*, 1817; *A geological map on a reduced scale*, 1819; and *New geological atlas of England and Wales*, 1819–1824.

James Hutton (1726–97), of Edinburgh, qualified in medicine, but immediately turned to farming, and became interested in geology. He published several books on physics and metaphysics, and a lengthy manuscript on agriculture from his pen is now in possession of the Geological Society of Edinburgh. His *Theory of the earth* was first read before the Royal Society of Edinburgh in 1785 and was published as a small pamphlet entitled *Abstract of a dissertation read in the Royal Society of Edinburgh, upon the seventh of March, and fourth of April, M,DCC,LXXXV, concerning the system of the earth, its duration, and stability*. This was anonymous, but Sir Edward Bailey has suggested that it was compiled by John Playfair. It was distributed by Hutton to his friends in 1785, and the *Theory* was printed in the first volume of the *Transactions*, 1788. A revised version of the *Theory* was published in two volumes, 1795.[2] Chapters four to ten of a third volume were found in manuscript in the Geological

[1] Eyles, V. A. John Woodward, F.R.S. (1665–1728), physician and geologist. *Nature*, 206, 1965, pp. 868–870.
[2] Reprinted in *Historiae Naturalis Classica*, T.1, 1959.

Society of London Library, and were edited by Sir Archibald Geikie, and published by the Society in 1899. Sir Edward Bailey[1] has examined and elucidated each chapter of the three volumes of the *Theory of the earth*, and provided a biography of Hutton. Hutton's other main works were *Dissertations on the philosophy of light, heat and fire*, Edinburgh, 1794; *An investigation of the principles of knowledge, and of the progress of reason from sense to science and philosophy*, three volumes, 1794; but his *Theory of the earth* is of major importance, and laid the foundations of modern geology.[2] A collection of his letters in the British Museum dating between 1768 to 1774 indicate Hutton's knowledge of geology and of fossils at that period, and some of these have been published.[3] Hutton's theories were advanced, and more adequately explained by John Playfair (1748–1819) in his *Illustrations of the Huttonian theory*, 1802. Another friend of Hutton, Sir James Hall (1761–1832), has been called the father of experimental geology. He was President of the Royal Society of Edinburgh, and accounts of his experiments are contained in the Society's *Transactions*. V. A. Eyles[4] has contributed a study of Hall which illustrates the debt owed to chemistry by geology, and mentions many letters from the Dunglass muniments in the General Register House, Edinburgh.

The biological sciences were very adequately represented in eighteenth-century literature, and the names of Buffon, Haller, Gilbert White, John Hunter, Spallanzani, Erasmus Darwin and Jenner suggest that it had more than its share of eminent figures. René Antoine Ferchault de Réaumur (1683-1757), a native of La Rochelle, studied mathematics and physics in Paris, but also contributed to other branches of science. He was the author of one of the greatest entomological publications, his *Mémoires pour servir a l'histoire des insectes*, six volumes, Paris, 1734–42, containing 250 plates and about five thousand illustrations. The work was incomplete at his death, but volume seven was published in Paris in 1928. This is

[1] Bailey, Sir Edward Battersby. *James Hutton—the founder of modern geology*, Amsterdam, [*etc.*], 1967.

[2] See Eyles, V. A. James Hutton, 1726–97. Commemoration of the 150th anniversary of his death. Note on the original publication of Hutton's *Theory of the earth*, and on the subsequent forms in which it was issued. *Proc. Roy. Soc. Edin.*, 63B, 1950, pp. 377–386; Eyles, V. A. Bibliography and the history of science. *J. Soc. Bib. Nat. Hist.*, 3, 1955, pp. 63–71; Eyles, V. A. A bibliographical note on the earliest printed version of James Hutton's *Theory of the earth*, its form and date of publication. *J. Soc. Bib. Nat. Hist.*, 3, 1955, pp. 105–108; and Gerstner, Patsy A. James Hutton's theory of the earth and his theory of matter. *Isis*, 59, 1968, pp. 26–31.

[3] See Eyles, V. A., and Joan M. Some geological correspondence of James Hutton. *Ann. Sci.*, 7, 1951, pp. 316–339; also Davies, Gordon L. The eighteenth-century derivation dilemma and the Huttonian theory of the earth. *Ann. Sci.*, 22, 1966, pp. 129–138.

[4] Eyles, V. A. The evolution of a chemist. Sir James Hall, Bt., F.R.S., P.R.S.E., of Dunglass, Haddingtonshire, (1761–1832), and his relations with Joseph Black, Antoine Lavoisier, and other scientists of the period. *Ann. Sci.*, 19, 1963 [1965], pp. 153–182; see also Chaldecott, J. A. Scientific activities in Paris in 1791. Evidence from the diaries of Sir James Hall for 1791 and other contemporary records. *Ann. Sci.*, 24, 1968, pp. 21–52.

concerned with ants, and was previously published in an English translation by William Morton Wheeler.[1] Réaumur was also the author of an important book on artificial incubation, *Art de faire éclore et d'élever en toutes saisons des oiseaux domestiques de toutes espèces*, two volumes, Paris, 1749, which went into a second edition in 1751. This was translated into English by Abraham Trembley as *The art of hatching and bringing up domestic fowls, by means of artificial heat*, London, 1750, and a German translation by J. C. Thenn, in two volumes, was published at Augsburg between 1767 and 1768. Particulars of the life and writings of Réaumur are contained in a series of publications by Jean Torlais,[2] and Jean Chaïa[3] has edited some of Réaumur's correspondence.[4]

George Edwards (1694–1773) was a native of Stratford, Essex, and after apprenticeship to a tradesman travelled in Holland, Norway and France. On his return to England, Edwards studied natural history, drawing and colouring illustrations of animals. In December, 1733, Sir Hans Sloane, then President of the College of Physicians, appointed Edwards as Library Keeper. He contributed numerous papers to the *Philosophical Transactions*, gained the Copley Medal in 1750, and was elected F.R.S. The writings of Edwards are bibliographically complicated, some having been issued in irregularly published parts, but his main works are as follow: *A natural history of birds*, [*etc.*], three parts, London, 1743–50, which was translated into French; *A natural history of uncommon birds, and of some other rare and undescribed animals, quadrupeds, reptiles, fishes, insects*, [*etc.*], four parts, 1743–51; and *Gleanings of natural history*, [*etc.*], three parts, London, 1758–64.[5] Edwards was supplied with specimens of birds by Mark Catesby (1682/3–1749), a self-taught artist who drew his birds against realistic

[1] *The natural history of ants, from an unpublished manuscript in the Archives of the Academy of Sciences in Paris by René Antoine Ferchault de Réaumur*, [*etc.*], New York, London, 1926.

[2] Torlais, Jean. *Réaumur d'après des documents inédits*, Paris, 1961; Torlais, Jean. Chronologie de la vie et des oeuvres de René-Antoine Ferchault de Réaumur. *Rev. Hist. Sci., Paris*, 11, 1958, pp. 1–12; Torlais, Jean. Réaumur philosophe. *Rev. Hist. Sci., Paris*, 11, 1958, pp. 13–33; Torlais, Jean. Réaumur, de Geer et la création de l'entomologie. *Progr. Méd. Paris*, 86, 1958, pp. 143–145; Torlais, Jean. Une rivalité célèbre: Réaumur et Buffon. *Presse méd., Paris*, 66, 1958, pp. 1057–1058.

[3] Chaïa, Jean. Sur une correspondence inédite de Réaumur avec Artur, premier médecin du Roy à Cayenne. *Episteme*, 2, 1968, pp. 36–57, 121–138.

[4] See also Birembaut, Arthur. Fontenelle, Réaumur et le gaz naturel. *Rev. Hist. Sci., Paris*, 11, 1958, pp. 82–48; Birembaut, Arthur. Réaumur et l'élaboration des produits ferreux. *Rev. Hist. Sci., Paris*, 11, 1958, pp. 138–166; Davy de Virville, Ad. Réaumur botaniste. *Rev. Hist. Sci., Paris*, 11, 1958, pp. 134–137; McKie, Douglas. René Antoine Ferchault de Réaumur (1683–1757), the Pliny of the eighteenth century. *Sci. Progr.*, 45, 1957, pp. 619–627; Rostand, Jean. Réaumur et la resistance des insectes à la congélation. *Rev. Hist. Sci., Paris*, 15, 1962, pp. 71–72; Rostand, Jean. Réaumur, embryologiste et geneticien. *Rev. Hist. Sci., Paris*, 11, 1958, pp. 34–50; Taton, René. Réaumur mathématicien. *Rev. Hist. Sci., Paris*, 11, 1958, pp. 130–133; and Théoridès, Jean. Réaumur (1683–1757) et les insectes sociaux. *Janus*, 48, 1959, pp. 62–76.

[5] See McAtee, W. L. The North American birds of George Edwards. *J. Soc. Bib. Nat. Hist.*, 2, vi, 1950, pp. 194–205.

botanical backgrounds. Born at Castle Hedingham, Essex, Catesby was in Virginia from 1712–19, and returned to South Carolina in 1722 for almost three years, before proceeding to the Bahamas. Upon returning to England he began his monumental work, for which he engraved his own plates. This was published in eleven parts as *The natural history of Carolina, Florida, and the Bahama Islands*, two volumes, London, 1731–43 [1729–47], of which later editions edited by George Edwards were published in 1754 and 1771. Catesby was elected F.R.S. in 1732, and was the friend of Ray and many other eminent naturalists.[1]

A miniature painter of Nuremberg, August Johann Roesel von Rosenhof (1705–59) studied the habits and structure of insects. His writings on the subject appeared in a popular monthly entitled *Insekten-Belustigungen*, and later in volume form between 1746 and 1761. Rosenhof studied the development of frogs, and his *Die natürliche Historie der Frosche*, Nuremberg, 1758, represents an outstanding publication on natural history. All the illustrations in his writings were engraved by Rosenhof himself, and he also studied *hydra*, or fresh-water polyps. Henry Baker (1698–1774), a bookseller, poet and naturalist, was also interested in *hydra*, and published *A natural history of the polype*, London, 1743. Baker also wrote *The microscope made easy*, [*etc.*], 1743, which went into several editions, including a French translation published in 1754. Abraham Trembley (1710–84) corresponded with Baker, and observed natural regeneration in *hydra* by budding; he was the author of *Mémoires pour servir a l'histoire d'un genre de polypes d'eau douce à bras en forme de cornes*, Leyden, 1744, and although Baker published his book in the previous year, most of his material was based upon Trembley's researches.[2] Tom Gibson[3] has recognized Abraham Trembley as a pioneer in the history of homografts, for he united one polyp inside another, reunited halves of a divided polyp and of different polyps of the same species.

Pieter Lyonet (1706–89) was a Dutch lawyer and amateur naturalist who studied the minute anatomy of the caterpillar. His *Traité anatomique de la chenille qui rouge le bois de saule*, published in 1760, is a monument of outstanding research. Lyonet owned fine collections of paintings and sea shells, and made his own drawings and engravings of his insect studies. He had an extensive correspondence with other scientists, and was a member of many learned societies, including the Royal Society. The bulk of his work was published in 1832, many years after his death, as *Recherches*

[1] See Frick, George Frederick, and Stearns, Raymond Phineas. *Mark Catesby: the Colonial Audubon*, Urbana, 1961; Frick, George Frederick. Mark Catesby: the discovery of a naturalist. *Pap. Bib. Soc. Amer.*, 54, 1960, pp. 163–175; and Allen, Elsa G. New light on Mark Catesby. *Auk*, 54, 1937, pp. 349–363.

[2] See Baker, John R. *Abraham Trembley of Geneva: scientist and philosopher*, 1710–1784, 1952.

[3] Gibson, Tom. The first homografts: Trembley and the polyps. *Brit. J. plast. Surg.*, 19, 1966, pp. 301–307.

sur l'anatomie, et les métamorphoses des différentes espèces d'insectes, [*etc.*], Paris & London, 1832, and a catalogue of his collection was issued as *Catalogue raisonné du célèbre cabinet de coquilles de feu P. Lyonet*, [The Hague, 1796].[1]

George Louis Leclerc, Comte de Buffon (1707–88), was born near Dijon, and after an early interest in mathematics and physics, turned to biology. In 1735 he translated Hales' *Vegetable statics* into French. Buffon became director of the Jardin du Roi in 1739, and his monumental *Histoire naturelle*, [*etc.*], was published in forty-four volumes between 1749 and 1804. The first three volumes were published in 1749, and the last eight appeared between 1788 and 1804. Several eminent zoologists completed the work after Buffon's death, and there have been several editions, translations and abridgements of the *Histoire*.[2] Buffon destroyed the early drafts of his publications, but incomplete drafts of several parts of his *Les époques de la nature*, 1778, have survived, enabling Jacques Roger[3] to print the manuscript versions with the relevant passages of the printed edition. This publication includes an extensive introduction, notes, bibliography, and a study of Buffon's cosmology and religious beliefs. The *Oeuvres complètes de Monsieur le Comte de Buffon* appeared in fourteen volumes, Paris, 1774–89, and the *Oeuvres philosophiques de Buffon*, edited by Jean Piveteau, was published in Paris in 1955.[4]

The versatile Albrecht von Haller (1708–77) wrote extensively on anatomy, physiology and medicine, while his bibliographical contributions are of primary significance.[5] He graduated from Leyden in 1727, where he was the pupil of Boerhaave and Albinus, and became professor of anatomy, botany and surgery at the University of Göttingen. Haller and Conrad Gesner planned to compile a complete flora of Switzerland but the latter withdrew from the project after accomplishing some of the work, and Haller continued alone. Fourteen years after his botanical journey in Switzerland in 1728, he published *Enumeratio methodica stirpium indegenarum Helvetiae*, Göttingen, 1742, and forty years after the journey produced his great work *Historia stirpium indigenarum Helvetiae inchoata*, Berne, 1768, a catalogue of 2,486 species of Swiss plants. Between 1728 and 1771 Haller and Gesner exchanged about six hundred letters, most of Haller's having been published by H. E. Sigerist in 1923, and a translated selection of the passages relating to the collecting of plants is

[1] See Van Seters, W. H. *Pierre Lyonet, 1706–1789: sa vie, ses collections de coquillages et de tableaux, ses recherches entomologiques*, The Hague, 1962.

[2] A modern re-examination of the *Histoire naturelle* is provided by Wohl, Robert. Buffon and his project for a new science. *Isis*, 51, 1960, pp. 186–199.

[3] *Mémoires du Muséum national d'Histoire naturelle*. Serie C: Science de la terre, Tome X: Buffon, *Les époques de la nature. Edition critique avec le manuscrit, une introduction et des notes*, Paris, 1962.

[4] See Roule, Louis. *Buffon et la description de la nature*, Paris, 1924.

[5] See Thornton, pp. 139–140, 244–245.

provided in a paper by Sir Gavin de Beer.[1] Haller retired to his native Berne in 1753 and commenced writing the innumerable papers and books that bear his name. His *Bibliotheca anatomica*, [*etc*.], two volumes, Zürich, 1774–77; and *Bibliotheca botanica*, [*etc*.], two volumes, Berne and Basle, 1771–72, with *Index emendatus perfecit J. Christian Bay*, Berne, 1908, which were reprinted together in two volumes, 1968, are of special interest to scientists. His anatomical and physiological writings include *Icones anatomicae*, [*etc*.], eight parts, Göttingen, 1743–56; *De respiratione experimenta anatomica*, Göttingen, 1746; *Disputationes anatomicae*, seven volumes, Göttingen, 1746–52; *Primae lineae physiologicae in usum praelectionum academicarium*, Göttingen, 1747; *Elementa physiologiae corporis humani*, eight volumes, 1757–66 (Plate 9), one of the most important books in the history of physiology,[2] and *De partium corporis humani praecipuarum fabrica et functionibus*, [*etc*.], eight volumes, Berne and Lausanne, 1778. Several of these were translated, and went into numerous editions. Haller's lesser works were collected together as *Opera minora emendata, aucta et renovata*, [*etc*.], three volumes, Lausanne, 1762–68. His work on the nervous system and on muscle fibres earned for him the title of father of experimental physiology.[3] Many of his manuscripts, including unpublished material, and sixty-seven bound volumes of correspondence addressed to him are preserved in the Stadt-Bibliothek at Berne.

Born in Geneva of French parents, Charles Bonnet (1720–93) was influenced by the writings of Réaumur, and followed his experimental methods. Bonnet studied aphids, and regeneration of lost parts in *hydra*, and the respiration of butterflies and caterpillars. His publications are contained in *Oeuvres d'histoire naturelle, et de philosophie*, eighteen volumes, Neuchâtel, 1779–83.

It must be most unusual for a book on natural history to be in print over one hundred and eighty years after its first publication, and to have gone into about two hundred editions during that period. Yet such has happened to White's *Natural history of Selborne*. Gilbert White was born on 18th July, 1820, at Selborne, and was educated at Oriel College, Oxford, where he graduated B.A. in 1743 and M.A. three years later. He became a curate, first at Swarraton and then of Selborne (1751). He occupied a similar position at Durley, near Bishop's Waltham, in 1753 for eighteen months, but finally returned to Selborne, and remained

[1] de Beer, Sir Gavin R. Haller's *Historia stirpium*. *Ann. Sci.*, 9, 1953, pp. 1–46.

[2] See Buess, Heinrich. Albrecht von Haller and his "Elementa physiologiae" as the beginning of pathological physiology. *Med. Hist.*, 3, 1959, pp. 123–131; and Buess, Heinrich. Zur Entstehung der *Elementa physiologiae* Albrecht Hallers (1708–1777). *Gesnerus*, 15, 1958, pp. 17–35.

[3] See also Bay, J. Christian. Albrecht von Haller, medical encyclopedist. *Bull. Med. Lib. Ass.* 48, 1960, pp. 393–403.

there as curate for the rest of his life. White studied the natural history of the neighbourhood, and corresponded with several eminent naturalists. Gilbert White's brother Benjamin, a natural history publisher, printed the first edition of *The natural history and antiquities of Selborne*, [*etc.*], London, 1789, the title-page of which does not bear the name of the author. The work consists of forty-four letters addressed to Thomas Pennant, sixty-four letters to the Hon. Daines Barrington, all written from Selborne, and twenty-six letters on the antiquities of Selborne. An abbreviated German translation was published in Berlin in 1792, and in 1802 appeared the second English edition in Gilbert White's *Works*, edited by W. Marwick, two volumes, London, printed by John, the son of Benjamin White. Edward A. Martin[1] has compiled an exhaustive bibliography of White's writings, from which the following notes are derived. Innumerable editions, issues, selections and translations of the *Natural history* have appeared, and the following outstanding editors have associated themselves with the book. John Mitford's edition was first published in 1813, fifty copies of this edition being on large paper; Sir William Jardine's name first appeared on an edition from Edinburgh in 1829; J. Rennie (1833); Thomas Brown (Edinburgh, 1833); Edward Blyth (1836); Edward Turner Bennett (1837); Leonard Jenyns [1843]; J. G. Wood (1854); Frank Buckland (1875); Thomas Bell (1877), who lived in Gilbert White's house at Selborne for over thirty years, and included much new material on White; G. Christopher Davies [1879]; Grant Allen (1900); L. C. Miall and W. Warde Fowler, reproducing the original edition of 1789 (1901); Richard Kearton (1902); E. M. Nicholson (1930); Nonesuch Press (1938); James Fisher (1947); R. M. Lockley (Everyman's Library, 1949); Lutterworth Press (1951); and W. Sidney Scott (Folio Society, 1962). It has also been translated into Danish (1951), Swedish (1963), French (1949), and Japanese (twice in 1949, and again in 1958). Most editions have been reprinted on several occasions. *A naturalist's calendar* was extracted from White's manuscripts by J. Aikin, Gilbert White having died on 26th June, 1793. It was published as a separate work in 1795, but was added to the *Natural history* in the second edition, published in 1802. The manuscript of White's Journal has been in the British Museum since 1881, and has been published as *Journals of Gilbert White. Edited by Walter Johnson*, London, 1931. Claude A. Prance[2] has provided bibliographical details of the major editions, White's other writings are listed, and the illustrations of his books are discussed. The *Natural history of Selborne* is a delightful introduction to nature. Written

[1] Martin, Edward A. *A bibliography of Gilbert White, the naturalist and antiquarian of Selborne*, [*etc.*], 1934; the first edition was published in [1897].

[2] Prance, Claude A. Some uncollected authors. XLIII. Gilbert White, 1720–93. *Book Collector*, 17, 1968, pp. 300–321; see also Scott, W. S. Gilbert White's The natural history of Selborne. *Book Collector*, 18, 1969, pp. 89–90.

by a modest, observant amateur, it echoes his spirit of adventure in revealing to others the marvels that he discovered in the countryside around him.[1]

Thomas Pennant (1726–98) was born at Downing, Flintshire, and attributed his early interest in natural history to a copy of Francis Willughby's *Ornithology*, 1678, given to him at the age of twelve. Pennant was elected F.R.S., and, at the instance of Linnæus, to the Royal Society of Uppsala. Among other publications he was the author of *The British zoology*, 1761–66; *Indian zoology*, 1769 and 1790; *Genera of birds*, 1773 and 1781; *Natural history of the turkey*, 1781; *Arctic zoology*, two volumes, London, 1784–85; and *The literary life of the late Thomas Pennant, Esq. By Himself*, London, published in 1793, despite the fact that he lived until 1798.[2]

Charles de Geer (1720–78), a native of Sweden, followed up the work of René Antoine Ferchault de Réaumur (1683–1757),[3] and published a work with the same title as that of Réaumur, *Mémoires pour servir à l'histoire des insectes*, seven volumes, 1752–78. This was largely influenced by the writings of Linnæus. Petrus Camper (1722–89), who spent most of his life in Leyden, made important contributions to comparative anatomy, particularly in his study of the apes. He was the author of *Naturkundige Verhandelingen over den Orang Outang*, [*etc.*], Amsterdam, 1782; and *Dissertation sur les variétés naturelles qui caracterisent la physionomie des hommes*, [*etc.*], translated from the Dutch by H. J. Jansen, Paris, 1791. Camper was also the author of a book on vision, which has been translated into English with a facsimile of the original Latin text as *Optical dissertation on vision*, 1746, Nieuwkoop, 1962, with an introduction by G. Ten Doesschate.[4]

Although mainly remembered for his contributions to surgery, John Hunter[5] (1728–93) must receive consideration for his important contributions to comparative anatomy. He investigated the electric organs of the torpedo fish, the structure of whales, air-sacs of birds, and in fact collected any rarity in natural history. Many of his papers were published in the *Philosophical Transactions*, but at his death he left a large collection of manuscripts, many of which were destroyed by his brother-in-law, Sir Everard Home (1763–1832), who had used much of the material. John Hunter's books have been the subject of a handlist compiled by

[1] See Holt-White, Rashleigh. *The life and letters of Gilbert White of Selborne. Written and edited by his great-grand-nephew.* [*etc.*], 2 vols., 1901; Johnson, Walter. *Gilbert White, pioneer, poet, and stylist*, 1928; and Scott, Walter S. *White of Selborne*, 1950.

[2] See McAtee, W. L. The North American birds of Thomas Pennant. *J. Soc. Bib. Nat. Hist.*, 4, 1963, pp. 100–124.

[3] See pp. 175–176.

[4] See also Thornton, pp. 116–117.

[5] See Thornton, pp. 118–120, etc.

W. R. LeFanu,[1] and the following represent the non-medical items: *Observations on certain parts of the animal œconomy*, printed in Hunter's house in 1786, with a second edition in 1792, and another, edited by Sir Richard Owen, in 1837. This contains a selection of Hunter's papers from the *Philosophical Transactions*, together with other material. A German translation was printed at Braunschweig in 1813. Hunter's *Works*, edited by James F. Palmer, were published between 1835 and 1837 in four volumes, with an extra volume of plates. *Observations and reflections on geology*, [*etc.*], London, 1859,[2] was followed by *Essays and observations on natural history, anatomy, physiology, psychology and geology*, [*etc.*], two volumes, London, 1861, this being edited by Sir Richard Owen. Jessie Dobson[3] has contributed a biography of Hunter, which includes a chronological list of his writings, and a useful list of source material. Thomas R. Forbes[4] has investigated Hunter's two papers on spontaneous intersexuality in vertebrates, based on material at the Royal College of Surgeons of England.[5]

George Stubbs (1724–1806) achieved fame as the painter of animals, but he made a particular study of anatomy. Born in Liverpool, he settled in York in 1745 and a surgeon procured cadavers for him to dissect, and requested him to lecture to students. Stubbs dissected, drew and engraved the plates for his book, *The anatomy of the horse*, [*etc.*], London, 1766, which was printed for the author, and is recognized as a masterpiece; the latest edition was published in 1966, with a modern veterinary paraphrase by J. McCunn and C. H. Ottaway. Recently rediscovered anatomical drawings by Stubbs have resulted in a re-examination of his achievements. During the recataloguing of the Free Public Library, Worcester, Mass., Thurston Taylor, the Librarian, discovered four volumes of manuscript material relating to Stubbs' work on comparative anatomy, together with 124 of his drawings. These were once in the possession of Thomas Bell (1792–1880), and went to the Free Library through Dr. John Green, who presented his collection of 7,000 volumes to form the basis of the collection. The text of the manuscript is in French, but some of the plates were printed before the death of Stubbs, who had commenced work on

[1] Le Fanu, William Richard. *John Hunter: a list of his books*, for the Royal College of Surgeons of England, 1946; see also *Medical Classics*, January, 1940, which contains selections from Hunter's writings, a bibliography, etc.

[2] This was written about 1790. See Jones, F. Wood. John Hunter as a geologist. [*etc.*]. *Ann. Roy. Coll. Surg. Engl.*, 12, 1953, pp. 219–245.

[3] Dobson, Jessie. *John Hunter*, Edinburgh, London, 1969; see also Dobson, Jessie. John Hunter's animals. *J. Hist. Med.*, 17, 1962, pp. 479–486; and Dobson, Jessie. John Hunter's museum. *St. Bart's Hosp. J.*, 66, 1962, pp. 146–150.

[4] Forbes, Thomas R. John Hunter on spontaneous intersexuality. *Amer. J. Anat.*, 116, 1965, pp. 269–300.

[5] See also Peachey, George C. *A memoir of William & John Hunter*, Plymouth, *for the author*, 1924; Gloyne, S. Roodhouse. *John Hunter*, Edinburgh, 1950; and Kobler, John. *The reluctant surgeon: the life of John Hunter*, London, [*etc.*], 1960.

the book in 1795. Entitled *A comparative anatomical exposition of a human body with that of a tiger and a common fowl*, tables i–x were printed in 1804, and tables xi–xv probably between then and 1806. Edward Orme published all fifteen plates with the text in 1817, but the text differs widely from that discovered at Worcester, Mass. This material was exhibited in London by the Arts Council of Great Britain in 1958, and the illustrated catalogue compiled for that exhibition contains interesting information on Stubbs, and on his work as a comparative anatomist.[1] He painted a baboon and other animals for John Hunter, and was the first accurately to produce a representation of the adult rhinoceros.[2]

The early voyages of discovery are of special significance in natural history, as they often contain the first descriptions of birds, mammals and plants, in addition to geographical and geological material. The extensive journeys of Captain James Cook (1728–79) are no exception. Cook was born on 27th October, 1728, at Marton, Cleveland, Yorkshire, and after being apprenticed to a coal-shipper and studying navigation, he volunteered for the Royal Navy in 1755. In 1763, and again from 1764 to 1767, he was engaged in surveys of the St. Lawrence and the coast of Newfoundland. He observed an eclipse of the sun near Cape Ray, and his remarks were printed in the *Philosophical Transactions* for 1766. Cook's first voyage round the world was accomplished between 1768 and 1771 in his ship the *Endeavour*. The voyage was chiefly scientific, and Green, the astronomer, and Joseph Banks went with Cook. They charted New Zealand, took possession of the east coast of Australia, and confirmed the existence of the Torres Strait. On the second voyage, between 1772 and 1775, there were two ships, with two naturalists and two astronomers, and on this voyage Cook succeeded in mastering scurvy. His classic paper on the prevention of that disease was first printed in Sir John Pringle's *A discourse upon some late improvements of the means of preserving the health of mariners*, [*etc.*], London, 1776, and was then published in the *Philosophical Transactions*[3]; Cook was awarded the Copley Medal of the Royal Society for this work. His third voyage took place between 1776 and 1779, in which latter year Cook was killed. He had been elected F.R.S. in 1776. Cook was the originator of scientific marine surveying, and contributed to geography, navigation and medicine. Sir Maurice Holmes[4] has compiled a bibliography of Captain James Cook, and there

[1] See *George Stubbs. Rediscovered anatomical drawings from the Free Public Library, Worcester, Massachusetts*, 1958.

[2] See also Fountain, R. B. George Stubbs (1724–1806) as an anatomist. *Proc. Roy. Soc. Med.*, 61, 1968, pp. 639–646.

[3] *Phil. Trans.*, 66, 1776, pp. 402–406. See also Thrower, W. R. Contributions to medicine of Captain James Cook, F.R.S., R.N. *Lancet*, 1951, II, pp. 215–219.

[4] Holmes, Sir Maurice Gerald. *Captain James Cook, R.N., F.R.S.: a bibliographical excursion*, 1952; reprinted Franklin, New York, 1968.

is a similar list of material in the Mitchell Library and other libraries in Australia, issued by the New South Wales Public Library.[1] He wrote *Directions for navigating on part of the south coast of Newfoundland, [etc.]*, London, 1766, and *Directions for navigating the west-coast of Newfoundland [etc.]*, London, St. John's, Newfoundland, 1768. There are several accounts of Cook's voyages, the official account of the second voyage, by Cook himself, appearing as *A voyage towards the South Pole, and round the world. Performed in His Majesty's Ships the Resolution and Adventure. In the years 1771, 1773, 1774 and 1775*: two quarto volumes, 1777. The official account of the third voyage, by James Cook and James King, was entitled *A voyage to the Pacific Ocean. Undertaken by the Command of His Majesty, for making discoveries in the Northern Hemisphere. . . . In the years 1776, 1777, 1778, 1779 and 1780*, three volumes, London, 1784, the first two volumes being by Cook, and the third by King. This work was sold out on the third day after publication. A third edition was published in 1785. The original journal of Cook's first voyage is in the Mitchell Library, Sydney, and was published as *Captain Cook's Journal during his first voyage round the world made in H.M. Bark "Endeavour", 1768–71*, [etc.], edited by W. F. L. Wharton, London, 1893. The Hakluyt Society is publishing the original journals of all three voyages, several unofficial collections of these having been previously published. Volumes 1–3 have been published (1955–67), and a fourth is to follow. Accounts of Cook's voyages, both official and unofficial, went into numerous editions and translations, and a bibliography of the French translations has been compiled by Stanley Roberts.[2,3]

The main purpose of Cook's voyage was astronomical investigation, but this yielded comparatively little. Banks and Solander concentrated on collecting botanical specimens, and 1,300 new species were included, but Banks' scheme for comprehensive lavish publication was never realized. Solander has been blamed as being lazy, but Banks owned the specimens, which together with the notes, descriptions, drawings and engravings are now housed in the British Museum (Natural History). Daniel Carl Solander (1733–82) was born at Piteå, in northern Sweden, and was a student of Linnaeus at Uppsala University. In 1760 he came to

[1] New South Wales Public Library. *Bibliography of Captain James Cook, R.N., F.R.S., circumnavigator*, Sydney, 1928.

[2] Roberts, Stanley. Captain Cook's voyages: a bibliography of the French translations, 1772–1800. *J. Document.*, 3, 1947–48, pp. 160–176. See also Spence, Sydney Alfred. *Captain James Cook, R.N. (1728–1779): a bibliography of his voyages, to which is added other works relating to his life, conduct and nautical achievements*, Mitcham, 1960; and Merrill, Elmer Drew. The botany of Cook's voyages and its unexpected significance in relation to anthropology, biogeography and history. *Chron. Bot.*, 14, 1954, pp. 163–383.

[3] See also Muir, John Reid. *The life and achievements of Captain James Cook, R.N., F.R.S., explorer, navigator, surveyor, and physician*, London, Glasgow, 1939; and Holmes, Sir Maurice. Captain James Cook, R.N., F.R.S. *Endeavour*, 8, 1949, pp. 11–17.

England intending to stay for a year or two, but he travelled extensively, joined the staff of the British Museum, and catalogued various collections there. Elected F.R.S. in 1764, he became a great friend of Banks, and sailed with him in the *Endeavour*. In 1773 Solander became Keeper of the British Museum, and was also librarian and curator to Banks. Solander contributed widely to the writings of others, but wrote little himself except a condensation of Linnaeus' *General botany* under the title *Elementa botanica*, Uppsala, 1756. John Cawte Beaglehole has edited *The "Endeavour" journal of Joseph Banks, 1768–71*, two volumes, London, 1962, and Roy Anthony Rauschenberg[1] has contributed seveıal well-documented studies of Solander, based mainly on manuscripts, journals, diaries, logs and letters.[2]

Among others who accompanied Cook on his voyages we must mention William Anderson (1748–78), who went on two of the journeys and made extensive natural history collections, which he bequeathed to Sir Joseph Banks. Anderson also left a journal in manuscript, which is now in the Public Record Office.[3] Sidney Parkinson (*c.* 1745–71) was one of the artists engaged by Banks to accompany him on Cook's first voyage, during which Parkinson died. His botanical drawings are widely known, but his zoological work is less so. The Zoology Department of the British Museum (Natural History) possesses a volume containing forty water-colours by Parkinson, all dated 1767,[4] and some of these drawings, together with those by other artists, have been published by the British Museum (Natural History), with a text by P. J. P. Whitehead, as *Forty drawings of fishes made by the artists who accompanied Captain James Cook on his three voyages to the Pacific 1768–71, 1771–75, 1776–80*, London, 1968.

Friedrich Heinrich Alexander von Humboldt (1769–1859) made contributions of outstanding importance in many fields of learning. In the sciences his work in geography, meteorology, geophysics, astronomy, zoology and botany should be mentioned. Born in Berlin on 14th September, 1769, he died in the same city on 6th May, 1859, in his ninetieth year. He studied at the universities of Frankfurt-an-der-Oder and Göttingen, and at the Freiberg Bergakademie, becoming an inspector

[1] Rauschenberg, Roy Anthony. Daniel Carl Solander, naturalist on the "Endeavour". *Trans. Amer. Philos. Soc.*, N.S. 58, iii, 1968; Rauschenberg, Roy Anthony. Daniel Carl Solander, naturalist on the *Endeavour* voyage. *Isis*, 58, 1967, pp. 367–374; and Rauschenberg, Roy Anthony. A letter of Sir Joseph Banks describing the life of Daniel Solander. *Isis*, 55, 1964, pp. 62–67.

[2] See also Stearn, William T. The botanical results of the *Endeavour* voyage. *Endeavour*, 27 1968 pp. 3–10.

[3] See Keevil, J. J. William Anderson, 1748–1778, master surgeon, Royal Navy. *Ann. med. Hist.*, 2nd Ser., 5, 1933, pp. 511–524.

[4] See Sawyer, F. C. Some natural history drawings made during Captain Cook's first voyage round the world. *J. Soc. Bib. Nat. Hist.*, 2, vi, 1950, pp. 190–193.

of mines in Prussia. In 1795 he resigned this position to devote himself to scientific travel, probably inspired in this by Georg Forster[1] who had sailed on Cook's second voyage, and with whom Humboldt had travelled in Europe. His great voyage to South America from 1799 to 1804 was to provide the material for the greater part of his scientific work, but he had written a work on the vegetation of Freiburg, *Flora Fribergensis specimen, [etc.]*, Berlin, 1793, and *Versuche über die gereizte Muskel-und Nervenfasser*, (1797), on physiology.[2] Humboldt sailed in 1799 with the botanist Aimé Bonpland (1773–1858).[3] Owing to an outbreak of typhoid aboard ship Humboldt and Bonpland landed at Cumana, Venezuela. During their great journey the course of the Orinoco was explored, Chimborazo was climbed, a great collection of plants was made, meteorological data was recorded, and the zones of vegetation studied. The travellers returned to Europe, and settled in Paris to arrange and publish the results of the voyage. Humboldt delivered in 1805 a lecture which was to be published in 1807 as *Essai sur la géographie des plantes*, Tübingen. The title page of the *Essai* is sometimes dated 1805, but other copies bear the correct publication date of 1807. This pioneer scientific study of plant geography was later supplemented in an important essay Humboldt prefixed to C. S. Kunth's *Nova genera et species plantarum*, seven volumes, Paris, 1816–25, in which the relationship of plant distribution to temperature and altitude is discussed, and statistical analysis is introduced to the study and comparison of floras.[4] A German version was issued from Tübingen as *Ideen zur einer Geographie der Pflanzen*, 1807, with a second German edition from Vienna, 1811.[5] The results of the American expedition were published in the *Voyage aux régions équinoxiales du Nouveau Continent fait en 1799–1804*, twenty-four volumes, Paris, 1805–37, which includes the *Nova genera et species plantarum*, mentioned above, in which C. S. Kunth, continuing the work of Aimé Bonpland after the latter's emigration to the Argentine, described over three thousand new plants. A section of Part I of the *Voyage* was published as *Relation historique du voyage aux régions équinoxiales du nouveau continent*, three volumes, 1814, and this

[1] See Hoare, M. E. "Cook the discoverer": an essay by Georg Forster 1787. *Rec. Austral. Acad. Sci.*, 1, iv, 1969, pp. 7–16.

[2] Rothschuh, K. E. Alexander von Humboldt und die Physiologie seiner Zeit. *Sudhoffs Arch. Gesch. Med.*, 43, 1959, pp. 97–113.

[3] Schulz, Wilhelm. Aimé Bonpland. Alexander von Humboldts Begleiter auf d. Amerikanreise, 1799–1804. Sein Leben und Wirken besonders nach 1817 in Argentinien. *Akad. Wiss. Lit., Abh. math.-nat. Kl.*, Wiesbaden, 1960, Nr. 9.

[4] Stearn, W. T. Alexander von Humboldt (1769–1859) and plant geography. *New Scientist*, 5, 1959, pp. 957–959, and Stearn, W. T. Humboldt's "Essai sur la géographie des plantes". *J. Soc. Bib. Nat. Hist.*, 3, 1960, pp. 351–357. A facsimile of the "Essai" was issued as Sherborn Fund Facsimile number 1, 1959, by the Society for the Bibliography of Natural History.

[5] Reprinted as *Ostwald's Klassiker*, No. 248, 1960.

was reprinted in 1966. It was translated into English by H. M. Williams as Humboldt and Bonpland's *Personal narrative of travels to the equinoctal regions of the new continent during . . . 1799–1804*, four volumes [in three], London 1818–19; and also translated by T. Ross, three volumes, London, 1852–53. Humboldt's geophysical work, particularly the discovery of the decrease in intensity of the earth's magnetic force from the poles to the equator, has been described by Hanno Beck,[1] who is also the author of an extensive biography of Humboldt.[2] Among the many other publications of Humboldt we mention *Kosmos, Entwurf einer physischen Weltbeschriebung*, five volumes, Stuttgart and Tübingen, 1845–62, the fifth posthumous volume being edited by E. Buschmann, and translated into English as *Cosmos*, London, 1848–58; and the *Personal narrative of travels . . . 1799–1804*, [*etc.*], translated by H. M. Williams, four volumes, 1818–19. The correspondence of Humboldt was enormous, and Der Deutschen Akademie der Wissenschaften zu Berlin has proposed to publish a definitive edition, which will prove invaluable to historians of science.[3]

Lazzaro Spallanzani (1729–99), of Scandiano, near Modena, made significant experiments to disprove the doctrine of spontaneous generation, respiration and circulation. He was successively professor at Reggio, Modena and Pavia (1769–99), and wrote the following books:[4] *Saggio di osservazione microscopiche relative al sistema della generazione*, Modena, 1767; *Prodromi sulla riproduzione animale. Riproduzione della coda del girino*, Modena, 1768, which was translated into English, London, 1769; *De' fenomeni della circolazione osservata nel giro universale de' vasi*, [*etc.*], Modena, 1773, translated into French (Paris, An. VIII [1800]), and into English as *Experiments upon the circulation of the blood*, [*etc.*], London, 1801; *Opusculi di fisica animale e vegetabile*, [*etc.*], three volumes, Venice, 1782; and *Mémoires sur la respiration*, [*etc.*], Geneva, An. XI [1803], which was translated into English, London, 1805. Spallanzani is sometimes called the founder of experimental physiology,[5] but he also contributed to zoology by his work on sponges, rays, eels and crabs. In addition he was the author of *Chimico esame*, [*etc.*], Modena, 1796. Between 1793 and 1794 he

[1] Beck, Hanno. Das literarische Testament Alexander von Humboldts, 1799. *Forschungen und Fortschritte*, 31, 1957, pp. 65–70.

[2] Beck, Hanno. *Alexander von Hamboldt. Bd. I. Von der Bildungsreise zur Forschungsreise 1769–1804*. Wiesbaden, 1959; *Bd. II. Von Reisewerk zur "Kosmos" 1804–1859*, Wiesbaden, 1961; see also Kellner, L. *Alexander von Humboldt*, 1963; and Scurla, Herbert. *Alexander von Humboldt. Leben und Wirken. 2 Auflage*, Berlin, 1959.

[3] See Biermann, Kurt R., and Lange, Fritz G. Die Alexander von Humboldt-Briefausgabe. *Forsch. und Fortschr.*, 36, 1962, pp. 225–230.

[4] See Prandi, Dino. *Bibliografia delle Opere di Lazzaro Spallanzani: delle traduzione e degli scritti su di lui*. [*etc.*], Florence, 1951.

[5] Tortonese, E. Lazaro Spallanzani: founder of experimental physiology. *Endeavour*, 7, 1948, pp. 92–96.

conducted experiments on the ability of bats to avoid obstacles during flight, but the results of many of his later experiments were not published; they are recorded in MSS. in the Biblioteca Municipale, Reggio nell' Emilia, Italy, where some of Spallanzani's notebooks are also preserved.[1] An edition of his letters in four volumes was published as *Epistolario*, Florence, 1958–59. There is a modern collected edition of his works, *Le opere*, five volumes in six, Milan, 1932–36.[2]

A Danish field naturalist, Otto Friderich Müller (1730–84), conducted important research on the taxonomy of invertebrates, and was also a keen botanist. He was the author of *Zoologiae Danicae prodromus*, 1776, a preliminary to his *Zoologia Danica*, of which Part I was published in 1777, Part 2 in 1784, while Parts 3 and 4, 1789–1806, were published after Müller's death, and were largely the work of P. C. Abildgaard and others.[3] Müller succeeded Georg Christian Oeder (1728–91) as editor of the *Flora Danica*, which was issued in fifty-four parts, containing 3,240 plates, and appeared from 1761–1883.[4]

Erasmus Darwin (1731–1802) was born at Elston Hall, near Derby, and took a degree in medicine at Cambridge in 1755, before settling down for a time at Lichfield. In 1781 he moved to Derby, and became friendly with James Watt, Josiah Wedgwood, James Keir and Thomas Day. Erasmus Darwin was interested in chemistry and mechanics, but he also anticipated the theory to be advanced fifty years after his death by his grandson, Charles Robert Darwin. Erasmus Darwin was the author of several books, but his most important was *Zoonomia; or, the laws of organic life*, two volumes, London, 1794–96, which contains a system of pathology, a treatise on generation, and hints at the theory of evolution later advanced with overwhelming scientific evidence by his grandson. This went into numerous editions and translations, including printings at New York, 1796 (three volumes); Philadelphia, 1797 and 1818; Dublin, 1800 (two volumes); Boston, 1803 and 1809; a second English edition in two volumes, London, 1796; a third English edition in four volumes, 1801, which was translated into Italian, four volumes, Venice, 1803–06. Erasmus Darwin's *The Botanic Garden. Poem in two parts with philosophical notes*, contains as Part I *The economy of vegetation*, London,

[1] See Dijkgraaf, Sven. Spallanzani's unpublished experiments on the sensory basis of object perception in bats. *Isis*, 51, 1960, pp. 9–20.

[2] See also Doetsch, Raymond N., *ed. Microbiology: historical contributions from 1776 to 1908 by Spallanzani, Schwann, Pasteur, Cohn, Tyndall, Koch, Lister, Schloesing, Burrill, Ehrlich, Winogradsky, Warington, Beijerinck, Smith, Orla-Jensen*, New Brunswick, New Jersey, 1960. Contains original accounts of contributions in chronological order, with references at the end of each section.

[3] See Anker, Jean. *Otto Friderich Müller's "Zoologia Danica"*, Copenhagen, 1950. (Library Research Monographs, published by the University Library, Scientific and Medical Department, Vol. 1.)

[4] See Anker, Jean. From the early history of the Flora Danica. *Libri*, I, 1951, pp. 334–350.

1791, and Part II *The loves of the plants*, Lichfield, 1789, the latter having been published first. This went into several editions, and was translated into French, Italian and Portuguese.[1] He was also the author of *Phytologia, or the philosophy of agriculture and gardening*, [*etc.*], London, 1799, and *Temple of nature*, [*etc.*], published posthumously in 1803.[2] A Russian translation of this by N. A. Cholodkovsky, edited with a commentary by E. N. Pavlovsky, was published by the Academy of Sciences of Russia in Moscow, 1954. The translation had originally appeared in 1911 in the *Journal of the Ministry of Education*, St. Petersburg.[3] Desmond King-Hele[4] has written a biography of Erasmus Darwin, and also edited a selection of his writings.[5]

Although his theories were opposed by both Haller and Bonnet, Caspar Friedrich Wolff (1734–94) is regarded as the pioneer of modern embryology. Born in Berlin, Wolff went to St. Petersburg as a member of the Russian Academy of Sciences, and spent the last thirty years of his life there after the hostile reception of his theories had compelled him to resign his teaching post in Berlin. His "De formatione intestinonum" first appeared in 1768 in a journal,[6] but was translated by Johann Friedrich Meckel (1761–1833) into German as *Über die Bildung des Darmkanals in bebruteten Hühnchen*, Halle, 1812. This work on the development of the intestines in the chick is of primary significance in the history of biology, and became known mainly through the German translation. Wolff's *Theoria generationis*, Halle, [1759], was translated into German by the author as *Theorie von der Generationen*, Berlin, 1764, which was reprinted with an introduction by Robert Herrlinger, Hildesheim, 1966. A second edition of the Latin version appeared in

[1] See Emery, Clark. Scientific theory in Erasmus Darwin's *The Botanic Garden*, 1789–1791. *Isis*, 33, 1941–42, pp. 313–325; and Robinson, Eric. Erasmus Darwin's Botanical Garden and contemporary opinion. *Ann. Sci.*, 10, 1954, pp. 314–320.

[2] See Primer, Irwin. Erasmus Darwin's *Temple of nature:* progress, evolution, and the Eleusinian mysteries. *J. Hist. Ideas*, 25, 1964, pp. 58–76.

[3] See Hoare, Cecil A. Erasmus Darwin in Russia. *Ann. Sci.*, 11, 1956, pp. 255–256.

[4] King-Hele, Desmond. *Erasmus Darwin*, 1963; and *The essential writings of Erasmus Darwin. Chosen and edited with a linking commentary by Desmond King-Hele*, 1968; see also King-Hele, Desmond. Dr. Erasmus Darwin and the theory of evolution. *Nature*, 200, 1963, pp. 304–306.

[5] See also Pearson, Hesketh. *Doctor Darwin*, London, Toronto, 1930, and Penguin Books, 1943; Abbatt, William. Dr. Erasmus Darwin, the author of "Zoonomia". *Ann. med. Hist.*, 3, 1921, pp. 387–390; Krumbhaar, E. B. The bicentenary of Erasmus Darwin and his relation to the doctrine of evolution. *Ann. med. Hist.*, N.S.3, 1931, pp. 487–500; Barlow, Nora. Erasmus Darwin, F.R.S. (1731–1802). *Notes Rec. Roy. Soc. Lond.*, 14, 1959, pp. 85–98; Cohen, *Lord* Henry. Erasmus Darwin. *Univ. Birmingham hist. J.*, 11, 1967, pp. 17–40; and Moody, J. W. T. Erasmus Darwin, M.D., F.R.S.: a biographical and iconographical note. *J. Soc. Bib. Nat. Hist.*, 4, 1964, pp. 210–213.

[6] *Novi commentarii academiae scientiarum imperialis Petropolitanae*, 12, 1768, pp. 403–507; 13, 1769, pp. 478–530.

1774, and a new German translation by Paul Samassa was published in 1896.[1,2]

Jean Baptiste Pierre Antoine de Monet, Chevalier de Lamarck (1744–1829), was born at Bazentin-le-Petit, Picardy, served in the Army, and followed other occupations before, at the age of twenty-four he devoted himself to science. He studied botany for ten years, and published *Flore française*, [*etc.*], three volumes, Paris, 1778, followed by a second edition in 1793, and a third, edited in collaboration with A. P. de Candolle, six volumes, Paris, 1805–15. In 1781–82 Lamarck visited Holland, Germany and Hungary with Buffon, and then became keeper of the herbarium at the Jardin du Roi, afterwards the Jardin des Plantes. He was the author of a little-known but very significant contribution to the development of modern geology which was printed at his own expense. Entitled *Hydrogéologie, ou recherches sur l'influence qu'ont les eaux sur la surface du globe terrestre; sur les causes de l'existence du bassin des mers*, [*etc.*], 1802 (An. X), it was translated into English by Albert V. Carozzi as *Hydrogeology*, Urbana, Ill., 1964.[3] When almost fifty years of age, Lamarck became very interested in invertebrate zoology, and his evolutionary hypothesis has attracted considerable attention. His most important publication was *Philosophie zoologique*, [*etc.*], two volumes, Paris, 1809, with a reprint of this printed in 1830, and a second edition, also in two volumes, Paris, 1873. A translation with an introduction by Hugh Elliot has been published in London as recently as 1914 with the title *Zoological philosophy: an exposition with regard to the natural history of animals*, [*etc.*]. A German translation of *Philosophie zoologique* was published at Jena in 1876. Lamarck was also the author of numerous pamphlets, articles in periodicals and encyclopaedias, and the following books: *Histoire naturelle des animaux sans vertèbres*, [*etc.*], seven volumes, Paris, 1815–22, with a second edition in eleven volumes, Paris, 1835–45; and *Système analytique des connaissance de l'homme restreintes à celles qui proviennent directement ou indirectement de l'observation*, Paris, 1820. A rare forerunner of the *Philosophie zoologique* was published as *Recherches sur l'organization des corps vivans, et particulièrement sur son origine*, [*etc.*], Paris, 1802. H. Graham Cannon[4] has suggested that criticisms of Lamarck's evolutionary theory have been unjustified, and were mostly based on Cuvier's eulogy of Lamarck, in which the latter's views were ridiculed by one who

[1] Ostwald's *Klassiker*, Nr. 84–85.

[2] See also Aulie, Richard P. Caspar Friedrich Wolff and his "Theoria generationis", 1759. *J. Hist. Med.*, 16, 1961, pp. 124–144; Herrlinger, Robert. C. F. Wolff's "Theoria generationis" (1759). Die Geschichte einer epochemachenden Dissertation. *Z. Anat. EntwGesch.*, 121, 1959, pp. 245–270; and Herrlinger, Robert. C. F. Wolff: Theoria generationis (1759). *Neue Z. ärztl. Fortbild.*, 48, 1959, pp. 954–955.

[3] See Carozzi, Albert V. Lamarck's theory of the earth: *Hydrogéologie*. *Isis*, 55, 1964, pp. 293–307.

[4] Cannon, H. Graham. *Lamarck and modern genetics*, Manchester, 1959.

did not appreciate them. Charles C. Gillispie[1] has also contributed papers presenting a modern evaluation of Lamarck's work.[2]

Vicq d'Azyr (1748–94) was a Paris physician who became interested in comparative anatomy and made special studies of the brain. He planned a *Traité d'anatomie et de physiologie*, Paris, 1786, only one part of which had been published at the time of his death, but his writings were collected together as *Oeuvres de Vicq-D'Azyr, recueillies et publiées avec les notes et un discours sur la vie et ses ouvrages, par Jacq. L. Moreau*, six volumes and a folio atlas, Paris, An. XIII, 1805.

Edward Jenner (1749–1823), of Berkeley, Gloucestershire, achieved fame for his discovery of vaccination,[3] but he was also a naturalist of some distinction. His association with John Hunter from 1770 to 1793 provided a stimulus that compelled him to busy himself with the problems of hibernation, migration, digestion and similar subjects. W. R. LeFanu[4] has complied a monumental bio-bibliography of Jenner, providing a *vade mecum* on his life and writings, from which the following facts are gleaned. Most of Jenner's publications were either in pamphlet form, or appeared in journals and were subsequently reprinted, and sometimes also translated. For example, his classic paper "Observations on the natural history of the cuckoo" first appeared in the *Philosophical Transactions*,[5] was offprinted to the extent of fifty copies, and was later translated into French [LeFanu 9] and Italian [LeFanu 10]. Jenner's paper on bird migration was not read to the Royal Society until six months after his death, and it was then printed in the *Philosophical Transactions* [LeFanu 113].[6] This was reset and issued separately, also in 1824, and again reprinted in Gloucestershire [LeFanu 115], and Chipping Sodbury, 1882 [LeFanu 116], with a German abstract published at Erfurt, 1824 [LeFanu 117].[7]

[1] Gillispie, Charles Coulston. The formation of Lamarck's evolutionary theory. *Arch. int. Hist. Sci.*, 9, 1956, pp. 323–338; and Gillispie, Charles Coulston. Lamarck and Darwin in the history of science. *Amer. Scient.*, 46, 1958, pp. 388–409.

[2] See Packard, Alpheus S. *Lamarck, the founder of evolution: his life and work. With translations of his writings on organic evolution.* New York, [*etc.*], 1901. Bibliography of Lamarck's writings, pp. 425–445; *The Lamarck manuscripts at Harvard, edited by William Morton Wheeler and Thomas Barbour*, Cambridge, Mass., 1933. This contains six holograph manuscripts given in the original and also translated; and Newth, D. R. Lamarck in 1800. A lecture on the invertebrate animals and a note on fossils taken from the *Système des animaux sans vertèbres* by J. B. Lamarck. Translated and annotated by D. R. Newth. *Ann. Sci.*, 8, 1952, pp. 229–254.

[3] See Thornton, pp. 131–133.

[4] LeFanu, William Richard. *A bio-bibliography of Edward Jenner, 1749–1823*, 1951.

[5] *Phil. Trans.*, 78, 1778, Part 2, pp. 219–237.

[6] *Ibid.*, 114, 1824, pp. 12–44.

[7] See Baron, John. *The life of Edward Jenner*, 2 vols., 1827–38; Cameron, Sir Gordon Roy. Edward Jenner, F.R.S., 1749–1823. *Notes Rec. Roy. Soc. Lond.*, 7, 1949–50, pp. 43–53; Francis, W. W., and Stevenson, Lloyd G. Three unpublished letters of Edward Jenner. *J. Hist. Med.*, 10, 1955, pp. 359–368; Thomas, K. Bryn. A Jenner letter. *J. Hist. Med.*, 12, 1957, pp. 449–458; and LeFanu, William Richard. Letters from Edward Jenner. *Ann. Roy. Coll. Surg. Engl.*, 39, 1966, pp. 370–372.

Eighteenth-century botanical science is of vital significance, if only for the work of the immortal Linnæus. But several other botanists demand consideration, including Stephen Hales (1677–1761), a doctor of divinity who conducted numerous experiments on plant physiology. Hales was born at Bekesbourne, Kent, graduated at Cambridge, and with William Stukeley (1687–1765) studied botany, dissected animals and became interested in chemistry. Hales then concentrated on botany, and after his appointment as curate at Teddington he performed all his experiments there. The results of these were communicated to the Royal Society, which awarded him the Copley Medal in 1739. In 1725 Hales had read a paper which was printed as *Vegetable staticks; or an account of some statical experiments on the sap in vegetables*, [*etc.*], 1727, which is the first part of *Statical essays*. A reprint, with a foreword and notes by M. A. Hoskin, was published in 1961. A second issue appeared in 1731, and Buffon translated it into French (1735). Part 2 was entitled *Haemastaticks*, [*etc.*], 1733, and was translated into French by De Sauvages (1744). Both parts were translated into German (1748), Dutch (1750) and Italian (1756). In 1741 Hales read a paper to the Royal Society on ventilation in prisons, which was published as *A description of ventilators and a treatise on ventilators*, two parts, London, 1743–58, and his other main work was entitled *Philosophical experiments*, [*etc.*], London, 1739. Stephen Hales, a clergyman without a medical degree, contributed to medical history by his experiments on the circulation,[1] on ventilation[2] and hygiene, but his botanical investigations were outstanding, not only for their ingenuity, but because they initiated a new branch of science.[3]

Herman Boerhaave (1668–1738)[4] was a versatile individual whose botanical work has tended to be overlooked. He was in constant contact with botanists and collectors throughout the world, and was responsible for the publication of Vaillant's *Discours sur la structure des fleurs*, Leyden, 1718, to which Boerhaave contributed an unsigned preface, and also completed and saw through the press Vaillant's *Botanicon Parisiense*, Leyden and Amsterdam, 1727, for which he also wrote an extensive preface. Boerhaave helped Linnæus considerably, and the latter dedicated to him his *Genera plantarum*, 1737.[5] Boerhaave's own *Index alter plantarum*

[1] See Hoff, H. E., and Geddes, L. A. The contribution of the horse to knowledge of the heart and circulation. 1. Stephen Hales and the measurement of blood pressure. *Conn. Med.*, 29, 1965, pp. 795–800.

[2] See Harris, D. Fraser. Stephen Hales, the pioneer in the hygiene of ventilation. *Sci. Month.*, November, 1916, pp. 440–454; and Foregger, Richard. Two types of respiratory apparatus of Stephen Hales. *Anaesthesia*, 11, 1956, pp. 235–240.

[3] See Clark-Kennedy, A. E. *Stephen Hales, D.D., F.R.S.: an eighteenth-century biography*, Cambridge, 1929; and Burget, G. E. Stephen Hales (1677–1761). *Ann. med. Hist.*, 7, 1925, pp. 109–116.

[4] See also pp. 156–158 for his chemical writings and additional references.

[5] See Lindeboom, G. A. Linnæus and Boerhaave. *Janus*, 47, 1957, pp. 264–274.

quae in Horto Academico Lugduno-Batavo alantur, was published in Leyden in 1720 and 1727, and an unauthorized account of his botanical lectures appeared under the title *Historia plantarum* in 1727.[1]

Carl Linnæus, afterwards von Linné (1707–78), was born in the Swedish province of Småland, and studied at Lund and Uppsala, where he became lecturer and demonstrator in the university botanical garden. He was given a grant to make a scientific journey to Lapland, as the result of which he published his first major work, *Flora Lapponica*, [*etc.*], Amsterdam, 1737, which is in Latin and appeared in a second edition, edited by J. E. Smith, London, 1792. The diary that Linnæus kept on this tour was not published until 1811, when it appeared in two volumes translated into English by J. E. Smith as *Lachesis Lapponica, or A tour in Lapland*, [*etc.*], London. In 1735 Linnæus went to Holland and took a medical degree at Hardewijk. Thence he went to Leyden, to become friendly with Boerhaave, who introduced him to a rich banker. The latter employed Linnæus as superintendent of his zoological and botanical gardens at Hortecamp. He now produced *Genera plantarum*[2] and *Critica botanica*, an English translation of the latter by Sir Arthur Hort being published by the Ray Society in 1938. Linnæus visited England and France in 1736, but two years later he was again in Stockholm, where he practised as a physician. He then became professor of medicine and natural history at Uppsala, and in 1768 was ennobled and assumed the style of "von Linné". He made several journeys through Sweden, and persuaded a number of his pupils to make extensive botanical expeditions to foreign lands. The literary output of Linnæus, both in Latin and Swedish, was enormous, and there are several bibliographies of his writings. The best is that by Johan Markus Hulth, and is based on the unique collection in the Royal University of Uppsala. Entitled *Bibliographia Linnaeana*, [*etc.*], 1907, this was to have covered three volumes, but is still incomplete.[3] The collection at the British Museum (Natural History) is also outstanding, and its catalogue represents a bibliography of some 3,874 entries.[4] We note the following as being among the more important of the writings by Linnæus, many of which went into several editions and translations. *Bibliotheca botanica*, [*etc.*], Amsterdam, 1736, an early work which was the second important

[1] See Gibbs, F. W. Boerhaave and the botanist. *Ann. Sci.*, 13, 1957, pp. 47–61; and Thornton, pp. 122–123.

[2] A facsimile of the 5th edition, 1754, was published in the series *Historiae Naturalis Classica*, 3, 1961, with an introduction by William T. Stearn.

[3] See also *A catalogue of the works of Linnæus issued in commemoration of the 250th anniversary of the birthday of Carolus Linnæus, 1707–78*, Stockholm, 1957; and Williams, Terrence. *A checklist of Linnæus 1735–1835 in the University of Kansas Libraries*, Lawrence, 1964.

[4] *A catalogue of the works of Linnæus (and publications more immediately relating thereto) preserved in the libraries of the British Museum (Bloomsbury) and the British Museum (Natural History) (South Kensington), Second edition*, 1933; pp. 231–236 lists bibliographies of the writings of Linnæus, 1730–1932.

bibliography of botany to be printed, and which contains *Fundamenta botanica, Pars* I, which has its own title-page, and is separately paginated. A reprint was published as *Bibliotheca et fundamenta botanica*, Munich, 1968; *Classes plantarum*, 1738; *Flora Svecica, sistens animalia Sveciae regni*, [*etc.*], Stockholm, 1746, which went into a second edition in 1755; *Fauna Svecica*, [*etc.*], 1746; *Curiositas naturalis*, 1748; *Philosophia botanica*, Stockholm, 1751, some copies of which have a variant title-page with Amsterdam added to the imprint.[1] This work was reprinted in 1966[2], and the book was also printed in Vienna, 1755 and 1763; *Species plantarum*, 1753,[3] a facsimile, with an introduction by W. T. Stearn being published by the Ray Society in two volumes, 1957–59. A second, considerably enlarged edition was published in Stockholm, 1762–63, and a reprint of this, with errors corrected, was printed a year later in Vienna, and is usually known as the third edition. A fourth edition was published between 1797 and 1830. Also, *Politia naturae*, 1760; and the most important, in which he introduced his system of binomial nomenclature, *Systema naturae regnum vegetabile*, Leyden, 1735. A facsimile of this first edition, with an introduction and a first English translation of the "Observationes" by M. S. J. Engel-Ledeboer and H. Engel, was published in Nieuwkoop, 1964. This work grew with each edition, and the tenth (two volumes, Stockholm, 1758–59) is of special significance because it has been taken as the basis of modern botanical and zoological binomial nomenclature. This was published in facsimile at Leipzig, 1894, and by the British Museum in 1939. The collections of Linnæus found their way to London, and are now at the Linnean Society. His classification of plants by their sexual organs is not only of historical importance but influenced all subsequent botanists, although discarded in favour of the "natural order". His descriptions of plants are concise, and his disciples followed his example, which rendered signal service to the development of botany as a science.[4]

Michel Adanson (1727–1806) was born at Aix-en-Provence, and in

[1] See Guédès, Michel. L'Edition originale de la *Philosophia botanica* de Linné (1751). *J. Soc. Bib. Nat. Hist.*, 4, 1968, pp. 385–389.

[2] *Historia Naturalis Classica*, 48.

[3] See Papers commemorating the bicentenary of the publication of the *Species plantarum*. *Proc. Linn. Soc.*, 165, 1955, pp. 151–166.

[4] See Gourlie, Norah. *The prince of botanists: Carl Linnæus*, 1953; Hagberg, Knut. *Carl Linnæus. . . . Translated from the Swedish by Alan Blair*, 1952; Jackson, Benjamin Daydon. *Linnæus (afterwards Carl von Linné): the story of his life, adapted from the Swedish of Theodor Magnus Fries, . . . and brought down to the present time in the light of recent research*, 1923; Larsson, B. Hjalmar. Carolus Linnæus, physician and botanist. *Ann. med. Hist.*, 2nd Ser., 10, 1938, pp. 197–214; Smith, J. E. *A selection of the correspondence of Linnæus and other naturalists*, [*etc.*], 2 vols., 1821; Baehni, Charles. Les grands systèmes botaniques depuis Linné. *Gesnerus*, 14, 1957, pp. 83–93; James, W. O. Linnæus (1707–1778). *Endeavour*, 16, 1956, pp. 107–112; Poynter, F. N. L. Linnæus, naturalist and doctor, 1707–78. *Brit. med. J.*, 1957, I, pp. 1359–1361; Uggla, Arvid Hj. *Linnæus*, Stockholm, 1957; Larson, James L. Linnæus and the natural method. *Isis*, 58, 1967, pp. 304–320; and Larson, James L. The species concept of Linnæus. *Isis*, 59, 1968, pp. 291–299.

1749 visited Senegal making collections in every branch of natural history. He wrote *Histoire naturelle du Sénégal. Coquillages. Avec la relation abrégée d'un voyage fait en ce pays, pendant . . . 1749–53*, Paris, 1757, part of which was translated into English as *A voyage to Senegal, the Isle of Goree, and the River Gambia*, in J. Pinkerton's *A general collection of . . . voyages*, [*etc.*], Vol. 16, 1814. Adanson was elected F.R.S. and Membre de la Légion d'Honneur. His classic publication in plant taxonomy was published as *Familles des plantes*, Paris, volume 1, 1764, volume 2, 1763, and was reprinted in Lehre, 1966. The Hunt Library contains Adanson's botanical library, letters and manuscripts, and has published a collective two-volume study to celebrate the bicentennial of this work. This contains a biography of Adanson, a bibliography of his writings and papers on his works.[1]

Jan Ingen Housz (1730–99) continued Priestley's experiments on plant physiology, demonstrating that plants give off a little carbon dioxide in darkness, and in light fix the free carbon dioxide in the air. He was born at Breda in Holland, and visited England on two occasions before settling there in 1789. Ingen Housz discovered photosynthesis, and appreciated the fact that plants derive food from the atmosphere. He was the author of *Experiments on vegetables, discovering their great power of purifying the common air in sunshine, but injuring it in the shade or at night*, London, 1779; *On the nutrition of plants*, 1796; *Vermischte Schriften*, Vienna, 1782, a German translation of his French manuscript; and his collected essays were published as *Nouvelle experiences et observations sur divers objects de physique*, Paris, 1785, and were dedicated to Benjamin Franklin.[2]

The experiments on pollination conducted by Joseph Koelreuter 1733–1806) form the basis of our knowledge on the subject. His *Vorlaufige Nachricht von einigen des Geschlecht der Pflanzen betreffenden Versuchen und Beobachtungen*, Leipzig, 1761, sets out his experiments and observations in a remarkably clear manner. Koelreuter's views on pollination and fertilization were further advanced by Konrad Sprengel (1750–1816) in his *Das neu entdeckte Geheimnis das Natur in Bau und Befruchtung der Blumen*, 1793.

A pupil of Linnæus, Carl Peter Thunberg (1743–1828), a Swedish botanist, discovered nearly 1,900 new plants. He left Sweden in 1770 and returned nine years later, having visited the Cape, Japan, Java and Ceylon. Thunberg published 293 medical and scientific works, including *Resa uti Europa, Africa, Asia förätad ären 1770–1779*, Uppsala, 1788–93,

[1] Carnegie Institute of Technology. The Hunt Botanical Library. *Adanson. The bicentennial of Michel Adanson's "Familles des plantes"*, Parts 1–2, Pittsburgh, 1963–64.

[2] See Foregger, Richard. Jan Ingen Housz and Joseph Priestley on carbon dioxide absorption, *Anaesthesiology*, 17, 1956, pp. 511–522; and Reed, Howard S. Jan Ingenhousz, plant physiologist. With a history of the discovery of photo-synthesis. *Chron. Bot.*, 11, 1949 pp. 285–393.

which was translated into English, French and German; *Flora Japonica*, [*etc.*], Leipzig, 1784; *Prodromus plantarum Capensium*, [*etc.*], Uppsala, 1794–1800; and *Flora Capensis*, [*etc.*], Uppsala, 1807.

William Withering (1741–99) was born at Wellington, Shropshire, and qualified in medicine at Edinburgh.[1] In 1775 he settled in Birmingham and wrote the first complete flora of the British Isles, *A botanical arrangement of all the vegetables naturally growing in Great Britain, with descriptions of the genera and species according to Linnæus*, Birmingham, 1776, of which numerous editions were published, including printings in 1787 (two volumes), with a third volume in 1792; 1796 (four volumes) and editions produced by his son in 1801,[2] 1812, 1818 and 1830. The fourteenth edition was published in 1877, and the titles alter in the various editions. Withering's principles for the use of digitalis are contained in his *An account of the foxglove, and some of its medical uses*, [*etc.*], Birmingham, London, 1785, a rare book which was reprinted in 1949. Withering was in Lisbon for health reasons for six months, and while there wrote *A chemical analysis of the water at Caldas da Rainha*, Lisbon, 1795. The *Miscellaneous tracts* of William Withering were edited by his son and published in two volumes, London, 1822. Birmingham Central Library houses the Withering family letters.[3]

The *Botanical Magazine*, founded in 1787 by William Curtis (1746–99), has continued publication down to the present day, and keeps the name of its founder before us. Curtis was born at Alton, Hampshire, and set up as an apothecary in London, finally to give his entire time to botany. In 1772 he was appointed *Praefectus Horti* at the Society of Apothecaries' Garden at Chelsea, and held that position until 1777. The first part of his *Flora Londinensis* was published in May, 1775, and parts appeared periodically until 1798. A two-volume edition was published 1777–98. There is a difference of opinion regarding the number of plates in this first edition, but F. Cardew[4] states that there were 432, and the second edition edited by Hooker, five volumes, 1817–28, contains 647. Three hundred copies of each part were printed, and the plates were hand-coloured by some thirty colourists. The text also is invaluable. Curtis opened his London Botanic Garden at Lambeth in 1779, and was the author of several books of lasting interest. These include *Instructions for collecting and preserving insects*,

[1] See also Thornton, pp. 127–128.

[2] See Greene, S. W. The publication date of William Withering's A systematic arrangement of British plants (Ed. 4), London, 1801. *J. Soc. Bib. Nat. Hist.*, 4, i, 1962, pp. 66–67.

[3] See Peck, Thomas Whitmore, and Wilkinson, Kenneth Douglas. *William Withering of Birmingham, M.D., F.R.S., F.L.S.*, Bristol, London, 1950; McMichael, Sir John. William Withering in perspective. *Univ. Birmingham hist. J.*, 11, 1967, pp. 41–50; and Estes, J. Worth, and White, Paul Dudley. William Withering and the purple foxglove. *Sci. Amer.*, 212, 1965, pp. 110–116, 119.

[4] Cardew, F. A note on the number of plates in Curtis's "Flora Londinensis", 1777, and Hooker's enlarged edition, 1817–1828. *J. Soc. Bib. Nat. Hist.*, 2, vi, 1950, pp. 223–224.

particularly moths and butterflies, [*etc.*], London, for the author, 1771, his first publication; a translation of Linnæus' *Fundamenta entomologiae,* [*etc.*], 1772; *Linnæus' system of botany,* 1777; *A short history of the brown-tail moth,* 1782, which was reprinted in facsimile in 1969 as the first of the series Classica Entomologica; and *General observations on the advantage which may result from the introduction of the seeds of our best grasses,* 1787, which went into several editions. The anonymously published *A catalogue of the plants growing wild in the environs of London,* London, 1774, has been attributed to William Curtis, but J. S. L. Gilmour[1] gives several reasons for doubting this.[2] James Sowerby (1757–1822) made some of the drawings for Curtis's *Botanical Magazine* and for the *Flora Londinensis,* in addition to several other botanical works. Sowerby wrote *An easy introduction to drawing flowers according to nature,* 1788, and *Flora luxurians,* 1789–91. He was responsible for the 2,590 drawings contained in the thirty-six volumes of *English botany,* 1790–1814; the original drawings are now in the British Museum (Natural History). Sir James Edward Smith provided the text for the *English botany,* although it is generally associated with the name of the artist.[3]

Antoine Laurent de Jussieu (1748–1836) continued the work of Linnæus in advancing a theory of classification, and his *Genera plantarum,* 1789, described a system of classification devised by his uncle Bernard de Jussieu (1699–1777). The younger de Jussieu later produced several monographs on families of plants.

Johann Wolfgang von Goethe (1749–1832), poet, dramatist and philosopher, was throughout his life a keen student of science and nature. His chief published contribution to science is his *Versuch, die Metamorphose der Pflanzen zu erklaren,* Gotha, 1790, regarding the value of which there is considerable dissension.[4] This work was reprinted in the same year, and French translations appeared in 1829 by F. Gingins-Lassarez, in 1831 by F. J. Soret and in 1837 by C. F. Martins. The first English translation, by Emily M. Cox, appeared in 1863.[5] Goethe proposed to publish a more detailed work on the subject and amassed a collection of notes and many drawings, but it is only in recent years that these have been collected and published.[6] He studied anatomy at Jena and wrote a short treatise on the

[1] Gilmour, J. S. L. A "Catalogue of London plants" attributed to William Curtis. *J. Soc. Bib. Nat. Hist.,* 2, v, 1949, pp. 181–182.

[2] See also Curtis, W. Hugh. *William Curtis 1746–1799. Fellow of the Linnean Society, botanist and entomologist,* [*etc.*], Winchester, 1941; Gilmour, J. S. L. William Curtis. *Nature,* 157, 1946, pp. 14–15; Savage, S. Curtis as naturalist and humanist. *Nature,* 157, 1946, pp. 15–16.

[3] See also p. 346.

[4] See Arber, Agnes. Goethe's botany. The Metamorphosis of plants (1790) and Tobler's Ode to nature (1782). With an introduction and translations. *Chron. Bot.,* 10, 1946–47, pp. 63–126.

[5] See *Journal of Botany,* 1, 1863, pp. 327–345, 360–374.

[6] See particularly Troll, W. *Goethe's Morphologische Schriften,* Jena, 1926, quoted by Arber.

intermaxillary bone in 1784.[1] The theoretical views at which he arrived were written up as *Erster Entwurf einer allgemeinen Einleitung in die vergleichende Anatomie* in 1795, which was circulated in manuscript, but it was not printed until 1820 when it was published in *Zur Morphologie*, and the plates were not published until 1831. In his work on colour, *Beiträge zur Optik*, 1791, and the work he regarded as his most important contribution to science, *Farbenlehre*, 1808, Goethe was opposed to the physicists' views. This was translated by Charles Eastlake as *Theory of colours*, 1840, and reprinted in 1968. A recent study of this work by J. R. Kantor[2] attempts to justify it as a genuine work of science, in contrast to Sir Charles Sherrington's[3] assessment, and M. H. Wilson[4] has repeated Goethe's experiments, concluding that his views were in advance of his period, and fully consistent with modern theories. Under the auspices of the Deutsche Akademie der Naturforscher (Leopoldina) and with a group of distinguished editors, the publication commenced in 1947 of the *Leopoldina-Ausgaben von Goethes Schriften zur Naturwissenschaft*, Weimar. This will be issued in three sections, the first comprising the text, to be in twelve volumes, while the second will comprise a complementary series of volumes of supplementary material and notes. The third section will aim at including all Goethe's correspondence on scientific matters, both letters and replies.[5]

Robert John Thornton (*c.* 1768–1837) has been described as a pompous and vain individual, but his name is also associated with a luxuriously produced botanical work. He was lecturer in medical botany at Guy's Hospital, and exhibited paintings and drawings in Dr. Thornton's Linnæan Gallery in Bond Street. He compiled a catalogue of this exhibition which went into four editions. Thornton issued the *New illustration of the sexual system of Carolus Linnæus* between 1799 and 1809, the concluding part of which became famous as *The temple of Flora*. Subscriptions to the parts of the work were disappointing, and in 1811 an Act was passed permitting a lottery to be held in order that Thornton might recover his expenses. This was a failure, and he was ruined financially. The original parts were issued in folio, and in book form in 1807, the artists being

[1] Wells, George A. Goethe and the intermaxillary bone. *Brit. J. Hist. Sci.*, 3, 1967, pp. 348–361.

[2] Kantor, J. R. Goethe's place in modern science. *Goethe bicentennial studies by members of the Faculty of Indiana University. Edited by H. J. Meesen*, 1950, pp. 61–82.

[3] Sherrington, Sir Charles. *Goethe on nature and on science*, Cambridge, 1942.

[4] Wilson, M. H., and Brocklebank, R. W. Goethe's colour experiments. *Yr. Bk. Phys. Soc.*, 1958, pp. 12–21.

[5] See Gray, Ronald D. *Goethe the alchemist: a study of alchemical symbolism in Goethe's literary and scientific works*, Cambridge, 1952; *Gesnerus*, 6, iii–iv, contains several articles on Goethe, including Baehni, Charles. M. de Goethe, botaniste, pp. 110–128; Steiner, Hans. Goethe und die vergleichende Anatomie, pp. 129–143; Schüepp, Otto. Goethe als Botaniker, pp. 144–158; see also Larson, James L. Goethe and Linnæus. *J. Hist. Ideas*, 28, 1967, pp. 590–596; and Wells, George A. Goethe and evolution. *J. Hist. Ideas*, 28, 1967, pp. 537–550.

Peter Henderson, Philip Reinagle (1749–1833), Abraham Pether (1756–1812) and Sydenham Teast Edwards. The lottery issue of the book, which appeared in 1811, has a title-page dated 1810, and a quarto edition appeared in 1812. It has been stated that no two copies of the original work are alike, and the *Temple of Flora* has been reprinted in recent years.[1] This contains useful bibliographical information by Handasyde Buchanan. Although Thornton's work has little botanical value, it does represent a magnificently produced work. He was also the author of *The elements of botany*, two volumes, 1812; *Practical botany*, 1808; *Botanical extracts, or philosophy of botany*, two volumes, 1810; *A new family herbal*, 1810; *Outlines of botany*, 1812; *An easy introduction to the science of botany*, [*etc.*], 1823, and several other books.[2]

[1] Thornton, Robert John. *Temple of Flora: with plates faithfully reproduced from the original engravings and the work described by Geoffrey Grigson; with bibliographical notes by Handasyde Buchanan*, London, 1951.

[2] See Blunt, Wilfrid. *The art of botanical illustration*, [*etc.*], 1950; Chapter 18. Thornton and the Temple of Flora, pp. 203–208. Also Hill, Brian. The best laid schemes: Bonnell Thornton, M.B. (1724–1768), and Robert James Thornton, M.D. (1768–1837). *Practitioner*, 200, 1968, pp. 722–726.

CHAPTER VI

Scientific Books of the Nineteenth Century

"The names of the prime-movers of science disappear gradually in a general fusion, and the more a science advances, the more impersonal and detached it becomes."

CLAUDE BERNARD

The growth of scientific literature becomes overwhelming as we approach the present century, and selection of the more important publications in each major branch of science becomes an increasingly difficult problem. Furthermore, as these branches split into even more minute specialities, each having its own enormous weight of literature, it becomes impossible, within a single chapter, to do justice to every aspect of scientific writing. It must also be appreciated that many important contributions to science were not contained in books, but were, in increasing numbers, the subject of articles in periodicals. This might imply that for the purpose of this book certain major contributors to the development of science are ignored because they did not write books, but a few of these figures, Mendel for example, are included. Today, most major advances in knowledge are first announced in journals, and sometimes take a considerable time after this event to appear in textbooks. Monographs on highly specialized topics are decreasing in number, chiefly due to the economics of the publishing and book trades, and it is probable that in the future there will be little room for scientific literature other than periodicals and students' textbooks.

The nineteenth century saw a vast increase in the output of printed material on all branches of science, and there was much re-evaluation in addition to continued research. Several branches of science were reborn in their modern conceptions during this period.[1] Chemistry advanced by strides that are as yet undiminished despite the vast progress made within a century and a half; biology was stimulated by the writings of Charles Darwin, which evoked a wealth of literature from disciples and opponents; and the related subjects each had its outstanding champions, not all of whom have retained their premier positions in the light of modern knowledge.

Classics of science may not be recognized until a considerable period

[1] See Dingle, Herbert, *ed. A century of science, 1851–1951. Written by specialist authors under the editorship of Herbert Dingle*, London, [*etc.*], 1951.

after publication, and best-sellers of today may be entirely neglected a few years after their appearance. It is therefore difficult to assess the value of comparatively modern literature, and in this chapter we mention merely the highlights in each major branch of science.

In some subjects it is impossible to indicate certain books as outstanding contributions to scientific thought, either because the more eminent exponents did not write books, or their discoveries were announced in the periodical press. Astronomy, for example, made extraordinary progress, largely as the result of great improvements in the size of the telescopes employed. Yet few eminent books devoted to the subject were published in the nineteenth century.[1] This despite the fact that several prominent astronomers, both amateur and professional, were recording important discoveries in the heavens.

Friedrich Wilhelm August Argelander (1799–1875), a native of Memel, became assistant to Friedrich Wilhelm Bessel (1784–1846) at Königsberg, and in 1823 was director of the observatory at Abo, and later at Bonn (1836). Argelander compiled *Uranometria nova*, 1843, a chart of all naked-eye stars visible in Europe, and carried out a scheme of Bessel's for charting all stars down to the ninth magnitude. He worked on this from 1852 to 1863, and with E. Schönfield and A. Krüger produced the *Bonner Durchmusterung*, 1863. This gives the position of 324,198 stars. Several other catalogues deserve mention: the *Harvard photometry*, 1884, of Edward Charles Pickering (1846–1919), professor of astronomy and director of the observatory at Harvard, which gives the magnitudes of 4,260 stars; Charles Pritchard (1808–93), president of the Royal Astronomical Society (1866), and Savilian professor of astronomy at Oxford (1870), was responsible for *Uranometria nova Oxoniensis*, 1885, giving the magnitudes of 2,784 naked-eye stars; and G. Müller and P. Kempf were responsible for the *Potsdamer photometrische Durchmusterung*, 1894–1907, containing the magnitudes of 14,199 stars.

One of the most prominent British astronomers of this period was Sir Joseph Norman Lockyer (1836–1920), a native of Rugby, who studied all aspects of astronomy, and was elected F.R.S. in 1869. He built the Norman Lockyer Observatory at Salcombe Regis, near Sidmouth, and is also remembered as an editor of *Nature*. Lockyer was the author of numerous books, several of an elementary nature, but including *Solar physics*, 1873; *The chemistry of the sun*, 1887; *The movements of the earth*, 1887; *The meteoritic hypothesis*, 1890; *The sun's place in nature*, 1897; *Recent and coming eclipses*, 1897; and *Inorganic evolution*, 1900.[2]

[1] See Waterfield, Reginald L. *A hundred years of astronomy*, 1938; and Armitage, Angus. *A century of astronomy*, 1950.

[2] See Lockyer, T. Mary, and others. *Life and work of Sir Norman Lockyer. By T. Mary Lockyer and Winifred L. Lockyer, with the assistance of H. Dingle,* [*etc.*], 1928.

The close relationship between astronomy and mathematics is evident from the work of Carl Friedrich Gauss (1777–1855), a native of Brunswick, who was closely connected with Göttingen both as student and as professor of mathematics. Gauss was eminent as a theoretical astronomer, in addition to being one of the greatest mathematicians and physicists of the nineteenth century. His *Disquisitiones arithmeticae*, Leipzig, 1801, following his doctoral dissertation of 1799, elaborated his theory of numbers. Gauss was also the author of a book contributing to the development of practical astronomy, *Theoria motus corporum coelestium*, Hamburg, 1809, which was translated into English by C. H. Davis as *Theory of the motion of the heavenly bodies moving about the sun in conic sections*, Boston, 1857. A reprint was published in New York, 1963. His collected writings were published as *Werke*, [*etc.*], Göttingen, twelve volumes, 1863–1929.[1]

Jean Baptiste Joseph Fourier (1768–1830) was born at Auxerre, where he was educated at the military school, to become teacher of mathematics there in 1784. He later taught at the École Normale, and became professor of analysis at the École Polytechnique. Fourier conducted investigations on the conduction of heat, and in 1822 appeared his *Théorie analytique de la chaleur*, which was published in Paris. He left an unfinished manuscript on determinate equations which was edited by Navier, and published as *Analyse de équations determinées*, Paris, 1831. This contains "Fourier's theorem". His works, edited by G. Darboux, were published in two volumes, Paris, 1888–90.

A great friend of Lagrange and Laplace, Siméon Denis Poisson (1781–1840) was the author of a large number of papers published in periodicals, and wrote a memoir on finite difference at the age of eighteen. His main treatises include *Traité de méchanique*, two volumes, Paris, 1811, of which a second enlarged edition was published in 1833; *Mémoires sur les surfaces élastiques*, Paris, 1814; *Théorie nouvelle de l'action capillaire*, Paris, 1831; *Théorie mathématique de la chaleur*, Paris, 1835, to which a supplement was published in 1837; *Recherches sur la probabilité de jugements*, [*etc.*], Paris, 1837; and *Recherches sur le mouvement des projectiles dans l'air*, [*etc.*], Paris, 1839.

Baron Augustin Louis Cauchy (1789–1857), a native of Paris, elaborated a theory of algebraic equivalences, and made several contributions to mechanics. His major treatises include *Cours d'analyse de l'École Polytechnique*, 1821; *Le calcul infinitésimal*, 1823; and *Leçons sur les applications du calcul infinitésimal a la géometrie*, 1826–28.[2] His works were published by the French Minister of Public Instruction in twenty-seven volumes, 1882–1938.

[1] See Dunnington, G. Waldo. *Carl Friedrich Gauss: titan of science. A study of his life and work*, New York, 1955.

[2] See Dubbey, J. M. Cauchy's contribution to the establishment of the calculus. *Ann. Sci.* 22, 1966, pp. 61–67.

Although Jacob Steiner (1796–1863), a native of Switzerland, was rather backward in early life, he later became professor of geometry at Berlin University. He was the author of numerous publications, but his most prominent, *Systematische Entwickelung der Abhängigkeit geometrischer Gestalten von einander*, 1832, was of primary significance in geometry.

Rudolf Friedrich Alfred Clebsch (1833–72) was born at Königsburg and became professor of theoretical mechanics at Karlsruhe (1858), professor of mathematics at Giessen (1863) and professor of applied mathematics at Göttingen (1868), where he died in 1872. His book on the theory of elasticity was first published as *Theorie der Elastizität Korper*, 1862, and a greatly expanded French translation by Alfred-Arné Flamant (1839–1913), with editorial annotations by Saint-Venant, was issued as *Théorie de l'élasticité des corps solides*, Paris, 1883 (reprinted 1966). Clebsch was also the author of *Theorie der Abelischen Funktion*, written with W. P. Gordon, 1866; *Theorie der binären algebraischen Formen*, 1872; and numerous papers. With C. Neumann he founded *Mathematischen Annalen* in 1868.

Two figures attained prominence in mathematics despite the fact that they died at an early age. Niels Henrik Abel (1802–29), a Norwegian by birth, conducted research into the theory of elliptical functions, and there have been two editions of his collected works, one published by the Swedish Government in 1839, and a more complete edition published in 1881.[1] Evariste Galois (1811–32) died as the result of a duel at the age of twenty, and although he published only five short articles in periodicals, his main work was his contribution to the group theory, or the theory of symmetry. This was elaborated by later mathematicians, and is recognized as the work of a genius.[2]

The calculating machine of Charles Babbage (1791–1871) was a forerunner of the computer, for the building of which he received a government grant, but died before his work was finished. Babbage was born near Teignmouth, Devon, on 26th December, 1791, and was educated at Cambridge, where he was professor of mathematics from 1828 to 1839. He was a founder of the Analytic Society of Cambridge, the Royal Astronomical Society and of the Statistical Society, and helped to establish the postal system in England in 1840. Babbage invented the first speedometer and the ophthalmoscope, and was the author of several books and articles in the *Philosophical Transactions*. His publications include *The comparative view of the various institutions for the assurance of lives*, London, 1826, reprinted New York, 1967; *The table of logarithms of the natural*

[1] See Ore, Oystein. *Niels Henrik Abel: mathematician extraordinary*, Minneapolis, London, 1957.

[2] See Birkhoff, Garrett. Galois and group theroy. *Osiris*, 3, 1938, pp. 260–268; and Sarton, George. Evariste Galois. *Osiris*, 3, 1938, pp. 241–259.

numbers from 1 to 108000, London, 1827, second edition 1831; *Decline of science in England*, 1830; *Economy of machines and manufactures*, London, 1834; *The ninth Bridgewater treatise. A fragment*, London, 1837, second edition, 1838 (reprinted London, 1967 and New York, 1968), and Philadelphia, 1841; and *Passages from the life of a philosopher*, London, 1864, reprinted London, 1968. Philip and Emily Morrison have edited a selection of Babbage's writings under the title *Charles Babbage and his calculating engines*, New York, London, 1961 and New York, 1962.[1]

Sir William Rowan Hamilton (1805–65) was born in Dublin of Scottish parents, and at the age of seventeen discovered an error in Laplace's *Mécanique celeste*. Five years later Hamilton was appointed professor of astronomy at Dublin University. In 1843 he invented a new method of mathematical analysis known as quaternions, which is contained in his posthumously published *Elements of quaternions*, 1866. Hamilton had published his *Theory of systems of rays*, and was also the author of numerous mathematical papers which have been published as *The mathematical papers of Sir William Rowan Hamilton. Edited for the Royal Irish Academy by A. W. Conway and J. L. Synge*, two volumes, Cambridge, 1931–40.

A native of Lincoln, George Boole (1815–64) became professor of mathematics at Queen's College, Cork, and was one of the originators of mathematical logic. Boole's work has gained greater recognition in recent years, and his main works are represented by *An investigation of the laws of thought, on which are founded the mathematical theories of logic and probabilities*, New York, 1854; *A treatise on the calculus of finite differences*, Cambridge, 1860, and London, 1880, still a standard work on the subject; *The mathematical analysis of logic*, Oxford, 1948; and *Studies in logic and probabilities*, London, (1952).[2]

William Kingdon Clifford (1845–79) was professor of mathematics at University College London from 1871 until his death, and was a philosopher as well as a mathematician and geometrician. He left several unfinished manuscripts which were published posthumously, those of mathematical interest being *Lectures and essays*, 1879; *Mathematical fragments*, [*etc.*], 1881; and *Mathematical papers*, [*etc.*], 1882, which contains a bibliography of Clifford's writings.

Physics progressed rapidly at the hands of numerous scientists, but the history of the subject in the first half of the nineteenth century has been

[1] See Bowden, B. V. *Faster than thought. A symposium on digital computing machines*, London, 1953; Moseley, M. *Irascible genius: a life of Charles Babbage, inventor*, London, 1964; and Hollindale, S. H. Charles Babbage and Lady Lovelace—two 19th century mathematicians. *Bull. Inst. Math. Applic.*, 2, 1966, pp. 2–15.

[2] See Kneale, W. Boole and the algebra of logic. *Notes Rec. Roy. Soc. Lond.*, 12, 1956, pp. 53–63; and Taylor, Sir Geoffrey. George Boole, F.R.S., 1815–1864. *Notes Rec. Roy. Soc. Lond.*, 12, 1956, pp. 44–52.

badly neglected, as has been pointed out by L. Pearce Williams,[1] who also indicates numerous sources of original manuscript material. The major figures contributing to the subject follow in roughly chronological sequence. Thomas Young (1773–1839) was an infant prodigy, who excelled in several subjects. He was a physician, and compiled a bibliography devoted to medicine;[2] as an Egyptologist he deciphered the demotic text of the Rosetta Stone; he founded the science of physiological optics,[3] and performed important experiments on light and sound. Young correctly explained capillary phenomena, and his wave theory of light was a major step in science. His dissertation was published as *De corporis humani viribus conservatricibus dissertatio*, Göttingen, 1796, and is very rare and little known. He wrote sixty-one articles for the first edition of *Encyclopaedia Britannica*, the earlier ones appearing over the initials "O.R." at Young's request, and he contributed numerous papers to the *Philosophical Transactions*, *Nicholson's Journal*, the *Philosophical Magazine*, etc. His *Course of lectures on natural philosophy and the mechanical arts* was published in two large quarto volumes, 1807, while a second, octavo, edition appeared in 1845. Young's writings were published as *Miscellaneous works*, three volumes, 1855, the first two volumes including his scientific memoirs, being edited by George Peacock, while Volume 3 contains his hieroglyphical essays and correspondence, and was edited by John Leitch. Mention must also be made of *Oeuvres optalmologiques de Thomas Young. Traduites et annotées par M. Tscheming*, [*etc.*], Copenhagen, 1894.[4]

Hans Christian Oersted (1777–1861) discovered electromagnetism in 1820, and announced his preliminary studies in a four-page pamphlet "Experimenta circa effectum conflictus electrici in acum magneticam", which was promptly translated into French, Italian, German, English and Danish. It was originally distributed privately, and a facsimile with an English translation were published in 1928.[5] Oersted was a student at Copenhagen University, and his electrochemical system was first described in *Ansicht der chemischen Naturgesetze durch die neueren Entdeckungen gewonnen*, Berlin, 1812, of which a greatly revised edition was published

[1] Williams, L. Pearce. The physical sciences in the first half of the nineteenth century: problems and sources. *History of Science*, 1, 1962, pp. 1–15.

[2] See Thornton, pp. 246–247.

[3] Observations on vision. *Phil. Trans.*, 83, 1793, pp. 169–181; and, On the mechanism of the eye. *Phil. Trans.*, 91, 1801, pp. 23–88; see also Millington, E. C. History of the Young-Helmholtz theory of colour vision. *Ann. Sci.*, 5, 1941–47, pp. 167–176.

[4] See Wood, Alexander, and Oldham, Frank. *Thomas Young, natural philosopher, 1773–1829*, [*etc.*], Cambridge, 1954. Appendix III (pp. 346–350) consists of a bibliography of Young's writings. See also Peacock, George. *The life of Thomas Young*, 1855; and Robinson, Herbert Spencer. Thomas Young: a chronology and a bibliography with estimates of his work and character. *Med. Life*, 36, 1929, pp. 227–540.

[5] *Isis*, 10, 1928, pp. 437–444. This consists of facsimiles of the original article, and of the English translation from *Annals of Philosophy*, 16, 1820, pp. 273–276.

as *Recherches sur l'identité des forces chimiques et électriques*, Paris, 1813. A collection of his lectures and essays on the relationship between science and religion, aesthetics and poetry and other topics was published as *The soul in nature*, London, 1852, and was reprinted in 1966 with a biographical introduction by P. L. Moeller. Oersted's scientific papers were published as *Naturvidenskabelige skrifter. . . . Scientific papers*, edited by Kirstine Meyer, three volumes, Copenhagen, 1920, which contains a biography of Oersted.[1]

The science of electro-magnetism was also considerably developed by André-Marie Ampère (1775–1836), a native of Polémieux, near Lyons. He was professor of mathematics at the Lyons Lyceum, and later at the École Polytechnique, Paris. Ampère is regarded as the founder of the science of electro-dynamics, and he enunciated the laws of force between a magnet and a constant electric current. His name is perpetuated in the unit of current. Ampère's original papers appeared in periodicals, but he was also the author of *Théorie des phénomènes électrodynamiques uniquement déduite de l'expérience*, Paris, 1826.[2]

The unit of electrical resistance is named after Georg Simon Ohm (1787–1854), a native of Erlangen, who discovered the law relating resistance to current and voltage. Ohm also studied acoustics and crystal interference, and eventually became professor of physics and mathematics at Munich. He was the author of numerous papers on mathematics, but little of significance except a pamphlet entitled *Die galvanische Kette mathematisch bearbeitet*, Berlin, 1827, which summarizes Ohm's law, previously outlined in a journal article.

Augustin Jean Fresnel (1788–1827) was a native of Normandy and became a civil engineer, devoting his leisure to the study of physical optics. He explained the theory of the interference of polarized light, and extended Young's undulatory theory of light. Fresnel's papers were mainly presented to the Académie des Sciences, but many were not printed until some years after his death.

The pioneer work on kinetics of John Herapath (1790–1868) was largely neglected, mainly because his earliest work on the subject was rejected by the Royal Society. Herapath was born in Bristol, and taught mathematics. His papers on the kinetic theory were published between 1816 and 1822 in *Annals of Philosophy*, without Herapath knowing of Daniel

[1] See Dibner, Bern. *Oersted and the discovery of electromagnetism*, New York, 1963; and Stauffer, Robert C. Speculation and experiment in the background of Oersted's discovery of electromagnetism. *Isis*, 48, 1957, pp. 33–50.

[2] See Valson, C. A. *Etude sur la vie et les ouvrages d'Ampère*, 1885; André-Marie Ampère, 1775–1836. L'Oeuvre et la vie d'Ampère. *Revue générale de l'électricité*, 6e ann. Nov. 1922. Numéro special. (Contains list "Publications d'Ampère sur l'électrodynamique", pp. 78–80); and Gardiner, K. R., and D. L. André-Marie Ampère and his English acquaintances. *Brit. J. Hist. Sci.*, 2, 1965, pp. 235–245.

Bernoulli's earlier quantitative, mathematical account of the theory as published in 1738 in his *Hydrodynamica*. Herapath was proprietor and manager of *Railway Magazine* (1836), later *Herapath's Railway and Commercial Journal*, published weekly. Joule revived Herapath's kinetic theory of gases, and later Bernoulli's earlier theory was re-discovered, and Herapath's work was neglected. In addition to several other papers, he was the author of *Mathematical physics*, two volumes, London, 1847, the first volume of which consists mainly of his articles in *Annals of Philosophy*.[1]

Although Michael Faraday (1791–1867) held the chair of chemistry at the Royal Institution, much of his research work lay within the domains of physics. He was born at Newington Butts, Surrey, and became assistant to Davy at the Royal Institution, with which body he was associated for most of his life. In chemistry, Faraday discovered benzol and butylene, and in physics his main work was connected with electromagnetic rotations. A. E. Jeffreys[2] has compiled a bibliography of Faraday's writings, excluding translations and American reprints, but giving locations of various letters and manuscripts. Jeffreys records in strict chronological order all books and separate publications, articles in periodicals, lectures whether printed or not, and manuscript lecture notes. Faraday's earliest papers were based on lectures given at the City Philosophical Society between 1816 and 1819, and his first book was published as *Chemical manipulations: being instructions to students in chemistry on the methods of performing experiments of demonstration or of research, with accuracy and success*, London, [*etc.*], 1827 and 1829, with a new edition published in 1830, and a third edition in 1842. His *Experimental researches in electricity*, [*etc.*], three volumes, London, 1839–55, contains articles reprinted from the *Philosophical Transactions*, the *Quarterly Journal of Science*, the *Philosophical Magazine*, and *Proceedings of the Royal Institution*. A second edition of volume one was published in London [1878–82?]; and a selection of material from the first edition appeared in Everyman's Library in [1912], and was reprinted six times between 1922 and 1951. A further collection of Faraday's articles in periodicals was published as *Experimental researches in chemistry and physics*, [*etc.*], London, 1859. Numerous reprints were issued of his *A course of six lectures on the chemical history of the candle*, [*etc.*], London, 1861, including a basic English edition in 1933. The manuscript of Faraday's diary has been in the possession of

[1] See Brush, S. G. The development of the kinetic theory of gases. 1. Herapath. *Ann. Sci.*, 13, 1957 [1959], pp. 188–198; Mendoza, Eric. The surprising history of the kinetic theory of gases. *Memoirs and Proceedings, Manchester Lit. & Phil. Soc.*, 105, 1962, pp. 15–28; and Talbot, G. R., and Pacey, A. T. Some early kinetic theories of gases: Herapath and his predecessors. *Brit. J. Hist. Sci.*, 3, 1966, pp. 133–149.

[2] Jeffreys, Alan Edward. *Michael Faraday: a list of his lectures and published writings*, [*etc.*], 1960. The original version of this useful bibliography was submitted in part requirement for the Diploma in Librarianship, University College, London. It also contains a list of the main biographies of Faraday (pp. xxv–xxvi).

the Royal Institution since his death in 1867, and it has been published as *Faraday's Diary: being the various philosophical notes of experimental investigation . . . during the years 1820–1862*, eight volumes, 1932–36. A definitive biography by L. Pearce Williams[1] has been published, and there have been several other biographical studies, and investigations into various aspects of Faraday's work.[2]

Sadi Carnot (1796–1832), a French engineer, was the author of only one scientific publication, which was concerned with the theory of heat and the efficiency of heat engines. Entitled *Réflexions sur la puissance motrice du feu et sur les machines propres à develloper cette puissance*, Paris, 1824, this founded the modern science of thermodynamics. An English translation by R. H. Thurston was published in 1890, and was reprinted, with additional material by Eric Mendoza, New York (Dover Books), 1960. A facsimile of the 1824 edition was issued in 1966 [1967]. A paper by Benoit-Paul-Emile Clapeyron (1799–1864) probably saved Carnot's book from oblivion, and led to the incorporation of his contribution into the history of thermodynamics. This was "Mémoire sur la puissance motrice de la chaleur" (*J. de l'école polytechn.*, 14, 1834, pp. 153–190), and it also appeared in German (1843) and English (1837) translations. Clapeyron contributed to the mathematical theory of the elasticity of solid bodies, and was the author of *Mémoire sur la force motrice de la chaleur*, 1843.[3]

The English physicist James Prescott Joule (1818–89) studied under Dalton, and wrote several important papers on electricity and magnetism. He conducted experiments on the relation of work and heat, and his *On the calorific effects of magneto-electricity, and on the mechanical value of heat*, 1843, sets out "Joule's equivalent". He later published in a Manchester weekly paper a vital article entitled "Matter, living force and heat". Joule followed up the work of Carnot, and the joule, the practical unit of energy, is so named in his honour. Joule's experiments

[1] Williams, L. Pearce. *Michael Faraday*, 1965.

[2] See Bence-Jones, Henry. *The life and letters of Faraday*, 2 vols., 1870; Thompson, Silvanus P. *Michael Faraday, his life and work*, London, [*etc.*], 1898; Kendall, James. *Michael Faraday, man of simplicity*, 1955; Hartley, Sir Harold. Michael Faraday as a physical chemist. *Trans. Faraday Soc.*, 49, 1953, pp. 473–488; Cragg, R. H. "Work, finish and publish". The chemistry of Michael Faraday 1791–1867. *Chemistry in Britain*, 3, 1967, pp. 482–486; Levere, T. H. Faraday, matter, and natural theology—reflections on an unpublished manuscript. *Brit. J. Hist. Sci.*, 4, 1968, pp. 95–107; Ross, S. Faraday consults the scholars: the origins of the terms of electrochemistry. *Notes Rec. Roy. Soc. Lond.*, 16, 1961, pp. 187–220; Seeger, R. J. Michael Faraday: his scientific insights and philosophical outlook. *Physis*, 8, 1966, pp. 220–231; and Gorman, Mel. Faraday on lightning rods. *Isis*, 58, 1967, pp. 96–98.

[3] See Kuhn, Thomas S. Sadi Carnot and the Cagnard engine. *Isis*, 52, 1961, pp. 567–574; Mendoza, Eric. Sadi Carnot and the Cagnard engine. *Isis*, 54, 1963, pp. 262–263; Nash, Leonard K. The Carnot cycle and Maxwell's relations. *J. chem. Educ.*, 41, 1964, pp. 368–372; Frank, F. C. Reflections on Sadi Carnot. *Physics Education*, 1, 1966, pp. 11–18; and Kerker, Milton. Sadi Carnot. *Sci. Mon.*, 85, 1957, pp. 143–149.

on the expansion of air[1] have been the subject of a paper by A. P. Hatton and L. Rosenfeld,[2] who added results from Joule's original notebook, now in the Joule Museum at Salford, that were not published in his original paper. Joule was elected a Fellow of the Royal Society in 1850, and ten years later received the Copley Medal. His *Scientific Papers* were published in two volumes, 1884–87, and these were reprinted in 1963.

John Tyndall (1820–93), a native of Leighlin Bridge, County Carlow, was a teacher of mathematics and surveying, and also studied chemistry at Marburg under Bunsen before he confined his attention to physics. He became professor of natural philosophy at the Royal Institution in 1853,[3] and in 1867 succeeded Faraday as director, having been elected F.R.S. in 1852. He studied radiant heat in relation to gases and vapours, and conducted researches on sound. Tyndall was the author of *Heat considered as a mode of motion*, 1863, with a second edition in 1865; *Sound*, 1867; *Faraday as a discoverer*, 1868, *Fragments of science*, [*etc.*], 1871, which went into several editions; *Forms of water*, 1872; and *Contributions to molecular physics in the domain of radiant heat*, London, 1872, containing his collected papers.[4]

After holding chairs of anatomy, physiology and pathology at various universities, Herman Ludwig Ferdinand von Helmholtz (1821–94) became a professor of physics at Berlin in 1871 and devoted himself entirely to physics. Helmholtz invented an ophthalmoscope, and made other important contributions to physiological optics and acoustics. He was the author of numerous works, including *Über die Erhaltung der Kraft*, [*etc.*], Berlin, 1847, on the conservation of energy; *Beschreibung eines Augenspiegels zur Untersuchung der Netzhaut im lebenden Auge*, Berlin, 1851, which was translated into English by Thomas Hall Shastid, Chicago, 1916; *Die Lehre von den Tonempfindungen als physiologische Grundlage für die Theorie der Musik*, Braunschweig, 1863, which went into several editions, including a French translation, 1868–74, and an English translation by A. J. Ellis, 1875, of which a second edition was published in 1885, and reprinted, New York (Dover Books), 1954; *Handbuch der physiologischen Optik*, Leipzig, 1867, with a second edition published at Hamburg in 1896, and an English translation in three volumes, 1924–25, the third edition of which has been reprinted in two volumes, edited by James P. C.

[1] "On the changes of temperature produced by the rarefaction and condensation of air." *Phil. Mag.*, 26, 1845, pp. 369–383.

[2] Hatton, A. P., and Rosenfeld, L. An analysis of Joule's experiments on the expansion of air. (Including original results not previously published). *Centaurus*, 4, 1955–56, pp. 311–318.

[3] See Thompson, D. John Tyndall and the Royal Institution. *Ann. Sci.*, 13, 1957, pp. 9–22.

[4] See [Young, H.] *A record of the scientific work of John Tyndall, D.C.L., L.L.D., F.R.S., 1850–1888*, for private circulation, 1935; and Eve, A. S., and Creasey, C. H. *Life and work of John Tyndall. . . . With a chapter on Tyndall as a mountaineer by Lord Schuster, and a preface by Granville Proby*, 1945.

Southall, New York (Dover), 1963; and *Wissenschaftliche Abhandlungen*, three volumes, Leipzig, 1882–95. An Italian translation, edited by V. Cappelleti was published as *Opere*, Turin, 1967 in the *Classici della Scienza* series.[1]

Among numerous other significant contributions to physics, chemistry and electricity, Clausius established the kinetic theory of gases in a series of papers published in *Annalen der Physik* from 1857 onwards, and formulated the second law of thermodynamics. One of the most original theoretical physicists of the nineteenth century, Rudolf Julius Emanuel Clausius (1822–88) was born in Koslin, Prussia (now Koszalin, Poland), and studied at Berlin and Halle. He became professor of physics at Zürich (1855–67), Würzburg (1867–69), and Bonn (1869–88), and was elected F.R.S. in 1868. Clausius was the author of several books and numerous papers, including *Die mechanische Wärmetheorie*, two volumes, 1864–67, and three volumes, 1876–91, which was translated into French, two volumes, Paris 1868. Many of his papers were published in the *Annalen*, and most appeared as translations in the *Philosophical Magazine*.[2]

For some years Heinrich Rudolf Hertz (1857–94), a native of Hamburg, served as assistant to Helmholtz, and in 1889 Hertz was appointed professor of physics at Bonn. He conducted researches on electric waves, his papers being published in the *Annalen der Physik* between 1887 and 1890. These were collected together and published in 1892, while an English translation of the writings of Hertz by D. E. Jones was issued in three volumes as *Electric waves*, 1893; *Miscellaneous papers*, 1896; and *Principles of mechanics*, 1899. The latter had first appeared as *Die Principien der Mechanik*, Leipzig, 1894. Hertz is remembered eponymously by Hertzian waves.[3]

Another British physicist, William Thomson, later Lord Kelvin (1824–1907), was professor of natural philosophy at Glasgow University from 1846 to 1899, and contributed usefully to mathematics, electricity, magnetism and geology, while he also supervised the making of the first Atlantic cable. Kelvin's writings include *Treatise on natural philosophy*, 1867; *Elements of natural philosophy*, 1873; *Reprint of papers on electrostatics and magnetism*, 1872; and *Popular lectures and addresses*, three volumes, 1889 to 1894. In 1844 Kelvin delivered a series of lectures on "Molecular dynamics and the wave theory of light" at Johns Hopkins

[1] See Warren, R. M. *Helmholtz on perception, its physiology and development*, New York, 1968; M'Kendrick, John Gray. *Hermann Ludwig Ferdinand von Helmholtz*, 1899; and Crombie, A. C. Helmholtz. *Sci. Amer.*, 198, 1958, pp. 94–102.

[2] See Brush, S. G. The development of the kinetic theory of gases. III. Clausius. *Ann. Sci.*, 14, 1958, pp. 185–196.

[3] See Süsskind, Charles. Hertz and the technological significance of electromagnetic waves. *Isis*, 56, 1965, pp. 342–345.

University, Baltimore, which were published in 1904. His *Collected mathematical and physical papers*, six volumes, were published 1882–1911.[1]

James Clerk Maxwell (1831–79), a native of Edinburgh, became professor of natural philosophy at Aberdeen (1856), at King's College, London (1860), and professor of experimental physics at Cambridge (1871), where he planned the Cavendish Laboratory. He continued the work of Faraday in electricity, founding the electrodynamic theory of light, and also contributed to mathematics and other branches of science. Clerk Maxwell was the author of *Treatise on electricity and magnetism*, 1873, his greatest work, of which a third edition, edited by J. J. Thomson was published in two volumes, Oxford, 1892, and reprinted in Oxford, 1904, Stanford, 1953, and New York, 1954; *Theory of heat*, 1871; *Matter and motion*, 1876; and he edited the works of Cavendish as *Electrical researches of the Hon. Henry Cavendish*, 1879. A memorial edition of his works was published by the Cambridge University Press in 1890 as *The Scientific papers of James Clark Maxwell*, two volumes, which were reprinted in Paris, 1927, and New York (Dover Books), 1952.[2]

Josiah Willard Gibbs (1839–1903) was born at New Haven, and educated at Yale, where he later became professor of mathematical physics. Willard Gibbs created the science of chemical thermodynamics, and made important contributions to electrochemistry and physical chemistry in general. Most of his writings appeared in periodicals, but certain of his lectures were published by E. B. Wilson as *Vector analysis*, [*etc.*], 1901, and his *Elementary principles of statistical mechanics* was issued in 1902. *The scientific papers of J. Willard Gibbs*, 1906, were later reprinted with the addition of his *Elementary principles* as *The collected works of J. Willard Gibbs*, two volumes, 1928.[3]

[1] See Gray, Andrew. *Lord Kelvin: an account of his scientific life and work*, London, New York, 1908, (English Men of Science Series); Thompson, S. P. *Life of Lord Kelvin*, 2 vols., 1910; and Silliman, Robert H. William Thomson: smoke rings and nineteenth-century atomism. *Isis*, 54, 1963, pp. 461–474.

[2] See Campbell, L., and Garnett, W. *The life of James Clerk Maxwell, with a selection from his correspondence and occasional writings*, [*etc.*,] 1882; Glazebrook, R. T. *James Clerk Maxwell and modern physics*, London, [*etc.*] 1896, (Century Science Series); Hopley, I. B. Maxwell's work on electrical resistance. *Ann. Sci.*, 13, 1957, pp. 265–272; 14, 1958, pp. 197–210; 15, 1959, pp. 51–55; Hopley, I. B. Maxwell's determination of the number of electrostatic units of one electromagnetic unit of electricity, *Ann. Sci.*, 15, 1959, pp. 91–108; Hopley, I. B. Clerk Maxwell's experiments on colour vision. *Sci. Progr.* 48, 1960, pp. 46–66; Brush S. G. The development of the kinetic theory of gases. IV. Maxwell. *Ann. Sci.*, 14, 1958, pp. 243–255; Bernstein, Henry T. J. Clerk Maxwell on the history of the kinetic theory of gases, 1871. *Isis*, 54, 1963, pp. 206–216; and Bork, Alfred M. Maxwell and the vector potential. *Isis*, 58, 1967, pp. 210–222.

[3] See Donnan, F. G., and Haas, Arthur, *eds. A commentary on the scientific writings of J. Willard Gibbs*, [*etc.*], 2 vols., New Haven, London, 1938; and Rukeyser, Muriel. *Willard Gibbs*, New York, 1942.

A pioneer of telegraphy, John Joseph Fahie (1846–1934) was a native of Tipperary, served with the Electric and International Telegraph Company in London (1865–67), and in 1867 went abroad in the service of the Indo-European Government Telegraph Department. He spent much of his life in Persia until his retirement in 1898, and wrote numerous articles on telegraphy. His *Faults in submarine cables* was privately published by the Society of Telegraph Engineers in 1874, to be followed ten years later by *A history of electric telegraphy, chiefly compiled from original sources and hitherto unpublished documents*, London, New York, 1884. Fahie's other scientific books include *A history of wireless telegraphy, including some bare-wire proposals for subaqueous telegraphs*, Edinburgh, London, 1899; *Galileo: his life and work*, 1903; and *Memorials of Galileo Galilei (1564–1642)*, 1929.[1]

The discoverer of the electron (1897), and pioneer in numerous discoveries in electricity and magnetism, Sir Joseph John Thomson (1856–1940) was awarded many honours, including the Nobel Prize in Physics (1906) and the Order of Merit (1912). His numerous publications include *A treatise on the motion of vortex rings*, 1883, reprinted 1968; *Applications of dynamics to physics and chemistry*, 1888, reprinted 1968; *Notes on recent researches in electricity and magnetism*, 1893, reprinted 1968; *Elements of mathematical theory of electricity and magnetism*, 1895; *Discharge of electricity through gases*, 1897; *Conduction of electricity through gases*, 1903; *The structure of light: the corpuscular theory of matter*, 1907; *Rays of positive electricity*, 1913; *The electron and chemistry*, 1923; and *Recollections and reflections*, 1936. Physics made progress at the hands of numerous other scientists, certain of them being considered below for their contributions to chemistry and other related subjects, while others, particularly those of the later period, have been dealt with in a book by William Wilson.[2]

Probably no branch of science has made more progress during the past one hundred and fifty years than has chemistry. Its literature is enormous, and continues to grow rapidly, for certain aspects of the field, biochemistry for example, are comparatively recent branches of the subject. During the nineteenth century, chemistry attracted the attention of some of our greatest scientists, and their discoveries sometimes opened vast new fields for investigation. A brief paper in a periodical might thus prove the origin of a new industry, and few books devoted to chemistry published during this period contained the results of up-to-date experiments. Handbooks, textbooks and dictionaries lagged behind knowledge that daily increased, to be recorded in journals that greatly increased in size to accommodate the spate of

[1] Whitehead, E. S. *A short account of the life and work of John Joseph Fahie*, Liverpool, London, 1939.
[2] Wilson, William. *A hundred years of physics*, 1950.

papers. The systems of chemistry that have attempted to synthesize current knowledge have been sadly out of date on publication, and the subject has so developed that the chemist cannot hope to keep abreast of advances save in a specialized branch of the subject.

J. R. Partington's[1] outstanding histories of the subject give one some idea of the tremendous strides in chemistry, and contain invaluable information on the chief personalties in well-documented surveys. Alexander Findlay[2] performs a similar service for the past hundred years in a volume providing much bibliographical and biographical material. Both these authorities provide information which emphasizes the fact that few of the eminent chemists of the period under consideration were the authors of books.[3] This being the criteria for inclusion here, the following selection is provided as a guide to the more prominent chemical books published in the nineteenth century.

Friedrich Accum (1769–1838) was a native of Westphalia, who spent some time in England promoting the adoption of gas lighting. He was the author of numerous books, including the following: *System of theoretical and practical chemistry*, two volumes, 1803, with a second edition in 1807; *A practical essay on the analysis of minerals*, [*etc.*], two volumes, London, 1804, of which a second edition was published in 1808 with the title *A manual of analytical mineralogy; Analysis of a course of lectures on mineralogy*, [*etc.*], London, 1809; *Manual of a course of lectures on experimental chemistry and on mineralogy*, [*etc.*], London, 1810; *A practical treatise on gaslight*, [*etc.*], London, 1815 (two editions), with a third English edition in 1816, a German translation, Berlin, 1816, and also a French version, Paris, 1816; *A practical essay on chemical reagents or tests*, [*etc.*], London, 1815, with a second edition in 1818, a third in 1828, and an Italian translation, Milan, 1819; *Chemical amusements*, London, 1817 (two editions), with a third in 1818, a fourth in 1819, and a Spanish translation, Paris, 1836; and *A treatise on adulteration of food and culinary poisons*, [*etc.*], 1820 (two editions).[4]

The first Regius Professor of Chemistry at Glasgow University, Thomas Thomson (1773–1852) was an early worker on the atomic theory, and founded one of the first chemical laboratories in Great Britain. He wrote numerous articles in periodicals as the result of his research work and became widely known for his constantly revised

[1] Partington, J. R. *A history of chemistry*, Vols. 2–4, 1961–64, (Volume 1, Part 1, 1970; and *A short history of chemistry. . . . Third edition*, 1957.

[2] Findlay, Alexander. *A hundred years of chemistry. . . . Third edition, revised by T. I. Williams*, 1965.

[3] See also Thornton, John L., and Tully, R. I. J. History and biography of chemistry. *In*, Bottle, R. T., *ed. The use of chemical literature. Second edition*, 1969, pp. 235–250.

[4] See Cole, R. J. Friedrich Accum (1760–1838). A biographical study. *Ann. Sci.*, 7, 1951. pp. 128–143.

textbook which went into numerous editions and translations. Entitled *A system of chemistry*, the first edition was published in four volumes at Edinburgh in 1802, and was followed by a second edition, four volumes, 1804; third edition, five volumes, 1807; fourth edition, five volumes, 1810; fifth edition, four volumes, London, 1817; sixth edition, four volumes, 1820; and a seventh edition bearing the title *A system of chemistry of inorganic bodies*, two volumes, London, 1831. It also appeared in French, German and American editions, and the first exposition of Dalton's atomic theory appeared in the third edition (1807). Thomson founded and edited *Annals of Philosophy* (1813–21), and was also the author of *Elements of chemistry*, 1810; *History of the Royal Society from its institution to the end of the eighteenth century*, London, 1812; *The history of chemistry*, two volumes, London, 1830–31, of which an undated "second edition" in one volume was printed without even correction of misprints[1]; *An attempt to establish the first principles of chemistry by experiment*, two volumes, London, 1825; and *Chemistry of organic bodies. Vegetables*, London, 1838.[2]

Although nearly forty years were to pass before the general acceptance of Avogadro's hypothesis put forward in his famous paper "Essai d'une manière de déterminer les masses rélatives des molécules des corps, [*etc.*]"[3] his work has been described as forming the basis of the whole of theoretical chemistry, and especially the structure theory of chemistry.[4] Lorenzo Romano Amedeo Carlo Avogadro, Count of Quaregna and Cerreto (1776–1856), was born in Turin on 9th August, 1776. Initially trained for the law, he studied mathematics and physics, becoming professor in these subjects at Vercelli, and in 1820, first professor of mathematical physics at Turin. The "Essai" was published in 1811 and was followed by other important papers in 1814[5] and 1821.[6] He also published *Fisica de' corpi ponderabili*, four volumes, Turin, 1837–41, and his selected works were published as *Opere scelte*, Turin, 1911.[7]

[1] See Kent, Andrew. Thomas Thomson (1773–1852), historian of chemistry. *Brit. J. Hist. Sci.*, 2, 1964–65, pp. 59–63.

[2] See Partington, J. R. Thomas Thomson, 1773–1852. *Ann. Sci.*, 6, 1948–50, pp. 115–126; and Klickstein, Herbert S. Thomas Thomson: pioneer historian of chemistry. *Chymia*, 1, 1948, pp. 37–53.

[3] *Journal de Physique*, 73, 1811, pp. 58–76; English translation in Alembic Club reprint No. 4. Foundations of the molecular theory, 1893, pp. 28–51.

[4] Pauling, Linus. *In*, Hinshelwood, Cyril N., and Pauling, Linus. Amedeo Avogadro. *Science*, 124, 1956, pp. 708–713.

[5] Mémoire sur les masses rélatives des corps simples, [*etc.*]. *Journal de Physique*, 78, 1814, pp. 131–156.

[6] Nouvelles considérations sur la théorie des proportions déterminées dans les combinaisons et sur la détermination des masses des molécules des corps. *Memorie della Reale Accademia delle Scienze di Torino*, 26, 1821, pp. 1–162.

[7] See also Meldrum, Andrew N. *Avogadro and Dalton. The standing in chemistry of their hypothesis*, Edinburgh, 1904; and Coley, N. G. The physico-chemical studies of Amedeo Avogadro. *Ann. Sci.*, 20, 1964 [1965], pp. 195–210.

ELEMENTA

PHYSIOLOGIÆ

CORPORIS HUMANI.

AUCTORE

ALBERTO v. HALLER,

Præside Societatis Reg. Scient. Götting.
Sodali Acadd. Reg. Scient. Paris. Reg. Chir. Gall.
Imper. Berolin. Suecic. Bononiens.
Societt. Scient. Britann. Upsal. Helvet.
In Senatu Supremo Bernensi Ducentumviro.

TOMUS PRIMUS.

FIBRA. VASA. CIRCUITUS SANGUINIS. COR.

LAUSANNÆ,

Sumptibus Marci-Michael. Bousquet & Sociorum.

MDCCLVII.

Plate 9. Title-page of volume one of Albrecht von Haller's *Elementa physiologiae corporis humani*, Lausanne, 1757.

LECTURES

ON

PHYSIOLOGY, ZOOLOGY,

AND THE

Natural History of Man,

DELIVERED AT

THE ROYAL COLLEGE OF SURGEONS

BY

W. LAWRENCE, F.R.S.

PROFESSOR OF ANATOMY AND SURGERY TO THE COLLEGE, ASSISTANT SURGEON TO ST. BARTHOLOMEW'S HOSPITAL, SURGEON TO BRIDEWELL AND BETHLEM HOSPITALS, AND TO THE LONDON INFIRMARY FOR DISEASES OF THE EYE.

WITH TWELVE ENGRAVINGS.

LONDON:

PRINTED FOR J. CALLOW, MEDICAL BOOKSELLER, PRINCES STREET, CORNER OF GERARD STREET, SOHO.

1819.

PLATE 10. Title-page of Sir William Lawrence's *Lectures on physiology, zoology, and the natural history of man*, London, 1819.

Louis Jacques Thenard (1777–1857) was a French chemist who became a professor at the Sorbonne. He discovered peroxide of hydrogen, invented a dye known as Thenard's blue, and made numerous important investigations in organic chemistry. He was the author of *Traité de chimie élémentaire*, 1813–16, which went into several editions, a portion of the work being translated into English by John George Children as *An essay on chemical analysis*, [*etc.*], London, 1819. In the previous year had appeared *A treatise on the general principles of chemical analysis*, [*etc.*], translated by Arnold Merrick from the French, London, 1818. Thenard worked with Joseph Louis Gay-Lussac (1778–1850) on chlorine and the alkali metals, and the two jointly produced *Recherches physico-chimiques*, [*etc.*], two volumes, Paris, 1811. Gay-Lussac held several important posts in Paris, having been professor of physics at the Sorbonne, professor of practical chemistry at the École Polytechnique and professor of general chemistry at the Jardin des Plantes. He was assistant to Berthollet and to Fourcroy, in 1804 he ascended in a balloon with Biot to conduct experiments, and in 1805 he made a scientific expedition with Alexander von Humboldt into Italy and Germany. Gay-Lussac discovered cyanogen and its compounds, and investigated the properties of iodine, writing an important paper entitled "Mémoire sur l'iode", [etc.],[1] which was also separately published, and was translated into German in 1889. His other important discoveries are recorded in the above publication written with Thenard.[2]

Sir Humphry Davy (1778–1829) was born in Penzance, the son of a wood-carver, and after a brief apprenticeship to an apothecary was released from his indentures to work with Thomas Beddoes at the Pneumatic Institute. He later became professor of chemistry at the Royal Institution, where Michael Faraday became his assistant. Davy discovered the anaesthetic properties of nitrous oxide, or "laughing gas", which he described in *Researches chemical and philosophical, chiefly concerning nitrous oxide, or dephlogisticated nitrous air, and its respiration*, London 1800. This edition has been described as "excessively rare". An annotated bibliography of Davy's published writings has been compiled by June Z. Fullmer[3], who has contributed several other studies on various aspects of his life and writings[4], and is preparing an edition of his correspondence. This is based on a collection of 330 letters housed

[1] In *Ann. de Chim.*, 91, 1814, pp. 5–161.
[2] See also Webb, K. R. Gay-Lussac (1778–1850) as chemist. *Endeavour*, 9, 1950, pp. 207–210; and Crosland, M. P. The origins of Gay-Lussac's law of combining volumes of gases. *Ann. Sci.*, 17, 1961 [1963], pp. 1–26.
[3] Fullmer, June Z. *Sir Humphry Davy's published works*, Cambridge, Mass., 1969.
[4] These include: Fullmer, June Z. Humphry Davy's adversaries. *Chymia*, 8, 1962, pp. 147–164; Davy's sketches of his contemporaries. *Chymia*, 12, 1967, pp. 127–150; and Humphry Davy and the gunpowder manufactory. *Ann. Sci.*, 20, 1964 [1965], pp. 165–194.

at the Royal Institution, and 500 items from other sources, and will form the basis of a biography of Davy. The Royal Institution also contains several of Davy's laboratory notebooks, and other material associated with him. He had taken an interest in tanning and leather manufacture, and this aspect of his early work has been described by C. H. Spiers,[1] while Robert Siegfried[2] has devoted a paper to Davy's little-known work on the diamond. Davy conducted research on electrolysis,[3] on chlorine, and on oxides of fluorine, but he is probably best remembered for his invention of the miner's safety lamp, first described in print in *On the safety lamp for coal miners: with some researches on flame*, London, 1818, but which had previously been described in a letter to John Buddle dated 14th February, 1817.[4] Other publications by Sir Humphry Davy include *Syllabus of a course of lectures*, 1802; *Elements of chemical philosophy*, Volume 1, Part 1, London, 1812, no further parts having been published; and *Elements of agricultural chemistry*, [*etc.*], London, Edinburgh, 1813, of which a sixth edition was published in 1839. Davy contributed forty-six papers to the Royal Society, and many of his articles were reprinted in other journals, either fully or in abstract. He published *Discourses to the Royal Society*, 1827, and an incomplete collection of his writings were also issued as *Collected works of Sir Humphry Davy. Edited by his brother John Davy*, nine volumes, London, 1839–40, of which Volume 1 is biographical. Davy was elected F.R.S. in 1803, knighted in 1812, created a baronet six years later, and served as President of the Royal Society from 1820 to 1827. While seriously ill towards the end of his life, Davy wrote a book on fly-fishing, which also contains observations on natural history, together with some philosophical discussions. Entitled *Salmonia: or days of fly fishing. In a series of conversations. With some account of the habits of fishes belonging to the genus Salmo. By an angler*, London, 1828, this went into several editions, the fifth appearing in 1869.[5] He died at Geneva on 29th May, 1829. Several biographies of Sir Humphry have been published,[6] and

[1] Spiers, C. H. Sir Humphry Davy and the leather industry. *Ann. Sci.*, 24, 1968, pp. 99–113.

[2] Siegfried, Robert. Sir Humphry Davy on the nature of the diamond. *Isis*, 57, 1966, pp. 325–335.

[3] Russell, Colin A. The electrochemical theory of Sir Humphry Davy. Part I. The Voltaic pile and electrolysis. Part II: Electrical interpretations of chemistry. *Ann. Sci.*, 15, 1959, pp. 1–25; Part III. The evidence of the Royal Institution manuscripts. *Ann. Sci.*, 19, 1963, pp. 255–271.

[4] See Weil, E. An unpublished letter by Davy on the safety-lamp. *Ann. Sci.*, 6, 1948–50, pp. 306–307; and Keys, Thomas E. Sir Humphry Davy and his safety-lamp for coal miners. *Mayo Clinic Proc.*, 43, 1968, pp. 865–891.

[5] See Keys, Thomas E. The "Salmonia" of Sir Humphry Davy. *Bull. Med. Libr. Ass.*, 44, 1956, pp. 431–442.

[6] See Davy, John. *Memoirs of the life of Sir Humphry Davy, Bart. By his brother*, 2 vols., 1836; Gregory, Joshua C. *The scientific achievements of Sir Humphry Davy*, 1930; Paris, J. A., *The life of Sir Humphry Davy*, 1831; and Thorpe, Sir Thomas Edward, *Humphry Davy, poet and philosopher*, 1896.

most are unsatisfactory for various reasons,[1] particularly those by J. A. Paris and by his brother John Davy. Two modern studies by Sir Harold Hartley[2] and Anne Treneer[3] are reliable sources, but a definitive biography remains desirable. The bibliography by June Z. Fullmer contains information on Davy's publications, including translations and items not published separately.

The Swedish chemist, Jons Jakob Berzelius (1779–1848), was professor of botany and pharmacy at Stockholm University, later professor of chemistry at the Chirurgico-Medical Institute, Stockholm, and determined the atomic and molecular weights of numerous chemical substances. He discovered several elements, and devised the symbols still employed for these. The most important publication by Berzelius in book form was his *Treatise on chemistry*. This was first published as *Lärbok i kemien*, six parts, Stockholm, 1808–30, of which Parts 1–2 were published in enlarged editions, 1817–22. There were numerous translations of this book, including eleven German editions, two Dutch, three Italian, one Spanish and four French. The second French, the definitive edition, was published as *Traité de chimie, minerale, végétale et animale*, [*etc.*], six volumes, Paris, 1845–50. Volume 4 (1831) of this repeats the first detailed exposition of the electrochemical theory provided by Berzelius in *Essai sur la théorie des proportions chimiques de l'électricité*, Paris, 1819[4,5]

A British chemist, William Prout (1785–1850), was the first to suggest that the atomic weights of all elements are multiples of that of hydrogen, and he was also the first to obtain urea in a pure form. Prout's *On the relation between the specific gravities of bodies in their gaseous state and the weight of their atoms*, [*etc.*], London, 1815–16, was first published anonymously in *Annals of Philosophy*.[6]

Leopold Gmelin (1788–1853) came from a family of famous chemists, and was professor of chemistry at Heidelberg from 1817 to 1851. He made

[1] See Fullmer, June Z. Davy's biographers: notes on scientific biography. *Science*, 155, 1967, pp. 285–291.

[2] Hartley, Sir Harold, *Humphry Davy*, 1966, (*British Men of Science Series*).

[3] Treneer, Anne. *The mercurial chemist. A life of Sir Humphry Davy*, 1963.

[4] See Russell, Colin A. The electrochemical theory of Berzelius. Part 1, Origins of the theory. Part 2, An electrochemical view of matter. *Ann. Sci.*, 19, 1963, pp. 117–145.

[5] See also Jorpes, J. E. *Jac. Berzelius: his life and work. Translated by B. Steele*, Stockholm, 1966. (*Bidrag till K. Svenska Vetenskapsakademiens Historia*, VII); Soloviev, I. I. New materials for the scientific biography of J. J. Berzelius—the scientific relations of Berzelius with Russian scholars, from unpublished letters. *Chymia*, 7, 1961, pp. 109–125; Holmberg, Arne. *Bibliographie de J. J. Berzelius*, [*etc.*], 3 parts, Stockholm, 1933–36; Robinson, Victor, Jons Jakob Berzelius (1779–1848). *Med. Life*, 35, 1928, pp. 165–185; Tilden, Sir William A. A sketch of Berzelius, *Med. Life*, 35, 1928, pp. 187–209; and Hartley, Sir Harold. The place of Berzelius in the history of chemistry, [*etc.*], *K. Svenska Vetenskapsakademiens Arsbok för år 1948, Bilaga*, pp. 31–50.

[6] Also reprinted as Alembic Club Reprint No. 20.

several important discoveries in chemistry, and his *Handbuch der theoretischen Chemie*, 1817–19, was reprinted and translated on numerous occasions, including an English translation in nineteen volumes, 1848–72. Gmelin also introduced the names 'ester' and 'ketone'.

Friedrich Wöhler (1800–82), German chemist, was the author of over three hundred papers, but is known chiefly for "Über kunstliche Bildung von Harnstoff", 1828.[1] Mainly because of an over-eulogistic obituary notice by A. W. Hofmann,[2] it has been held that this paper described the synthesis of urea from inorganic compounds and was thus responsible for the overthrow of vitalism. Douglas McKie[3] showed in 1944 that this was not so, and that the vitalistic theory continued in use until Berthelot's work in 1860; nevertheless the legend continues to be stated without qualification.[4] Wöhler was also the author of *Grundriss der anorganische Chemie*, 1831, and *Grundriss der organische Chemie*, Berlin, 1840.

Following apprenticeship to an apothecary, Jean Baptiste André Dumas (1800–84) conducted research in physiology and physiological chemistry before going to Paris in 1823. There he became assistant to Thenard, and was later to hold the position of professor of chemistry at the Sorbonne, the École Polytechnique and the École de Medecine. Dumas investigated the atomic theory, vapour densities, the composition of blood, among other important subjects, and became one of the most influential French chemists of the nineteenth century. His writings were admired for their lucid style, and the following represent his more important books: *Mémoire sur les substances vegetales qui se rapprochement du camphre, et sur quelques huiles essentielles*, [*etc.*], Paris, 1832; *Leçons sur la philosophie chimique*, [*etc.*], Paris, 1837; and *Traité de chimie appliquée aux arts*, eight volumes, Paris, Brussels, 1828–46.

Justus Liebig (1803–73) (who became von Liebig in 1845) acquired an early interest in chemistry. He was born at Darmstadt, and at the age of nineteen went to Paris as a pupil of Gay-Lussac. Liebig became professor of chemistry at Giessen (1825), and later at Munich (1852), which latter position he held until his death. He was eminent as a teacher, and became interested in agricultural and physiological chemistry,

[1] *Annalen der Physik und Chemie*, 12, 1828, pp. 253–256; French translation in *Annales de Chimie*, 37, 1828, pp. 330–334; English version in *Quarterly Journal of Science*, 1, 1828, pp. 491–492.

[2] *Berichte der Deutschen Chemischen Gesellschaft*, 15, 1882, pp. 3127–3290; includes a list of Wöhler's works.

[3] McKie, Douglas. Wöhler's synthetic urea and the rejection of vitalism: a chemical legend. *Nature*, 153, 1944, pp. 608–610.

[4] See also Campaigne, Ernest J. Wöhler and the overthrow of vitalism. *J. chem. Educ.*, 32, 1955, p. 403; Kurzer, Frederick, and Sanderson, Phyllis M. Urea in the history of organic chemistry. *J. chem. Educ.*, 33, 1956, pp. 452–459; Hartman, L. Wöhler and the vital force. *J. chem. Educ.*, 34, 1957, pp. 141–142.

acting as pioneer in these two subjects. Liebig introduced the well-known method of organic analysis, and edited the *Annalen der Pharmacie* (1832–39), later *Annalen der Chemie und Pharmacie* (1840–75), which became the *Annalen der Chemie*, and is now known as *Justus Liebig's Annalen der Chemie*.[1] His writings in book form include *Die Chemie in ihrer Anwendung auf Agricultur und Physiologie*, Braunschweig, 1840, which was translated into English by Lyon Playfair as *Chemistry in its application to agriculture and physiology*, 1840. This was followed by *Die Thierchemie oder der organische Chemie in ihrer Anwendung auf Physiologie und Pathologie*, Braunschweig, 1842; *Handbuch der organischen Chemie mit Rücksicht auf Pharmacie*, Heidelberg, 1843; *Chemische Briefe*, Heidelberg, 1844; *Chemische Untersuchungen über das Fleisch und seine Zubereitung, zum Nahrungsmittel*, Heidelberg, 1847; *Die Grundsätze der Agricultur-Chemie*, Braunschweig, 1855; *Über Theorie und Praxis in der Landwirthschaft*, Braunschweig, 1856; *Naturwissenschaftliche Briefe über die moderne Landwirthschaft*, 1859; and *Reden und Abhandlungen*, Leipzig, 1874 (reprinted 1965), consisting of a selection of his papers. Several of these works have appeared in English and French translations.[2]

A pioneer of physical chemistry, and founder of colloid chemistry Thomas Graham (1805–69) was born in Glasgow, where he became professor of chemistry (1830–37), and occupied the same post at University College, London (1837–55). He was Master of the Mint from 1855–69, having been elected F.R.S. in 1836, and serving as President of the Chemical Society in 1840. He formulated "Graham's Law" (the speeds of diffusion of different gases are inversely proportional to the square roots of their densities) (1834), and introduced the term osmosis. Graham was the author of numerous scientific papers, and *Elements of chemistry*, 1842, with an edition in two volumes, 1850, and of volume two again in 1858. His *Researches on the arsenates, phosphates, and modifications of phosphoric acid* was published in the Alembic Club reprints series, Edinburgh, 1904.[3]

An assistant to Dumas, August Laurent (1807–53), was born on 14th November, 1807, at La Folie, near Langres, and qualified as a mining engineer before turning to chemistry. He occupied the chair of chemistry at Bordeaux (1838), but after seven years returned to Paris, and in 1848 was appointed assayer at the Mint. His experimental and theoretical work in organic chemistry were of particular significance, as also was his work on atomic weights. In his later work Laurent was associated with Charles

[1] See Phillips, J. P. Liebig and Kolbe, critical editors. *Chymia*, 11, 1966, pp. 89–97.

[2] See also Holmes, Frederic L. Elementary analysis and the origins of physiological chemistry. *Isis*, 54, 1963, pp. 50–81; and Lipman, Timothy O. Vitalism and reductionism in Liebig's physiological thought. *Isis*, 58, 1967, pp. 167–185.

[3] See Mokrushin, S. G. Thomas Graham and the definition of colloids. *Nature*, 195, 1962, p. 861; and *The life and works of Thomas Graham, D.C.L., F.R.S., illustrated by 64 unpublished letters. Prepared by R. Angus Smith*, 1884.

Frederic Gerhardt (1816–56), with whom he was responsible for the *Comptes rendus mensuels des travaux chimiques. Années 1845–1851*, seven volumes, 1846–59. Laurent was also the author of the posthumously published *Méthode de chimie*, 1854, and for *Chemical method, notation, classification and nomenclature*, translated by William Odling, and published by the Cavendish Society in 1855. This latter work contains Laurent's substitution theory. Charles Gerhardt had studied under Liebig at Giessen, and had been lecture-assistant to Dumas, before becoming professor of chemistry at Montpellier (1841–48). After working for some years with Laurent, Gerhardt was appointed professor of chemistry and pharmacy at Strassburg in 1855, a year before his death. He made numerous important contributions to organic chemistry, and was the author of *Précis de chimie organique*, two volumes, Paris, 1844–45, and *Traité de chimie organique*, 1853–56.[1]

The German chemist Robert Wilhelm Bunsen (1811–99) was born at Göttingen, and became professor of chemistry at Heidelberg (1852–99). Bunsen excelled as a teacher, and was essentially practical, the Bunsen cell, Bunsen burner and the Bunsen battery being some of his inventions. He discovered the elements caesium and rubidium. His writings include *Gasometrische Methoden*, 1857, and *Chemical analysis by spectral analysis*, 1860, the result of his work with G. R. Kirchhoff.[2] His collected works, edited by Ernst August Max Bodenstein and Wilhelm Ostwald, were published as *Gesammelte Abhandlungen*, three volumes, 1904.

Jean Servais Stas (1813–91) was professor of chemistry at the École Royale Militaire, Brussels, and worked on the determination of atomic weights. His writings were published as *Oeuvres complètes*, three volumes, Brussels, 1894.

George Fownes (1815–49), professor of chemistry to the Pharmaceutical Society, and later professor of practical chemistry at University College, London, was elected F.R.S. in 1845. He was the author of *Manual of elementary chemistry, theoretical and practical*, 1844, which went into numerous editions; *Chemical tables*, 1846; and *Introduction to qualitative analysis*, 1846.

Another student under Liebig, Hermann Kopp (1817–92), became professor of chemistry at Heidelberg. With Liebig he founded the *Jahresbericht der Chemie*, and although Kopp is mainly remembered for his contributions to the history of chemistry, he also conducted important work in physical chemistry. His main writings were *Geschichte der Chemie*, four volumes, Braunschweig, 1843–47; *Beiträge zur Geschichte*

[1] See Potter, Owen. August Laurent's contributions to chemistry. *Ann. Sci.*, 9, 1953, pp. 271–286.

[2] See Schiek, O. Centenary of spectral analysis. In commemoration of its discovery by Robert Kirchoff and Wilhelm Bunsen. *Jena Review*, 4, 1959, pp. 143–146.

der Chemie, three parts in one, Braunschweig, 1869–75; *Die Entwicklung der Chemie in der neueren Zeit*, Munich, 1873 (reprint announced, 1963); and *Die Alchemie in alterer und neuerer Zeit*, [*etc.*], Heidelberg, 1886.

August Wilhelm von Hofmann (1818–92) was born at Giessen, and also studied under Liebig before coming to London, where he was head of the Royal College of Chemistry from 1845 to 1865,[1] when he returned to Germany as professor at the Prussian Academy, Berlin. Hofmann was elected F.R.S. in 1851, was President of the Chemical Society in 1861, and was the author of many reports and papers, some written with Thomas Graham and Edward Frankland. Hofmann's writings have been the subject of a bibliography by Kathleen Mary Hammond,[2] who lists 377 numbered items, including *Harrogate and its resources. Chemical analysis of its medicinal waters*, [Harrogate, 1854]; *On mauve and magenta*, [*etc.*], London, 1862; and *Introduction to modern chemistry, experimental and theoretic*, [*etc.*], 1865. This went into several translations, including German (Braunschweig, 1866), with a sixth edition in 1877; Dutch (Utrecht, 1866); Italian (Torino, 1869); Polish (Cracow, 1875); and Russian, 1866. Hofmann studied the organic bases in coal tar, and particularly influenced the development of the dye industry.

Although Louis Pasteur (1822–95) is looked upon as the founder of modern bacteriology, he was also an eminent chemist. He occupied chairs of chemistry at Strassburg (1848), Lille (1854) and the Sorbonne (1867), and spent the last years of his life at the Institut Pasteur, which was erected in his honour. Pasteur was born at Dôle, and became assistant to Balard at the École Normale. In 1853 he was awarded the Prix de la Société de Pharmacie de Paris for the synthesis of racemic acid, and four years later the Royal Society awarded him the Rumford Medal for his studies on crystallography. Pasteur is renowned for his experiments disproving spontaneous generation, for his investigations into the causes of silk-worm disease, of diseases of wine, of chicken cholera and of hydrophobia. Many of Pasteur's writings were published in the *Annales de Chimie*, and his major publications include *Mémoire sur la fermentation alcoolique*, 1860; *Mémoire sur la fermentation acétique*, 1864; *Études sur le vin*, [*etc.*], Paris, 1866; *Études sur le vinaigre*, [*etc.*], Paris, 1868; *Études sur les maladies des vers a soie*, [*etc.*], two volumes, Paris, 1870; *La théorie des germes et ses applications à la médecine et à la chirurgie*, 1878; *Sur les maladies virulentes et en particulier sur la maladie appelée vulgairement choléra des poules*, 1880; *De la possibilité de rendre les moutons réfracteures au charbon par la méthode des inoculations préventives*, 1881. His writings were collected

[1] See Beer, John J. A. W. Hofmann and the founding of the Royal College of Chemistry. *J. chem. Educ.*, 37, 1960, pp. 248–251.

[2] Hammond, Kathleen Mary. *August Wilhelm von Hofmann: bibliography submitted in part requirement for University of London Diploma in Librarianship*, 1967. [Typescript.]

together and published in two volumes, Paris, 1922, and as *Oeuvres de Pasteur, réunies par Pasteur Vallery-Radot*, Paris, 1932–39. Extracts have been published as *Pasteur: extraits de ses oeuvres présentés par R. Dujarric de la Rivière*, Paris, 1967.[1]

The Italian chemist Stanislao Cannizzaro (1826–1910) was born at Palermo, Sicily, and became professor of chemistry at Genoa (1858), at Palermo (1861), and Rome (1872), and conducted important researches in organic chemistry. His work on the cyanides and the alcohol groups was of special significance, and he recognized the importance of Avogadro's hypothesis. Cannizzaro's *Sunto di un corso di filosofia chimica fatto nella R. Universita di Genova. Nota sulle condensazioni di vapore dell autore stesso*, Pisa, 1858, a very rare work on the determination of atomic weights, has been reproduced on two occasions,[2] and is based on a course of lectures in which Cannizzaro attempted to rationalize the conceptions of atom and molecule and of atomic and molecular weight.[3]

The French chemist Pierre Eugène Marcellin Berthelot (1827–1907) (generally known as Marcellin Berthelot) was a native of Paris, and became professor of organic chemistry at the Collège de France. He was a pioneer of thermo-chemistry, and conducted research on explosives and on the synthetic preparation of organic compounds. He later held political appointments, and also wrote several works on the history of chemistry. His other important works include *Chimie organique fondée sur la synthèse*, Paris, 1860 (reprinted Brussels, 1966); *Sur la force de la poudre et des matières explosives*, 1872; *Mécanique chimique*, 1878; *Thermochimie*, 1897; and *Les carbures d'hydrogène*, 1901.[4]

The classification of the elements attracted the attention of several eminent chemists, and one of the most ambitious schemes was that advanced by William Odling (1829–1921). Born in Southwark, he became Medical Officer of Health to Lambeth[5] (1856–62), lecturer on chemistry at Guy's Hospital (1857), at St. Bartholomew's Hospital

[1] See Vallery-Radot, René. *The life of Pasteur. . . . Translated from the French. By Mrs. R. L. Devonshire. With a foreword by Sir William Osler*, 2 vols., 1911; Dubos, René J. *Louis Pasteur, free lance of science*, 1951; Nicolle, Jacques, *Louis Pasteur: a master of scientific enquiry*, 1961 (French edition, Paris, 1953; German edition, Berlin, 1959); Mann, J. H. *Louis Pasteur*, New York, 1964; Malpas, J. S. The life and works of Louis Pasteur. Part I [–II]. *St. Bart's Hosp. J.*, 59, 1955, pp. 150–159, 185–194; and Portmann, Georges. Louis Pasteur. *Archives of Laryngology*, 90, 1969, pp. 800–812.

[2] Alembic Club Reprint No. 18; Ostwald's *Klassiker*, No. 30.

[3] See Hartley, Sir Harold. Stanislao Cannizzaro, F.R.S. (1826–1910) and the first International Chemical Conference at Karlsruhe in 1860. *Notes Rec. Roy. Soc. Lond.*, 21, 1966, pp. 56–63.

[4] See Virtanen, R. *Marcelin Berthelot: a study of a scientist's public role*, Lincoln, Nebraska, 1965. (*University of Nebraska Studies*, N.S. No. 31); and Centenaire de la création d'un enseignement de chimie organique au Collège de France. *Bull. Soc. chim. Française*, 1964, pp. 1681–1689.

[5] See Brown, John R., and Thornton, John L. William Odling as Medical Officer of Health to Lambeth 1856–1860. *Med. Officer*, 102, 1959, pp. 77–78.

(1863–70); Fullerian Professor in the Royal Institution (1868), and Waynflete Professor of Chemistry at Oxford (1872–1912). He was the author of *A course of practical chemistry*, [*etc.*], London, 1854, which went into a fifth edition in 1876, and was translated into Russian in 1867. His translation of Auguste Laurent's *Chemical method* was published by the Cavendish Society in 1855. Odling's *A manual of chemistry, descriptive and theoretical*, Part 1, 1861, was translated into Russian (1863), German (1865), and French (1868), but sections only of Part 2 were published, and appeared in *Chemical News*. He also wrote *Lectures on animal chemistry, delivered at the Royal College of Physicians*, 1866, translated into Russian, 1867; *Lectures on the chemical changes of carbon*, 1869, translated into French, 1870; and *Outlines of chemistry*, 1870. Odling was elected F.R.S. and F.R.C.P. in 1859, and a full list of his publications is provided by J. L. Thornton and Anna Wiles.[1]

Friedrich August Kekulé (1829–96) was a native of Darmstadt, and after studying architecture at Giessen came under the influence of Liebig and other chemists, and found his true vocation. He worked in Paris and London, where he met A. W. Williamson, William Odling, and others, and worked as assistant to John Stenhouse at St. Bartholomew's Hospital. In 1856 he went to Heidelberg as a teacher of organic chemistry, and two years later Kekulé became professor of chemistry at Ghent, and during this period developed his benzene theory. He went to Bonn in 1867, and occupied the chair of chemistry there until his death. His hexagon formula for benzene was first published in a paper in 1865,[2] and this formula has become of primary importance in organic chemistry.[3] This was but one of his numerous contributions to science, and he contributed a large number of papers to periodicals. Kekulé's *Lehrbuch der organischen Chemie, oder der Chemie der Kohlenstoffverbindungen* was not completed, only Volumes 1 to 3 and Part 1 of Volume 4 having been published, 1861–87.[4]

[1] Thornton, John L., and Wiles, Anna. William Odling, 1829–1921. *Ann. Sci.*, 12, 1956 [1957], pp. 288–295; see also Smeaton, W. A. Centenary of the law of octaves: Döbereiner, Newlands, Odling and others. *Roy. Instn. Chem. J.*, 88, 1964, pp. 271–274.

[2] *Bull. Soc. Chim. Fr.*, 3, 1865, p. 98; and *Bull. Acad. Roy. Belg.* (*Cl. Sc.*), 19, 1865, p. 551. See also Benfey, O. Theodor, *trs.* August Kekulé and the birth of the structural theory of organic chemistry in 1858. *J. chem. Educ.*, 35, 1958, pp. 21–23.

[3] See Walker, Oswald J. August Kekulé and the benzene problem. *Ann. Sci.*, 4, 1939, pp. 34–46; McKie, Douglas. Kekulé and the benzene ring. *Pharmaceut. J.*, 195, 1965, pp. 197–200; Russell, C. A. Kekulé and benzene. *Chemistry in Britain*, 1, 1965, pp. 141–142; and Brown, H. C. Foundations of the structural theory. *J. chem. Educ.*, 36, 1959, pp. 104–110; see also pp. 319–339.

[4] See *Kekulé centennial. A symposium co-sponsored by the Division of History of Chemistry, the Division of Organic Chemistry, and the Division of Chemical Education at the 150th meeting of the American Chemical Society, . . . 1965*. Advances in Chemistry, Series 61, Washington, 1966; Ruske, Walter. August Kekulé und die Entwicklung der chemischen Strukturtheorie. *Naturwissenschaften*, 52, 1965, pp. 485–489; and Darmstaedter, Ludwig, and Oesper, Ralph E. August Kekulé. *J. chem. Educ.*, 4, 1927, pp. 697–702.

Johann Ludwig Wilhelm Thudichum (1829–1901) was born on 27th August, 1829, at Büdingen, and studied under Bunsen at Heidelberg, and under Liebig at Giessen. In 1853 he emigrated to London, qualifying M.R.C.S., and becoming naturalized in 1859. Thudichum published over two hundred scientific articles, and carried out extensive work on the chemistry of the brain. He conducted experiments on haemoglobin, urinary pigments, and the chemistry of wine and beer. He became the first Director of the Laboratory of Chemistry and Pathology at St. Thomas's Hospital, London, in 1865, but resigned after six years and became a leading otolaryngologist. He was the author of *A treatise on the pathology of the urine, including a complete guide to its analysis*, London, 1858, which went into a second edition in 1877; *A treatise on gall-stones: their chemistry, pathology and treatment*, London, 1863; *A treatise on the origin, nature and varieties of wine*, [*etc.*], London, New York, 1872; *A manual of chemical physiology, including its points of contact with pathology*, London, 1872; *A treatise on the chemical constitution of the brain*, [*etc.*], London, 1884, reprinted with an historical introduction by David L. Drabkin, Hamden, Conn., 1962, of which a Russian translation was published, 1885–86, and a greatly enlarged German edition was issued as his last publication with the title *Die chemische Konstitution des Gehirns des Menschen und der Tiere*, [*etc.*], Tübingen, 1901; *Grundzüge der anatomischen und klinischen Chemie*, [*etc.*], Berlin, 1886; and *Briefe über öffentliche Gesundheitspflege ihre bisherigen Leistungen und heutigen Aufgaben*, Tübingen, 1898. Thudichum's notebooks and chemical preparations were discovered in the stable of his former home by Otto Rosenheim (1871–1955), and are preserved at the National Institute for Medical Research, Hampstead.[1]

A Londoner, Sir William Crookes (1832–1919), studied chemistry under Hofmann, and was founder and editor of *Chemical News* (1859–1906). Crookes conducted research in spectroscopy and radioactivity, inventing several instruments, including the radiometer[2] and the spinthariscope. In 1861 he discovered the new metal, thallium, and Crookes also stressed the necessity of providing artificial fertilizers as sources of nitrogen. He was the author of *Select methods in chemical analysis*, 1871; *Practical handbook of dyeing and calico printing*, 1874; *The wheat problem*, 1899; and *Diamonds*, 1909. Knighted in 1897, Crookes was awarded the Order of Merit in 1910.

Carl Schorlemmer (1834–92) became professor of organic chemistry at Manchester, where he was associated with Sir Henry Enfield Roscoe

[1] See Drabkin, David L. *Thudichum: chemist of the brain*, Philadelphia, 1958. This contains an annotated bibliography of Thudichum's writings, pp. 209–234; and Debuch, H., and Dawson, R. M. C. Prof. J. L. W. Thudichum (1829–1901). *Nature*, 207, 1965, p. 814. See also Thornton, pp. 182–183.

[2] See Woodruff, A. E. William Crookes and the radiometer. *Isis*, 57, 1966, pp. 188–198.

(1833–1915). Schorlemmer conducted research on the paraffin hydrocarbons, on the alcohols, on aurin and on suberone. He was the author of *A manual of the chemistry of the carbon compounds*, London, 1874; and *Der Ursprung der organischen Chemie*, Braunschweig, 1889. With Roscoe he wrote *A treatise on chemistry*, two volumes in three, London, 1878–1911, the English edition of which was not completed, although the German translation was concluded by Brühl.

Dmitri Ivanovich Mendeleev or Mendeléeff (1834–1907) was born at Tobolsk in Siberia, and became professor of chemistry at the University of St. Petersburg (Leningrad) (1866–1907). Mendeleev carried out important investigations into the properties of solutions, the thermal expansion of liquids, and in 1869 formulated the periodic law in a paper on the classification of the elements.[1,2] He wrote a textbook on chemistry, published at St. Petersburg, 1868–70, of which the first English edition, from the fifth Russian edition, was published as *The principles of chemistry*, London, 1891, third edition, two volumes, 1905. His periodic law was embodied in *La loi périodique des éléments chimiques*, 1879; and his *Attempt towards a chemical conception of the ether* was also translated from the Russian by G. Kamensky.[3] A collection of Mendeleev's writings was published by the U.S.S.R. Academy of Sciences as *Sochineniya*, twenty-five volumes, Leningrad, 1934–54, and his library and personal records are in Leningrad University.[4]

Although primarily considered as an historian of chemistry, Sir Thomas Edward Thorpe (1845–1925) also made notable contributions to our knowledge of the derivatives of fluorine and phosphorus, and of the paraffin hydrocarbons. Born near Manchester, he became professor of chemistry at the Andersonian Institution, Glasgow (1870), at Yorkshire College, Leeds (1874), and at the Imperial College of Science, South Kensington (1885). He was knighted in 1909. Thorpe was the author of *A history of chemistry*, two volumes, London, 1909–10; *Essays in historical chemistry*, third edition, London, 1911; *Alcoholometric tables*, 1915; and *Dictionary of applied chemistry*, three volumes, 1890–93, of which the fourth edition, edited by J. F. Thorpe and M. A. Whiteley, appeared in twelve volumes, 1937–56.

Jacobus Henricus Van't Hoff (1852–1911) was a native of Amsterdam,

[1] First expounded in *Russkoe fiziko-khimicheskoe obshchestvo zhurnal*, I, 1869, and in Liebig's *Annalen der Chemie und Pharmacie*, Supplementband VIII, 1871; both published in Ostwald's *Klassiker*, No. 68, 1895.

[2] See Leicester, Henry M. Factors which led Mendeleev to the periodic law. *Chymia*, I, 1948, pp. 67–74; Greenaway, Frank. A pattern of chemistry. Hundred years of the periodic tables. *Chemistry in Britain*, 5, 1969, pp. 97–99; and Vucinich, Alexander. Mendeleev's views on science and society. *Isis*, 58, 1967, pp. 342–351.

[3] See Posin, Daniel Q. *Mendeleyev. The story of the great scientist*, New York, 1948.

[4] See Krotikov, V. A. The Mendeleev Archives and Museum of the Leningrad University. *J. chem. Educ.*, 37, 1960, pp. 625–628.

and studied under Kekulé at Bonn, before becoming lecturer (1877) and then professor of chemistry at the University of Amsterdam. In 1896 he was appointed honorary professor at Berlin University. Van't Hoff was the most eminent physical chemist of his period, and worked upon the stereochemistry of carbon, formulated the osmotic theory of solutions, conducted research upon the crystallization of salts from solutions, and applied the principles of mathematics and thermodynamics to chemical equilibria. His *Dix années dans l'histoire d'une théorie. Deuxième édition de "La chimie dans l'espace"*, Rotterdam, 1887, formed the basis of the science of stereochemistry, and had first appeared in 1875. Van't Hoff's other epoch-making work was his *Études de dynamique chimique*, Amsterdam, 1884, which was revised and enlarged by Ernest Cohen, translated by Thomas Ewen, and published in Amsterdam and London, 1896.

Sir William Ramsay (1852–1916) was a native of Glasgow, and occupied chairs of chemistry at University College, Bristol (1880–87), and at University College London (1887–1912). Ramsay was knighted in 1902 and was awarded the Nobel Prize for chemistry in 1904. He discovered argon (with Lord Rayleigh), helium, krypton, neon and xenon, and in 1910 estimated the molecular and atomic weights of niton, a radium emanation. Sir William Ramsay was the author of *Investigations on the toluic and nitroluic acids*, [*etc.*], Tübingen, 1872, a rare dissertation prepared for the degree of doctor of natural sciences; *Gases of the atmosphere: the history of their discovery*, 1896, of which a fourth edition was published in 1915; *Modern chemistry*, 1900; *Essays biographical and chemical*, 1908; *Elements and electrons*, 1912; and *The life and letters of Joseph Black*, [*etc.*], London, 1918.[1]

Wilhelm Ostwald (1853–1932) was born at Riga, and was professor of physical chemistry at Leipzig from 1887 to 1906, being awarded the Nobel Prize for chemistry in 1909. Ostwald was one of the most eminent chemists of his period, and his researches were chiefly in electro-chemistry. He discovered the method of oxidizing ammonia to form oxides of nitrogen, and founded the *Zeitschrift für physiologische Chemie*. Ostwald's writings include *Lehrbuch der allgemeine Chemie*, of which a second edition was published in Leipzig, 1885–87; *Grundlinien der anorganischen Chemie*, which went into a fifth edition, Leipzig, 1922; and *Die wissenschaftlichen Grundlagen der analytischen Chemie*, Leipzig, 1894.

Edmund O. von Lippman (born 1857) was a native of Vienna, and in addition to being an eminent historian of science was also a great

[1] See Travers, Morris W. *A life of Sir William Ramsay*, 1956; Ramsay, Sir William. *The life and letters of Joseph Black, M.D. . . . With an introduction dealing with the life and work of Sir William Ramsay by F. G. Donnan*, 1918; and Tilden, Sir William A. *Sir William Ramsay, K.C.B., F.R.S.: memorials of his life and work*, 1918.

chemist. In 1877 he was employed in a sugar factory, and much of his work was devoted to the sugars. Lippman was the author of a large number of papers, together with the following books: *Der Zucker, seine Derivate und sein Nachweis*, Vienna, 1878; *Die Zuckerarten und ihre Derivate*, Braunschweig, 1882; *Die Chemie der Zuckerarten*, 1890, with a third edition in 1904; *Geschichte des Zuckers*, [*etc.*], Leipzig, 1890, with a new edition, Berlin, 1929; *Rohstoffe*, [*etc.*], 1900, which went into an eighth edition, Berlin, 1934; *Abhandlungen und Vorträge zur Geschichte der Naturwissenschaften*, two volumes, Leipzig, 1906–13; *Entstehung und Ausbreitung der Alchemie*, two volumes, Berlin, 1919–31; and among others, *Beiträge zur Geschichte der Naturwissenschaften und Technik*, Berlin, 1923.[1]

Although he graduated in physics, and held several important posts devoted to that subject, Hermann Walther Nernst (1864–1941) later became professor of physical chemistry at Göttingen (1895) and at Berlin (1905–22, 1924–34). He formulated the Nernst theorem, or Third Law of Thermodynamics, and was awarded the Nobel Prize for chemistry in 1920. Nernst advanced a theory of galvanic cells and the "atom chain-reaction" theory in photo-chemistry, among other contributions to chemistry, and invented the Nernst incandescent electric lamp. His *Theoretische Chemie*, 1893, was translated into English by Charles Skeele Palmer, London, 1895, a fifth edition, revised by L. W. Codd, having been published in 1923, while his *Die theoretischen und experimentellen Grundlagen des neuen Warmesatzes*, 1918, also appeared in an English translation as *The new heat theorem*, translated by G. Barr, 1926. Other translations of his books include *Experimental and theoretical applications of thermodynamics to chemistry*, 1907; and *Theory of the solid state*, 1914.

Outstanding among nineteenth-century geologists was Sir Charles Lyell (1797–1875), whose work influenced the development of the theory of evolution, and who later became a close friend of Charles Darwin. Lyell was first interested in entomology, but was attracted to geology while at Oxford. He travelled extensively in Europe observing the denudation of the land by river and sea, and the results of erosion and earthquakes. He was elected F.R.S. in 1826, and two years later travelled on the Continent with Murchison. Lyell was the author of numerous papers, and his most important work, *The principles of geology*, [*etc.*], was first published in three volumes, 1830–33. Volume 2, 1852, deals particularly with biology.[2] This work was reprinted in 1834 in four volumes, the reprint being styled the third edition, Volumes 1 and 2 of the first edition having been previously reissued. There were many revisions of this textbook, a one-volume edition having appeared in

[1] See Partington, J. R. Edmund O. von Lippman. *Osris*, 3, 1938, pp. 5–21.
[2] See Coleman, William. Lyell and the "reality" of species: 1830–33. *Isis*, 53, 1962, pp. 325–338.

1853; a tenth in two volumes, 1866–68, contains Lyell's adoption of organic evolution and a twelfth edition was published in 1875. The last part of Lyell's *Principles* was published separately in 1838 as *Elements of geology*, the sixth edition of this appearing in 1865, while editions three to five bore the title *A manual of elementary geology*. Lyell was also the author of *Travels in North America, with geological observations*, two volumes, 1845; *A second visit to the United States of North America*, two volumes, 1849; *The antiquity of man*, 1863, which went into a fourth edition in 1873; and *The student's elements of geology*, 1871, this also going into four editions. Knighted in 1848, and created a baronet in 1864, Lyell greatly influenced the development of geology in his time. He developed Hutton's theory, and his writings overshadowed similar works in their field.[1] Lyell's friend Sir Roderick Impey Murchison (1792–1871) had an early Army career before becoming interested in science. He attended lectures at the Royal Institution, and later made extensive geological explorations in Europe, and was elected F.R.S. in 1826. Murchison became Director-General of the Geological Survey, and founded a chair of geology at Edinburgh. He was the author of *The Silurian system*, 1838, which was rewritten with the title of *Siluria* and published in 1854. With Van Keyserling and de Verneuil he wrote *The geology of Russia and the Ural Mountains*, 1845.[2]

The geologist and engineer William Smith (1769–1839) was responsible for many geological maps, pamphlets and lithographs, and his manuscript papers are preserved in the Department of Geology and Mineralogy, Oxford University. His publications include *A memoir to the map and delineation of the strata of England and Wales, with part of Scotland*, London, 1815; *Strata identified by organized fossils*, [*etc.*], four parts, London, 1816–1819; *Stratigraphical system of organized fossils*, [*etc.*], London, 1817; and *A new geological map of England and Wales*, [*etc.*], London, 1820.[3]

The biological sciences made extensive progress during the nineteenth century, and became even more specialized within their respective branches.[4] Comparative anatomy progressed rapidly at the hands of Owen, Mivart, Ray Lankester and others. Embryology and histology began their development as independent sciences, and it was during the

[1] See Bonney, Thomas George. *Charles Lyell and modern geology*, 1895; and Eiseley, Loren C. Charles Lyell. *Sci. Amer.*, 201, 1959, pp. 98–106; and Bailey, Sir Edward. Charles Lyell, F.R.S. (1797–1875). *Notes Rec. Roy. Soc. Lond.*, 14, 1959, pp. 121–138.

[2] Geikie, Sir Archibald. *Life of Sir Roderick Impey Murchison*, [*etc.*], 2 vols, 1875.

[3] See Eyles, Joan M. William Smith (1769–1839): a bibliography of his published writings, maps and geological sections, printed and lithographed. *J. Soc. Bib. Nat. Hist.*, 5, 1969, pp. 87–109.

[4] See Dawes, Ben. *A hundred years of biology*, 1952; and Mendelsohn, Everett. The biological sciences in the nineteenth century: some problems and sources. *History of Science*, 3, 1964, pp. 39–59

early nineteenth century that physiology was established as a distinct science.[1] Several important theories of evolution were advanced, the work of Darwin being outstanding as a signal influence on the future development of the subject. The travels of several explorers produced remarkable results in the form of travel books, zoological specimens, and descriptions of new species. The Challenger Expedition organized at the suggestion of the Royal Society by the British Government explored scientifically the physical, chemical, geological and biological conditions of the great ocean basins. The collections made were housed in the British Museum, and the scientific results were published mainly under the direction of Sir John Murray as *Reports of the scientific results of the voyage of the "Challenger" during 1873–1876*, fifty volumes, London, 1880–1895, reprinted in 1965. Books on the mammalia or birds of individual countries or specific areas became more numerous, and several sumptuously produced volumes appeared, such as the publications of Audubon and John Gould.

Karl Ernst von Baer (1792–1876), a native of Estonia, studied medicine and zoology, and his researches in embryology were of primary significance in the history of that subject. He discovered the mammalian ovum, and greatly influenced the work of his successors. Von Baer's writings include *De ovi mammalium et hominis genesi*, Leipzig, 1827;[2] reprinted Brussels, 1960; *Über Entwickelungsgeschichte der Thiere: Beobachtung und Reflexion*, two volumes, Königsberg, 1828–37; *Reden gehalten in wissenschaftlichen Versammlungen und kleinere Aufsätze vermischen Inhalts*, three volumes, St. Petersburg; and his autobiography, *Nachrichten über Leben und Schriften des Herrn Geheimraths Dr. Karl Ernst von Baer, mitgetheilt von ihm selbst*, [*etc.*], St. Petersburg, 1866.[3] In addition to his physiological work, Theodor Schwann (1810–82) was a pioneer in the development of the cell theory. He was a professor of anatomy at Louvain and later at Liège (1847–80), and wrote the classic *Mikroskopische Untersuchungen über die Uebereinstimmung in der Struktur und dem Wach-*

[1] See Franklin, K. J. History of physiology, 1851–1951. *Advancement of Science*, 8, 1951–52, pp. 293–302; Franklin, K. J. Physiology and histology. *In*, Dingle, Herbert, *ed. A century of science*, 1951, Chapter XV, pp. 222–238; Schiller, J. Physiology's struggle for independence in the first half of the nineteenth century. *History of Science*, 7, 1968 [1969], pp. 64–89; and Olmsted, J. M. D. French contributions to physiology in the XIX century. *Tex. Rep. Biol. Med.*, 13, 1955, pp. 306–316. This last paper describes the discoveries made by the brilliant French physiologists between 1810 and 1860.

[2] See Sarton, George. The discovery of the mammalian egg and the foundation of modern embryology. With a complete facsimile (no. xiii) of K. E. von Baer's fundamental memoir: De ovi mammalium et hominis genesi (Leipzig, 1827). *Isis*, 16, 1931, pp. 315–379. This was followed up by a translation into English as, On the genesis of the ovum of mammals and of man, by Karl Ernst von Baer, translated into English by Charles Donald O'Malley. Introduction by I. Bernard Cohen. *Isis*, 47, 1956, pp. 117–153.

[3] See Raikov, B. E. *Karl Ernst von Baer, 1792–1876. Sein Leben und Werk. Deutsche Übersetzung mit Anmerkungen von H.v. Knorre*, Leipzig, 1969. (*Acta Historica Leopoldina*, Nr. 5).

sthum der Thiere und Pflanzen, 1839, of which an English translation by Henry Smith was published by the Sydenham Society in 1847.[1,2]

The German embryologist Oscar Hertwig (1849–1922) was professor of anatomy at Jena (1878) and director of the institute of anatomy and biology in Berlin (1888–1921). He demonstrated the fusion of the nuclei of sperm and ovum in fertilization, and his embryological researches led to the foundation of the germ-layer theory. With his brother Richard (1850–1920), Oscar Hertwig was the author of *Studien zur Blättertheorie*, 1879–83; *Die Cölomtheorie*, 1881; and *Entwicklung des mittleren Keimblattes der Wirbelthiere*, 1883. He also wrote several works himself, including *Lehrbuch der Entwicklungsgeschichte*, 1886, which went into a ninth edition in 1910, with an English translation in 1892; *Die Zelle und die Gewebe*, two volumes, 1893–98, which was also translated into English; *Der Kampf um Kernfragen der Entwicklungs und Vererbungslehre*, 1909, with a second edition in 1920; and an edition of the *Handbuch der Entwicklungslehre der Wirbelthiere*, three volumes, 1901–06.[3]

The Cambridge embryologist, Francis Maitland Balfour (1851–82), although dying at an early age while climbing in the Alps, was foremost among the founders of modern embryology. He developed certain of Haeckel's embryonic theories, and his work was continued by a school of thought that still flourishes. Balfour's monumental textbook was entitled *A treatise on comparative embryology*, two volumes, 1880–81, which was reprinted in 1885. A memorial edition of his writings was published as *The works of Francis Maitland Balfour*, [*etc.*], and was edited by Michael Foster and Adam Sedgwick in four volumes, 1885.

John James Audubon (*c.* 1785–1851) was born at Les Cayes, Santo Domingo, and educated in France. Returning to America, he explored the forest, producing bird and animal drawings which he afterwards published in parts. These part issues have become rare. Audubon's *The birds of America, from original drawings by John James Audubon*,[4] appeared in four double elephant folio volumes consisting of eighty-seven parts published between 1827 and 1838. These were printed in

[1] See Florkin, Marcel. Un manuscrit de Schwann montrant que les "Mikroskopische Untersuchungen" constituent le premier volume d'une oeuvre intitullée "Theorie der Zellen", dont le second volume est constitué par une série de manuscrits encore inédits. *Bull. Acad. roy. Belg.* (*Cl. Sci.*), Ser. 5, 10, pp. 1054–1061.

[2] See also Causey, Gilbert. *The cell of Schwann*, [*etc.*], Edinburgh, London, 1960; Florkin, Marcel. *Naissance et déviation de la théorie cellulaire dans l'oeuvre de Theodore Schwann*, Paris, Liège, 1960; *Lettres de Théodore Schwann*, Liège, 1961; Waterman, Rembert. *Theodor Schwann. Leben und Werk*, Dusseldorf, 1960. Marcel Florkin contributed a valuable series of studies on Schwann and his work to *Rev. Méd. Liège* between 1957 and 1960.

[3] See Weissenberg, Richard, *Oscar Hertwig, 1849–1922. Leben und Werk eines deutschen Biologen*, Leipzig, 1959. (Lebensdarst. deutscher Naturforsch., 7. Mit Zeittafel, Bibliographie der Veröffentlichungen von und über Hertwig, von Rudolph Zaunick.)

[4] Some 400 of the original water-colours for this are in the New York Historical Library, and others are in Harvard University Library.

ON

THE ORIGIN OF SPECIES

BY MEANS OF NATURAL SELECTION,

OR THE

PRESERVATION OF FAVOURED RACES IN THE STRUGGLE FOR LIFE.

BY CHARLES DARWIN, M.A.,

FELLOW OF THE ROYAL, GEOLOGICAL, LINNÆAN, ETC., SOCIETIES;
AUTHOR OF 'JOURNAL OF RESEARCHES DURING H. M. S. BEAGLE'S VOYAGE ROUND THE WORLD.'

LONDON:
JOHN MURRAY, ALBEMARLE STREET.
1859.

The right of Translation is reserved.

PLATE 11. Title-page of the first edition of Charles Darwin's *Origin of species*, London, 1859.

seen climbing branches, almost like a creeper; it often, like a shrike, kills small birds by blows on the head; and I have many times seen and heard it hammering the seeds of the yew on a branch, and thus breaking them like a nuthatch. In North America the black bear was seen by Hearne swimming for hours with widely open mouth, thus catching, like a whale, insects in the water. Even in so extreme a case as this, if the supply of insects were constant, and if better adapted competitors did not already exist in the country, I can see no difficulty in a race of bears being rendered, by natural selection, more and more aquatic in their structure and habits, with larger and larger mouths, till a creature was produced as monstrous as a whale.

As we sometimes see individuals of a species following habits widely different from those both of their own species and of the other species of the same genus, we might expect, on my theory, that such individuals would occasionally have given rise to new species, having anomalous habits, and with their structure either slightly or considerably modified from that of their proper type. And such instances do occur in nature. Can a more striking instance of adaptation be given than that of a woodpecker for climbing trees and for seizing insects in the chinks of the bark? Yet in North America there are woodpeckers which feed largely on fruit, and others with elongated wings which chase insects on the wing; and on the plains of La Plata, where not a tree grows, there is a woodpecker, which in every essential part of its organisation, even in its colouring, in the harsh tone of its voice, and undulatory flight, told me plainly of its close blood-relationship to our common species; yet it is a woodpecker which never climbs a tree!

Petrels are the most aërial and oceanic of birds, yet in the quiet Sounds of Tierra del Fuego, the Puffinuria

Plate 12. The page of Darwin's *Origin of species*, 1859, containing the story of the bear catching insects in the water. This was omitted from later printings.

London for the author, and the text later appeared with the title *Ornithological biography, or an account of the habits of the birds of the United States of America,* [*etc.*], five volumes, Edinburgh, 1831–39. Later editions combining text and plates appeared, including *The birds of America,* [*etc.*], seven volumes in octavo, New York and Philadelphia, 1840–44, reissued in 1861, and in eight volumes, New York (1870). With John Bachman, Audubon was the author of *The viviparous quadrupeds of North America*, three volumes, New York, published in thirty parts from 1845 to 1848, followed by three volumes of text, and a supplementary work, *The quadrupeds of North America*, three volumes, 1854. Audubon was also responsible for *A synopsis of the birds of North America*, Edinburgh, London, 1839. Other items of interest include Maria R. Audubon's *Audubon and his journals, with zoological and other notes by Elliott Coues,* two volumes, New York, 1898; *Journal of John James Audubon during his trip to New Orleans in 1820–1821. Edited by Howard Corning*, Boston, 1929, published by the Club of Odd Volumes, which also published *Letters of John James Audubon, made while obtaining subscriptions to his "Birds of America", 1840–1843. Edited by Howard Corning*, Boston, 1930. Audubon's plates have been reissued on numerous occasions, and in 1951 appeared *Audubon's Animals. Compiled and edited by Alice Ford*, London, which is a republication of *Viviparous quadrupeds of North America.*[1]

Another brilliant artist, John Gould (1804–81), was responsible for a remarkable series of zoological monographs containing about three thousand coloured plates. Several of his publications were not completed during Gould's lifetime, and others were incorporated in later works. His more important ones include *A century of birds from the Himalaya Mountains*, London, 1832, the first of his famous folios, most of the letterpress of which was written by N. A. Vigors; *The birds of Europe*, five volumes, London, issued in twenty-two parts between 1832 and 1837, of which the plates were hand-coloured by Mrs. Gould; *A monograph of the Ramphastidae, or family of toucans*, London, 1834, which was issued in a revised edition, 1852–54; *A synopis of the birds of Australia, and the adjacent islands*, four parts, London, 1837–38; *The birds of Australia*, seven volumes issued in thirty-six parts, London, 1840–48, which contains 600 hand-coloured plates, and to which there was a supplement in five parts, London, 1851–69, with a text, two volumes, 1865; *Mammals of Australia*, three volumes, London, [1845]–63, containing 182 coloured plates; *A monograph of the Trochilidae, or family of humming-birds*, five volumes, 1849–61, issued in twenty-five parts, to which there was a

[1] See Adams, Alexander B. *John James Audubon: a biography*, London, 1967; Ford, A. E. *John James Audubon*, Norman, Okla., 1964. Herrick, Francis Hobart. *Audubon the naturalist; a history of his life and time. . . . Second edition. Two volumes in one*, New York, London, 1938; and Rourke, Constance. *Audubon. . . . With 12 colored plates from original Audubon prints; black and white illustrations by James MacDonald*, London, [*etc.*], 1936.

supplementary volume published 1880–87, completed by R. Bowdler Sharpe, as Gould had died in 1881; *The birds of Asia*, [*etc.*], seven volumes, London, 1850–83, issued in thirty-five parts, the last three of which were completed by R. Bowdler Sharpe, and which contain 530 coloured plates; *The birds of Great Britain*, five volumes, London, 1862–73, issued in twenty-five parts, with 367 hand-coloured plates; *The birds of New Guinea and the adjacent Papuan Islands, including many new species recently discovered in Australia*, five volumes, 1875–88, published in twenty-five parts, completed by R. Bowdler Sharpe, and containing 320 hand-coloured plates; and *A monograph of the Trogonidae, or family of trogons* London, 1838, with a second edition in four parts, London, 1858–75. Many editions of *Birds of Australia*, *Birds of Asia*, and *Birds of Europe*, have been published.

Baron Georges Léopold Chrétien Fréderic Dagobert Cuvier (1769–1832) was born on 23rd August, 1769, at Montbéliard, and was educated at Stuttgart. In 1795 he became assistant to the professor of comparative anatomy at the Muséum d'Histoire Naturelle, and read his first paper at the opening of L'Institut National in 1796. This was later published as *Mémoires sur les espèces d'éléphants vivants et fossiles*, 1800. An abridgement of his course of lectures at the École du Panthéon appeared as *Tableau élémentaire de l'histoire naturelle des animaux*, Paris, 1798, the first general statement of Cuvier's classification of the animal kingdom. In 1799 he was appointed professor of natural history in the Collège de France, and the following year appeared *Leçons d'anatomie comparée*, five volumes, 1800, in the preparation of which he was assisted by A. M. C. Dumeril and G. L. Duvernoy. Cuvier held several other official appointments, and in 1831 was made a peer of France, but he died on 13th May, 1832, in Paris. Cuvier has been regarded as the founder of the science of palaeontology, and his influence was widespread. He was the author of: *Recherches sur les ossements fossiles de quadrupèdes*, four volumes, 1812, consisting of a series of separately paged memoirs, which went into a second edition, five volumes, Paris, 1821–24; a third edition, five volumes, Paris, 1825; and a fourth edition in ten volumes, with a two-volume atlas, Paris, 1834–36. This work appeared in English as *Researches on fossil bones*, fourth edition, two parts, London, 1834–35. *Le règne animal distribué après son organisation*, [*etc.*], four volumes, Paris, 1817;[1] second edition, five volumes, 1829–30; third edition, seventeen volumes, Paris, 1836–49, which was issued in 262 sections, with contributions by several other eminent zoologists. An English translation of this was also published in more than forty parts between 1824 and 1835 as *The animal kingdom*, [*etc.*], sixteen volumes, London, 1827–35. This

[1] See Whitehead, P. J. P. The dating of the 1st edition of Cuvier's Le règne animal distribué d'après son organisation. *J. Soc. Bib. Nat. Hist.*, 4, 1967, pp. 300–301.

contains much additional material by E. Griffith and others.[1] Several abridged editions of this appeared, including printings in 1840, 1849, 1851 and 1863. With A. Valenciennes, Cuvier was responsible for *Histoire naturelle des poissons*, twenty-two volumes, Paris, 1828–49, which was issued in four separate editions, the octavo version with black and white plates being the most quoted. A reprint of this was announced in 1969. Numerous other writings came from his pen. Cuvier was one of the most eminent naturalists of the early nineteenth century, and his method of classification forms the basis of our modern system.[2]

The marine biological work of John V. Thompson (1779–1847) has been largely overlooked, but he revolutionized some aspects of zoological thought. His early interests were botanical, and he produced a *Catalogue of plants growing in the vicinity of Berwick-on-Tweed* in 1807. Qualifying as surgeon, he travelled with the Army abroad for some years, until in 1835 he was sent as medical officer to the convict settlement of New South Wales, where he settled for life. He wrote several articles between 1830 and 1836, and contributed a series of six privately printed memoirs (in five numbers) with the title of *Zoological researches and illustrations; or, natural history of nondescript or imperfectly known animals in a series of memoirs*, Nos. I–V, Cork, 1828–34. A facsimile of this work, with an introduction by Alwyne Wheeler, was published by the Society for the Bibliography of Natural History in 1968 (Sherborn Fund Facsimile, No. 2). Thompson's early work on the cirrepedes and on barnacle development was of particular significance.

The British zoologist William Yarrell (1784–1856) was a voluminous writer, but his literature was scientific as well as popular. In addition to his articles in periodicals, he was the author of *A history of British fishes*, two volumes, London, 1836, first issued in nineteen parts, 1835–36, at half-a-crown each, and of which other editions appeared in 1841 and 1859. This last was edited by Sir John Richardson and contains a bibliography of Yarrell's writings, together with a biographical sketch by John Van Voorst, his friend and publisher. Yarrell's *A history of British birds*, three volumes, London, 1843, issued in thirty-seven parts, 1837–43, went into a second edition in 1845 together with a supplement, a third edition

[1] See Cowan, C. F. Notes on Griffith's "Animal kingdom" of Cuvier [1824–35]. *J. Soc. Bib. Nat. Hist.*, 5, 1969, pp. 137–140.

[2] See Coleman, William. *Georges Cuvier, zoologist. A study in the history of evolution theory*, Cambridge, Mass., 1964; Coleman, William. A note on the early relationship between Georges Cuvier and Louis Agassiz. *J. Hist. Med.*, 18, 1963, pp. 51–63; Théodoridés, Jean. Quelques documents inédits ou peu connus relatifs à Georges Cuvier, à sa famille et à son salon. *Stendahl Club*, No. 33, 1966, pp. 55–64; No. 34, 1967, pp. 179–188; Petit, Georges, and Théodoridès, Jean. Les cahiers de notes de Georges Cuvier et son premier *Diarium zoologicum. Comptes rendus des séances de l'Académie des Sciences*, 246, 1958, pp. 352–354; and Swinton, W. E. Early history of comparative anatomy. *Endeavour*, 19, 1960, pp. 209–214.

with a second supplement in 1856, and a fourth edition, revised and enlarged by Alfred Newton and Howard Saunders, four volumes, 1871–85, originally issued in thirty parts. A condensed text with a selection of the illustrations to this work was published by Saunders in 1889 with the title *An illustrated manual of British birds.*[1]

Another semi-popular writer on zoology was William Swainson (1789–1855), author of *Zoological illustrations, or original figures and descriptions of new, rare, or interesting animals*, first series, three volumes, London 1820–23 with a second series, also in three volumes, London, 1829–33; *A preliminary discourse on the study of natural history*, London, 1833; *Exotic conchology*, four parts, 1821–22, which was re-issued with two parts added (1834–35?), with a second edition edited by S. Hanley, 1841, and of which a facsimile of the first edition, edited by R. Rucker Abbott, was announced in 1968[2]; *A treatise on the geography and classification of animals*, London, 1835; *On the natural history and classification of birds*, two volumes, London, 1836–37, with a similarly dated edition in Lardner's Cabinet Cyclopaedia; *The natural history of the birds of Western Africa*, two volumes, 1837, in Jardine's Naturalist Library; *Animals in menageries*, London, also in Lardner's Cabinet Cyclopaedia, one issue not having a date on the title-page, the other, published in the same year, being dated 1838; *The natural arrangement and relations of the family of flycatchers, or musicicapidae*, Edinburgh, 1838, in Jardine's Naturalist Library; *The natural history and classification of fishes, amphibians, and reptiles, or monocardian animals*, two volumes, London, 1838–39, in Lardner's Cabinet Cyclopaedia; *A selection of the birds of Brazil and Mexico. The drawings by William Swainson*, London, 1841, originally issued in six parts, 1834–36, with the title *Ornithological drawings*, without text.

The greatest comparative anatomist of the nineteenth century was Sir Richard Owen (1804–92), a native of Lancaster, who became curator of the Museum at the Royal College of Surgeons of England and then superintendent of the Natural History Department of the British Museum. Owen was elected F.R.S. in 1834 and knighted in 1884. He acquired a remarkable knowledge of extinct mammals, and was eminent as a zoologist, although in later years he was engaged in several controversies regarding evolution.[3] Owen was the author of numerous zoological papers and sections of scientific publications. His books include *Memoir on the pearly nautilus (Nautilus Pompilius, Linn.)*, [*etc.*], London, 1832; *Odontography; or, a treatise on the comparative anatomy of the teeth*, [*etc.*],

[1] See Forbes, Thomas R. William Yarrell, British naturalist. *Proc. Amer. Philos. Soc.*, 106, 1962, pp. 505–515.

[2] See McMillan, Nora F. William Swainson's *Exotic conchology*. *J. Soc. Bib. Nat. Hist.*, 4, 1963, pp. 198–199.

[3] See MacLeod, Roy M. Evolutionism and Richard Owen, 1830–68: an episode in Darwin's century. *Isis*, 56, 1965, pp. 259–280.

two volumes, 1840–45; *Lectures on the comparative anatomy and physiology of the invertebrate animals*, [*etc.*], London, 1843, of which a second edition was published in 1855; *A history of British fossil mammals and birds*, London, 1846; *Lectures on the comparative anatomy and physiology of the vertebrate animals, Part* 1. *Fishes*, London, 1846, this being the only part published; *On the archetype and homologies of the vertebrate skeleton*, London, 1848; *A history of British fossil reptiles*, four volumes, 1849–84; *On the nature of limbs*, London, 1849; *Descriptive catalogue of the osteological series contained in the museum of the Royal College of Surgeons*, two volumes, London, 1853; *On the classification and geographical distribution of the mammalia*, [*etc.*], London, 1859; *Palaeontology; or, a systematic summary of extinct animals and their geological relations*, Edinburgh, 1860, second edition, 1861; *On the extent and aims of a national museum of natural history*, London, 1862; *Monograph on the aye-aye*, London, 1863; *Memoir on the gorilla*, London, 1865; *On the anatomy of vertebrates*, three volumes, 1866–68; *Researches on the fossil remains of the extinct mammals of Australia, with a notice of the extinct marsupials of England*, two volumes, 1877–78; *Memoirs of the extinct wingless birds of New Zealand; with an appendix on those of England, Australia, Newfoundland, Mauritius, and Rodriguez*, two volumes, London, 1879.[1]

George Bennett (1804–93) has been described as "the greatest of the physician naturalists of Australia", and was the author of a large number of books and articles devoted to botany and zoology. Bennett was born at Plymouth on 31st January, 1804, and at the age of fifteen went on a voyage to Ceylon and Mauritius. Returning to England in 1821, he studied medicine, and qualified in 1828. He became acquainted with Richard Owen, with whom he corresponded, and to whom he sent specimens, throughout his career. Between 1828 and 1835 Bennett made a voyage to Australia and the South Seas as surgeon-naturalist, and he brought back a large collection of plants which he presented to the British Museum. He also brought back a pearly nautilus, which Richard Owen described in a memoir published by the Royal Society. In 1832 Bennett sailed for Sydney, whence he made extensive journeys into the interior, and on his return to England published his *Wanderings in New South Wales, Batavia, Pedir Coast, Singapore and China; being the journal of a naturalist in these countries, during 1832, 1833 and 1834.* two volumes, London, 1834. In April, 1835, Bennett left England to return to Australia, and was away for over twenty years. He catalogued the Australian Museum, his work being published as *A catalogue of specimens of natural history and miscellaneous curiosities deposited in the Australian Museum*,

[1] See Owen, Richard. *The life of Richard Owen. By his grandson Richard Owen. . . . Also an essay on Owen's position in anatomical science by T. H. Huxley*, [*etc.*], 2 vols., 1894; Cave, A. J. E. Sir Richard Owen. *St. Bart's Hosp. J.*, 68, 1964, pp. 71–73, 76, Cave, A. J. E. The glands of Owen. *St. Bart's Hosp. J.*, 57, 1953, pp. 131–133; and Cave, A. J. E. The muscles of Owen, *St. Bart's Hosp. J.*, 61, 1957, pp. 138–140.

Sydney, 1837. Visiting England in 1859 he prepared for publication *The gatherings of a naturalist in Australasia*, [*etc.*], London, 1860, which includes detailed descriptions of the flora and fauna of Australia. Bennett died on 29th September, 1893, and many of his letters and other documents are in the Royal College of Surgeons of England Library, and in the Owen Collection at the British Museum.[1]

Louis Jean Rodolphe Agassiz (1807–73) was a native of Switzerland, and after studying medicine became professor of natural history at Neuchâtel (1832–46). Following a lecture tour in America, he became professor of natural history at Harvard, where he founded the Museum of Comparative Zoology. Agassiz was a disciple of Cuvier, and his most important work was on fossil fishes, *Recherches sur les poissons fossiles*, three volumes and supplement, Neuchâtel, 1833–44. Agassiz visited Britain in 1834, 1835 and 1840, on the last occasion with the purpose of discovering evidence of the glacial action.[2] He wrote *Études sur les glaciers*, two volumes, Neuchâtel, 1840, which was reprinted in 1966; and *Nomenclator zoologicus*, issued in twelve fascicules, Solothurn, 1842–46 [1847].[3] He was also the author of several bibliographies,[4] and of *Lake Superior: its physical character, vegetation and animals, compared with those of other and similar regions*, [*etc.*], Boston, 1850; *Contributions to the natural history of the United States of America*, Boston, London, 1857–62, four volumes of the projected ten being published; *An essay on classification*, London, 1859; *The structure of animal life*, New York, 1866; (with Augustus A. Gould) *Principles of zoology*, [*etc.*], Boston, 1848, of which a revised and enlarged edition appeared in 1855, with a reprint in 1856; and (with Elizabeth Agassiz) *A journey in Brazil*, Boston, 1886, with a further edition in 1888.[5]

Although mainly remembered for his theory of natural selection as expounded in his *Origin of species*, Charles Darwin made numerous other vital contributions to natural history and to geology. In fact, there were many predecessors of Darwin whose work influenced the development of his theory. Numerous papers have been written upon individuals whose writings contained hints that were later collected together and used as evidence by Darwin, and books by H. B. Glass, Owsei Temkin

[1] See Coppleson, V. M. The life and times of George Bennett, *Bull. Post-Grad. Comm. Med. Univ. Sydney*, 11, ix, 1955, pp. 207–264. Bibliography of his writings (115 items), pp. 232–236.

[2] Davies, Gordon L. A tour of the British Isles made by Louis Agassiz in 1840. *Ann. Sci.*, 24, 1968, pp. 131–146.

[3] See Bowley, Donovan R., and Smith, Hobart M. The dates of publication of Louis Agassiz's *Nomenclator zoologicus*. *J. Soc. Bib. Nat. Hist.*, 5, 1958, pp. 35–36.

[4] See p. 236.

[5] See Lurie, Edward. *Louis Agassiz: a life in science*, Chicago, 1960, abridged edition, 1966; Robinson, Mabel L. *Runner of the mountain tops. The life of Louis Agassiz*, New York, 1939; Baron, Walter. Zu Louis Agassiz's Beurteilung des Darwinismus. *Sudhoffs Arch. Gesch. Med.*, 40, 1956, pp. 259–277; and Weir, J. A. Agassiz, Mendel, and heredity. *J. Hist. Biol.*, 1, 1968, pp. 179–203.

and W. L. Straus,[1] Loren C. Eiseley,[2] and Francis C. Haber,[3] in particular, survey forerunners of Darwin in extensive studies. Two articles by Conway Zirkle[4] perform a similar function in briefer form.[5] James Parkinson (1755–1824), author of numerous popular writings on medicine, on politics, and notable for his classic description of paralysis agitans, also wrote *Organic remains of a former world*, three volumes, London, 1804–11, the third volume of which deals with vertebrate and invertebrate fossils. Parkinson was an original member of the Geological Society, and several of his papers published in the Society's *Transactions* deal with fossils.[6] Sir William Lawrence (1783–1867), a noted surgeon, delivered several series of lectures at the Royal College of Surgeons, which were published as *An introduction to comparative anatomy and physiology*, [*etc.*], London, 1816, and *Lectures on physiology, zoology and the natural history of man*, [*etc.*], London, 1819 (Plate 10). The latter in particular advanced views that offended the religious scruples of the period, and Lawrence was induced to suppress the book. However, without his sanction, it was republished as *Lectures on comparative anatomy, physiology and the natural history of man*, [*etc.*], 1822, and went into numerous editions, the final printing of the ninth edition appearing in 1848. Lawrence was a disciple of Cuvier, and did not discuss evolution. In fact he was opposed to the suggestion that man evolved from the animal kingdom, yet it was suggested that his writings contradicted the scriptures and were blasphemous.[7] Possibly the advanced thinking of both Parkinson and Lawrence was due to the fact that they had attended lectures on comparative anatomy by James Macartney (1770–1843). He gave these courses at St Bartholomew's Hospital, London, and later at Trinity College, Dublin, where he became professor of anatomy and surgery. The main biological writings of Macartney consist of thirteen articles in Abraham Rees' *Cyclopaedia*, these containing the results of much original research.[8]

[1] Glass, Hiram Bentley, Temkin, Owsei, and Straus, William L., *eds. Forerunners of Darwin, 1745–1859*, Baltimore, 1959; reprinted 1968.

[2] Eiseley, Loren C. *Darwin's centenary. Evolution and the men who discovered it*, 1958.

[3] Haber, Francis C. *The age of the world. Moses to Darwin*, Baltimore, 1959.

[4] Zirkle, Conway. Natural selection before the "Origin of species". *Proc. Amer. Philos. Soc.*, 84, 1941, pp. 71–123; and Zirkle, Conway. Species before Darwin. *Proc. Amer. Philos Soc.*, 103, 1959, pp. 636–644.

[5] See also Burrow, J. W. *Evolution and society: a study in Victorian social theory*, Cambridge 1966.

[6] See Eyles, Joan M. James Parkinson (1755–1824). *Nature*, 176, 1955, pp. 580–581; see also Thornton, pp. 163–164.

[7] See Goodfield-Toulmin, June. Blasphemy and biology. *Rockefeller Univ. Rev.*, Sept.-Oct., 1966, pp. 9–18; and Mudford, Peter G. William Lawrence and *The natural history of man*. *J. Hist. Ideas*, 29, 1968, pp. 430–436.

[8] See Macalister, Alexander. *James Macartney, . . . a memoir*, 1900; Thornton, John L. A diary of James Macartney (1770–1843), with notes on his writings. *Med. Hist.*, 12, 1968, pp. 164–173; and Thornton, John L. James Macartney (1770–1843). *St. Bart's Hosp. J.*, 65, 1961, pp. 121–123.

Vestiges of the natural history of creation, London, 1844, was first published anonymously, but was written by Robert Chambers (1802–71), and created a stir in religious and scientific circles, Chambers was a naturalist and a founder of the publishing firm of W. & R. Chambers. He was the author of several works, mainly on Scotland, but the *Vestiges* attracted most attention. Within seven months of publication four editions were printed, and five more appeared within the next six years. In 1884 the twelfth edition was published with an introduction by Alexander Ireland, who revealed the name of the author for the first time. There had been much speculation regarding this, the authorship having been attributed to various distinguished scientists and philosophers.[1]

Born on 12th February, 1809, and christened Charles Robert, Charles Darwin was a grandson of Erasmus Darwin (1731–1802), author of *Zoonomia*, 1794–96, which contained a germ of the theory to be advanced by Charles. It had been intended that Darwin should study medicine, but after a short period at Edinburgh he gave up the idea and proceeded to Cambridge. His future career was determined when he was invited to accompany Captain Robert Fitzroy in the *Beagle* on a voyage which was to last five years. Several accounts of the voyage have been published, including books and articles by A. Moorehead,[2] Bern Dibner,[3] Howard E. and Valami Gruber[4] in an article based on Darwin's manuscript notebooks kept during the voyage, and now preserved in Cambridge University Library, P. D. F. Murray,[5] and *Charles Darwin and the voyage of the Beagle. Edited with an introduction by Nora Barlow*, London, 1945. An account of the expedition was published as *Narrative of the surveying voyages of Her Majesty's Ships Adventure and Beagle between the years 1826 and 1836*, [*etc.*], *Volume III. Journal and remarks, 1832–1836, 1839.* This third volume, for which Charles Darwin was responsible, was issued separately in 1839 and 1840, and a revised popular edition appeared in Murray's Home and Colonial Library as *Journal of researches into the natural history and geology of the countries visited during the voyage of H.M.S. "Beagle"*, [*etc.*], 1845.This edition was first published in three separate half-crown parts, and then in one volume during the same year. The 1860 (third) edition bears the title *A naturalist's voyage*, [*etc.*], and the first edition appeared in a facsimile reprint in 1952. A German translation of the 1840 edition was published in 1844, the first of Darwin's works to be translated. Five volumes devoted to the zoology of the expedition, intended solely for specialists, were published with the aid of a Government

[1] See Millhauser, M. *Just before Darwin: Robert Chambers and 'Vestiges'*, Middletown, 1968.
[2] Moorehead, A. *Darwin and the Beagle*, 1969.
[3] Dibner, Bern. *Darwin of the Beagle. Second edition*, New York, 1964.
[4] Gruber, Howard E., and Gruber, Valmai. The eye of reason: Darwin's development during the *Beagle* voyage. *Isis*, 53, 1962, pp. 186–200.
[5] Murray, P. D. F. The voyage of the "Beagle". *New Biology*, No. 28, 1959, pp. 7–24.

grant as *Zoology of the voyage of H.M.S. "Beagle". . . . Edited and superintended by Charles Darwin*, [*etc.*]. Part 1, on fossil mammalia by Richard Owen, contains a geological introduction by Darwin, and was published in 1840, a year after the second part. Part 2, dealing with mammalia by George R. Waterhouse, has a notice of their habits and ranges by Charles Darwin, and appeared in 1839. Part 3 deals with birds, by John Gould, 1841; Part 4, fish, by Leonard Jenyns, 1842; and Part 5, reptiles, by Thomas Bell, 1843. These five parts were issued as nineteen numbers between February 1838 and October 1843. The geology of the voyage was described by Darwin in a publication issued in three parts as *The structure and distribution of coral reefs*, [*etc.*], 1842, with a second edition in 1874; *Geological observations on the volcanic Islands*, [*etc.*], 1844; and *Geological observations on South America*, [*etc.*], 1846, the two latter appearing in a second edition in one volume, 1876. Before the appearance of his *magnum opus* Charles Darwin published the following four monographs: *A monograph on the fossil lepadidae; or, pendunculated cirripedes of Great Britain*, 1851; *A monograph on the sub-class cirripedia. . . . The lepadidae*, [*etc.*], 1851, and *A monograph on the sub-class cirripedia. . . . The balanidae*, [*etc.*], 1854, these two publications of the Ray Society having been reprinted together in one volume, Weinheim, 1964; and *A monograph on the fossil balanidae and verrucidae of Great Britain*, 1854. Probably in 1839 Charles Darwin printed a questionnaire regarding the breeding of animals, containing twenty-one lengthy questions on eight pages. The original was known only by a copy in the British Museum (Natural History), but it has been reprinted by the Society for the Bibliography of Natural History as *Questions about the breeding of animals* [*1840*]. *With an introduction by Sir Gavin de Beer*, London, 1968 (Sherborne Fund Facsimile No. 3). A further copy of the original, with answers and the date May 1839 in Darwin's writing was found in September, 1968 in the Darwin archive at Cambridge University.[1]

In 1838 Charles Darwin read the *Essay on the principle of population*, [*etc.*], by Thomas Robert Malthus (1766–1834),[2] which had first appeared anonymously in 1798, and in 1803 with the name of the author. Darwin had already commenced a notebook on the subject of natural selection, and prepared drafts of his views on the subject in 1842 and 1844.[3] Before

[1] See Freeman, R. B., and Gautrey, P. J. Darwin's *Questions about the breeding of animals*, with a note on *Queries about expression*. *J. Soc. Bib. Nat. Hist.*, 5, 1969, pp. 220–225.

[2] See Habakkuk, H. J. Thomas Robert Malthus, F.R.S. (1766–1834). *Notes Rec. Roy. Soc. Lond.*, 14, 1959, pp. 99–108; *The travel diaries of Thomas Robert Malthus. Edited by Patricia D. James*, 1966; and Robert Malthus. *Brit. med. J.*, 1966, ii, pp. 698–699.

[3] See Stauffer, Robert Clinton, "On the origin of species": an unpublished version. *Science* 130, 1959, pp. 1449–1452; de Beer, Sir Gavin. Darwin's notebooks on transmutation of species. Part 1 [–6]. *Bull. Brit. Mus.* (*Nat. Hist.*) *Hist. Ser.*, 2, 1960–61, pp. 23–73, 75–118, 119–150, 151–183, 185–200; *Hist. Ser.*, 3, 1969, pp. 129–176; de Beer, Sir Gavin. Darwin's journal. *Bull. Brit. Mus.* (*Nat. Hist.*) *Hist. Ser.*, 2, 1967, pp. 3–21, a transcript of Darwin MS.

these were published, however, he received a paper setting forth similar ideas from Alfred Russel Wallace (1823–1913), who, incidentally, had been similarly influenced by Malthus. At the instance of his friends, Darwin presented a joint paper to the Linnean Society bearing the names of himself and Wallace, and entitled "On the tendency of species to form varieties; and on the perpetuation of varieties and species by natural means of selection". This was communicated by Sir Charles Lyell and J. D. Hooker, and was read on 1st July, 1858.[1] Darwin now hastened to finish his book, and *On the origin of species by means of natural selection, or the preservation of favoured races in the struggle for life* was published on 24th November, 1859 (Plates 11–12). A facsimile of this first edition, with an introduction by Ernst Mayr, has been published, Cambridge, Mass., 1964. Of the original edition 1,250 copies were printed, 1,192 being for sale, and the trade sale resulted in 1,500 copies being ordered on the first day, and a new printing of 3,000 copies appeared on 7th January, 1860, with a third of 2,000 copies in April, 1861. Editions followed in 1866, 1869 and 1872, and 1876, which should be regarded as the definitive edition, and the book has been reprinted on numerous occasions, including a reprint of the first edition, with a foreword by C. D. Darlington, in 1950. The centenary of the publication of the *Origin of species* was celebrated by a spate of books and articles on Darwin,[2] and a variorum text by Morse Peckham[3] was issued. This covers all the variants in the six texts published between 1859 and 1872, including additions, deletions, changes in style and punctuation. It indicates the development of Darwin's

140[2], University of Cambridge; Smith, Sydney. The origin of the "Origin" as discerned from Charles Darwin's notebooks and his annotations in the books he read between 1837 and 1842. *Advanc. Sci.*, 16, 1960, pp. 391–401; de Beer, Sir Gavin. The origins of Darwin's ideas on evolution and natural selection. The Wilkins Lecture. *Proc. Roy. Soc.*, 155B, 1962, pp. 321–338; Barlow, Lady Nora. Darwin's ornithological notes. [Cambridge University Handlist, (1960) No. 29 (ii)]. Edited with an introduction, notes and appendix. *Bull. Brit. Mus. (Nat. Hist.) Hist. Ser.*, 2, 1963, pp. 201–278; and Stauffer, Robert Clinton. Ecology in the long manuscript version of Darwin's *Origin of species* and Linnæus' *Oeconomy of nature*. *Proc. Amer. Philos. Soc.*, 104, 1960, pp. 235–241.

[1] See Darwin, Charles, and Wallace, Alfred Russel. *Evolution by natural selection*, Cambridge, 1958, which contains Darwin's Sketch of 1842, his Essay of 1844, and the joint paper of 1858, etc.; Sarton, George. Discovery of the theory of natural selection. With facsimile reproductions (Nos. viii–ix) of Darwin's and Wallace's earliest publications on the subject (1859). *Isis*, 14, 1930, pp. 133–154; and Poynter, F. N. L. Centenary of the Darwin-Wallace paper on natural selection. *Brit. med. J.*, 1958, I, pp. 1538–1540, which mentions the earlier paper by Wallace, "On the law which has regulated the introduction of new species". *Ann. & Mag. Nat. Hist.*, 16, 1855, pp. 184–196, to which Lyell drew Darwin's attention in 1856. See also the Darwin-Wallace Centenary issue of *J. Linn. Soc. Bot.*, 56 (365); *Zool.*, 44 (295), 1958, pp. 1–152.

[2] Good surveys of this literature are provided by Cushing, H. B. Darwin centennial publications. *Catholic Library World*, 31, 1959–60, pp. 423–429, 456–457; and by Fleming, Donald. The centenary of the *Origin of species*. *J. Hist. Ideas*, 20, 1959, pp. 437–446.

[3] Darwin, Charles. *The origin of species. . . . A variorum text. Edited by Morse Peckham*, Philadelphia, 1959. This contains a transcript of the publisher's records (pp. 775–785), and bibliographical descriptions (pp. 787–792).

thesis as revealed by successive editions of the *Origin*, except that published in 1876, which Peckham had not seen, but the method of arrangement is not easy to follow, and the book cannot be read as a text.[1]

Darwin devoted his entire life to the study of natural history and to recording his observations. His books include the following: *On the various contrivances by which orchids are fertilized by insects*, 1862, of which a second edition was published in 1877; *The movements and habits of climbing plants*, 1865, which had first appeared in the *Journal* of the Linnean Society, and of which a second edition was printed in 1875, with a final text in 1882, reprinted 1937; *The variation of animals and plants under domestication*, two volumes, 1868,[2] with a second revised edition in 1875; *The descent of man, and selection in relation to sex*, two volumes, 1871 (reprinted, Brussels, 1969), which went into a second edition in 1874; *The expressions of the emotions in man and animals*, 1872,[3] reprinted with a preface by Konrad Lorenz, Chicago, 1965; *Insectivorous plants*, 1875; *The effects of cross and self fertilization in the vegetable kingdom*, 1876, with a second edition in 1878; *The different forms of flowers on plants of the same species*, 1877; which was printed in a second edition in 1880; *The power of movement in plants*, 1880; and *The formation of vegetable mould, through the action of worms, with observations on their habits*, 1881. These books were the result of many years of patient experiment and thought, and are still worthy of careful study. Most of these were reprinted in Brussels in 1969. Other Darwin items of interest are: *The life and letters of Charles Darwin, including an autobiographical chapter. Edited by his son, Francis Darwin*, three volumes, 1887, of which an abridged, revised edition was published in 1892 as *Charles Darwin: his life told in an autobiographical chapter, and in a selected series of his published letters*, [*etc.*], 1892; *More letters of Charles Darwin*, [*etc.*], edited by Francis Darwin and A. C. Seward, two volumes, 1903; his correspondence with the Rev. J. Brodie Innes,[4] and other letters[5]; *The foundations of the origin of species: two essays written in 1842 and 1844 by Charles Darwin. Edited by his son, Francis Darwin*, 1909; and *Charles Darwin's diary of the voyage of H.M.S. "Beagle"*, [*etc.*], edited from the manuscript by Lady Nora Barlow, a granddaughter of Darwin, 1933. Lady Barlow has also edited Darwin's autobiography,

[1] See also Ellegård, Alvar. *Darwin and the general reader. The reception of Darwin's theory of evolution in the British periodical press, 1859–1872*, Göteborg, 1958.

[2] See Olby, R. C. Charles Darwin's manuscript of "Pangenesis". *Brit. J. Hist. Sci.*, 1, 1963, pp. 251–263.

[3] See Swisher, Charles N. Charles Darwin on the origins of behaviour. *Bull. Hist. Med.*, 41, 1967, pp. 24–43.

[4] See Stecher, R. M. The Darwin-Innes letters: the correspondence of an evolutionist with his vicar, 1848–1884. *Ann. Sci.*, 17, 1961 [1964], pp. 201–258.

[5] See de Beer, Sir Gavin. Some unpublished letters of Charles Darwin. *Notes Rec. Roy. Soc. Lond.*, 14, 1959, pp. 12–66; and de Beer, Sir Gavin. Further unpublished letters of Charles Darwin. *Ann. Sci.*, 14, 1958, pp. 83–115; and de Beer, Sir Gavin. The Darwin letters of Shrewsbury School. *Notes Rec. Roy. Soc. Lond.*, 23, 1968, pp. 68–85.

restoring the omissions from the 1887 and subsequent editions,[1] and his correspondence with John Stevens Henslow.[2] A list of Darwin manuscripts at Cambridge has been published as *Handlist of Darwin papers at the University Library, Cambridge*, 1960. B. J. Loewenburg[3] has provided an excellent review of recent studies on Darwin, Sir Gavin de Beer[4] has written a readable, but undocumented, survey of Darwin's life and work, and Julian Huxley and H. B. D. Kettlewell[5] have published a similar, authoritative biographical study for the general reader.[6] Charles Darwin died at Down House, Downe, on 19th April, 1882, and was buried in Westminster Abbey. A handlist of his writings has been compiled by R. B. Freeman,[7] excluding work published exclusively in periodicals. Forty-six works are listed, producing over five hundred items, and the handlist also contains an item by Darwin not previously published in English. Entitled "On the flight-paths of male humble bees", it first appeared in 1886 in a German collection of Darwin's shorter works published after his death, and bore the title "Über die Wege der Hummel-Männchen". This was the first detailed description of the flight paths and 'buzzing-places' of humble bees.

We have noted that Darwin's first public pronouncement on natural selection was contained in a paper published jointly with A. R. Wallace. At that time Wallace was in Borneo and it is significant that at no time did he and Darwin argue about priority in advancing the theory. Alfred Russel Wallace (1823–1913) accompanied H. W. Bates to the Amazon in 1848, and from 1854 to 1862 was engaged in studying the zoology of the Malay Archipelago. He was content to follow Darwin as his disciple, but later developed rather divergent theories. Wallace's writings include *A narrative of travels on the Amazon and Rio Negro*, [*etc.*], London, 1853; *The Malay Archipelago*, [*etc.*], two volumes, London, 1869, a second

[1] *The autobiography of Charles Darwin, 1809–1882. With original omissions restored. Edited with appendix and notes by his grand-daughter, Nora Barlow*, 1958.

[2] *Darwin and Henslow: the growth of an idea. Letters, 1831–1860, edited by Nora Barlow*, 1967.

[3] Loewenberg, Bert James. Darwin and Darwin studies, 1959–63. *History of Science*, 4, 1965, pp. 15–54.

[4] de Beer, Sir Gavin. *Charles Darwin. Evolution by natural selection*, 1963. (*British Men of Science Series*.)

[5] Huxley, Julian, and Kettlewell, H. B. D. *Charles Darwin and his world*, 1965.

[6] See also Darlington, C. D. *Darwin's place in history*, Oxford, 1959; Himmelfarb, Gertrude. *Darwin and the Darwinian revolution*, 1959; Keith, Sir Arthur. *Darwin revalued*, 1955; Wichler, Gerhard. *Charles Darwin: the founder of the theory of evolution and natural selection*, Oxford, [*etc.*], 1961; Barnett, S. A., *ed. A century of Darwin*, London, [*etc.*], 1958, containing fifteen chapters by various authors on aspects of Darwin's work; Bettany, G. T. *Life of Darwin*, 1887; Atkins, Hedley. The Darwin tradition. Thomas Vicary Lecture delivered at the Royal College of Surgeons of England on 29th October, 1964. *Ann. Roy. Coll. Surg.*, 36, 1965, pp. 1–25.

[7] Freeman, R. B. *The works of Charles Darwin. An annotated bibliographical handlist*, 1965; see also Freeman, R. B. On the origin of species, 1859. *Book Collector*, 16, 1967, pp. 340–344; and Nethery, Wallace. "On the origin of species", 1859. [Notes on a copy at the University of Southern California.] *Book Collector*, 17, 1968, p. 216.

edition of which was published in America the same year, with further editions in 1872, 1886, 1890 and 1902, and two separate reprints in 1962; *Contributions to the theory of natural selection*, London, 1870, with a second edition in 1871; *The geographical distribution of animals*, [*etc.*], two volumes, London, 1876, with an American reprint issued from New York in the same year and another in 1962, and a German translation by A. B. Meyer in two volumes, Dresden, 1876; *Tropical nature, and other essays*, London, 1878; *Island life; the phenomena and causes of insular faunas and floras*, 1880, which went into a second edition in 1892, there having been an American edition published in New York, 1881; *Darwinism; an exposition of the theory of natural selection, with some of its applications*, 1889; *Studies scientific and social*, two volumes, London, 1900, containing reprints of articles from periodicals; and *Man's place in the universe*, London, 1903. There are important unpublished notebooks and journals kept by Wallace preserved in the Linnean Society, London and Barbara G. Beddall[1] and H. Lewis McKinney[2] have contributed well-documented articles based on manuscript material dealing with Wallace and natural selection. Wilma George[3] and Amabel Williams-Ellis[4] have published biographical studies of A. R. Wallace, the second being intended for the general reader rather than the scientist.[5]

Although Darwin's theories attracted considerable attention and controversy, he was not physically capable of replying to his critics, but found an effectual champion in Thomas Henry Huxley (1825–95). The latter was born at Ealing on 4th May, 1825, and after studying medicine entered the Naval Medical Service, in which he joined the surveying ship *Rattlesnake*.[6] Huxley became professor of natural history and palaeontology at the Royal School of Mines (1854), Inspector of Salmon Fisheries (1881), professor of biology and dean of the Royal College of Science, and received many honours during his distinguished career. In 1845 he published a paper on the layer in hair,[7] which became

[1] Beddall, Barbara G. Wallace, Darwin, and the theory of natural selection: a study in the development of ideas and attitudes. *J. Hist. Biol.*, 1, 1968, pp. 261–323.

[2] McKinney, H. Lewis. Alfred Russel Wallace and the discovery of natural selection. *J. Hist. Med.*, 21, 1966, pp. 333–357.

[3] George, Wilma. *Biologist philosopher: a study of the life and writings of Alfred Russel Wallace*, London, New York, 1964.

[4] Williams-Ellis, Amabel. *Darwin's moon. A biography of Alfred Russel Wallace*, 1966.

[5] See Wallace, Alfred Russel. *My life: a record of events and opinions*, [*etc.*], 2 vols., 1905; Marchant, James. *Alfred Russel Wallace: letters and reminiscences*, 2 vols., 1916; Eiseley, Loren C. Alfred Russel Wallace. *Sci. Amer.*, 200, 1959, pp. 70–84; and Roberts, Ffrangcon. Alfred Russel Wallace (1823–1913). *Lancet*, 1958, II, pp. 580–581; and Pantin, C. F. A. Alfred Russel Wallace, F.R.S., and his essays of 1858 and 1855. *Notes Rec. Roy. Soc. Lond.*, 14, 1959, pp. 67–84.

[6] See *T. H. Huxley's diary of the voyage of H.M.S. Rattlesnake. Edited from the unpublished MS. by Julian Huxley*, 1935.

[7] On a hitherto undescribed structure in the human hair sheath. *Lond. med. Gaz.*, N.S.1, 1845, pp. 1340–1341.

known as "Huxley's layer", and during his four years' voyage in the *Rattlesnake* wrote several papers, one of which, "On the anatomy and affinities of the family of medusae", was published by the Royal Society, and he was elected a Fellow of the Society in 1851, receiving its gold medal in the following year. Huxley wrote several valuable papers, and in 1859, on the publication of Darwin's *Origin of species*, came out as its main supporter, but without accepting Darwin's views in their entirety. Huxley gave a series of lectures between 1859 and 1862 on the comparative anatomy of man and the higher apes, which were published as *Evidence as to man's place in nature*, London, 1863. This was his most famous biological publication. His other writings include *Lectures on the elements of comparative anatomy. On the classification of animals and on the vertebrate skull*, London, 1864; *An introduction to the classification of animals*, London, 1869; *A manual of the anatomy of vertebrated animals*, London 1871, a valuable systematic work which went into further editions from New York in 1872 and 1878; *Lectures on evolution*, [*etc.*], New York, 1882, containing his lectures delivered in New York and London in 1876; *Manual of the anatomy of invertebrated animals*, London, 1877; and *The crayfish*, 1880. Collections of Huxley's papers were issued as *The scientific memoirs of Thomas Henry Huxley. Edited by Michael Foster and E. Ray Lankester*, four volumes and a supplementary volume, London, 1898–1903; *Collected essays*, nine volumes, London, 1893–1901; and *Darwiniana*, [*etc.*], London, 1899. A collection of letters and manuscripts relating to Huxley was acquired by the Imperial College of Science and Technology in 1937, a catalogue of these having been published in 1946.[1] The College also houses a collection of Huxley's notebooks, drawings and other papers, a list of which has been prepared by J. Pingree,[2] and published by Imperial College.[3]

The science of heredity originated in the experimental investigations of Gregor Johann Mendel (1822–84), a native of Heinzendorff, in Moravia. In 1843 he joined the Augustinian monastery at Brünn. There he formed a botanical garden and a mineralogical collection, and in 1846 commenced his experiments on garden peas, which lasted for seven years, and he also studied hereditary characteristics. Mendel's law of heredity was first

[1] Dawson, Warren R. *The Huxley papers: a descriptive catalogue of the correspondence, manuscripts and miscellaneous papers of the Rt. Hon. Thomas Henry Huxley, P.C., D.C.L., F.R.S., preserved in the Imperial College of Science and Technology*, 1946.

[2] Pingree, J. *Thomas Henry Huxley: a list of his scientific notebooks, drawings and other papers, preserved in the College Archives*, 1968.

[3] See also Bibby, Cyril. *T. H. Huxley: scientist, humanist and educator*, 1959, in which pp. 260–266 contain a select list of Huxley's publications; Huxley, Leonard. *The life and letters of Thomas Henry Huxley*, 2 vols., 1900, which contains a list of his publications. Vol. 2, pp. 453–470; Peterson, Houston. *Huxley, prophet of science*, 1932; Ayres, Clarence. *Huxley*, New York, 1932; Irvine, William, *Apes, angels, and Victorians: a joint biography of Darwin and Huxley*, 1955; Clark, R. W. *The Huxleys*, 1968; and Bibby, Cyril. Huxley and the reception of the "Origin". *Vict. Stud.*, 3, 1959, pp. 76–86.

announced in 1865 and published as "Versuche über Pflanzen-Hybriden"[1] and a second paper appeared as "Über einige aus künstlicher Befruchtung gewonnene Hieracium-Bastarde" in 1869,[2] but his work was largely ignored until it was rediscovered many years later and the two papers were printed together.[3] These two articles have since been reprinted and translated several times.[4] Robert C. Olby and Peter Gautrey[5] discovered eleven references to Mendel before 1900, but none of the writers really understood Mendel's theory. His work was not known to Darwin, although the *Verhandlungen* went to 115 libraries and institutions on exchange, including the Royal Society, and forty reprints of the original article were distributed. Peter J. Vorzimmer[6] has mentioned the possibilities of Darwin becoming acquainted with Mendel's work, and Sir Ronald A. Fisher[7] has suggested that some of his results were too good to be true by statistical theory. Robert C. Olby[8] has written extensively on Mendel, his precursors and successors, and Sir Gavin de Beer's[9] papers on Mendelism suggest that it was not until the 1930's that the true significance of Mendel's work was appreciated. The rediscovery and partial appreciation of Mendel's work in 1900 independently by Hugo de Vries (1848–1935), Carl Correns (1864–1933) and Erik von Tschermak (1871–1962) has been investigated by Conway Zirkle,[10] and by J. S. Wilkie[11] who outlines the factors which favoured the rapid growth of Mendelian genetics after the rediscovery of Mendel's work. Curt Stern and Eva R. Sherwood[12] have edited a Mendel source book which contains a new translation from the original German of Mendel's two papers on hybridization, his letters to Carl Nägeli, papers by Hugo de Vries and by Carl Correns, letters from de Vries and Correns to H. F. Roberts, and papers

[1] In *Verhandlungen des Naturforschenden Vereins in Brünn*, 4, 1866, pp. 3–47. (A facsimile appeared in *J. Hered.*, 42, 1951, pp. 3–47.)

[2] *Ibid.*, 8, 1869, pp. 26–31.

[3] Ostwald's *Klassiker*, No. 121, 1901.

[4] See Van der Pas, Peter W. A note on the bibliography of Gregor Mendel. *Med. Hist.*, 3, 1959, pp. 331–333.

[5] Olby, Robert C., and Gautrey, Peter. Eleven references to Mendel before 1900. *Ann. Sci.*, 24, 1968, pp. 7–20.

[6] Vorzimmer, Peter J. Darwin and Mendel: the historical connection. *Isis*, 59, 1968, pp. 77–82.

[7] Fisher, Ronald A. Has Mendel's work been rediscovered? *Ann. Sci.*, 1, 1936, pp. 115–137.

[8] Olby, Robert C. *Origins of Mendelism*, 1966; and Olby, Robert C. The Mendel centenary. *Brit. J. Hist. Sci.*, 2, 1965, pp. 343–349.

[9] de Beer, Sir Gavin. Mendel, Darwin, and Fisher (1865–1965). *Notes Rec. Roy. Soc. Lond.*, 19, 1964, pp. 192–226; and de Beer, Sir Gavin. Mendel, Darwin, and Fisher: addendum. *Notes Rec. Roy. Soc. Lond.*, 21, 1966, pp. 64–71.

[10] Zirkle, Conway. The role of Liberty Hyde Bailey and Hugo de Vries in the rediscovery of mendelism. *J. Hist. Biol.*, 1, 1968, pp. 205–218; see also Zirkle, Conway. Some oddities in the delayed discovery of Mendelism. *J. Heredity*, 55, 1964, pp. 65–72.

[11] Wilkie, J. S. Some reasons for the rediscovery and appreciation of Mendel's work in first years of the present century. *Brit. J. Hist. Sci.*, 1, 1962, pp. 5–17.

[12] Stern, Curt, and Sherwood, Eva R., *eds. The origin of genetics. A Mendel source book*, San Francisco, London, 1966.

on Mendel's work by R. A. Fisher and by Sewall Wright. J. H. Bennett has edited Gregor Mendel's *Experiments in plant hybridization*, 1965[1] containing a translation of Mendel's original paper, a biography of Mendel by Bateson, and writings on Mendel's work by Sir Ronald Fisher. Mendel left an autobiography, which has been translated by Anne Iltis,[2] and printed with a facsimile of part of his manuscript. Mendel conducted other experiments on hybridization with bees,[3] and was also interested in hybridization of apple trees, gaining a medal for new varieties of apples and pears.[4] The centenary of the publication of Mendel's paper on heredity was widely celebrated and the proceedings of several symposia were published.[5] M. Jakubíček and J. Kubíček[6] have published a bibliography of Mendel's writings, and there are several studies of his life and work.[7] The Moravian Museum, Brno, Czechoslovakia, has published *Iconographia Mendeliana*, 1965, which contains reproductions of pictorial and written documents characterizing the life and work of Mendel; and *Folia Mendeliana*, first published in November 1966, contains an account of the 1965 centennial celebrations in Brno, and articles on Mendel.

Another eminent contributor to heredity was Sir Francis Galton (1822–1911), a grandson of Erasmus Darwin through his second wife, Charles Darwin bearing the same relationship by the first marriage of Erasmus. From 1846 to 1850 he was exploring in the Sudan and Damaraland, after which he became interested in meteorology, and formulated the theory of anticyclones. Galton was responsible for laying down the principles of eugenics, introduced composite portraiture, and systematized

[1] Mendel, Gregor. *Experiments in plant hybridisation. Mendel's original paper in English translation with commentary and assessment by the late Sir Ronald A. Fisher, together with a reprint of W. Bateson's biographical notice of Mendel. Edited by J. H. Bennett*, Edinburgh, London, 1965.

[2] Iltis, Anne. Gregor Mendel's autobiography. *J. Hered.*, 45, 1954, pp. 231–234.

[3] See Vecerek, O. Johann Gregor Mendel as a beekeeper. *Bee World*, 46, 1965, pp. 86–96.

[4] See Orel, V., and Vávra, M. Mendel's program for the hybridization of apple trees. *J. Hist. Biol.*, 1, 1968, pp. 219–224.

[5] See Sosna, Milan, *ed. G. Mendel memorial symposium, 1865–1965. Proceedings held in Brno, Aug. 4–7, 1965. Collected papers*, Prague, 1967; *Mendel centenary: genetics, development and evolution. Proceedings of a symposium, 1965, edited by R. M. Nardone*, Washington, 1968; Brink, Royal Alexander, and Styles, E. Derek, *eds. Heritage from Mendel. Proceedings of the Mendel Centennial Symposium, 1965, sponsored by the Genetics Society of America*, Madison, 1967; Commemoration of the publication of Gregor Mendel's pioneer experiments in genetics. *Proc. Amer. Philos. Soc.*, 109, 1965, pp. 189–284; and The birth of genetics. Mendel-De Vries-Correns-Tscherman in English translation. *Genetics*, 35, No. 5, ii, Supplement.

[6] Jakubíček, M., and Kubíček, J. *Bibliographia Mendeliana*, Brno, 1965.

[7] See Iltis, Hugo. *Life of Mendel. . . . Translated by Eden and Cedar Paul*, [*etc.*], 1932, reprinted New York, 1966; Posner, E., and Skutil, J. Darwin and Mendel. *Midland med. Rev.*, 2, 1967, pp. 112–118; Sorsby, Arnold. Gregor Mendel. *Brit. med. J.*, 1965, i, pp. 333–338; Posner, E. The enigmatic Mendel. *Bull. Hist. Med.*, 40, 1966, pp. 430–440; Gasking, Elizabeth B. Why was Mendel's work ignored? *J. Hist. Ideas.*, 20, 1959, pp 60–84; and Stomps, Th. J. On the rediscovery of Mendel's work by Hugo de Vries. *J. Hered.*, 45, 1954, pp. 293–294.

methods of finger-prints. Galton was knighted in 1909, and was the author of *Narrative of an explorer in tropical South Africa, being an account of a visit to Damaraland in 1851*, first issued in 1853 with the title *Tropical South Africa*, and which went into several editions; *Hereditary genius*, 1869; *English men of science*, 1874; *Inquiries into human faculty and its development*, 1883; *Natural inheritance*, 1889, in which he formulated his law of ancestral inheritance, having published a tentative formulation in 1885, and proceeding to a more advanced one in 1897[1]; and *Noteworthy families*, 1906. Galton's scientific study of finger-prints led to these being used as an invaluable method of crime detection, and his work on the subject is described in *Finger prints*, 1893; *Blurred finger prints*, 1893; and *Finger print directory*, 1895.[2]

Eminent as a palaeontologist, Joseph Leidy (1823–91) was also keenly interested in anatomy, botany, geology, mineralogy and zoology. A native of Philadelphia, Leidy studied medicine and lectured on anatomy. He conducted exhaustive researches with the microscope, and became professor of anatomy at Pennsylvania. He studied fossil molluscs, bacteria, parasitic worms, and in 1846 detected *trichina spiralis* in ham. From 1871 to 1883 he occupied the chair of natural history at Swarthmore College, and later was professor of zoology and comparative anatomy at Pennsylvania, and also professor of zoology at the Academy of Natural Sciences. Leidy was the author of *Ancient fauna of Nebraska*, 1854; he published an American edition of Quain's *Anatomy* in 1849; *Elementary treatise on human anatomy*, 1861, with a second edition in 1889; *Freshwater rhizopods of North America*, 1879; *The extinct mammalian fauna of Dakota and Nebraska*, [*etc.*], 1869, and numerous other contributions to science.[3]

The author of several semi-popular works on natural history, many of which can still be read with profit, and some of which remain useful to the professional scientist, Philip Henry Gosse (1810–88) illustrated many of his own writings. He was elected F.R.S. in 1856, and his books include *Introduction to zoology*, 1844; *The ocean*, 1845; *Birds of Jamaica*, 1847; *A naturalist's sojourn in Jamaica*, 1851; *The aquarium: an unveiling of the wonders of the deep sea*, 1854; *Manual of marine zoology*, two volumes, 1855–56; *Omphalos*, 1857; *Actinologia Britannica: a history of the British sea anemones and corals*, twelve parts, 1858–60; and *The rotifera*, two

[1] See Swinburne, R. G. Galton's law-formulation and development. *Ann. Sci.*, 21, 1965, pp. 15–31.

[2] See Pearson, Karl. *The life, letters and labours of Francis Galton*, 3 vols. [in 4], Cambridge, 1914–30; Wilkie, J. S. Galton's contribution to the theory of evolution, with special reference to the use of models and metaphors. *Ann. Sci.*, 11, 1956, pp. 194–205; Platt, Sir Robert. Darwin, Mendel, and Galton. *Med. Hist.*, 3, 1959, pp. 87–99.

[3] See Middleton, William Shainline, Joseph Leidy, scientist. *Ann. med. Hist.*, 5, 1923, pp. 100–112; and Osborn, Henry Fairfield. Joseph Leidy, founder of vertebrate paleontology in America, [*etc.*], *Science*, 59, 1924, pp. 173–176.

volumes, 1886, written with C. T. Hudson.[1] Peter Stageman has published a bibliography devoted to the first editions of Gosse.[2]

Francis Trevelyan Buckland (1826–80), generally known as Frank Buckland, was a surgeon who became prominent as a writer on natural history. He became Inspector of Fisheries and his official annual reports led to legislation promoting purer rivers and the conservation of fish. Buckland was a pioneer of wild life conservation, and he established a museum. A true naturalist and an extensive writer, he was also eccentric, and tested the culinary possibilities of a wide variety of animals. His books include *Curiosities of natural history*, 1857, with a second edition in 1858, a fourth in 1859, and a new series in two volumes in 1866; *Log-book of a fisherman and zoologist*, 1875; *Natural history of British fishes*, 1880; and *Notes and jottings from animal life*, 1882.[3]

Carl Gegenbaur (1826–1903) has been called the founder of modern comparative anatomy. A native of Würzburg, he was professor of anatomy at Jena (1855–73) and Heidelberg (1873–91), and contributed to our knowledge of the evolution of the skull and to embryology. His important textbook was entitled *Grundzüge der vergleichenden Anatomie*, Leipzig, 1874, of which a second edition was published in 1870, and an English translation in 1878; he also wrote *Vergleichende Anatomie der Wirbelthiere mit Berücksichtigung der Wirbellosen*, two volumes, 1898–1901. In 1875 he founded the *Morphologisches Jahrbuch*.

The British zoologist Alfred Newton (1829–1907) made a special study of birds in various parts of the world, and in 1866 was appointed professor of comparative anatomy at Cambridge. His most important publication was written with the aid of several other authorities and published as *A dictionary of birds*, [*etc.*], London, in four continuously paginated parts. 1893–96. A second edition appeared in 1899. With Howard Saunders, Newton revised the fourth edition of William Yarrell's *A history of British birds*, foui volumes, London, 1871–85, and also compiled a catalogue of the egg collection of John Wolley, who died in 1859 at the age of thirty-six. Entitled *Ootheca Wolleyana*, this also includes Newton's own collection, and was published in four parts, part 1, 1864; part 2, 1902; part 3, 1905; and part 4, 1907. A biography by

[1] See Andrews, John S. Philip Henry Gosse, F.R.S. (1810–88). *Lib. Assn. Rec.*, 63, 1961, pp. 197–201; see also Hedgpeth, Joel W. Fishes of the Murex. (Notes for a bibliography of marine natural history.) *Isis*, 37, 1947, pp. 26–32.

[2] Stageman, Peter. *A bibliography of the first editions of Philip Henry Gosse, F.R.S. With introductory essays by Sacheverell Sitwell and Geoffrey Lapage*, Cambridge, 1955. See also review of this work by Denys W. Tucker. *J. Soc. Bib. Nat. Hist.*, 3, 1957, pp. 223–229; and Sawyer, F. C. Some additions to P. Stageman's "Bibliography of the first editions of Philip Henry Gosse" (1955). *J. Soc. Bib. Nat. Hist.*, 3, 1957, pp. 221–222.

[3] See Burgess, G. H. O. *The curious world of Frank Buckland*, 1967; Snell, William E. Frank Buckland—medical naturalist. *Proc. Roy. Soc. Med.*, 60, 1967, pp. 291–296; and Gordon, Mervyn. Memories of Frank Buckland and extracts from some of his records. *Ann. Roy. Coll. Surg. Engl.*, 10, 1952, pp. 133–139.

A. F. R. Wollaston[1] contains a short-title list of Newton's published papers, and is based on his correspondence preserved over a period of more than fifty years.

St. George Jackson Mivart (1827–1900) was born in London, and after being lecturer in anatomy at St. Mary's Hospital became professor at Louvain. He was in conflict with the evolutionists regarding the evolution of the human mind, but wrote several outstanding books, including *Man and apes, an exposition of structural resemblances and differences bearing upon questions of affinity and origin*, London, 1873; *Lessons from nature, as manifested in mind and matter*, London, 1876; *The cat. An introduction to the study of backboned animals, especially mammals*, London, 1881; *Dogs, jackals, wolves, and foxes: a monograph of the Canidae*, London, 1890; *Birds; the elements of ornithology*, London, 1892; *Types of animal life*, London, 1893, with a reprint in 1894; and *A monograph of the lories, or brush-tongued parrots, composing the family Loriidae*, London, 1896.

Another English zoologist, Sir William Henry Flower (1831–99), first served as a surgeon during the Crimean War, and as curator of the museum of the Royal College of Surgeons of England, before becoming director of the British Museum (Natural History) at South Kensington. Flower was prominent in making popular the study of natural history, but his publications were eminently scientific, and include *Diagrams of the nerves of the human body*, 1861; *An introduction to the osteology of the mammalia*, London, 1870, the third edition of which was edited by Hans Gadow, and published in 1885; *An introduction to the study of mammals living and extinct*, written with Richard Lydekker (1849–1915), London, 1891; and *Essays on museums and other subjects connected with natural history*, London, 1898.[2]

The Swiss scientist Wilhelm His (1831–1904) became professor of anatomy at Leipzig, and studied embryology by means of serial sections cut with the microtome. He investigated in particular the development of the nervous system and the origin of nerve fibres. His writings include *Die erste Entwicklung des Hühnchens im Ei: Untersuchungen über das Ei und die Eientwicklung bei Knochenfischen*, [*etc.*], Leipzig, 1873; and *Unsere Körperform und das physiologische Problem ihre Entstehung. Briefe an einer befreundeten Naturforscher*, Leipzig, 1874.[3]

An early convert to the Darwinian theory was Ernst Heinrich Haeckel (1834–1919), a physician who became professor of comparative anatomy

[1] Wollaston, A. F. R. *Life of Alfred Newton*, [*etc.*], 1921.

[2] See Dobson, Jessie, Conservators of the Hunterian Museum. IV. William Henry Flower. *Ann. Roy. Coll. Surg. Engl.*, 30, 1962, pp. 383–391.

[3] See *Wilhelm His, der Ältere. Lebenserinnerungen und ausgewählte Schriften. Zusammengestellt und herausgegeben von E. Ludwig*, Berne, Stuttgart, 1965. (*Hubers Klassiker der Medizin und der Naturwissenschaften*, Bd. 6); and Picken, Laurence. The fate of Wilhelm His. *Nature*, 178, 1956, pp. 1162–1165.

and then of zoology at Jena. Haeckel became Darwin's chief disciple in Germany, and his writings on evolution were translated into several languages. His invaluable researches on the medusae, the sponges and the radiolaria are outstanding, but unfortunately much of his writing is obscured by pseudo-scientific theory. Haekel was the author of *Generelle Morphologie der Organismen. Allgemeine Grundzüge der organischen Formen-Wissenschaft, mechanisch begründet durch die von Charles Darwin reformirte Descendenz-Theorie*, two volumes, Berlin, 1866, which has never been translated into English or any other language, and in which the term 'ecology' was first used.[1] The text of his course of popular lectures on evolution was published as *Naturliche Schöpfungsgeschichte*, Berlin, 1868, the ninth edition of which appeared in 1897, by which time it had been translated into ten European languages, and into Japanese. Among other books Haeckel was the author of *Systematischen Phylogenie. Entwurf eines natürlichen Systems der Organismen auf Grund ihrer Stammesgeschichte*, three volumes, Berlin, 1894–96; *Kuntsformen der Natur*, three volumes, Leipzig, [1904]; and *Prinzipien der generellen Morphologie der Organismen*, [*etc.*], Berlin, 1906. He founded the Phyletisches Museum, which contains his collections and drawings, and also Haeckel Haus, to which he transferred his archives, library, drawings and paintings.[2]

Richard Bowdler Sharpe (1847–1909) was an eminent British naturalist whose writings dominated their field over a considerable period. Many of these were written in collaboration with other zoologists, and the following are his more important books: *A monograph of the Alcedinidae: or family of kingfishers*, published in fifteen parts, London, 1868–71; *Catalogue of the birds in the British Museum*, twenty-seven volumes, 1874–98, of which Bowdler Sharpe himself wrote fourteen volumes, and edited the remainder (with Claude Wilmott Wyatt); *A monograph of the Hirundinidae or family of swallows*, two volumes, issued in twenty parts, London, 1885–94; *Monograph of the Paradiseidae, or birds of paradise and Ptilonorhynchidae, or bower-birds*, two volumes, issued in eight parts, London, 1891–98; *A hand-book to the birds of Great Britain*, four volumes, London, 1894–97, in Allen's Naturalist's Library, and again in Lloyd's Natural History, four volumes, 1896–97; *Hand-list of the genera and species of birds in the British Museum*, five volumes, London, 1899–1909, with an index by W. R. Ogilvie-Grant published in 1912; and a *Catalogue of the collection of birds' eggs in the British Museum*, five volumes, London, 1901–12.

Sir Edwin Ray Lankester (1847–1929), an eminent biologist, was successively Linacre Professor of Comparative Anatomy at Oxford,

[1] See Smit, P. Ernst Haeckel and his "Generelle Morphologie": an evaluation. *Janus*, 1967 pp. 236–252; and Stauffer, Robert C. Haeckel, Darwin, & ecology. *Quart. Rev. Biol.*, 32, 1957, pp. 138–144.

[2] See also De Grood, D. H. *Haeckel's theory of the unity of nature*, Boston, 1965.

director and then keeper of zoology at the British Museum (Natural History) at South Kensington. He was elected F.R.S. in 1875 and knighted in 1907. Lankester accepted the Darwinian theory, and he popularized the study of evolution. Several of his writings were of a popular nature, and went into numerous editions, They include *The uses of animals in relation to the industry of man*, [*etc.*], London, 1860; *On comparative longevity in man and the lower animals*, London, 1870; *A treatise on zoology*, nine volumes, London, 1900–09; *Extinct animals*, New York, 1905; *Science from an esay chair*, 1910, with a fourteenth edition, 1919; and *More science from an easy chair*, 1920. Most of his major contributions to biology were published in the form of articles in scientific periodicals, and are of special significance on account of their number and the quality of the scholarship displayed.

The brothers Franz Andreas Bauer (1758–1840) and Ferdinand Lukas Bauer (1760–1826), "set standards in botanical illustration that have never been surpassed". Both were born in Austria, and in 1786 John Sibthorp (1758–96) visited Vienna and took Ferdinand Bauer on a journey to the Near East. They returned to England together in 1787, and Sibthorp died before his *Flora Graeca* was finished. This was published in ten folio volumes between 1806 and 1840, the cost exceeded £30,000, and only twenty-eight complete sets were originally issued. The full title is *Flora Graeca; sive Plantarum rariorum historia quas in provinciis aut insulis Graeciae legit, investigavit et depingi curavit J. Sibthorp. . . . Characteres omnium descriptiones et synonyme elaboravit* [Volumes I–VII] *J. E. Smith* [Volumes VIII–X] *J. Lindley*, ten volumes, London, 1806–40, and the work was reprinted, Graz, 1967. Henry Bohn had reissued forty copies only between 1845 and 1856. In 1801 Ferdinand Bauer and Robert Brown sailed to Australia with Matthew Flinders, and remained there until 1805. Ferdinand made 2,064 drawings, and later published *Illustrationes florae Novae Hollandiae*, Parts 1–3, 1806–13, but it was a financial failure. He returned to Austria in 1814 and died twelve years later. His brother Franz was persuaded by Sir Joseph Banks in 1790 to settle at Kew to paint flowers, and he stayed there until his death in 1840. He was keenly interested in plant anatomy and histology, and he was the author of *Delineations of exotic plants*, 1796–1803; *Strelitzia depicta*, 1818; and *Illustrations of orchidaceous plants*, 1830–38. He illustrated numerous other books and articles in periodicals. Many of the brothers' original drawings are in the British Museum (Natural History) and other institutions.[1]

Botanical science also progressed at the hands of numerous eminent scholars during this century. Travellers brought home new species from far afield, and books were published dealing with specific localities

[1] Stearn, W. T. Franz and Ferdinand Bauer, masters of botanical illustration. *Endeavour*, 19, 1960, pp. 27–35.

and with individual species, while the microscope became an important new vehicle to botanists such as Robert Brown (1773–1858). He was born at Montrose and educated at Marischal College, Aberdeen, and Edinburgh University. He took an early interest in natural history, and published his first paper at the age of eighteen. From 1795 to 1799 he served mainly in Ireland as assistant surgeon to a Scottish infantry regiment, and then met Sir Joseph Banks, who offered him the post of naturalist to an expedition to New Holland led by Captain Matthew Flinders. Brown was abroad from 1801 to 1805, and on his return became librarian to the Linnean Society. He published the results of his travels as *Prodromus florae Novae Hollandiae et insulae Van Diemen exhibens characteres plantarum quas annis 1802–5 per oras utrusque insulae collegit et descripsit Robertus Brown*, Volume 1, London 1810, to which a supplement appeared in 1830, no further volumes being published. Also in 1810, Brown succeeded Dryander as librarian to Sir Joseph Banks, and on the death of the latter in 1820 was left the library and collections with a house for life. But in 1827 Brown transferred the books and specimens to the British Museum, and became keeper of the Botanical Department there. Brown discovered the nucleus of the vegetable cell and the methods of fertilization in certain plants. He was a pioneer in the study of the anatomy of fossil plants, and published several papers in the *Transactions of the Linnean Society*. Elected F.R.S. in 1811, Brown was awarded the Copley Medal in 1839. He observed the so-called "Brownian movements" of small particles suspended in liquids, which is of special significance in physics. This work was described in a privately printed pamphlet entitled *A brief account of microscopical observations . . . on the general existence of active molecules in organic and inorganic bodies*, 1828. This limited edition is very rare, and ranks Brown among the great experimenters of the period. He followed up this pamphlet by a sequel entitled *Additional remarks on active molecules*, 1829. Brown was the author of several botanical studies, and his collected works were published in German by Nees von Esenbeck as *Vermischte botanische Schriften*, 1825–34, and by the Ray Society as *The miscellaneous botanical works of Robert Brown*, two volumes, 1866–67, with a third folio volume consisting of an atlas of plates, 1868. His researches led to him being described as "perhaps the greatest figure in the whole history of British botany".[1]

George Bentham (1800–84), the son of Sir Samuel Bentham and a nephew of Jeremy Bentham, was privately educated and acquired a wide knowledge of languages. He lived on the Continent for long periods and inherited a keen interest in botany from his mother. His first botanical work was *Catalogue des plantes indigènes des Pyrénées et de*

[1] See Green, J. H. S. Robert Brown (1773–1858) and the Brownian movement. *Research*, 11, 1958, pp. 290–291.

Bas-Languedoc, avec des notes et observations, Paris, 1826. Bentham was the author of numerous papers and memoirs, including *Outlines of a new system of logic*, 1827; *Plantae Hartwegiana*, London, 1839–57; *Labiatarum genera et species*, 1832–36, and *Scrophularineae Indicae*, 1835, both on the genera and natural orders of Indian plants; *The botany of the voyage of H.M.S. Sulphur, under the command of Captain Sir Eduard Belcher during the years 1836–42*, 1844, reprinted 1965; *Flora Hongkongenis*, 1861; *Flora Australiensis*, seven volumes, 1863–78; *Handbook of the British flora*, 1858, with an enlarged, illustrated edition in two volumes, 1863–65, and a seventh edition by A. B. Rendle in 1924. This latter book was compiled in collaboration with J. D. Hooker, as was also *Genera plantarum*, three volumes, 1862–83. Sir Joseph Dalton Hooker (1817–1911) was the younger son of Sir William Jackson Hooker (1785–1865), and succeeded his father as director of Kew Botanical Gardens. Sir William was the author of *The botany of Capt. Beechey's voyage; comprising an account of the plants collected by Messrs. Lay and Collie . . . during the voyage to the Pacific and Bering's Strait, performed in H.M.S. Blossom . . . 1825–28*, (written in collaboration with G. A. W. Walker-Arnott), London, [1830–] 1841, of which a reprint was announced from Weinheim, 1965; and *Niger flora; or an enumeration of the plants of western tropical Africa, collected by Theodor Vogel, botanist to the voyage of the expedition sent . . . to the River Niger in 1841 under the command of Capt. H. D. Trotter, including Spicelegia Gorgonea by P. Barker Webb and Flora Nigritiana, by . . . J. D. Hooker and G. Bentham, with a sketch of the life of Dr. Vogel*, London, 1849, of which a reprint with a bibliographical note by Frans A. Stafleu was published in the series *Historiae Naturalis Classica*, L, Lehre, 1966. W. J. Hooker was also the author of *Icones plantarum; or figures, with brief descriptive characters and remarks, of new or rare plants selected from the author's herbarium*, series 1 and 2, ten volumes, London, [1836] 1837–54; series 3, volumes 11–20, 1867–91, edited by J. D. Hooker and D. Oliver. All these volumes were reprinted in Lehre, 1966–67, and again in New York, 1967. J. D. Hooker graduated in medicine, but was particularly interested in botany, and accompanied Sir James Clark Ross in the *Erebus* on his Arctic expedition of 1839–43. Hooker published the botanical results in six volumes as *Flora Antarctica*, two volumes, 1844–47; *Flora Novae Zelandiae*, two volumes, 1853–55; and *Flora Tasmaniae*, two volumes, 1855–60. These six volumes were reprinted in four, with a separate atlas, in New York, 1964. He became a great friend of Charles Darwin, and was a keen traveller, visiting Eastern Nepal and Sikkim, Palestine, Morocco, Syria and the United States of America. He discovered numerous new plants, and was the author of a large number of papers. Hooker's major publications include *Himalayan journals*, 1854, with a second edition in the following year; *Illustrations of Sikkim-Himalayan plants*, 1855; *Flora Indica*, only Volume 1 of which was published, 1855; *Introductory essay on the flora of Tasmania*, 1859; *Outlines of the distribution of Arctic plants*, 1862; *Handbook of the New Zealand flora*, 1867, and *Flora of British India*, seven volumes, 1897. Hooker also revised the *Index Kewensis*.[1]

A French agricultural chemist, Jean Baptiste Boussingault (1802–87), conducted extensive researches on the absorption of nitrogen by plants, proving that it is derived from nitrates in the soil. His work is recorded in *Agronomie, chimie agricole et physiologie*, 1860–72, which went into a second edition in 1884, and which was also translated.

Mathias Jacob Schleiden (1804–81) was a native of Hamburg, and became director of the Botanical Garden in Jena. His special branch of research was the plant cell, and he was the author of many botanical papers. Schleiden was the author of *Grundzügen der Botanik*, two volumes Leipzig, 1842–43, of which a fourth edition appeared in 1861, and which was translated into English by Edwin Lankester as *Principles of scientific botany, or botany as an inductive science*, 1849. A facsimile reprint of this has been published, with a new introduction by Jacob Lorch, New York, London, 1969.

Another German botanist, Wilhelm Friedrich Benedict Hofmeister (1824–77), was professor at Heidelberg (1863–71) and then at Tübingen. He contributed extensively to plant morphology, his main work being on the sexuality of liverworts, mosses, ferns and other plants, and on the embryology of Bryophytes and Pteridophytes. Hofmeister's *Vergleichende Untersuchungen der Keimung Entfaltung und Fruchtbildung hoherer Kryptogamen und der Samenbildung der Coniferen*, 1851, was translated into English in 1862.

Julius von Sachs (1832–97) was born at Breslau, and after being left an orphan was invited to Prague by J. E. Purkinje, for whom Sachs' father had made drawings. While attending the university Sachs worked as assistant and artist to Purkinje, and graduated doctor of philosophy in 1856, eventually to become professor of botany at Freiburg (1867) and at Würzburg. He was keenly interested in problems of plant nutrition, and contributed many papers to all branches of botany, especially to plant physiology. These were collected together as *Gesammelte Abhandlungen*, 1892–93. Sachs was also the author of *Handbuch der Experimentalphysiologie der Pflanzen*, 1865, of which a French edition was published in 1868; *Lehrbuch der Botanik*, 1868, with a French translation in 1874, and an English translation in 1875 and 1882; *Vorlesungen über Pflanzen-*

[1] See Allan, Mea. *The Hookers of Kew, 1785–1911*, 1967; Turrill, W. B. *Joseph Dalton Hooker*, 1964 (*British Men of Science Series*); Huxley, Leonard. *Life and letters of Sir Joseph Dalton Hooker*, [*etc.*], 2 vols., 1918; list of Hooker's writings, vol. 2, pp. 486–506. Also Turrill W. B. Joseph Dalton Hooker, F.R.S. (1817–1911). *Notes Rec. Roy. Soc. Lond.*, 14, 1959, pp. 109–120; and Turrill, W. B. Pioneer plant geography. The phytogeographical researches of Sir Joseph Dalton Hooker. *Lotsya*, 4, 1953.

physiologie, two volumes, Leipzig, 1882, with a second edition in 1887, and an English translation by H. M. Ward published at Oxford, also in 1887; and *Geschichte der Botanik*, 1875, a reprint of which was announced in 1968, with an English translation in 1890, a reprint of this having appeared in two volumes, New York, 1967.[1]

[1] See Nēmec, B. Julius Sachs in Prague. In, *Science, medicine and history*, Vol. 2, 1953, pp. 211–216.

CHAPTER VII

The Rise of the Scientific Societies

"The best thinkers and scientists had long since been penetrated with the conviction that the work done in universities was valueless and has as little to do with them as possible. Hence arose the necessity for some organized plan for fostering organized research, bringing scientific men together and keeping them in touch with one another by means of publications. This haven of research and experiment dreamed of by Bacon in his *House of Solomon*, was found in the scientific society."

FIELDING HUDSON GARRISON

Societies originated as the result of man's tendency to congregate, thus forming groups for the discussion of subjects of mutual interest. From obscure beginnings, with gatherings of committee size in small communities, these sometimes grew to assume national, and even international importance. Sometimes development was initiated by organizations which had grown large and strong enough to attract membership other than local, so that branches and sections were formed to cater for those residing at a distance from the parent body. Both methods of growth are promptly recognized in societies existing today, but most have probably been founded in emulation of bodies with similar interests, or to cater for rapidly developing branches of knowledge hitherto unprovided for.

Learned societies are known under seven headings, notably schools, museums, lyceums, academies, societies, universities and congresses, and according to J. L. Myers[1] came into existence in this order. Science had its beginnings in the earliest civilizations of China, Greece, Egypt and Arabia, and there is firm evidence of the existence of academies in those times. The word 'academy' has undoubtedly been given to the world from the school founded by Plato, but associations of scholars existed before Plato. Little research has been done on the history of these early academies, the best general survey being that by Edward Conradi.[2]

Within the limits of a single chapter we cannot hope to incorporate

[1] Myers, J. L. *Learned societies. A lecture delivered in the Library of the Department of Education at the University of Liverpool, 4th November, 1922.*

[2] Conradi, Edward. Learned societies and academies in early times. *Pedagogical Seminary*, 12, 1905, pp. 384–426.

historical details of all the scientific societies that have flourished during the past four centuries, and an attempt is therefore made to outline the development of the movement as a whole, to mention the pioneer societies in each special branch of science, and in particular to note their influence on the output of scientific literature. The references given as footnotes contain fuller information, and should be consulted for additional details.

Teachers attracted pupils, who became disciples of their schools of thought, and later, students grouped themselves together to discuss their activities. Then followed the phase in which scholars jointly conducted experiments, publishing the results of their research as the 'proceedings' of the particular society, without incorporating the names of individuals. We find societies at an early date publishing the results of their meetings as journals, memoirs, proceedings, transactions, etc., and certain of them published the writings of their members as monographs, exercising a form of censorship, as did the Royal Society at an early date. Later, the societies collected books to form libraries, and in certain instances societies were formed primarily as reading clubs. The early societies were very general in scope, incorporating all the arts and sciences as subjects for discussion, but the recent rapid development of specialization has led to the formation of societies devoted to branches of science previously considered quite inconspicuous. Scientific societies fostered the development of experimental science, for discussion prompted men to prove their reasoning, instead of accepting without question the opinions of their forerunners.

In Ancient Greece there were a number of schools such as those of Thales (*c.* 640–546 B.C.) who gathered around him a number of students of mathematical and natural science, Pythagoras (*c.* 582–497 B.C.), and that known as the Atomistic School. Thales is reputed to have been the first to demonstrate certain theorems later incorporated in Euclid's *Elements*, but none of his writings survive. The school of Pythagoras had a very close organization[1] and built up the study of arithmetic, particularly the theory of numbers. Of the literature of the Atomistic School also little remains. Plutarch and Archimedes mention the valuable work of Democritus (*c* 470–400), and Diogenes Laertius[2] gives a brief account of the life and work of Leucippus and Democritus, the founders of Atomism. According to Diogenes Laertius, on the authority of Thrasyllus, Democritus was the author of many works on a great variety of subjects including the following: *On the planets; Of nature; Of the different shapes of atoms; On geometry; Numbers; Geography;* and many

[1] Diels, H. Ueber die ältesten Philosophischenschulen der Griechen. *Philosophische Aufsätze*, Leipzig, 1887, pp. 241–260.

[2] Diogenes Laertius, *Lives of eminent philosophers*, IX, 30–49, translated by R. D. Hicks, Loeb Classical Library, 1925.

ethical, literary and musical works now unfortunately lost. The variety of these works is so great that, as Conradi says, "We must assume . . . that there was a masterful organization which collected this wealth of material and worked it over scientifically". The atomic theory for which this school is famed differs from the modern theory of the same name in being purely speculative, and not based on data obtained experimentally.

At the academy of Plato (427–347 B.C.) zoology, botany, geography, mathematics and astronomy were studied, in addition to philosophy, the chief contribution to science being the formation of study methods in mathematics and astronomy. Usener, quoted by Conradi, says it is wrong to consider these schools as a number of pupils flocking around a master. They were organizations for research as well as for learning and listening, so their affinity with the scientific society as we know it is veıy evident.

Around the museum and libraries of Alexandria founded by the Ptolemies many scholars congregated, and their association led to great advances, especially in physics and medicine. The most renowned men of science were accommodated in buildings attached to these institutions, being supported by the kings, and thus enabled to devote all their energies to study and discussion. Owing to the unhampered researches of these scholars, science now broke away from philosophy to some extent, and developed into a number of special studies. The activities of this 'academy' extended nearly a thousand years with fluctuating intellectual fertility, and the following works and names are directly connected with it. Probably the most famous is Euclid (*c* 330–260 B.C.), whose *Elements of geometry* still provides the basis of all geometrical teaching in our schools. The *Elements* survives in the Greek text to these days in a number of manuscripts. The first Latin translation was made by Adelard of Bath about 1120 from the Arabic, into which language it was translated in the eighth and ninth centuries. The first printed edition was published by Ratdolt, the pioneer of mathematical publishing, in Venice, 1482, rapidly followed by others in 1486 and 1491. Euclid wrote a number of other works, many of which are lost. Then came Aristarchus of Samos (*c.* 310–230 B.C.), and Archimedes (287–212 B.C.), whose extant works include *On floating bodies*, giving an account of his discoveries which are the basis of the idea of specific gravity; followed by Apollonius of Perga (*c.* 220 B.C.), Eratosthenes (*c.* 276–149 B.C.), a librarian, and later, Ptolemy of Alexandria (*c.* A.D. 170), whose *Almagest* Charles Singer[1] calls "one of the most influential of all scientific writings". Enough has been recorded above to show the importance of the Alexandrian Academy's contribution to science and scientific literature. Incidentally, the Alexandrian is said to have been the earliest museum known.

[1] Singer, Charles. *A short history of science to the nineteenth century*. Oxford, 1941. p. 84.

The Arabian societies carried on the traditions associated with the schools of Greece and Alexandria, and towards the end of the tenth century there existed a society in Basra called Ichwanal-sofa (Brethren of Purity) which closely resembled our present academies. This society produced an encyclopaedia covering philosophical, physical, mathematical, biological and theological matters, and its influence was widespread.

In contemporary Muslim Spain at Cordova, Islamic culture flourished under the Caliphs Mahomet I and 'Abd-ar-Rahman VII, and reached its height under Hakam II. Hakam II combined the various royal libraries into one tremendous collection, said to have contained between four hundred thousand and six hundred thousand manuscripts. Although there appears to have been little formal organization, an academy was formed by the learned men who flocked there from all over the world, making Cordova the intellectual centre of the West. Spain thus formed a channel for the transmission of the learning of Greece and the East to Western Europe, principally through the work of Jewish translators, and such men as Albucasis (*c.* 1000), Alpetragius (*c.* 1180), whose book on astronomy was used by Copernicus, and later Maimonides (1135–1204) and Averroës (1126–98). The glory of Cordova faded with the fall of the Caliphate and the destruction of the great library, but its influence, in spite of its short life, was considerable.

Although little is known of academies other than those already mentioned, between the suppression of the Platonic schools in A.D. 529 and their revival in Italy eight centuries later, it is not unlikely that other associations of scholars were active in this period. It is probable that their meetings were held in secret owing to opposition from governmental and religious authorities, until, with the Renaissance, it was once more possible to meet openly; not always, however, without opposition. The Church was particularly active in its persecution of those attempting scientific advancement, and this attitude was no doubt responsible for the early Italian academies being devoted primarily to literary studies.

Following the 'schools' of Ancient Greece, and the Alexandrian museum and library, we turn to Italy as the home of the earliest modern scientific societies. In 1560 Giambattista (or Giovanni Babtista) Della Porta (1536–1615) organized a group of friends in his home in Naples, the group being known as the Academia Secretorum Naturae or Accademia del Segreti, and the members as the 'Otiosi' ('the Idlers'). This society was short-lived, for Della Porta was accused of sorcery, and although acquitted was forced to discontinue his activities.[1] He

[1] See Armytage, W. H. G. Giambattista della Porta and the Segreti. *Brit. med. J.*, 1960, I, pp. 1129–1130; see also Hanlon, C. Rollins. The decline and fall of scientific societies. *Surgery*, 46, 1959, pp. 1–8.

wrote on telescopes, the force of steam, on distillation, and was the author of *Magiae naturalis, sive de miraculis rerum naturalium*, 1558, an English translation of which was published as *Natural magick* in 1658, and of which a facsimile edition was printed in 1957, 1958 and 1959[1]; *De humana physiognomia*, 1586; and *Phytognomonica*, [*etc.*], Naples, 1588, and Rouen, 1650, which made the first attempt to group plants by their geographic locale and distribution. This could be described as the beginnings of ecology. The Accademia dei Lincei ("the Lynxes", i.e., "the keen-eyed inquirers") was founded at Rome in 1603 by Prince Federigo Cesi, and its members included Della Porta and Galileo. This society was forced to close owing to the strange nature of the experiments conducted at its meetings, but it was revived in 1609, and in that year commenced recording the proceedings of its meetings as *Gesta Lynceorum*, the earliest recorded publication of scientific research carried out by a society.[2] The society planned to found centres for scientific research and discussion in other cities and countries, all to be united in their work. These centres were to comprise a museum, library, laboratories and printing office. Publications were authorized and printed, such as Galileo's *Il Saggiatore*, (Rome, 1623), and *Istoria e dimostrazioni intorno alle macchie solari e loro accidenti*, (Rome, 1613), Stelluti's account of the anatomy of bees (Rome, 1625), Colonna's observations on botany (1624), and the monumental edition of Fernandez de Oviedo's *Thesaurus Mexicanus* (Rome, 1651), annotated by the members, and printed at the expense of Cesi, although not published until long after his death in 1630. With the loss of his guidance, and the persecutions resultant from the condemnation of the Copernican teaching in 1615, support for the society dwindled, and its activities ceased in 1657. The name was revived with the formation of the present Accademia Nazionale dei Lincei in 1784.

Although the work of the Accademia della Crusca was almost entirely literary, it is worthy of inclusion in this survey, since, by its admission of foreign members—which included philosophers, archaeologists and natural scientists—it caused the idea of academies to spread to the rest of Europe. The Crusca, founded in 1582 at Florence, is famed for its lexicographical research.

The third scientific society having its origin in Italy was the Accademia del Cimento (1657–67), which was founded at Florence. Here nine scientists conducted experiments during the ten years of the society's existence, the results being published as *Saggi di Naturali esperienze fatte nell' Accademia del Cimento*. This appeared in 1666, with a second edition

[1] This is in the Collectors Series in Science, edited by Derek J. Price, who contributed a preface on "Giambattista della Porta and his Natural magick", pp. v–ix.

[2] Ornstein, Martha. *The rôle of scientific societies in the seventeenth century.* (*Third edition*). Chicago, 1938, p. 75.

in 1667, and was translated into English by Richard Waller (1684), into Latin by P. von Musschenbroek (1731), and into French (1755). A photo-litho reproduction of the editio princeps was published in 1957 to celebrate the tercentenary of the foundation of the society, and a reprint of the Waller translation was reprinted with an introduction by A. Rupert Hall, New York, 1964. The instruments and fittings of the Accademia are preserved in the Museo di Storia della Scienza in Florence.

The Accademico Investigante, or Academy of Investigators, was formed at Naples in 1665, and met in the palace of Andrea Concublet, Marquis of Arena. The Academy declined on several occasions, to be revived for the fourth and last time in 1735 by Stefano di Stefano; it finally ceased to exist in 1737.[1]

In England, Francis Bacon in the *New Atlantis*, and indeed throughout his writings, was stressing the necessity for an organization to co-ordinate and encourage the work of scientists. Contemporary writers, both English and Continental, frequently mention his work as being the inspiration of the academies of the seventeenth century. Outside Italy, the Societas Ereunetica, or Zetica, of Rostock, founded by Joachim Jungius (1587–1657) in 1622, is probably the earliest scientific academy. No publications are known to have resulted from its meetings, but it is of interest for its early date, and its encouragement of experimental methods. Jungius was a naturalist, mathematician and physician, had studied at Padua, and was a friend of Leibniz and Comenius. It is possible that the struggle now known as the Thirty Years' War was responsible for the termination of the society's activities in 1625.

In 1652 Johann Lorenz Bausch (1605–55), a physician of Schweinfurt, founded, probably as a result of his knowledge of the Lincei, a society the publications of which have been extensive, and which has existed without interruption from that date to the present time. This society, which was named the Academia Naturae Curiosorum, had as its objects "the advancement of medicine and pharmacy through observation, by presenting observations in monographs and communicating them to the members for correction and further elaboration".[2] It will be noted that its proposed function was fundamentally different from the academies already mentioned. The society having no settled home, its publications became the focus of its efforts, but its growth was slow and although a few monographs were published, it was not until 1670 that the first volume of *Miscellanea curiosa medico-physica* appeared. This dealt with anatomy, pathology, botany, chemistry, physics and zoology, and was based on the *Philosophical Transactions*, which had started publication in 1665. In 1677 the Emperor Leopold became patron, and the society grew

[1] See Fisch, Max H. The Academy of Investigators. In *Science, medicine and history*, Vol. 1, 1953, pp. 521–563.

[2] Ornstein, Martha, *Op. cit.*, p. 170 .

in importance. The *Miscellanea*, like the society, changed its title a number of times, and is of great importance for the history of seventeenth-century science. It helped to forward the growth of interest in science, not only in the German states, but in England, where the Royal Society discussed its papers, and in France, where its contributions were often noted in the *Journal des Sçavans* (*Savants*).

Although founded by Royal Charter on 15th July, 1662, the Royal Society had its beginnings in informal meetings which commenced about 1645 in London, and later in Oxford. The Oxford section expired about 1660,[1] but the London meetings held at Gresham College, although interrupted by political troubles in 1658, thrived and led to the formation of the present Society.[2] Publication of the *Philosophical Transactions* began on 6th March, 1665 (Plate 13). This periodical has the distinction of being the oldest periodical publication of a learned society, but the *Journal des Sçavans*, which began publication in Paris two months earlier, was the first journal to include scientific matter. The *Philosophical Transactions* were first published by Henry Oldenburg. the secretary of the Society, on his own initiative, although the President and the Council exercised some supervision. The career of Henry Oldenburg (1615?–77) has received considerable attention in recent years, largely as the result of the publication of his vast correspondence, edited by A. Rupert Hall and Marie Boas Hall.[3] This reveals many unknown facts regarding the life of Oldenburg, and the internal history of the Royal Society, including the disagreement between Newton and Hooke, and the editors have provided several supplementary, significant papers.[4] Although the idea of publish-

[1] See Gunther, R. T. *Early science in Oxford, Vol.* 4. *The Philosophical Society*, Oxford, *for the subscribers*, 1925.

[2] A review of available literature on the history of the Royal Society, with details of uninvestigated manuscript sources is provided by Hall, Marie Boas. Sources for the history of the Royal Society in the seventeenth century. *History of Science*, 5, 1966, pp. 62–76. Good accounts of the founding of the Society are provided by Purver, Margery. *The Royal Society: concept and creation*, [*etc.*], 1967; McKie, Douglas. The origins and foundation of the Royal Society of London. *Notes Rec. Roy. Soc. Lond.*, 15, 1960, pp. 1–37. The July 1960, issue of this journal is a special tercentenary number, and contains a wealth of papers on the early Fellows of the Society: it was separately published in book form as Hartley, Sir Harold. *The Royal Society—its origins and founders*, 1960. See also Andrade, E. N. da C. *A brief history of the Royal Society*, 1960; Andrade, E. N. da C., and Martin, D. C. The Royal Society and its foreign relations. *Endeavour*, 19, 1960, pp. 72–80; and Times, The. *The Royal Society tercentenary. Compiled from a special supplement of the Times, July, 1960*. 1961.

[3] *The correspondence of Henry Oldenburg, edited and translated by A. Rupert Hall and Marie Boas Hall*, Madison, Milwaukee and London, Vols. 1–6, 1965–69; to be completed in about ten volumes.

[4] See Hall, Alfred Rupert, and Hall, Marie Boas. Why blame Oldenburg? *Isis*, 53, 1962, pp. 482–491; Hall, Marie Boas. Oldenburg and the art of scientific communication. *Brit. J. Hist. Sci.*, 2, 1965, pp. 277–290; Hall, A. Rupert, and Hall, Marie Boas. Some hitherto unknown facts about the private career of Henry Oldenburg. *Notes Rec. Roy. Soc., Lond.* 18, 1963, pp. 94–103; and Hall, A. Rupert, and Hall, Marie Boas. Further notes on Henry Oldenburg. *Notes Rec. Roy. Soc. Lond.*, 23, 1968, pp. 33–42; see also Bluhm, R. K. Henry Oldenburg, F.R.S. (*c.* 1615–1677). *Notes Rec. Roy. Soc. Lond.*, 15, 1960, pp. 183–197.

ing a periodical was probably discussed by the members of the Society before the publication of the *Journal des Sçavans*, the connection of the *Philosophical Transactions* with this predecessor is shown by three of the ten papers in the first number being taken from the *Journal des Sçavans*.[1] The early issues contained papers contributed by the members, letters, reviews of books, and notes of interesting phenomena culled from Oldenburg's extensive correspondence with scientists in other countries. This connection with foreign scientists, notably Leeuwenhoek and Malpighi, and its consequent international character, greatly increased the value of the Society's publication. There is no need to stress the importance of this and the other periodical published by the Royal Society—The *Proceedings* (1832 to date). T. H. Huxley[2] has recorded this in the following words: "If all the books in the world except the Philosophical Transactions were destroyed, it is safe to say that the foundations of physical science would remain unshaken, and that the vast intellectual progress of the last two centuries would be largely, though incompletely, recorded."[3]

All the great names of English science are to be found in the pages of the *Philosophical Transactions*. Probably the period of most importance is the latter half of the eighteenth century, which contains contributions of great value from men such as Herschel, Priestley, Franklin, John Hunter, Rumford and Henry Cavendish.

The activity of the Royal Society in aiding the publication of the work of individual scientists is less widely known, yet many of the greatest scientific books of the seventeenth century appeared under its imprint, such as: John Evelyn, *Sylva, or a discourse of forest trees*, 1664 (the first book printed by the Royal Society); Robert Hooke, *Micrographia*, 1665; Thomas Sprat, *The history of the Royal Society*, 1667; Marcello Malpighi, *Anatome plantarum*, 1675; Francis Willughby, *Ornithologiae libri tres*, 1676 (completed by John Ray); Nehemiah Grew, *The anatomy of plants*, 1682; Isaac Newton, *Philosophiae naturalis principia mathematica*, 1687. The latter is generally regarded as the greatest scientific work ever published, and its influence on subsequent thought is incalculable. As Sir Henry Dale once stated, it "was one of the greatest intellectual achievements in the history of mankind . . . providing for more

[1] Brown, Harcourt. *Scientific organizations in seventeenth-century France, 1620–1680*, Baltimore, 1934, p. 199. Reprinted New York, 1967.

[2] Huxley, T. H. On the advisableness of improving natural knowledge. In, *Lay sermons, addresses and reviews*, 1, 1870, p. 4.

[3] See also the following series of articles: Rattansi, P. M. The intellectual origins of the Royal Society. *Notes Rec. Roy. Soc. Lond.*, 23, 1968, pp. 129–143; Hill, Christopher. The intellectual origins of the Royal Society—London or Oxford? *Notes Rec. Roy. Soc. Lond.*, 23, 1968, pp. 144–156; and Hall, Alfred Rupert, and Hall, Marie Boas. The intellectual origins of the Royal Society—London and Oxford. *Notes Rec. Roy. Soc. Lond.*, 23, 1968, pp. 157–168.

than two centuries a framework for the mechanical interpretation of the universe and a basis for the building of physical science, and therewith of the material structure of our modern civilization".

Another publication of the Royal Society of great value, especially to librarians, is the *Catalogue of scientific papers*. This general author index to papers in scientific periodicals covers the literature of 1800–1900, and was published, partly with Government funds, between 1867 and 1902. Three subject indexes were also published (1908–14). The work was later undertaken by an international committee, and published annually by the Royal Society until the Great War, 1914, as the *International catalogue of scientific literature*, 1901–14. A reprint was announced in 1969.[1] Since 1938 the Royal Society has published *Notes and Records of the Royal Society*, which was first issued as a house journal for Fellows of the Society. It has become one of the more important periodicals dealing with the history of science, and since 1949 has included notes on books relating to the history of the Society. This section was expanded in 1953 to include articles published in journals.

In 1967 the Royal Society moved from Burlington House to Carlton House Terrace, the new premises being described in an article by Lord Holford.[2]

The Académie Française, which devoted its labours to the improvement of the French language, was formed in 1653, and leading from this association informal meetings of philosophers and others led to the founding of the Académie des Sciences in 1666. The Académie had one fundamental difference from the Royal Society in the former having an official position in the State under the active patronage of the King, Louis XIV. Pensions were given to the members to enable them to devote their full time to research, and financial aid was also available for apparatus. Some foreign scientists were invited to become members, the number of which was limited. This dependence on the Court was a mixed blessing, as was seen when after 1686 less sympathy was shown to purely theoretical researches, and the members were called upon to solve the practical problems of the Court, such as calculating the odds in various Court games! At first, with the excellent *Journal des Sçavans* available, there was little inclination for the Académie to issue publications. However, even before its reorganization in 1699, following a period of lethargy, certain publications of value were issued

[1] The following provide additional information on the Royal Society: Sprat, Thomas. *The history of the Royal-Society of London, for the improving of natural knowledge*, 1667; Thomson, Thomas, *History of the Royal Society, from its institution to the end of the eighteenth century*, 1812; Weld, Charles Richard. *A history of the Royal Society*, [*etc.*], 2 vols., 1848.

[2] Holford, Lord. The new home of the Royal Society. *Notes Rec. Roy. Soc. Lond.*, 22, 1967, pp. 23–36; see also Martin, D. C. Former homes of the Royal Society. *Notes Rec. Roy. Soc. Lond.*, 22, 1967, pp. 12–19.

under its imprint, the first being Perrault's *Mémoires pour servir a l'histoire naturelle des animaux*, Paris, 1671(–76), recording the dissection of many beasts carried out by Perrault and other members of the Académie. This was subsequently translated into English by Alexander Pitfield and published by the Royal Society in 1688 as *The natural history of animals, containing the anatomical description of several creatures dissected by the Royal Academy of Science at Paris*. A more notable publication was Huygens' *Horologium oscillatorum*, Paris, 1673, containing the description of his pendulum clock, and later, his *Traité de la lumière*, 1690, the result of research undertaken while in Paris as a member of the Académie. Other early co-operative publications were *Mémoires pour servir a l'histoire naturelle des plantes*, edited by Denis Dodart, Paris, 1676, and *Recueil d'observations faites en plussieurs voyages par order de Sa Majesté, pour perfectionner l'astronomie et la géographie . . . par messieurs de l'Académie Royale des Sciences*, Paris, 1693.

Upon its reorganization, already mentioned, the Académie started a regular periodical publication, the *Histoire de l'Académie (Royale) des Sciences. Avec les mémoires de mathématiques et de physique*, (Paris, 1702–97). This publication covered the work of the Académie from 1699 until 1790. In 1793 the Académie was suppressed, being restarted as a section of the Institut National, its periodical publication then becoming *Mémoires de l'Institut national des sciences et arts. . . . Sciences mathématiques et physiques*, [1798]–1806, continued as *Mémoires de la classe des sciences mathématiques et physiques de l'Institut de France*, 1806–15, and later *Mémoires de l'Académie des Sciences de l'Institut de France*, Paris, 1816, etc. In 1733 the researches of the members made between the dates of foundation and reorganization, chiefly collected from the *Journal des Sçavans*, were published as *Histoire (Mémoires) de l'Académie Royale des Sciences depuis son établissement en 1666 jusqu'à 1699*, in eleven volumes. The *Comptes rendus hebdomadaire des séances de l'Académie des Sciences* commenced publication in 1835 and is still in progress. The periodical publications of this Académie contain many important papers, including the work of Laplace, Réaumur and Coulomb.[1]

In the latter half of the seventeenth century many scientific societies were founded, such as the Collegium Curiosum sive Experimentale of J. C. Sturm of Altdorf, which published two volumes of proceedings (1676–85); and the Accademia delle Scienze del'Istituto di Bologna, which the astronomer Manfredi formed in 1690.

[1] See McKie, Douglas. The early years of the *Académie des Sciences. Endeavour*, 25, 1966, pp. 100–103; Fauré-Fremiet, E. Les origines de l'Académie des Sciences de Paris. *Notes Rec. Roy. Soc. Lond.*, 21, 1966, pp. 20–31; George, Albert J. The genesis of the Académie des Sciences. *Ann. Sci.*, 3, 1938, pp. 372–401; and Watson, E. C. The early days of the Académie des Sciences as portrayed in the engravings of Sébastien Le Clerc. *Osiris*, 7, 1939, pp. 556–587.

As a result of energetic propaganda on the part of Gottfried Wilhelm Leibniz (1646–1716), who was a member of the Royal Society, the Societas Regia Scientiarum (Preussischen Akademie der Wissenschaften zu Berlin) was founded by the Elector Frederick III (later Frederick I of Prussia) in 1700. Leibniz became first President, and thus was launched another society which, with the British Society and the French Academy, form the great triumvirate of scientific societies. Leibniz envisaged a great academy of all the German states centred in Vienna, and the academy issued *Miscellanea Berolinensia*, 1710–44, published in French for the years 1745–1804 as *Histoire de l'Académie Royale des Sciences et des Belles-Lettres de Berlin* (1746–1807), and other French titles, finally becoming *Abhandlungen der k. (preussischen) Akademie der Wissenschaften zu Berlin*, 1804, [*etc.*], (1815, [*etc.*]), and later issued in two classes.[1]

The St. Petersburg Academy was established in 1725, largely as the result of conversations between Leibniz and Peter the Great, but neither lived to see the academy formally established. Its publications included *Commentarii Academiae Societarium Imperialis Petropolitanae*, 1728–48, when the title was changed to *Novi Commentarii*, [*etc.*], and in 1777 to *Acta Academiae*, [*etc.*], In Sweden an academy was established at Uppsala in 1710, and the Collegium Curiosorum, which first met in 1739 at Stockholm, became the Kongliga Svenska Vetenskaps Akademien in 1741. The Danish Kongelige Danske Videnskabernes Selskab was founded at Copenhagen in 1743 by Christian VI. In Austria the Österreichische Akademie der Wissenschaften, Wien was founded by the Emperor Ferdinand I by a statute of 14th May, 1847, to further scientific research and to issue publications. It was divided into two sections, the mathematical and natural sciences, and the philosophical and historical sciences. The Academy began publishing *Denkschriften* (1850) and *Sitzungsberichte* (1848), and these are of great historical significance, the earlier volumes having recently been reprinted.

An early society comparable with the Académie des Sciences and the Royal Society was the Society of Arcueil, based on a village on the outskirts of Paris. There Claude-Louis Berthollet and Pierre Simon Laplace provided hospitality to other eminent men of science such as Gay-Lussac, Arago, Chaptal, Humboldt, Thenard, Biot, De Candolle, Dulong, Malus, Bérard, A. B. Berthollet, Descotils and Poisson, among others. There were fifteen full members and several associates, and meetings were held at week-ends from 1807 to 1813. C.-L. Berthollet continued to work in his laboratory from 1814 to 1822, but the fall of Napoleon and subsequent events greatly affected the development of scientific investigation. Many distinguished foreign scientists visited Arcueil, and the society

[1] See Calinger, Ronald S. Frederick the Great and the Berlin Academy of Sciences (1740–66). *Ann. Sci.*, 24, 1968, pp. 239–249.

published *Mémoires de Physique et de Chimie de la Société d'Arcueil*, volume 1, 1807; volume 2, 1809; and volume 3, 1817. This final volume was in press in 1813, and a copy in the Bibliothèque Nationale, Paris has a title-page with this date. An excellent study of the Society of Arcueil has been published by Maurice Crosland[1], who provides a select bibliography of printed sources with references to other societies, and to individual members of the Society. Crosland also wrote an introduction to the facsimile edition of the *Mémoires*, announced in 1967.

In America, the short-lived Boston Philosophical Society was founded in 1683, and Benjamin Franklin, who in 1727 founded the Junto, a secret literary and scientific society, also stimulated the formation of a scientific society by the publication in 1743 of a pamphlet entitled *A proposal of promoting useful knowledge among the British plantations in America.* This led to the formation in Philadelphia of the American Philosophical Society in the same year, Franklin being President from 1769 to 1790. The Society published *Transactions* from 1771 and *Proceedings* from 1838. The American Academy of Arts and Sciences was founded at Boston in 1780, published *Memoirs* from 1785 and *Proceedings* from the year 1864. The first American chemical society was founded in the University of Pennsylvania by John Penington in 1789, but it lasted only one year. It was followed by the Chemical Society of Philadelphia, founded by John Redman Coxe and other medical students in 1792;[2] it survived for about fifteen years. In 1811 the Columbian Chemical Society[3] was formed in Philadelphia, and published a journal entitled *Memoirs of the Columbian Chemical Society*, volume one of which appeared in 1813. The Society seems to have disappeared in the following year. The Delaware Chemical and Geological Society was formed in 1821, and in 1861 Charles F. Chandler founded the Chemical Society of Union College, Schenectady, possibly the first college chemical society in America.[4] In 1863 the Beloit Chemical Society was formed.

Probably the largest scientific organization in the world, the American Chemical Society[5] was founded in New York in 1876, and is particularly renowned for its publications, including the *Proceedings of the American*

[1] Crosland, Maurice P. *The Society of Arcueil. A view of French science at the time of Napoleon I*, 1967.

[2] See Miles, Wyndham D. John Redman Coxe and the founding of the Chemical Society of Philadelphia in 1792. *Bull. Hist. Med.*, 30, 1956, pp. 469–472.

[3] See Miles, Wyndham D. The Columbian Chemical Society. *Chymia*, 5, 1959, pp. 145–154. See also Miles, Wyndham D. Early American chemical societies. 1. The 1789 Chemical Society of Philadelphia. 2. The Chemical Society of Philadelphia. *Chymia*, 3, 1950, pp. 95–113; and Bolton, Henry Carrington. *Chemical societies of the nineteenth century*, Washington, 1902.

[4] See Bacon, Egbert K. A precursor of the American Chemical Society—Chandler and the Society of Union College. *Chymia*, 10, 1965, pp. 183–197.

[5] See Browne, Charles Albert, and Weeks, Mary Elvira. *A history of the American Chemical Society: seventy-five eventful years*, Washington, 1952.

Chemical Society, (1876), which became the *Journal of the American Chemical Society* in 1879; *Chemical Abstracts*, published since 1907, and now also available in a number of sections; *Industrial and Engineering Chemistry*, (1909); *Analytical Chemistry*, (1948); *Chemical Titles*, published since 1960, and a growing list of other titles, now over twenty in number. The American Association for the Advancement of Science was founded in 1848 and comprises over 250 independent organizations, with a total membership exceeding two million. It grew out of the Association of American Geologists, founded in Philadelphia in 1840, and in 1842 the name was changed to the Association of American Geologists and Naturalists. At a meeting in Boston in 1847 it was decided to widen its scope, and in 1848 the first meeting of the American Association for the Advancement of Science took place in Philadelphia. *Proceedings* were issued until 1861, when publication was suspended until 1866. Among other publications the following are particularly noteworthy: *Science*, first published in 1883; *A.A.A.S. Bulletin*, 1942–46; *Biological Symposia*, issued since 1940; and *Science Books*, a quarterly review published since 1965 and consisting of authoritative, classified reviews of new books.[1]

The growth of scientific societies in the United States has been phenomenal since those early beginnings, and it is impossible to mention even a fraction of them. A list of these organizations edited by J. David Thompson[2] was published in 1909, and the National Academy of Sciences[3] has published a list providing information on scientific and technical societies in the United States. Entries are arranged alphabetically by title, and details are provided of addresses, dates of foundation, membership, meetings, libraries, funds, medals and publications. Yet another guide to American scientific societies has been compiled by Ralph S. Bates,[4] to emphasize the importance of these organizations on the development of science. The third edition brings the subject down to 1965 and includes the atomic and space ages, with the resultant formation of numerous new societies. It also includes "A chronology of science and technology in the United States" up to 1965 (pp. 237–244), and the Bibliography (pp. 245–293) lists general guides to American scientific societies, and references to individual societies arranged alphabetically under their names. This thoroughly documented study is an invaluable source of information on scientific societies in the United States.

There were numerous scientific societies in Scotland in the eighteenth

[1] See Fairchild, H. L. The history of the American Association for the Advancement of Science. *Science*, N.S.59, 1924, pp. 365–369, 385–390, 410–415.

[2] (Thompson, J. David, *Ed.*) *Handbook of learned societies and institutions. America*, (Carnegie Institution of Washington, Publication No. 39), Washington, 1909.

[3] National Academy of Sciences. *Scientific and technical societies of the United States*, 8th edition, Washington, 1968. Earlier editions included Canada.

[4] Bates, Ralph S. *Scientific societies in the United States. . . . Third edition*, Oxford, [*etc.*], 1965.

and early nineteenth centuries, including the Aberdeen Philosophical Society, 1758–73,[1] which was revived in 1840 and survived until 1939; the Aberdeen Natural History Association (1845–51); the Natural History and Antiquarian Society of Aberdeen, founded in 1863 and still flourishing; the Anatomical Society of Edinburgh (1833–45); the Natural History Society of Edinburgh, founded in 1782, united with the Royal Physical Society (1771), and still flourishing; the Scottish Microscopical Society (1889); and the Wernerian Natural History Society (1808).[2]

Edinburgh was the home of several early medical, philosophical and scientific societies, and in 1737 the Philosophical Society of Edinburgh was organized, to be chartered in 1783 as the Royal Society of Edinburgh.[3] The Society for the Investigation of Natural History, a students' society, was instituted in Edinburgh in 1782, and manuscripts relating to it are housed in the University Library.[4] Incidentally, a Chemical Society existed in the University of Edinburgh in 1785, a similar society is mentioned as having existed in the University of Glasgow in 1786, and James Kendall[5] provides details of certain other eighteenth-century chemical societies.

Ireland's first scientific institution, the Dublin Philosophical Society, was founded in 1684 by William Molyneux, and was the forerunner of the Royal Dublin Society. The original society seems to have disappeared about 1687, to be revived in 1693 but in 1731 a group met at Trinity College, Dublin, and formed the Dublin Society for "improving husbandry, manufactures and other useful arts", the "sciences" being added to its title at the second meeting. Its scientific work was particularly pronounced from 1792, when the Leskean collection of minerals was purchased and went to form the basis of a geological museum in 1795, with William Higgins as the curator. A number of professors were appointed, including Higgins (chemistry), Walter Wade (botany) and James Lynd (hydraulics, mechanics and experimental philosophy). *Transactions of the Dublin Society* were published from 1799 to 1810, the *Journal of the Royal Dublin Society* from 1856 to 1876, and *Scientific*

[1] See Fabian, Bernhard. David Skene and the Aberdeen Philosophical Society. *Bibliotheck*, 5, 1968, pp. 81–99.

[2] See Drummond, H. J. H. Records of medical and scientific societies in Scotland. I. Early medical and scientific societies of North-East Scotland. *Bibliotheck*, 1, ii, 1957, pp. 31–33; Finlayson, C. P. Records of scientific societies in Scotland. II. Records of scientific and medical societies preserved in the University Library, Edinburgh. *Bibliotheck*, 1, iii, 1958, pp. 14–19.

[3] See Davidson, J. N. The Royal Society of Edinburgh. *J. Roy. Inst. Chem.*, 78, 1954, pp. 562–566; and Kendall, James. The Royal Society of Edinburgh. *Endeavour*, 5, 1946, pp. 54–57.

[4] See McKie, Douglas. Some notes on a students' scientific society in eighteenth-century Edinburgh. *Sci. Progr.*, 49, 1961, pp. 228–241.

[5] Kendall, James. Some eighteenth-century chemical societies. *Endeavour*, 1, 1942, pp. 106–109.

Transactions and *Scientific Proceedings* began publication in 1877, the *Transactions* ceasing publication in 1909. Membership of the Society is limited to 9,000, and the library of 150,000 volumes is most outstanding.[1]

The eighteenth century saw the development of the specialist scientific society in England. Douglas McKie,[2] during the course of an interesting article devoted to scientific societies founded before the end of the eighteenth century, mentions the foundation of a Botanical Society in London in 1721 by John Martyn (1699–1768), which lasted for five years. This was apparently the first formally constituted body of any size specifically devoted to botany to be formed in Britain, but was preceded by the Temple Coffee House Botanic Club, which had a membership of forty, and survived from 1689 until at least 1713. The Botanical Society had as its first president Johann Jacob Dillenius (1681–1747), and meetings took place every Saturday evening, at first in the Rainbow Coffee House, Watling Street, and later in members' houses.[3] This was followed in 1745 by the Aurelian Society founded for the study of insects, but the Society's premises were destroyed by fire in 1748. It was succeeded by a second Society bearing the same name, and which existed between 1762 and 1766. A Society of Entomologists of London existed from 1780 to 1782, and in that year the Society for Promoting Natural History was also organized in London. Members of this Society left the parent body in 1788 to form the Linnean Society, which was founded by James Edward Smith (1759–1828), Samuel Goodenough (1743–1827) and Thomas Marsham (1747 or '48–1819). The property of the Society for Promoting Natural History was later handed over to the Linnean Society, the development of which has been ably recorded by A. T. Gage[4] in a well-documented, illustrated history. Among the numerous activities carried out in the early coffee-houses were the meetings of societies that were formed there. Many of these were short-lived, but examples include the Society for the Encouragement of Learning (1735), and the Aurelian Society, formed in the Swan Tavern, Cornhill, and later meeting at the York Coffee House in St. James Street.[5]

The genesis of provincial scientific societies is also found in the eighteenth century, and the Literary and Philosophical Society of Manchester may be taken as an outstanding example of such societies. It

[1] Crowley, Denis. The Royal Dublin Society. *J. Roy. Inst. Chem.*, 82, 1958, pp.10–18; and Hoppen, K. Theodore. The Royal Society and Ireland—William Molyneux, F.R.S. (1656–1698). *Notes Rec. Roy. Soc. Lond.*, 18, 1963, pp. 125–135.

[2] McKie, Douglas. Scientific societies to the end of the eighteenth century. *Phil. Mag.*, Commemoration issue. July, 1948, pp. 133–144.

[3] See Allen, D. E. John Martyn's Botanical Society: a biographical analysis of the membership. *Proc. Bot. Soc. Brit. Isles*, 6, 1967, pp. 305–324.

[4] Gage, A. T. *A history of the Linnean Society of London*, 1938.

[5] See Armytage, W. H. G. Coffee houses and science. *Brit. med. J.*, 1960, II, p. 213.

was founded in 1781, and soon possessed a library, publishing its *Memoirs* from the year 1785. In 1888 these *Memoirs* amalgamated with the *Proceedings*, hitherto published separately, to become the *Memoirs and Proceedings*, and this organ has contained major contributors to science from such outstanding names as Dalton, Joule, Rutherford, Bragg and Elliot Smith. The premises and valuable library of this Society were destroyed during the war.[1] The Lunar Society was a private and exclusive organization formed in Birmingham about 1765, but the name Lunar was not used until 1776. It had no organization, no records were kept, and its members became known as the "Lunatics" from their habit of meeting when the moon was full. Meetings continued until at least 1807. Distinguished scientists associated with the Society were Matthew Boulton, Erasmus Darwin, James Watt, Joseph Priestley, Thomas Day, Richard Lovell Edgeworth, James Keir,[2] Josiah Wedgwood, William Withering and others. Robert E. Schofield[3] has compiled a remarkable history of the Society, mainly from manuscript and other elusive sources, providing a "Lunar Society Bibliography" consisting of lists of books, pamphlets and papers by or about members of the Society.[4] The Derby Philosophical Society was founded about 1783 and survived until 1857. It had connections with the Lunar Society, probably through Erasmus Darwin, who, as President of the Derby Philosophical Society, was responsible for building up its library, a catalogue listing 275 items contained therein being preserved in Derby Borough Library.[5]

Among chemical societies founded in London before the Chemical Society, which was formed in 1841,[6] we note the Lunar Society's London-branch Chemical Society, which met during the 1780's at the Chapter Coffee House in Paternoster Row, and the Chemical Club, founded about 1782. This was connected with the Royal Society, as was also the

[1] See Ardern, L. L. The Manchester Literary and Philosophical Society. *J. chem. Educ.*, 39, 1962, pp. 264–265; Sheehan, Donal. The Manchester Literary and Philosophical Society. *Isis*, 33, 1941–42, pp. 519–523; also Fleure, H. J. The Manchester Literary and Philosophical Society. *Endeavour*, 6, 1947, pp. 147–151.

[2] See Smith, Barbara M. D., and Moilliet, J. L. James Keir of the Lunar Society. *Notes Rec. Roy. Soc. Lond.*, 22, 1967, pp. 144–154.

[3] Schofield, Robert E. *The Lunar Society of Birmingham. A social history of provincial science and industry in eighteenth-century England*, Oxford, 1963. See also Schofield, R. E. The Lunar Society of Birmingham: a bicentenary appraisal. *Notes Rec. Roy. Soc. Lond.*, 21, 1966, pp. 144–161; Schofield, Robert E. Membership of the Lunar Society of Birmingham. *Ann.* Sci., 12, 1956, pp. 118–136; Schofield, Robert E. The industrial orientation of science in the Lunar Society of Birmingham. *Isis*, 48, 1957, pp. 408–415; and King-Hele. The Lunar Society of Birmingham. *Nature*, 212, 1966, pp. 229–233.

[4] A special issue of the *Univ. Birm. hist. J.* (Vol. 11, No. 1, Oct., 1967) is devoted to the Lunar Society, and contains some very interesting papers.

[5] See Robinson, Eric. The Derby Philosophical Society. *Ann. Sci.*, 9, 1953, pp. 359–367.

[6] See Moore, Tom Sidney, and Philip, James Charles. *The Chemical Society, 1841–1941: a historical review*, 1947; and Gibson, C. S. The Chemical Society (of London) after one hundred years. *Endeavour*, 6, 1947, pp. 63–68.

Animal Chemistry Club. This latter resulted from the proposal of a Group of Fellows of the Royal Society to form a "Society for the Improvement of Animal Chemistry". A constitution was drawn up in 1808 and the Club existed until 1825, but it was more a dining club, with discussions after dinner. Papers communicated were published in the *Philosophical Transactions*.[1] The London Chemical Society was formed in 1824, with George Birkbeck as president, but the names of only ten members are recorded, none of these being well-known chemists of the period. The Society probably did not survive into 1825.[2]

In 1783 Bryan Higgins founded in London the Society for Philosophical Experiments and Conversations, which held weekly meetings when experiments were conducted, until 21st June, 1794. The Askesian Society (1796) and the British Mineralogical Society (1799) were also founded in London, the latter combining with the former in 1806. The Royal Institution, founded in 1799 by Count Rumford, was chartered in 1800, and maintains its original function of diffusing knowledge of the practical application of scientific research. K. D. C. Vernon[3] has published a detailed history of the early years of the Institution, based on manuscript minute books, archives and letters, and several biographical studies of Benjamin Thompson, Count Rumford (1753–1814) have been published.[4] His *Collected works*, to be in five volumes, began publication in 1964.

The Royal Horticultural Society, which has made important contributions to scientific botany as well as horticulture, was founded in 1804, to be followed by the Geological Society of London in 1807. This was the first learned society devoted entirely to geology, and it contributed significantly to the progress of the subject during the nineteenth century.[5] The Royal Astronomical Society was founded in 1820.[6] The latter

[1] Coley, N. G. The Animal Chemistry Club: assistant society to the Royal Society. *Notes Rec. Roy. Soc. Lond.*, 22, 1967, pp. 173–185.

[2] See Brock, W. H. The London Chemical Society, 1824. *Ambix.*, 14, 1967, pp. 133–139.

[3] Vernon, K. D. C. The foundation and early years of the Royal Institution. *Proc. Roy. Instn. G.B.*, 39, 1963, pp. 364–462. See also Martin, Thomas. Origins of the Royal Institution. *Brit. J. Hist. Sci.*, 1, 1962, pp. 49–63; Martin, Thomas. Early years at the Royal Institution. *Brit. J. Hist. Sci.*, 2, 1964, pp. 99–115; and Ironmonger, E. F. The Royal Institution and the teaching of science in the nineteenth century. *Proc. Roy. Inst. G.B.*, 37, 1958, pp. 139–158.

[4] See Bradley, D. *Count Rumford*, Princeton, N. J., 1967; Sparrow, W. J. *Knight of the White Eagle. A biography of Sir Benjamin Thompson, Count Romford (1753–1814)*, 1964. Chapters deal with him as founder of the Royal Institution. See also Brown, Sanborn C. Count Rumford as a scientist. *Proc. Roy. Instn. G.B.*, 39, 1963, pp. 583–620.

[5] See Woodward, Horace B. *The history of the Geological Society of London*, London, [*etc.*], 1908; and Rudwick, M. J. S. The foundation of the Geological Society of London: its scheme for co-operative research and its struggle for independence. *Brit. J. Hist. Sci.*, 1, 1963, pp. 325–355.

[6] See Dreyer, John Louis Emil, and Turner, Herbert Hall, *eds. History of the Royal Astronomical Society, 1820–1920*, The Society, 1923.

absorbed the Spitalfields Mathematical Society, which had existed from 1717 to 1846. The Zoological Society of London was founded in 1826,[1] the Royal Geographical Society in 1830, and the Entomological Society in 1833. The first meeting of the British Association for the Advancement of Science was held at York in 1831,[2] and the Statistical Society (1834), the London Mathematical Society (1865) and the Physical Society (1874) also demand notice among the host of scientific societies that were formed during the nineteenth century.

The Medico-Botanical Society of London was founded on 16th January, 1821, by John Frost (1803–40) for the study of the medicinal properties of plants of all countries, and it became the Royal Medico-Botanical Society in 1837. John Frost was born near Charing Cross, and intending to qualify in medicine, he became a pupil of Dr. Wright, apothecary to the Bethlem Hospital. However, Frost soon gave up medicine for botany, and founded this Society, of which he was director and lecturer in botany. He persuaded numerous eminent people to become Fellows and members, and a book containing the signatures of Fellows contains the names of twenty members of royal families, including Queen Victoria, George IV and William IV. The pages bearing the signatures are ornamented with wreaths of beautifully hand-painted flowers, and the book was missing for over a hundred years.[3] The Duke of Wellington, Sir Robert Peel and other public figures were represented, and Sir James McGrigor was the first president, to be followed by Lord Stanhope. Frost pestered everybody, and without any academic qualifications received several honours and appointments. In 1824 he became Secretary to the Royal Society, and was engaged in several ambitious schemes. He delivered an oration at every annual meeting of the Medico-Botanical Society, but he was arrogant, and became very unpopular with the members. At a meeting held on 8th January, 1830, Frost was deposed and expelled. Following financial difficulties, he went to live in Paris under an assumed name, and eventually went to Berlin, where he assumed the title Sir John Frost and practised medicine. He died on 17th March, 1840. The Society held meetings twice a month in Sackville Street, which were well attended, and to which Humphrey Gibbs sent collections of plants and flowers for decoration and for distribution to the audience. *Transactions* covering

[1] Stratton, G. Burder. The Zoological Society of London. *Brit. Book News*, 95, July, 1948, pp. 358–360.

[2] Howarth, O. J. R. *The British Association for the Advancement of Science: a retrospect, 1831–1931. . . . Centenary (second) edition*, London, 1931; Howarth, O. J. R. The British Association, *Endeavour*, 3, 1944, pp. 57–61; Lowe, D. N. The British Association for the Advancement of Science, *British Book News*, 93, May, 1948, pp. 232–234.

[3] Dr. Richard A. Hunter, who now possesses this volume, has kindly provided information regarding this Society, and intends to publish a study on the subject at a later date.

the years 1821 to 28th June, 1837, were published in four numbers, and number one was reprinted in the *Journal of Morbid Anatomy*, 1828. The Society's records continue to 1847–48, and it probably ceased to function about 1849.[1]

John Cohen and others[2] compiled a graph showing the cumulative number of learned and scientific national societies in Britain, and published tables showing the foundation of new societies arranged by subjects in twenty-year periods, and indicating the number of societies in relation to population in 1850, 1900 and 1952–53. The British Council[3] publishes *Scientific and learned societies of Great Britain*, the 1964 edition listing over eight hundred societies by subject, and providing such information as date of foundation; object; membership; subscription; meetings; library; and publications. There is also a more recent *Directory of British associations, Interests, activities and publications of trade associations, scientific and technical societies, professional institutes, learned societies, research organizations, chambers of trade and commerce, agricultural societies, trade unions, cultural, sports and welfare organizations in the United Kingdom and in the Republic of Ireland, Edition 2*, Beckenham, Kent, 1967–68, listing almost six thousand associations, with publications and subject indexes, and a useful list of initial abbreviations. A directory of natural history societies in Great Britain was published by the British Association for the Advancement of Science[4] in 1959, and gives details of aims; membership; meetings; amenities such as libraries, museums and equipment; and publications. Dates of foundation are provided, and there is an alphabetical list of publications and also a geographical index.

A special form of scientific society is that confining its activities to publishing. The Ray Society immediately comes to mind, and since its formation in 1844 it had by 1970 published some 145 volumes devoted to natural history. Richard Curle has described the foundation of the Society and provided much information gleaned from the Minute Books of the Society and other sources. His work[5] also includes a bibliography of the Society's publications. The arrangement is systematic and is followed by a list according to the year for which the publications were issued, the actual year of publication being given also. George Johnston (1797–1855)

[1] See Clarke, J. F., *Autobiographical recollections of the medical profession*, 1874, pp. 240–241, 267–282; and Hill, Brian. A Georgian careerist: John Frost (1803–1840). *Practitioner*, 188, 1962, pp. 262–266.

[2] Cohen, John, Hansel, C. E. M., and May, Edith F. Natural history of learned and scientific societies. *Nature*, 173, 1954, pp. 328–333.

[3] British Council. *Scientific and learned societies of Great Britain. A handbook compiled from official sources. 61st edition*, 1964.

[4] See Lysaght, Averil, *ed. Directory of natural history and other field study societies in Great Britain, including societies for archaeology, astronomy, meteorology, geology and cognate subjects*, 1959.

[5] Curle, Richard. *The Ray Society; a bibliographical history*, 1954.

was the founder of the Ray Society and he was also responsible for the formation of the Berwickshire Naturalists' Club in 1831. Johnston studied medicine at Edinburgh, and settled permanently at Berwick-on-Tweed, where he devoted all his spare time to natural history. He was the author of several books, none being of primary significance, but the Society continues its invaluable activities as a monument to his memory.[1] Founded just after the Ray Society, the Palaeontographical Society has published annual volumes since 1848 (for 1847), issued in the form of numbered monographs. Many of these were reprinted in 1966.

At a meeting of ornithologists held in London on 7th May, 1879, it was agreed to form an Association for reprinting certain ornithological works interesting for their utility or rarity. This Association, "The Willughby Society for the Reprinting of scarce Ornithological Works", issued its first publication, Marmaduke Tunstall's *Ornithologia Britannica*, 1771, edited were a preface by Alfred Newton, in 1880. A further eleven reprints were issued, the last, for the year 1883, being Johann Georg Wagler's *Six ornithological memoirs from the "Isis"*, (1829–32), edited by P. L. Sclater, 1884.

The Alembic Club was formed in the chemistry department at the University of Edinburgh in 1889, with five original members. One of its objects was to encourage the study of the history and literature of chemistry, and discussion meetings were held. In 1893 it published Joseph Black's *Experiments upon magnesia alba*, [*etc.*], 1755, as Alembic Club Reprint No. 1, which was followed by several other reprints of classic writings, over twenty having been printed. The Council of the Royal Society of Edinburgh took over the assets of the Club in 1952, and is now responsible for the continuance of its reprints.[2]

A survey of research on the history of scientific societies during the previous fifty years has been published by R. E. Schofield[3] in a comprehensive critical survey of the literature. Despite the impressive output, there is no major work discussing scientific societies from the end of the seventeenth century, and much remains to be accomplished in studying individual societies, those of specific countries, and those devoted to specialist subjects.

Most large cities house literary and philosophical, natural history and/or antiquarian societies, and these have encouraged the progress of science. National scientific societies, local institutions and groups of scientists interested in specialist subjects have all contributed to the development of science and its diffusion through the world. The growth

[1] See Yonge, C. M. George Johnston and the Ray Society. *Endeavour*, 14, 1955, pp. 136–139.
[2] See Kendall, James. The Alembic Club. *Endeavour*, 13, 1954, pp. 94–96.
[3] Schofield, R. E. Histories of scientific societies: needs and opportunities. *History of Science*, 2, 1963, pp. 70–83.

of these societies has been enormous, making a complete list of them an impossibility within the confines of a single volume. Their meetings have served for the interchange of information between kindred minds; their libraries have enabled members to study the extensive literature of their respective subjects; and their publications have spread scientific knowledge throughout the world.

CHAPTER VIII

The Growth of Scientific Periodical Literature

"In the history of science the academies completed the work of the universities in the Renaissance period and prepared the way for the laboratories of modern schools. Finally, it is important to note that it is to the academies that we owe the publication of the reports of their meetings, and these in turn gave rise to the first scientific journals."

ARTURO CASTIGLIONE

The development of the scientific periodical has proved of vital importance in the history of science. From small beginnings, and slow initial progress, the stream of journals of all types has become a flood that threatens to overwhelm those it was intended to benefit. The scientist expects speedy publication, and requires up-to-date information to keep him abreast of modern research. The journal has filled these rôles in the past, but modern conditions require even speedier publication and the spate of literature renders it more difficult for the specialist to sort out his requirements from the masses of papers that appear unendingly. But these conditions have taken three hundred years to develop, and the early years saw many false starts.

The seventeenth century saw the advent of the first scientific periodical, and before that period research work was published either as an essay (exercitatio), or as a separate treatise (tractatus). Before the art of printing had become firmly established, scholars exchanged information by letter, or by word of mouth, and we find scholars travelling widely between centres of learning, exchanging ideas, and by personal contact acquainting themselves with progress in their branches of learning. Printing enabled the written word to be widely disseminated, but it was not until scientists had grouped themselves into societies that we encounter the early journals, which were in fact accounts of the proceedings of those societies. Many of these early periodicals had correspondence sections, thus continuing the early method of propagating information, and even today the correspondence columns of *Nature*, for example, are an essential feature of that periodical.

The early news-letter sheets and newspapers, foreshadowed by the *Acta Diurna* of Ancient Rome, preceded the records of the proceedings of societies, which in turn grew into transactions, bulletins and journals.

Following these we find the general scientific periodicals, these often including medicine and allied subjects within their scope, and lastly the specialist journals, which become more specialized with the continued progress of scientific research. Periodicals are an essential feature of research work. They contain records of recent research, report lectures, contain reviews of books, abstracts of articles in other journals, reports of meetings, notices of forthcoming events, and correspondence, all these features proving of value to the scientist.

The overwhelming multitudes of scientific periodicals that have appeared for short or lengthy periods since their advent in the seventeenth century presents an interesting study for the historian. David A. Kronick[1], has written an extensive monograph on the origins and development of the scientific and technological press, covering the various types of periodicals such as journals, proceedings of societies, abstracts, reviews, almanacs, annuals, and collections of disputations, dissertations and similar academic writings. Numerous tables arranged by decade, country and subject, and an extensive bibliography are also included. Douglas McKie[2], has also contributed an important paper on the subject, and E. N. da C. Andrade[3] deals with it incidentally while covering a wider field, but two major sources of information are Henry C. Bolton[4] and Fielding H Garrison.[5] The latter deals with medical as well as scientific journals, and is particularly useful for the former, but a few remarks on his scientific material may prove interesting. Garrison notes that of the scientific periodicals of the seventeenth century six were French, three English, three German, three Italian and two Dutch. He lists these journals, together with details of editions, translations, abridgements, etc.

For the eighteenth century Garrison[6] gives a table (see below) indicating the country of origin of various groups of scientific journals.

[1] Kronick, David A. *A history of scientific and technical periodicals: the origins and development of the scientific and technological press, 1665–1790*, New York, 1962.

[2] McKie, Douglas, The scientific periodical from 1665 to 1798. *Phil. Mag.*, Commemoration issue, July, 1948, pp. 122–132.

[3] Andrade, E. N. da C. The presentation of scientific information. *Proc. Roy. Soc.*, A, 197, 1949–50, pp. 1–17; also in *Proc. Roy. Soc.*, B, 136, 1949–50, pp. 317–333; and *The Royal Society Scientific Information Conference, 21st June–2nd July, 1948. Report and papers submitted*, 1948, pp. 26–44.

[4] Bolton, Henry Carrington. *A catalogue of scientific and technical periodicals, 1665–1895. Together with chronological tables and a library check-list. . . . Second edition*, Washington, 1897. (Smithsonian Miscellaneous Collections, Vol. 40, 1898), reprinted New York, 1966; a list of American journals omitted by Bolton was compiled by William J. Fox (*Bull. of Bib.*, 5, 1908, pp. 82–85).

[5] Garrison, Fielding H. The medical and scientific periodicals of the seventeenth and eighteenth centuries. *Bull. Inst. Hist. Med.*, 2, 1934, pp. 285–343. (Supplement to *Bull. Johns Hopkins Hosp.*, 55, 1934); see also Kronick, David A. The Fielding H. Garrison list of medical and scientific periodicals of the 17th and 18th centuries; addenda et corrigenda. *Bull. Hist. Med.*, 32, 1958, pp. 456–474.

[6] Garrison, Fielding H. *Op. cit.*, p. 300.

PHILOSOPHICAL
TRANSACTIONS:
GIVING SOME
ACCOMPT
OF THE PRESENT
Undertakings, Studies, and Labours
OF THE
INGENIOUS
IN MANY
CONSIDERABLE PARTS
OF THE
WORLD.

Vol I.
For *Anno* 1665, and 1666.

In the *SAVOY*,
Printed by T. N. for *John Martyn* at the Bell, a little without *Temple-Bar*, and *James Allestry* in *Duck-Lane*, Printers to the *Royal Society*.

PLATE 13. Title-page of the first volume of *Philosophical Transactions*.

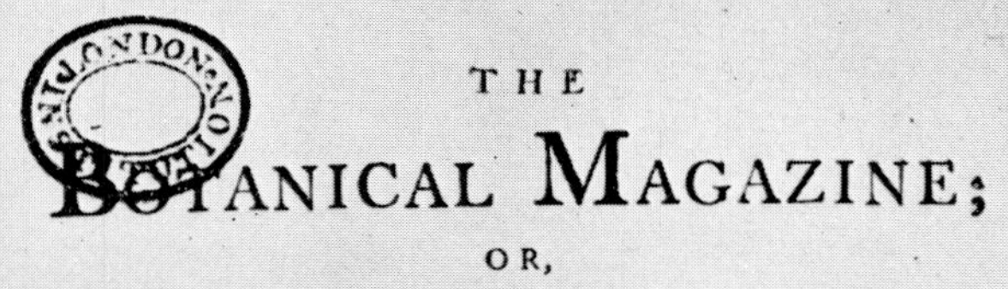

THE

BOTANICAL MAGAZINE;

OR,

Flower-Garden Diſplayed:

IN WHICH

The moſt Ornamental FOREIGN PLANTS, cultivated in the Open Ground, the Green-Houſe, and the Stove, are accurately repreſented in their natural Colours.

TO WHICH ARE ADDED,

Their Names, Claſs, Order, Generic and Specific Characters, according to the celebrated LINNÆUS; their Places of Growth, and Times of Flowering:

TOGETHER WITH

THE MOST APPROVED METHODS OF CULTURE.

A WORK

Intended for the Uſe of ſuch LADIES, GENTLEMEN, and GARDENERS, as wiſh to become ſcientifically acquainted with the Plants they cultivate.

By *WILLIAM CURTIS*,

Author of the FLORA LONDINENSIS.

VOL. I.

" A Garden is the pureſt of human Pleaſures."

VERULAM.

LONDON:

Printed by STEPHEN COUCHMAN, Throgmorton-Street,

For W. CURTIS, N° 3, *St. George's-Creſcent*, Black-Friars-Road;

And Sold by the principal Bookſellers in Great-Britain and Ireland.

MDCCXCIII.

PLATE 14. Title-page of the first volume of Curtis's *Botanical Magazine*.

	Scientific Societies	*Biology*	*Scientific Periodicals*	*Total*
Austria-Hungary			7	7
Belgium	1		1	2
Bohemia	1		1	2
Denmark	1	1	3	5
France	18	6	22	46
Germany	19	40	96	155
Great Britain	11	4	9	24
Holland	9		12	21
Italy	10		6	16
Mexico			2	2
Norway	1			1
Portugal	1			1
Russia	1		2	3
Spain		1	1	2
Sweden	6	5		11
Switzerland	5	9	5	19
United States	2			2
(Based on Garrison's Table)				319

He also further subdivides them by subject for the eighteenth century, listing twenty-nine devoted to biology (pp. 328–329), eighteen on botany (p. 329), seven dealing with zoology (pp. 329–330) and eight specializing in entomology (p. 330).

A table giving an idea of the subject distribution of modern scientific journals makes an interesting comparison, and is based on *Ulrich's Periodicals Directory*, 1959 and 1967, with figures in round brackets from the 1969 supplement.[1]

	1959	1967		1959	1967
Astronomy	32	63 (13)	Mathematics	119	226 (27)
Biology	136	194 (19)	Metallurgy	106	287 (35)
Botany	72	153 (38)	Meteorology	39	51 (15)
Chemistry	242	380 (47)	Ornithology	50	85 (8)
Electricity	126	314 (49) (Electricity and Electronics combined)	Physics	104	283 (37)
Electronics	58		Zoology	47	125 (21)
Entomology	27	80 (10)			

[1] *Ulrich's Periodicals Directory: a classified guide to a selected list of current periodicals, foreign and domestic. Ninth edition. Edited by Eileen C. Graves,* New York, 1959; and *Ulrich's International Periodicals Directory. Volume 1. Scientific, technical & medical. . . . Twelfth edition. Edited by Marietta Chicorel,* New York, London, 1967. Also, Third supplement, 1969. The 13th edition was published in 1969.

The 1969 Supplement contains entries for over 4,500 periodicals not previously included, and of these about 1,000 had been started since 1966. There is also a list of about 700 periodicals which had ceased publication or changed title.

An annotated bibliography of guides to scientific periodicals has been provided by Maureen J. Fowler in *Guides to scientific periodicals. Annotated bibliography*, London, 1966, which is based on the collection of guides held by the National Lending Library for Science and Technology. This also includes bibliographies and guides covering monographs as well as serials; directories listing the periodical publications of organizations mentioned; national bibliographies and other periodicals listing new serials; and abstracting journals publishing lists of serials abstracted. The national and other regional guides are divided into eleven geographical areas, which are subdivided into general lists; special subject lists in U.D.C. order; guides to abstracting journals; lists of government publications; guides to house journals; lists of new titles; directories; and union lists.

S. H. Scudder's *Catalogue of scientific serials of all countries, including the transactions of learned societies in the natural, physical and mathematical sciences, 1633–1876*, Cambridge, Mass., 1879 (reprinted New York, 1965), was originally issued as Harvard University Library, Special publication No. 1. It lists over 4,400 titles arranged by country and town of origin, with indexes of towns and titles, and a brief subject index. Scudder excludes periodicals devoted exclusively to medicine, agriculture and technology, and it is complementary to, but less complete than H. C. Bolton's *Catalogue* (see below).

A number of scientific items are mentioned in "An account of the periodical literary journals which were published in Great Britain and Ireland, from the year 1681 to the commencement of the Monthly Review, in the year 1749", by Samuel Parkes.[1] This gives details of title, editor, printer and publisher, with descriptions of type of contents.

Charles Harvey Brown[2] published in 1956 a survey of scientific periodical literature dealing with its acquisition, storage and discarding, with the increase in the number and cost of this form of literature, and listing the journals most frequently cited and abstracted in the various branches of science.

The *World list of scientific periodicals* is one of the largest lists of its kind; the first edition, published 1925–27, contains 25,000 entries, while the second edition, issued in 1934, lists 36,000 journals. The third

[1] *Quart. J. Sci. Lit., and the Arts*, 13, 1822, pp. 30–58, 289–312.

[2] Brown, Charles Harvey. *Scientific serials: characteristics and lists of most cited publications in mathematics, chemistry, geology, physiology, physics, botany, zoology and entomology*, [*etc.*], Chicago, 1956.

edition, published in 1953, covers the period 1900–50, and lists over 50,000 items. The fourth and last edition, 1963–65 is in three volumes, and lists over 60,000 items. It is maintained up to date at the National Central Library, which compiles *British Union-Catalogue of Periodicals, incorporating World List of Scientific Periodicals, New Periodical Titles*, published quarterly. The scientific items have also been published annually as *World List of Scientific Periodicals*, [*etc.*], since 1964, with a cumulative volume in 1970. The main purpose of the *World List* is that of giving the locations of items in British libraries. Sir Charles Sherrington[1] compiled the following interesting table indicating the language distribution of the entries in the second edition:

English scientific periodicals . . .	13,494
German ,, ,, . . .	6,186
French ,, ,, . . .	5,013
Russian ,, ,, . . .	1,833
Italian ,, ,, . . .	1,667

The first edition of Henry C. Bolton's[2] *Catalogue of scientific and technical periodicals* listing 5,105 journals, appeared in 1885, the second edition containing 8,603 entries. It includes all journals incorporating scientific material, but in the chronological tables entries are restricted to purely scientific periodicals. An analysis of these tables reveals only five purely scientific periodicals appearing before 1750, seventy-two between 1750 and 1800, 230 between 1801 and 1850, and 203 from 1851 to 1889. Many periodicals have survived for but a brief period, while others have relied upon changes of title and incorporation with other journals to prolong their careers. While some mention must be made of pioneers in the field of scientific periodical literature, we are mainly concerned with those that have usefully appeared over lengthy periods. The Royal Society[3] has published a list of British scientific periodicals containing 212 entries, and many of these have been in existence for over a century. The list is chiefly concerned with the publications of scientific societies, including medical items, but omits news journals, most commercially sponsored periodicals, and Government department publications. The main list is arranged alphabetically by names of societies or publishers, but there are indexes of names of journals and of general subjects. Entries include brief descriptions of the contents of each journal, of indexes, frequency, current volume numbers, approxi-

[1] Sherrington, Sir Charles S. Language distribution of scientific periodicals. *Nature*, 134, 1934, p. 625.

[2] Bolton, Henry Carrington. *Op. cit.*

[3] Royal Society. Information Services Committee. *A list of British scientific publications reporting original work or critical reviews*, 1950.

mate size, advertisements, directions to authors, and addresses of editors and of publishing organizations.

A chronological survey of scientific periodicals gives some idea of the gradual development of this material, and indicates the gradual development of specialization in the nature of journals. The first independent scientific periodical, the *Journal des Sçavans*, was published at Paris on 5th January, 1665, and appeared under that title until 1792. It was reprinted at Amsterdam, 1665–1792, and five volumes of a German translation by Friedrich Nitsche appeared between 1667 and 1671. The *Journal des Sçavans* was first edited by Denis de Sallo, under the pseudonym of Sieur de Hédouville, who was followed by the Abbé Gallois in 1666 and by Cousin in 1687. The periodical was suppressed during the Revolution, to resume in 1816 under the auspices of the Institut de France, with the title *Journal des Savants*.[1] The *Philosophical Transactions* of the Royal Society first appeared on 6th March, 1665 (Plate 13) and was based upon the *Journal des Sçavans*, containing similar features and even extracts from that periodical. The *Philosophical Transactions* was edited by Henry Oldenburg, secretary to the Royal Society, and was his own financial responsibility until 1676. After February 1678–79 (No. 142) publication stopped, to be resumed with No. 143 in January, 1682–83. During the interim Hooke published *Philosophical Collections* at the request of the Council of the Royal Society, but it resumed its original title in 1682–3, and has continued its invaluable career to the present day. There have been numerous abridgements of the *Philosophical Transactions*, listed elsewhere,[2] and also translations into Latin, French and German.[3] Volumes 1 to 70, 1665–1780, were reprinted in 1963–64. Since 1832 the Royal Society has also published *Proceedings*, which have been divided into Series A and B since 1905, devoted to Mathematical and Physical Sciences, and Biological Sciences respectively.

The *Acta Eruditorum* was first published in Latin at Leipzig in 1682, and was the first German scientific periodical. It was edited by Otto Mencke (1644–1707), and editorship remained in his family until 1754. It was published as the *Nova Acta Eruditorum* from 1732 to 1782 and contained many papers by Leibnitz. Reprints of these periodicals have been announced. A similar Dutch periodical, modelled on the *Journal des Sçavans*, the *Nouvelles de la république des lettres*, Amsterdam, 1684–1718, was edited by Pierre Bayle (1647–1706) during its early years. An Italian

[1] See Porter, J. R. The scientific journal—300th anniversary. *Bact. Rev.*, 28, 1964, pp. 211–230.
[2] Garrison, Fielding H. *Op. cit.*, Nos. 4–40.
[3] See Andrade, E. N. da C. The birth and early days of the "Philosophical Transactions". *Notes Rec. Roy. Soc. Lond.*, 20, 1965, pp. 9–27; Hulls, L. G. Phil. Trans.—the first fifty years. *Discovery*, 21, 1960, pp. 236–240; Wightman, W. P. D. "Philosophical Transactions" of the Royal Society. *Nature*, 192, 1961, pp. 23–24.

imitation of the *Journal des Sçavans* was published at Rome from 1668 to 1697, being first entitled *Giornale de Litterati*. Yet another series commencing publication in the seventeenth century was that published by the Académie Royale des Sciences under the titles *Histoire* and *Mémoires* in 164 volumes at Paris between 1698 and 1809, with various titles and supplementary volumes.

The eighteenth century saw the continued development of the scientific periodical, and towards the end of that period we find journals devoted to specific scientific subjects, including several that have survived to the present day.[1] Taken roughly in chronological order, the more important organs were the following: *Raccolta d'Opuscoli Scientifici e Filologici*, published in Venice between 1728 and 1757, when it was continued as *Nuova Raccolta d'Opuscoli Scientifici*, and published in forty-two volumes from 1755 to 1787; *Le Pour et le Contre*, [*etc.*], twenty volumes, Paris, 1733–40, of which another edition was published at The Hague; *Bibliothèque Britannique, ou Histoire des Ouvrages des Savans de Grande Bretagne*, twenty-five volumes, The Hague, 1733–47, which was followed by *Journal Britannique*, The Hague, in twenty-four volumes published between 1750 and 1757; *Göttingische Zeitung von gelehrten Sachen*, issued in fourteen volumes from 1739 to 1752, continued as *Göttingische Anzeigen von gelehrten Sachen unter Aufsicht der Königliche Gesellschaft der Wissenschaften*, 117 volumes, 1753–1801, and finally as *Göttingische gelehrte Anzeigen*, [*etc.*], 1802 onwards; *Hamburgisches Magazin*, in twenty-six volumes, 1747–67, and continued as *Neues Hamburgisches Magazin*, in twenty volumes, published at Hamburg and Leipzig from 1767 to 1780. *La Feuille Necéssaire, contenant divers sur les sciences, les lettres et les arts* was first published on 12th February, 1759, and was published weekly in forty-seven issues until 31st December, 1759, and then ceased to exist without warning. Published by Michel Lambert, it was succeeded by *L'Avant-Coureur. Feuille hebdomadaire, ou sont announcés les objets particuliers des sciences et des arts, le cours et les nouveautés des spectacles, & les livres nouveaux en tout genre*, which also appeared weekly under various publishers until the end of 1773, when it combined with *Gazette et Avant Coureur de litterature, des sciences et des arts*, which was published twice a week from 11th January, 1774, but after four issues was renamed the *Gazette de litterature, des sciences et des arts*, which continued as a separate publication until October, 1774. The early issues of *L'Avant-Coureur* contained several papers by Lavoisier.[2] The Abbé François Rozier (1734–93) was responsible for the monthly *Observations sur la Physique, sur l'Histoire Naturelle*,

[1] See Barnes, Sherman B. The editing of early learned journals. *Osiris*, 1, 1936, pp. 155–172.
[2] See Smeaton, W. A. L'Avant-Coureur. The journal in which some of Lavoisier's earliest research was reported. *Ann. Sci.*, 13, 1957, pp. 219–234.

et sur les Arts,[1] founded in July, 1771, and which appeared until 1823 under various changes of titles. This was the first appearance of a specialist periodical devoted to physics and chemistry; and *Chemische Journal, für die Freunde der Naturlehre, Gelahrtheit, Hauslhaltungskunst und Manufacturen*, in six volumes, Lemgo, 1778–81, when it continued with the title *Die Neuesten Entdeckungen, in der Chemie*, thirteen volumes, Leipzig, 1781–86. These were edited by Lorenz von Crell (1744–1816), a former pupil of Joseph Black, and who is regarded as the founder of chemical journalism. Among other periodicals devoted to chemistry Crell also edited *Chemisches Archiv* (1783); *Neues chemisches Archiv* (1784–91); *Neuestes chemisches Archiv* (1798); and *Chemische Annalen für die Freunde der Naturlehre, Arzneygelahrtheit, Haushaltungskunst und Manufacturen* (1784–1803).[2]

The year 1787 saw the foundation of *The Botanical Magazine* by William Curtis (1746–99) (Plate 14). The first fourteen volumes were published between 1787 and 1800, after which it became known as *Curtis's Botanical Magazine*, and has continued its useful career to the present day. It was originally published monthly, with a circulation of 2,000 copies, and was the first illustrated botanical periodical. The coloured plates were drawn by Sydenham Edwards until 1815, when he had produced almost 1,770 drawings, and the colourist was William Graves. These coloured plates have been an essential feature of *Curtis's Botanical Magazine* since the first number, and have no doubt heightened its popularity, although it has also maintained a strictly scientific outlook.[3]

The *Annales de Chimie* (*et de Physique*) is another invaluable scientific periodical with a lengthy history, having been published in various series and with minor changes in title since 1790. The first volume is dated 1789, but it did not appear until the following year. An excellent history of the *Annales* has been published by Marcel Delépine.[4] The first periodical strictly devoted to physics, the *Journal der Physik*, published at Halle and Leipzig, was published between 1790 and 1794. It then became *Neues Journal der Physik* (1795–97), *Annalen der Physik* (1799–1824), and finally *Annalen der Physik* (*und Chemie*), with slight variations in title, from 1824 onwards. The *Annalen* is associated with the name of Johann Christian Poggendorf (1796–1877), inventor of the galvanometer (1821), who edited 160 volumes of the *Annalen*. A biographical memoir, and a

[1] See McKie, Douglas. The "*Observations*" of the Abbé François Rozier (1734–93). *Ann. Sci.*, 13, 1957, pp. 73–89.

[2] See Yagello, Virginia E. Early history of the chemical periodical. *J. chem. Educ.*, 45, 1968, pp. 426–429.

[3] A bibliographical synopsis of the *Botanical Magazine* is contained in Blunt, Wilfrid. *The art of botanical illustration*, [*etc.*], 1950, p. 183; also, Chapter 15, (pp. 184–188) of that work is devoted to the *Botanical Magazine*. See also Hunkin, J. W. William Curtis, founder of the *Botanical Magazine*. *Endeavour*, 5, 1946, pp. 13–17.

[4] Delépine, Marcel. Les Annales de Chimie de leur fondation à la 173e année de leur parution. *Ann. Chim.*, 13e Sér., 7, 1962, pp. 1–11.

bibliography of Poggendorf's papers is contained in N.F.160, 1877, pp. vii–xxiv.

Towards the end of the century we must also note the appearance of yet another veteran that is still with us. *The Philosophical Magazine* appeared in a first series of forty-two volumes between 1798 and 1813, since when, with numerous amalgamations and changes of title, it has regularly appeared.[1] A reprint of the first 216 volumes was recently announced. Before leaving the eighteenth century we must also mention two *Transactions* of learned societies that are still current, those of the Royal Society of Edinburgh (1788) and of the Linnean Society of London (1791). *Nicholson's Journal* is also noteworthy, and appeared monthly from April, 1797 to 1813. It was edited by William Nicholson (1753–1815).[2]

Justus Liebig's Annalen der Chemie is one of the most important periodicals devoted to chemistry, and has been in continuous existence since 1832, when it was originally designated *Annalen der Pharmacie*. Liebig joined his colleague Philipp Lorenz Geiger (1785–1836) in editing the *Magazin für Pharmazie*, [*etc.*], in 1831, which underwent several changes of title and of editors, until in 1838 Liebig assumed complete control, and with the co-operation of Dumas in Paris and Thomas Graham in London, the *Annalen* was issued simultaneously in England, France and Germany. In 1840 the title became *Annalen der Chemie und Pharmacie*, until Liebig's death in 1873 when it was changed to *Justus Liebig's Annalen der Chemie und Pharmacie*, and the final two words were dropped a year later.[3]

The more important of the scientific periodicals founded in the first half of the nineteenth century are recorded in chronological order, that their development may be appreciated, but it will be recognized that numerous short-lived journals are not recorded, and that certain changes of title, etc., are omitted, these details being available in the bibliographical lists by Henry C. Bolton and by Fielding H. Garrison previously mentioned. In 1818 appeared the *American Journal of Science*, which reverted to the same title in 1880 after publication as the *American Journal of Science and Arts* from 1820 to 1879.[4] The *Memoirs of the Royal Astronomical Society* first appeared in 1821, and was followed in 1831 by the Society's *Monthly Notices*. *The Magazine of Natural History and Journal of Zoology, Botany, Mineralogy, Geology and Meteorology* was first issued in May, 1828, and in 1837 the title was changed to *The Magazine of Natural History*. In

[1] See Ferguson, Allan, and Ferguson, James. The Philosophical Magazine. *Phil. Mag.*, July, 1948, pp. 1–9; and Barr, E. Scott. Biographical material in the *Philosophical Magazine* to 1900. *Isis*, 55, 1964, pp. 88–90.

[2] See Lilley, S. "Nicholson's Journal" (1797–1813). *Ann. Sci.*, 6, 1948–50, pp. 78–101.

[3] See Van Klooster, H. S. The story of Liebig's Annalen der Chemie. *J. chem. Educ.*, 34, 1957, pp. 27–30.

[4] A classified list of scientific and chemical journals is contained in Crane, E. J., Patterson, Austin M., and Marr, Eleanor B. *A guide to the literature of chemistry. Second edition*, New York, London, 1957, pp. 76–123.

March, 1838, appeared the first issue of *The Annals of Natural History, or Magazine of Zoology, Botany and Geology*, and in 1840 the *Magazine* and the *Annals* were merged to form *The Annals and Magazine of Natural History, including Zoology, Botany, and Geology*, which has continued its distinguished career to the present day, but with another change of title in 1967 to *Journal of Natural History. An International Journal of Taxonomic and General Biology*. In 1830 appeared the *Proceedings of the Zoological Society of London*, which continued under that title until 1965 when it became the *Journal of Zoology*. *Botanical Miscellany* was also founded in 1830, and appeared under various titles for a large number of years. The *Annual Report of the British Association for the Advancement of Science* was published between 1831 and 1938, since when it has appeared as *The Advancement of Science*. The Zoological Society of London first issued its *Transactions* in 1833, in which year also appeared the *Entomological Magazine*, five volumes, 1833–38, followed by *The Entomologist*, 1840–42, which was revived in 1864 after a period of twenty-two years. The *Transactions of the (Royal) Entomological Society of London* first appeared in 1834. In 1837 Hooker's *Icones Plantarum* was first published, and the following year appeared the *Journal of the Royal Statistical Society* and the *Proceedings of the Linnean Society of London*, which in 1969 became *Biological Journal of the Linnean Society*.

The Chemical Society published its *Memoirs and Proceedings* from 1841 to 1847, after which the title was changed to *Journal of the Chemical Society*. From 1966–68 it was divided into sections A, B and C, with the addition of section D from 1969.

The Zoologist was first published in London in 1843 and continued until December, 1916, after which it was absorbed by *British Birds*, which had commenced publication in 1907. 1844 saw the first appearance of the *Transactions and Proceedings of the Botanical Society of Edinburgh* and of the *Transactions of the Royal Microscopical Society of London*, which appeared as the *Monthly Microscopical Journal* from 1869 to 1877, after which the title was changed to *Journal of the Royal Microscopical Society*, and in 1969 to *Journal of Microscopy*. The *Quarterly Journal of the Geological Society of London* first appeared in 1845, as did the *Scientific American*, to be followed in the same decade by the *Zeitschrift für Entomologie* (1847) and the *Zeitschrift für wissenschaftliche Zoologie* (1849).

In 1852 *Die Natur* first appeared in print, as did the *Quarterly Journal of Microscopical Science*, the title of which was changed to *Journal of Cell Science* in 1966, to be followed by the *Proceedings of the Royal Physical Society* (1854); *Journal of the Linnean Society of London* (Botany and Zoology, 1856), which divided into the *Botanical Journal of the Linnean Society* (1969), and *Zoological Journal of the Linnean Society*; and the *Zeitschrift für Mathematik und Physik* (1856). The British Ornithologists'

Union's organ *The Ibis* was first published in 1859, as was the *Proceedings of the Geologists' Association*. In 1862 appeared *Zeitschrift für analytische Chemie;* the *Journal of Botany, British and Foreign* (1863) became *Trimens Journal of Botany* in 1880, returning to its original title in 1883, and continuing publication until 1942 when it ceased to appear, but was later included in the *Annals and Magazine of Natural History*, already mentioned. The bi-monthly *Geological Magazine* first appeared in 1864, as did *The Record of Zoological Literature*, which became the *Zoological Record* in 1871. After several changes of title the *Quarterly Journal of Science* (1864–70) became the *Journal of Science and Annals of Astronomy, Biology, Geology,* [*etc.*], in 1880. The year 1865 saw the first appearance of the *Proceedings of the London Mathematical Society, Hardwicke's Science Gossip* and *Zeitschrift für Biologie*, the latter being published at Munich. In 1866 appeared the *Journal of the Royal Horticultural Society* and the *Meteorological Magazine*, published by the Meteorological Office. The *Journal of Anatomy and Physiology* was first published in 1867, and in 1916 became the *Journal of Anatomy*. Its centenary was celebrated by the publication of an excellent cumulative index.[1] That incomparable scientific periodical *Nature* was also founded in 1869, and maintains its position as the leading scientific periodical. From January, 1971, it was divided into three parts, each appearing weekly (Plate 15).[2]

The Physical Society first published its *Proceedings* in 1874, which became *Journal of Physics*, A, B, and C in 1969, and the *Quarterly Journal of the Royal Meteorological Society* first appeared in 1875. A periodical entitled *The Analyst* had been published between 1834 and 1840, but in 1877 appeared the journal which still survives bearing that title. The next decade saw the publication of the *Journal of the Society of Chemical Industry* (1882), which became the *Journal of Applied Chemistry* in 1951; and the *Zeitschrift für wissenschaftliche Mikroskopie* (1884).[3] *Annals of Botany, Journal of the Marine Biological Association, Kew Bulletin* and the *Zeitschrift für physikalische Chemie*, were all first issued in 1887. The *Journal of the British Astronomical Association* (1890), the *Bulletin of the British Ornithologists' Club* (1892), the *Geographical Journal* (1893), *Memoirs of the British Astronomical Association* (1893), *Physics Review* (1893),[4] later published by

[1] *Journal of Anatomy. Index to the first 100 years, 1866–1966. Compiled by Doreen Blake and Ruth E. M. Bowden*, 1968. See also Walls, E. W. The Journal of Anatomy, 1867–1966. *J. Anat.*, 100, 1966, pp. 1–4.

[2] See Gregory, Sir Richard. News and views from the scientific front. *Nature*, 151, 1943, pp. 517–519; and Barr, E. Scott. *Nature's* "Scientific worthies". *Isis*, 56, 1965, pp. 354–356.

[3] See Freund, Hugo. 80 Jahre Zeitschrift für wissenschaftliche Mikroskopie und mikroskopische Technik. Ihre Begründer und seiner Nachfolger. *Z. wissen. Mikrosk.*, 66, 1964, pp. 3–17.

[4] See Barr, E. Scott. Biographical material in the first series of the *Physical Review*. *Isis*, 58, 1967, pp. 245–246.

the American Physical Society, founded in 1899, *Proceedings of the Malacological Society of London* (1893), and the *Transactions of the British Mycological Society* (1898) were all inaugurated before the end of the century.

It has become increasingly obvious that the enormous output of scientific literature is overwhelming even the largest scientific libraries. Research workers cannot hope to read all the papers published even in limited fields of science, and the tracing of this material is also difficult. Abstracting journals have increased in size and number, even to the extent of badly overlapping in coverage, but the fact remains that a major problem awaits solution. Further co-operation between existing abstracting services, stricter editing, the ruthless pruning of articles, and various other ideas have been put forward, but practical methods have accomplished little. Abstracting periodicals such as *Biological Abstracts* and *Chemical Abstracts* are very useful, but their value is mitigated by the length of time elapsing before the appearance of abstracts, by the high cost of these organs, and the fact that in many cases, while they overlap, few attempt to be exhaustive.[1]

Tracing the location of scientific periodicals is an important matter, for no single library can hope to shelve every periodical required by research workers. Catalogues of periodicals contained in large libraries devoted to science, such as the Science Museum Library at South Kensington, London, the catalogues of more special collections, and joint catalogues of periodicals, such as that of journals contained in university libraries,[2] and the *British Union-catalogue of Periodicals*,[3] generally known as *BUCOP*, are all invaluable bibliographical tools. *BUCOP* is maintained up to date at the National Central Library, and published quarterly as *British Union-catalogue of Periodicals, incorporating World List of Scientific Periodicals. New Periodical Titles*, with an annual cumulation titled *World List of Scientific Periodicals*, [*etc.*], published since 1964, with a cumulative volume, 1970. The *World List* is of special value because it provides full titles and gives locations of copies, and is probably the location list most used in tracing required scientific journals in Britain.

Location lists of periodicals, and catalogues of the holdings of individual

[1] See Ditmas E. M. R. The co-ordination of abstracting services. Unesco's approach to the problem. *J. Document.*, 4, 1948–49, pp. 67–83. Appendix; Bibliography on abstracting of scientific periodicals, pp. 76–83; and Varossieau, W. W. A survey of scientific abstracting and indexing services. *Review of Documentation*, 16, 1949, pp. 25–46.

[2] Roupell, Marion G. *Union catalogue of the periodical publications in the university libraries of the British Isles* [*etc.*], 1937.

[3] *British Union-Catalogue of Periodicals: a record of the periodicals of the world, from the seventeenth century to the present day, in British libraries. . . . Edited . . . by James D. Stewart with Muriel E. Hammond and Erwin Saenger*, Vol. 1, A–C, 1955; 2, D–K, 1956; 3, L–R, 1957; 4, S–Z, 1958; supplementary volume, to 1960, 1962; see also Koster, C. J. A history of the British Union Catalogue of Periodicals, *Occasional News Letter. National Central Library*, No. 10, April, 1970, pp. 10–11.

libraries abound, and the following are typical examples of these tools. *Union list of serials in libraries of the United States and Canada*, 3rd edition, five volumes, New York, 1965, edited by Edna Brown Titus. This lists 156,499 titles held by 956 libraries, and is maintained up to date by *New Serial Titles*, published monthly by the Library of Congress, and cumulated annually. The Library of Congress also published Ruth S. Freitag's *Union lists of serials: a bibliography*, Washington, 1964, in which arrangement is geographical, with appropriate indexes and brief annotations.

The National Lending Library for Science and Technology has issued *A KWIC index to the English language abstracting and indexing publications currently being received by the National Lending Library*, 3rd edition, Boston Spa, 1969, in which titles are listed in alphabetical order of the subject key words. Over 8,000 titles are listed in *Periodical publications in the National Reference Library of Science and Invention. Part I. List of non-Slavonic titles in the Bayswater Division*, London, 1969, published by the British Museum. This supplements *Periodical publications in the Patent Office Library: list of current titles*, 3rd edition, 1965, representing some 9,000 titles. The International Association of Agricultural Librarians and Documentalists has published *Current agricultural serials. A world list of serials in agriculture and related subjects (excluding forestry and fisheries) current in 1964*, edited by D. H. Boalch, two volumes, Oxford, 1965–67. This includes many serials not in the *World List*, and also covers selected serials in chemistry, biology, physics, biochemistry, geology and zoology.

Periodicals are of primary importance to the modern research worker, and have done much to advance the cause of science. Their very number has overwhelmed scientists, who are frequently unable to trace required material which is buried beneath masses of related literature. Unless practical methods are adopted to control the publication of periodicals, adequately to abstract their contents, and make them freely available to research workers, the latter will be strangled by the very tools founded to facilitate the progress of scientific knowledge.

Estimates of the number of current scientific and technical periodicals vary considerably, and range from 100,000 to 30,000. K. P. Barr,[1] basing his estimate on National Lending Library figures, and excluding house journals and publishers' series, suggests 26,000 for December 1965. Of this number about 10,400 are in the English language. Obviously the numbers fluctuate with the advent of new titles, the decease of certain periodicals, and any estimate must depend upon precise definition and coverage. B. C. Vickery[2] has supplemented Barr's figures with information

[1] Barr, K. P. Estimates of the number of currently available scientific and technical periodicals. *J. Document.*, 23, 1967, pp. 110–116.

[2] Vickery, B. C. Statistics of scientific and technical articles. *J. Document.*, 24, 1968, pp. 192–195.

on the number and distribution of articles within these periodicals, showing the most used journals, and indicating that 2,400 periodicals might meet 90% of the demand. John Martyn and Alan Gilchrist[1] have published an evaluation of British scientific journals, providing some interesting tables. In one of these, journals are ranked by citations received, and it is estimated that the world's number of core science journals lies between 2,300 and 3,200. Many other publications concerned with the growth of periodical literature are available, including an article by Sir James Cook[2] and a survey by the National Science Foundation Office of Science Information Services.[3]

Lists of periodicals of primary importance in the fields of physics, the biological sciences, and the chemical sciences have been prepared by the International Council of Scientific Unions, Abstracting Board, and published as *Some characteristics of primary periodicals in the domain of the physical sciences* [*biological sciences; chemical sciences*], Paris, 1966[–1967]. These lists present a wealth of information regarding frequency of publication, numbers of pages, average number of words per page, number of papers published and their average length, average delay in publication, etc. A study of the literature of physics based on *Physics Abstracts* has also been published by Stella Keenan and Pauline Atherton,[4] an admirable review article on the growth of the literature of physics prepared by L. J. Anthony and others,[5] and publications on chemistry have been the subject of a comprehensive survey by R. S. Cahn.[6] This resulted in far-reaching recommendations for controlling the literature, some of which have been carried out.

Research newsletters contain much useful information, and many of them have a very limited circulation. Some provide details of research in progress, information on research material, and select bibliographies. H. V. Wyatt[7] has analysed the contents and scope of twenty-two such organs in the biological sciences.

Ralph H. Phelps and John P. Herlin[8] have made a survey of possible

[1] Martyn, John, and Gilchrist, Alan. *An evaluation of British scientific journals*, 1968. Aslib Occasional Publication No. 1.

[2] Cook, Sir James. The science information problem. *Adv. Sci.*, 23, 1966, pp. 305–309.

[3] National Science Foundation, Office of Science Information Services. *Characteristics of scientific journals, 1949–59*, Washington, 1964.

[4] Keenan, Stella, and Atherton, Pauline. *The journal literature of physics. A comprehensive study based on* Physics Abstracts (*Science Abstracts, Section A, 1961 issues*), New York, 1964.

[5] Anthony, L. J., East, H., and Slater, M. J. The growth of the literature of physics. *Reports on Progress in Physics*, 32, 1969, pp. 709–767.

[6] Cahn, R. S. *Survey of chemical publications and report to the Chemical Society*, London, 1965.

[7] Wyatt, H. V. Research newsletters in the biological sciences—a neglected literature service. *J. Document.*, 23, 1967, pp. 321–325.

[8] Phelps, Ralph H., and Herlin, John P. Alternatives to the scientific periodical: a report and bibliography. *Unesco Bull. Lib.*, 14, 1960, pp. 61–75. The bibliography contains 121 items.

alternatives to the scientific periodical, and they provide reasons for the necessity for such alternatives, but without finding a satisfactory solution. They mention delays in the publication of papers; restrictions on the length of articles; the fact that historical and theoretical material is sometimes omitted, and the number of references reduced by editors; that the multiplicity of journals results in the wide dissemination of papers; the high cost of journals compared with the number of articles of interest to individual subscribers; the high cost to societies issuing journals; and the waste of scientists' time in the unpaid editing and managing of journals. Various alternative suggestions include the provision of special radio and/or television stations for broadcasting science reports; and the use of tape recordings, microcards, depositories for material with notices in journals announcing availability, or the printing of summaries. The publication of separate papers as pamphlets, and the publication of journals as collections of separates, have been tried and abandoned, the cost of publishing and handling having proved excessive. W. S. Brown, J. R. Pierce and J. F. Traub[1] have described a computer-based system designed to permit subscribers to obtain "a personalized stream of papers", which would be issued separately by journals instead of being bound into issues. Although interesting, it does not appear to merit widespread practical application except within the framework of commercial firms functioning in closely defined fields of knowledge, and with substantial funds.

Estimates of the growth of scientific literature in recent years have been attempted by many authors, without any finding a solution to the problem. The numbers of books, journals, symposia, etc., are constantly growing, and indexes and abstracts fail to give comprehensive coverage despite an enormous increase in their own sizes. Mechanical methods of information storage and retrieval also fail to provide a solution to the problem, chiefly because of overlapping without complete coverage, bias in the selection of items indexed or abstracted, the time lag in providing lists of references, and the tremendous cost involved. Even the largest libraries find it impossible to absorb the increasing costs of essential bibliographical tools, yet the user of the small, isolated library expects a comprehensive service, and should be able to have access to the more sophisticated techniques. Derek J. de Solla Price[2] has examined the problem in some detail in several interesting publications. D. J. Urquhart[3]

[1] Brown, W. S., Pierce, J. R., and Traub, J. F. The future of scientific journals. A computer-based system will enable a subscriber to receive a personalized stream of papers. *Science*, 158, 1967, pp. 1153–1159.

[2] Price, Derek J. de Solla. *Science since Babylon*, New Haven, London, 1961; Price, Derek J. de Solla. *Little science, big science*, New York, London, Columbia Univ. Press, 1963; and Price, Derek J. de Solla. Networks of scientific papers. The patterns of bibliographic references indicates the nature of the scientific front. *Science*, 149, 1965, pp. 510–555.

[3] Urquhart, D. J. How the science information problem concerns YOU. *Adv. Sci.*, 23, 1966–67, p. 496.

suggests that scientists should be trained to understand the structure of scientific literature, since it will not be possible to rely upon information officers, librarians and computers to supply particular needs. Scientists should be trained to help themselves, for only the specialist in a particular field can decide what he wants for his research. In searching for it he can sift the wheat from the chaff at source, and also be usefully sidetracked to pursue previously unthought of lines of investigation. A large percentage of research time should be devoted to a thorough search of the literature, the compilation of an annotated bibliography of relevant material up to date, after which his research should add usefully to the knowledge gained by his predecessors. Too much money provided for research is wasted in the duplication of work already accomplished, sometimes way back in the past. A thorough knowledge of the history of any subject is needed before fresh chapters can be added to that history. Conflicting but interesting ideas on the problems of information retrieval, documentation and automation are provided in *Communication in science.*[1]

Scientific publishing in France is in a poor state, and public funds are made available to subsidize a large number of journals, many of which have a very small circulation.[2] Amalgamations and cessations of publication are suggested, and in the case of astronomy and astrophysics, a new monthly journal was proposed to be published in English, French and German as the result of co-operation between Scandinavian, Dutch, German and French scientists. A list of 600 French journals arranged within eighteen classified groups, with brief annotations in French, English and Spanish was published in a second edition by L'Union Nationale des Éditeurs-Exportateurs de Publications Françaises as *Catalogue de publications françaises. Scientifiques-techniques-professionelles-agricolés, 1967–1968*, Paris, 1969.

A survey of investigations into the information-gathering habits of scientists has been made by B. C. Vickery,[3] who refers to the scattering of scientific subjects over a wide range of journals, and investigates the forms of literature used by scientists. He mentions the heavy concentration on recent literature, but that in botany, zoology and entomology there is a large percentage of citations in journals to literature over fifty years old. We also note that the preparation of theses and of review articles often entails research into historical material, and that historians of all aspects of science must refer to the early issues of journals in most spheres of research.

The number of scientific periodicals is constantly growing, and

[1] See De Rueck, Anthony, and Knight, Julie, *eds. Communication in science: documentation and automation*, 1967; see also Maddox, John. Tape or type? *Lancet*, 1968, ii, p. 1071; and Maddox, John. Journals and the literature explosion. *Nature*, 221, 1969, pp. 128–130.

[2] See French journals in flux. *Nature*, 220, 1968, p. 221.

[3] See Vickery, B. C. The use of scientific literature. *Lib. Assn. Rec.*, 63, 1961, pp. 263–269.

certain publishers in particular are issuing streams of new journals on every conceivable subdivision of science. Furthermore, eminent scientists lend their names to these journals as editors, or serve on the editorial boards, and the contents of the journals deteriorate after the first few issues. These periodicals are preserved mainly in libraries, yet these are asked exorbitant subscriptions in excess of those expected from individuals who purchase the journals "for their own use". This leads to certain irregularities, and librarians are extremely critical of these unorthodox and unethical tactics. Unfortunately, few librarians are permitted to decide which journals they house, and which subscriptions should be cancelled, but they can advise their committees. If librarians could take joint action over these, and certain other matters, unscrupulous publishers would have a greatly decreased market for their wares, and certain periodicals would vanish, without leaving serious gaps.

Possibly one solution would be the introduction of greater control over scientific periodicals by societies, universities and institutions, who would not necessarily publish the journals, but would sponsor them through reputable publishing houses. They might also control the output of their members by discouraging the publication of material adding nothing to our knowledge of the subject. This, however, is a matter for national, and even international consideration, but the current trend is towards the opposite direction. Publication is encouraged; the use of grants must be justified by the number of articles printed, and the work of a university department tends to be judged by the weight of its literary output. We are faced with a gigantic suicidal pact, in which the advancement of science is hampered by the inability of research workers to trace new facts among the masses of literature that must eventually suffocate those in pursuit of true knowledge.

CHAPTER IX

Scientific Bibliographies and Bibliographers

"Every investigation must begin with a bibliography, and end with a better bibliography."

GEORGE SARTON

Bibliographies are keys to the vast accumulations of literature that are constantly growing, and can be compared with compasses guiding the traveller through unexplored regions. The objective may be clear, but the path tortuous, and much valuable material can be gleaned on the journey. Bibliographies devoted to scientific subjects are numerous, yet few specialist subjects are covered adequately. Current bibliographies on subdivisions of science abound, yet they tend to overlap, they are very expensive, they are not published soon enough to incorporate really up-to-date information, and complete coverage remains a dream. One cannot even list within a single chapter the bibliographies devoted to science in all its aspects. General scientific bibliographies attempt to cover the wide range of literature in all countries; others are restricted to certain chronological periods, to specific languages, countries or special branches of the subject, while others are restricted to certain types of material such as incunabula, manuscripts, or articles in periodicals. A. C. Townsend[1] has provided an excellent short survey of guides to scientific literature, and a monumental attempt at a complete list of published bibliographies was made by Theodore Besterman in his *A world bibliography of bibliographies, and of bibliographical catalogues, calendars, abstracts, digests, indexes, and the like. . . . Fourth edition,* [*etc.*], five volumes, Lausanne, 1965–66. This consists of an alphabetical dictionary arrangement of subjects, down to 1963 inclusive, and covers 117,187 bibliographies arranged under 15,829 headings. Entries from this work have been adapted to form a separately published series of subject bibliographies, which include agriculture; medicine; physical sciences; biological sciences; and geography.

A subject index of current bibliographies, both separately published, and in periodicals, is issued three times a year, the last (December) issue being a cumulative volume. *Bibliographic index. A cumulative bibliography*

[1] Townsend, A. C. Guides to scientific literature. *J. Document.*, 11, 1955, pp. 73–78.

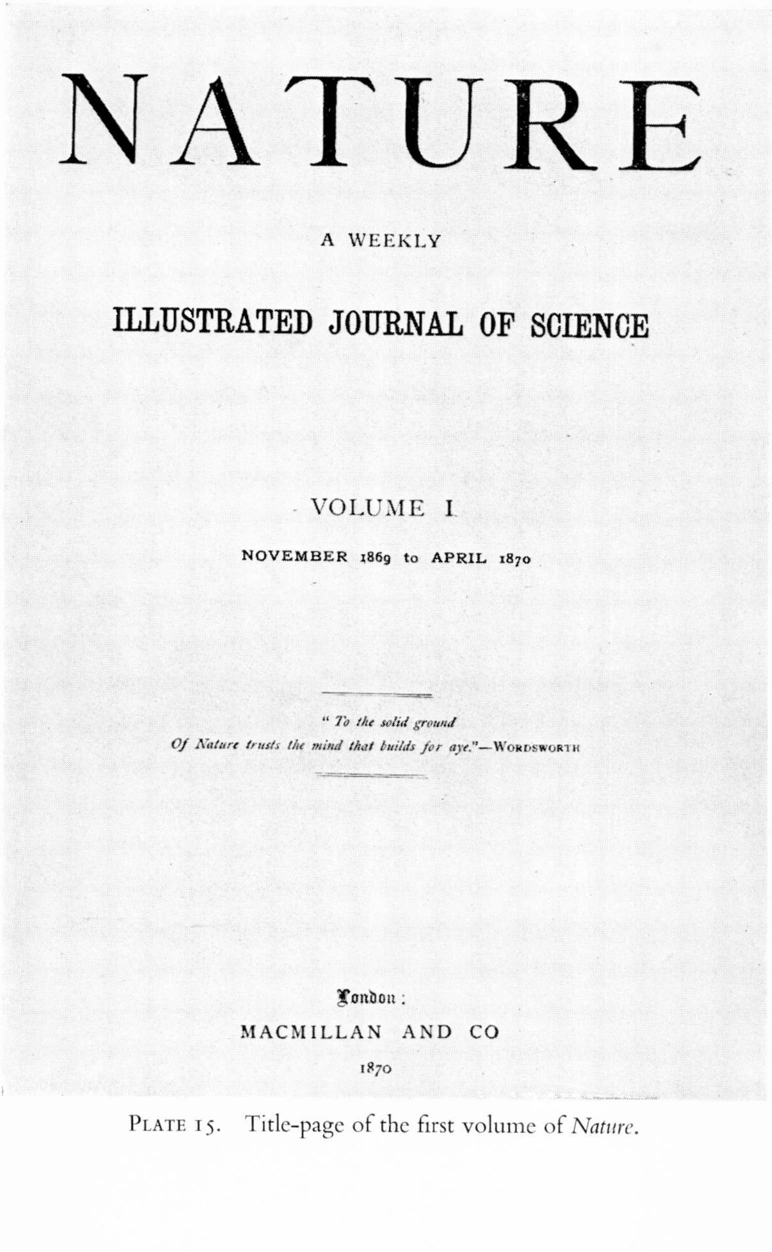

NATURE

A WEEKLY

ILLUSTRATED JOURNAL OF SCIENCE

VOLUME I

NOVEMBER 1869 to APRIL 1870

" To the solid ground
Of Nature trusts the mind that builds for aye."—WORDSWORTH

London:
MACMILLAN AND CO
1870

PLATE 15. Title-page of the first volume of *Nature*.

BIBLIOTHECA BOTANICA,

SIVE

CATALOGUS AUCTORUM ET LIBRORUM

omnium qui de Re Botanica, de Medicamentis ex Vegetabilibus paratis, de Re Rustica, & de Horticultura tractant,

A JOANNE-FRANCISCO SEGUIERIO

NEMAUSENSE DIGESTUS.

ACCESSIT

BIBLIOTHECA BOTANICA

J O. A N T. B U M A L D I,

SEU POTIUS

O V I D I I M O N T A L B A N I

B O N O N I E N S I S.

H A G Æ - C O M I T U M,

Apud J O A N N E M N E A U L M E,

M. D C C. X L.

Plate 16. Title-page of Jean François Seguier's *Bibliotheca botanica*, The Hague, 1740.

of bibliographies, New York, has been published since 1937, and contains a high percentage of scientific material. A survey of bibliographical tools was published in the April, 1967 issue of *Library Trends* devoted to "Bibliography: current state and future trends. Part 2." Among other subjects this contains chapters devoted to geography; history of science; general science; geology; biology; chemistry; physics; mathematics; agriculture; and medicine. Most of these sections are referred to elsewhere in this chapter, and should be consulted by those in search of a far more comprehensive survey of specialist subject bibliographies than can be attempted in one chapter.

Louise Noëlle Malclès'[1] *Les sources du travail bibliographique*, Volume 3, covers mathematics, astronomy, physics, mineralogy, chemistry, geology, anatomy, physiology, zoology, botany and medicine, listing histories, bibliographies, dictionaries, periodicals and reference books in these and other subjects. Providing a good bibliographical introduction, it is of particular interest to the non-specialist, and forms a basis for additional research.

A valuable work of reference exists in *Scientific, medical, and technical books published in the United States of America. A selected list of titles in print with annotations. Second edition. Books published to December 1956, [etc.]*, Washington, 1958. Edited by R. R. Hawkins, this was prepared under the direction of the National Academy of Sciences—National Research Council's Committee on Bibliography of American Scientific Books. It covers a very wide field, the contents of books are listed, entries are briefly annotated, there is an author and subject index, and also a directory of publishers. Approximately eight thousand titles are listed, covering books published since 1930 and still in print in 1958. It remains useful for publications covered by that period.

Gertrude Schutze has compiled a *Bibliography of guides to the S.T.M. literature, scientific-technical-medical*, New York, 1958, as Contributions to the S.T.M. Library Literature No. 1. This covers the period 1920–57, and the 654 citations include journal articles, papers, books, theses and similar material assisting one to evaluate basic sources and standard reference works. Entries are arranged under "general aids" and "topical aids", and a certain amount of foreign material is included.

Books listed under subject titles corresponding to article titles in the *McGraw-Hill Encyclopedia of science and technology* are contained in the *McGraw Hill Basic bibliography of science and technology*, [*etc.*], New York, [*etc.*], 1966. This contains about 8,000 entries, the emphasis being on basic books, these being additional to the reading lists in the *Encyclopedia*. The *Bibliographia scientiae naturalis Helvetica* is an excellent example of a

[1] Malclès, Louise Noëlle. *Les sources du travail bibliographique. Tome III. Bibliographies specialisées. (Sciences exactes et techniques.)*, [*etc.*], Geneva, Paris, 1958.

national bibliography of the sciences. Published annually since 1925, this title was assumed in 1948 when it was changed from *Bibliographie der Schweizerischen naturwissenschaftlichen und geographischen Literatur. Herausgegeben von der Schweizerischen Landsbibliothek*. This covers monographs and articles in periodicals published in Switzerland; concerning Swiss science; published abroad by Swiss authors; and by foreign writers working in Switzerland. Entries are arranged within nineteen classes, and prehistory, anthropology, geography and forestry are included among the usually recognized sciences.

The Science Museum Library publishes monthly a *List of accessions to the Library*, comprising complete catalogue references to all new books, journals, etc., and has also published a list of its current journals as *Current periodicals in the Science Museum Library. A hand-list compiled by H. D. Phippen*, ninth edition, London, 1965.

The Library issued also a *Bibliographical series* of specialist bibliographies. These appeared irregularly and were compiled in the Library in response to requests, over eight hundred being issued. The Royal Society was responsible for *A list of periodicals and bulletins containing abstracts published in Great Britain*, 1949, and *A list of British scientific publications reporting original work or critical reviews*, 1950. The Federation Internationale de Documentation has issued a directory of currently published abstracting and bibliographic services in its *Index bibliographicus*, the fourth edition of Volume 1 devoted to science and technology being published in 1959. Entries are arranged according to the Universal Decimal Classification with a subject and title index. Lists of bibliographies on particular subjects, of abstracting periodicals and catalogues of periodicals are examples of tools useful to those searching for requisite literature. The heading "bibliography" often includes various types of compilation which, strictly speaking, are not entitled to that honour. The *Bibliothecae*, for example, are seldom true bibliographies, but are either collections of writings or catalogues of collections. Similarly catalogues of libraries and personal collections are rarely exhaustive, yet they are of vital importance to scientific bibliography. Incomplete bibliographies are better than none at all, and perfection or complete coverage are goals most difficult of attainment.

Any attempt at tracing bibliographical information must commence with the most likely specialized source, proceeding back through more general subjects. For example, an inquirer desiring information in the literature devoted to ants would first consult a bibliography on ants, thence proceeding to bibliographies on entomology, zoology and natural history. For our purpose it is more logical to begin with general bibliographies, and then to deal with the main branches of science. Generally speaking, bibliographies contained in periodicals are omitted, and only the main branches of science are included, but sources of additional information are mentioned where possible. Similarly it has been found impossible to include many items covering restricted periods or localities, except where these are of outstanding significance.

An early subject index to articles appearing in the publications of scientific societies during the eighteenth century was compiled by Jeremias David Reuss as *Repertorium commentationum a societatibus litterariis editorum*, [*etc.*], Göttingen, 1801–21. This was published in sixteen volumes, Volume 1 covering natural history; Volume 2 deals with botany and mineralogy; Volume 3 is devoted to chemistry; Volume 4 covers physics; and Volumes 10 to 16 deal with science and medicine. This work emphasizes the need for looking beyond the specialized bibliographies, and consulting the larger, more general tools that cover extensive fields. A reprint was issued in 1961.

An example of the *Bibliothecae* above mentioned is E. A. Zuckold's *Bibliotheca historico-naturalis, physico-chemica et mathematica*, published at Göttingen in 1852. Another useful source of bibliographical information is Johann Christian Poggendorff's *Biographisch-literarisches Handwörterbuch zur Geschichte der exacten Wissenschaften*, [*etc.*], now in its seventh volume, which is appearing in parts, Berlin, 1858 onwards. Reprints of certain volumes have been published in Leipzig and Amsterdam. Volumes 1 and 2 cover the period up to 1858; Volume 3, 1858–83; Volume 4, 1883–1904; Volume 5, 1904–22; Volume 6, 1923–31; and Volume 7a, 1932–53. This was published as four volumes (in five), 1956–62. Poggendorff covers many branches of science (astronomy, chemistry, mathematics, physics, etc.), and is arranged alphabetically by scientists, providing brief biographies, with lists of publications, including both books and articles in periodicals. Hans Salié[1] has provided a brief biography of Johann Christian Poggendorff (1796–1877), who was born in Hamburg and trained as a pharmacist. He became editor of *Annalen der Physik und Chemie* in 1824, and held the post for fifty-two years. He lectured on the history of physics at Berlin University, and was the author of *Geschichte der Physik*, [*etc.*], Leipzig, 1879, which was reprinted in 1964. Salié gives a historical description of Poggendorff's *Handwörterbuch* and its continuation under various compilers, including proposals for its future development.

The Royal Society of London was responsible for the compilation and publication of the *Catalogue of scientific papers* which was published in four series in nineteen volumes at London (4th series, Cambridge), from 1866 to 1925. This consists of an author index of articles in over fifteen hundred scientific periodicals published between 1800 and 1900. A subject index (three volumes in four) appeared between 1908 and

[1] Salié, Hans. Poggendorff and *Poggendorff*. Translated by Ralph E. Oesper. *Isis.*, 57, 1966, pp. 389–392.

1914, but covers only mathematics, mechanics and physics, the remainder not having been published. The first series (Volumes 1 to 6) covers the years 1800 to 1863; the second series (Volume 7 to 8) covers 1864 to 1873; series three (Volumes 9 to 11) covers 1874 to 1883; Volume 12 is a supplementary volume devoted to 1800–83; and the fourth series (Volume 13 to 19) is devoted to the years 1884 to 1900. A reprint of twenty-three volumes bound in twenty-two was announced in 1969. This *Catalogue* was continued by the *International catalogue of scientific literature*, published by the Royal Society for the International Council of Scientific Workers. This was published in fourteen annual issues, each consisting of seventeen sections devoted to branches of science, covering the years 1901 to 1914, and published 1902–21. Books, and articles in periodicals are listed, and the subject subdivisions include mathematics, mechanics, physics, chemistry, astronomy, meteorology, geology, geography, palaeontology, biology, botany and zoology.[1] The volumes are arranged alphabetically by authors, with classified subject indexes. A reprint in thirty-two volumes instead of the original 238 was announced in 1969.

The library of Sir William Osler (1849–1919) is chiefly of medical significance, and has been considered elsewhere,[2] but the catalogue of the collection, published as *Bibliotheca Osleriana: a catalogue of books illustrating the history of medicine and science, collected, arranged and annotated by Sir William Osler, and bequeathed to McGill University*, Oxford, 1929, contains many items of purely scientific interest. About 7,600 bound volumes are listed therein, and the annotations render the catalogue eminently readable and of increased historical importance. Lloyd G. Stevenson contributed an introduction to the reprint of the *Bibliotheca* published by McGill University, Montreal, 1969. In the same category we must mention a bibliography by Fielding H. Garrison (1879–1935), which was revised by Leslie T. Morton and published as *A medical bibliography (Garrison and Morton): an annotated check-list of texts illustrating the history of medicine. . . . Third edition*, London, 1970. This contains 7,534 entries, classified and annotated, with adequate author and subject indexes. Biology, anthropology, microscopy, zoology and comparative anatomy, anatomy and physiology are among the subjects included that are of particular interest to historians of science. The *Bibliotheca Walleriana. The books illustrating the history of medicine and science collected by Dr. Erik Waller and bequeathed to the Royal University of Uppsala. A catalogue compiled by Hans Sallander*, two volumes, Stockholm, 1955, contains sections devoted to natural sciences; chemistry and alchemy; physics, botany,

[1] The Zoology volumes of the *International Catalogue* being the appropriate years of *The Zoological Record* (see p. 321).

[2] Thornton, pp. 288–289.

zoology, etc., and lists over 150 incunabula. These were collected by Eric Waller (1875–1955), a native of Sweden.

A fourth edition of Aslib's *Select list of standard British scientific and technical books* appeared in 1952, containing about one thousand textbooks classified by the Universal Decimal Classification. This was followed by *British scientific and technical books, 1935–52,* [*etc.*], London, 1956, and *British scientific books, 1953–57. A select list of recommended books published in Great Britain and the Commonwealth, edited by L. J. Anthony*, London, 1960, also published on behalf of Aslib, and based on the *Aslib Book List.* This was published quarterly from 1937, has appeared monthly since 1948, and represents a select list of recommended scientific and technical books grouped by main subjects, with author and subject indexes.

Although compiled as a list of books in a circulating library, the following represents one of the most useful, up-to-date bibliographical tools available to the scientist. The *Catalogue of Lewis's Medical, Scientific and Technical Lending Library. . . . New edition revised to Dec. 31, 1963,* Part I, Authors and titles, 1965, Part II, Classified index of subjects, 1966, is maintained up to date by means of quarterly supplements. This does not include the names of publishers, excludes paperback publications, and does not include material in foreign languages.

The American Association for the Advancement of Science has published since 1965 *Science books: a quarterly review*, Washington, which reviews critically textbooks and reference works in the pure and applied sciences. The School of Library Service at Columbia University has issued *Guide to the literature of science: for use in connection with the course in science literature. Prepared by Thomas P. Fleming. Second edition*, New York, 1957, a cyclostyled production listing the most important reference tools, book selection guides, encyclopaedias, dictionaries, standard texts, handbooks, histories, review publications, bibliographies, indexes and abstracts in each of the main subdivisions of science.

Harrison D. Horblit's *One hundred books famous in science*, New York, 1965, was published by the Grolier Club, and contains entries ranging from Pliny's *Historia naturalis*, 1469, to Einstein's *General theory of relativity*, 1916. Arranged alphabetically by authors, with a chronological index, entries have brief annotations, with collations and invaluable illustrations. A similar work containing 424 numbered entries was edited by John Carter and Percy H. Muir as *Printing and the mind of man. A descriptive catalogue illustrating the impact of print on the evolution of western civilization during five centuries*, London, [*etc.*], 1967. This contains extensive, scholarly annotations to each item, with useful biographical and bibliographical information, and many of the entries are of scientific interest.

A useful bibliography for early material is published as *L'Année philologique; bibliographie critique et analytique de l'antiquité gréco-latine.*

Première année 1927. . . . Bibliographie des années 1924–26 —, Paris, 1928 —. This includes a section "Sciences et métiers" in each annual volume, forming a useful bibliography of books and articles in periodicals on ancient science. Entries are classified as generalia (including general histories of science); astronomie; mathématiques; chimie et physique; sciences naturelles; médicine; géographie; exploitation du sol, etc. Another valuable current bibliography of French scientific publications was issued by the Ministère des affaires étrangères as *Informationes scientifiques*, Paris, 1956, and continued as *French Science News*, Paris, 1957–69. This was first issued six-monthly, but from 1959 appeared quarterly. Each issue contains notes on research, university news, congresses, films, etc., together with details of books published in France, entries being arranged by broad subjects, with brief accounts in English of each book. A number of French publishers presented a selection of scientific and technical books in French published between 1950 and 1960 in a publication by Cercle de la Librarie entitled *Catalogue collectif des livres françois de science et techniques, 1960–1962*, Paris, (1963). In this the books are classified methodically by subject-matter, followed by an index of key words in alphabetical order. There is an analytical index, and an alphabetical table of authors and translators. The Cercle de la Librarie has also published *Catalogue collectif des livres français de médecine et biologie, 1955–1965*, Paris, a classified list with an author index, and giving prices.

François Russo has compiled a selective bibliography of particular value to historians of science. Entitled *Éléments de bibliographie de l'histoire des sciences et des techniques. Deuxième édition*, [*etc.*], Paris, 1969, this contains a section devoted to the bibliography of the study of the history of science, including sections on methods; biographies of historians of science; societies and institutions arranged alphabetically by country; congresses, libraries and museums. The second part is concerned with printed sources, encyclopaedias, dictionaries, biographical dictionaries, bibliographies, catalogues of manuscripts, lists of scientific periodicals, followed by the actual titles and bibliographical details of periodicals on the history of science, and collective biographies. The various historical periods are then covered from antiquity, the Middle Ages, and the sixteenth to the twentieth centuries.

An invaluable guide to current literature is provided by the "Critical bibliography of the history of science and its cultural influences", published periodically in *Isis*. Compiled at the request of the Deutsche Gesellschaft für Geschichte der Medizin, Naturwissenschaft und Technik, volume I of *Index zur Geschichte der Medizin und Biologie* was edited by Walter Artelt, and covers 1945–48, having been published in 1953. Volume II, edited by Johannes Steudel, Wilfried Ricker and Claus Nissen, covers 1949–51/52, and was published in Munich, [*etc.*], 1966.

In this volume, the largest of the four sections is that devoted to medicine, but biology is adequately represented.

Marianne Winder[1] has published a bibliography of German astrological works printed between 1465 and 1600, providing locations of those located in London libraries. This includes writings in German, by German-speaking people, and translations of foreign works into German. The various editions of a work are arranged chronologically, and there are indexes of authors, printers and publishers.

An excellent guide to the literature of astronomy and astrophysics has been compiled by D. A. Kemp in the Macdonald Bibliographical Guides series. Entitled *Astronomy and astrophysics. A bibliographical guide*, London, 1970, this lists the major international works on these subjects, the authoritative textbooks, survey articles, progress reviews, reports of conferences and symposia, and papers reporting significant developments. Relevant specialized bibliographies are listed, and there are brief annotations. Bibliographies of astronomy are not numerous, and the next item to be mentioned is very specialized, covering only publications printed in Germany up to 1630. This is Ernst Zinner's *Geschichte und Bibliographie der astronomischen Literatur in Deutschland zur Zeit der Renaissance*, Leipzig, 1941, which contains 5,236 numbered entries. A similar bibliography listing 2,000 works printed up to the year 1650 is entitled *Astronomische chronologische Bibliographie. Einleitung zur mathematischen Bücherkenntniss*, [*etc.*], Breslau, 1784–98 [by Johann Ephraim Scheibel]. This was also published under the title *Astronomische Bibliographie*. Johann Friedrich Weidler's *Bibliographia astronomica* was published at Wittenberg in 1755 and contains 1,250 items. A French bibliography listing 5,000 entries was compiled by Joseph Jérôme Le Français de La Lande, and published at Paris in 1803 as *Bibliographie astronomique; avec l'histoire de l'astronomie depuis 1781 jusqu'à 1802*. This is arranged in chronological order, with author and subject indexes. On a larger scale we have a work by Jean Charles Houzeau (born 1820) and Albert Lancaster, published at Brussels from 1882 to 1889 as *Bibliographie générale de l'astronomie, ou catalogue méthodique des ouvrages, des mémoires et des observations astronomiques publiés depuis l'origine de l'imprimerie jusqu'en 1880*, two volumes [in three]. Volume 1, Part 1, is dated 1887; Part 2 bears the date 1889; and Volume 2 was published in 1882. Houzeau was director of the Observatory at Brussels, and his work is classified, with an author index. A list of periodicals is included, as is also a bibliography of astrology. In 1964 a new edition of this was published as two volumes in three, with additional material by D. W. Dewhirst. An adequate index of authors, a new introduction,

[1] Winder, Marianne. A bibliography of German astrological works printed between 1465 and 1600, with locations of those extant in London libraries. *Ann. Sci.*, 22, 1966 [1967], pp. 191–220.

short biographies of Houzeau and Lancaster, and a re-arrangement of the text facilitates reference to this primary bibliographical source. Two astronomical periodicals are also worthy of note from the bibliographical viewpoint. The first, *Jahrbuch der Astronomie und Geophysik*, [*etc.*], was published at Leipzig in twenty-three volumes from 1890 to 1912, and the twenty-fourth and final volume was entitled *Klein's Jahrbuch der Astronomie*, [*etc.*]. The *Astronomischer Jahresbericht*, Berlin, has appeared since 1899.

The bibliography of mathematics is adequate, and includes some interesting examples of the art. A. J. Lohwater[1] has provided a guide to current sources of information, and John E. Pemberton[2] has published a useful book under the title *How to find out in mathematics*, 1963. This deals with the organization of mathematical information; the use of libraries; inter-library co-operation; photocopying; catalogues; dictionaries; encyclopaedias; periodicals; abstracts; societies; history and biography among other subjects. Two other guides to the literature of the subject must be mentioned. G. A. Miller's *Historical introduction to mathematical literature*, New York, 1921, was first published in 1916, the author being professor of mathematics at the University of Illinois. The work is actually a general history of mathematics, but contains much biographical and bibliographical information. The second publication, Nathan Grier Parke's *Guide to the literature of mathematics and physics, including related works on engineering science. Second revised edition*, New York, 1958, is divided into two parts, the first dealing with the principles of reading and study; dictionaries; encyclopaedias, textbooks, etc.; tracing the literature; and periodicals. Part 2 deals with the literature of mathematics and physics, and contains over 5,000 entries.

The earliest bibliography of mathematics was that by Cornelius à Beughem, a magistrate and librarian at Emmerich, who compiled several other bibliographies.[3] His *Bibliographia mathematica et artificiosa novissima perpetuo continuanda, seu conspectus primus catalogi librorum mathematicorum . . . quotquot currente hoc semiseculo . . . in quavis lingua . . . typis prodierunt*, Amsterdam, 1685 and 1688, lists 3,000 entries. In the next century appeared *Einleitung zur mathematischen Bücherkenntnis*, [by Johann Ephraim Scheibel], published in eighteen parts at Breslau between 1769 and 1778, and containing 10,000 items. A four-volume work by Abraham Gotthelf Kästner, entitled *Geschichte der Mathematik. Geschichte der Künste und Wissenschaften*, Göttingen, 1796–1800, lists 5,000 mathematical works. Almost simultaneously appeared Friedrich Wilhelm August

[1] Lohwater, A. J. Mathematical literature. *Library Trends*, 15, 1967, pp. 852–867.
[2] Pemberton, John E. *How to find out in mathematics: a guide to sources of mathematical information arranged according to the Dewey Decimal Classification*, Oxford, 1963.
[3] See Thornton, p. 243.

Murhard's *Literatur der mathematischen Wissenschaften*, five volumes, Leipzig, 1797–1805, containing 10,000 entries. Several other mathematical works of bibliographical interest were published in the nineteenth century, the first two, by Johann Wolfgang Müller, being entitled *Auserlesene mathematische Bibliothek*, Nuremberg, 1820, and *Repertorium der mathematische Bibliothek*, two volumes, Augsburg, [*etc.*], 1822–25. Only the first part was published of J. Rogg's *Handbuch der mathematischen Literatur vom Anfange der Buchdruckerkunst. . . . Erste Abtheilung, welche die arithmetischen und geometrischen Wissenschaften enhält*, Tübingen, 1830, but it was continued by Ludwig Adolph Sohncke in his *Bibliotheca mathematica. Verzeichniss der Bücher über die gesammten Zweige der Mathematik . . . welche in Deutschland und dem Auslande von Jahre 1830 bis Mitte des Jahres 1854 erschienen sind*, Leipzig, 1854. This was also issued with a title-page in English.

James Orchard Halliwell (afterwards Halliwell-Phillipps) (1820–89) was a literary figure keenly interested in Shakespeareana and in mathematics. He made several important collections of books and pamphlets, which he sold at intervals, and his library of early mathematical and astronomical manuscripts was disposed of in 1840. Halliwell had previously published a catalogue of these in 1839, and he then appears to have concerned himself with literary work and controversy. He published *Rara mathematica* in 1839, and edited for the Historical Society of Science, which he founded in 1841, *A collection of letters illustrative of the progress of science in England from the reign of Queen Elizabeth to that of Charles the Second*, London, 1841. This was reprinted in 1965 together in one volume with *Popular treatise on science written during the middle ages in Anglo-Saxon, Anglo-Norman and English. Edited from the original manuscripts by Thomas Wright*. Halliwell was a precocious, rather notorious figure, and further information on his career is provided by A. N. L. Munby.[1]

A fascinating work, limited in usefulness by its small size, was compiled by a professor of mathematics at University College, London, Augustus De Morgan (1806–71),[2] who also wrote several mathematical works of significance. His bibliographical venture was published as *Arithmetical books from the invention of printing to the present time: being brief notices of a large number of works drawn up from actual inspection*, London, 1847, and lists 500 entries. No entry was taken from a catalogue, but every book was handled by the author, who contributed a lengthy introduction. Entries are in chronological order, and arranged by place and date of publication, author, title and size being indicated. Full annotations, the

[1] Munby, A. N. L. *The history and bibliography of science in England: the first phase, 1833–1845. To which is added a reprint of the Catalogue of scientific manuscripts in the possession of J. O. Halliwell, Esq.*, Berkeley, California, 1968.

[2] See also p. 114.

best feature of the work, are included, as is also an author index. This book was reprinted in 1966 with an introduction by A. Rupert Hall which throws light upon the character of De Morgan, a keen collector of books and a pioneer bibliographer.

Peter J. Wallis has issued a typescript "Check list of British mathematical writers up to 1850", Newcastle, 1967, with additions and corrections 1968, which is basically an interim author index to "A bibliography of British mathematics and its applications up to 1850" which is in preparation. It is estimated that about 5,000 mathematicians will be included and the completion of the work will be welcomed by historians of mathematics.

Although limited in scope, being confined to Germany and Italy respectively, the following two works are worthy of note. A. Erlecke's *Bibliotheca Mathematica. Systemat. Verzeichniss d. b. 1870 in Deutschland auf d. Gebieten der Arithmetik, Algebra, Analysis, Geometrie . . . Astronomie . . . Astrologie*, Halle, 1873; and Pietro Riccardi's *Biblioteca matematica italiana dalla origine della stampa ai primi anni del secolo XIX*, [*etc.*], Modena, 1893. The latter, a monumental work, is arranged alphabetically, with annotations, an appendix and chronological tables. Another Italian bibliography, *Bibliografia sui fondamenti della geometria*, by R. Bonola, was published in 1899.

Originally published as a series of articles,[1] D. Bierens de Haan's *Bibliographie néerlandaises historique-scientifique des ouvrages importants dont les auteurs sont nés aux 16e, 17e et 18e siècles sur les sciences mathématiques et physiques, avec leur applications* was published in Rome in 1883, and an unchanged reprint was published in Nieuwkoop in 1960. This contains 5,651 entries, with useful bibliographical descriptions of every work of importance.

The first important mathematical bibliography to appear in the twentieth century was E. Wölffing's *Mathematischer Bücherschatz*, [*etc.*], Leipzig, published in 1903. This is a classified bibliography of the most important literature on pure mathematics published during the nineteenth century, but excludes periodical literature. In 1908 appeared David Eugene Smith's *Rara arithmetica. A catalogue of the arithmetics written before the year MDCI, with a description of those in the library of George Arthur Plimpton of New York*, Boston and London, listing nearly 1,200 items. Full descriptions, with useful annotations and numerous facsimiles, are included, and Part 2 is devoted to manuscripts, indexes of dates, names, places and subjects. G. A. Plimpton died in 1936, and his collection is now in the Low Memorial Library at Columbia University. An addenda to the catalogue was published by David Eugene Smith as *Addenda to Rara arithmetica which described in 1908 such European arithmetics printed before*

[1] In *Bullettino di Bibliografia e de Storia delle Scienze Matematiche e Fisiche*, 1881–82.

1601 as were then in the library of the late George Arthur Plimpton, Boston and London, 1939.

Felix Müller's *Führer durch die mathematische Literatur*, Leipzig, 1909, places emphasis on the history of the subject, with reference to the more important books and articles in periodicals.

A chronological catalogue with extensive author and subject indexes was published for the University of St. Andrews, and was compiled by Duncan M. Y. Sommerville as *Bibliography of non-Euclidean geometry, including the theory of parallels, the foundations of geometry, and space of* n *dimensions*, London, 1911. This lists over 4,000 publications, but the entries are very brief and there are no annotations.

Two works of particular American appeal complete our books on mathematical bibliography. L. G. Simon's *Bibliography of early American textbooks on algebra published in the Colonies and the United States through 1850, together with a characterization of the 1st edition of each work*, New York, 1936; and Louis C. Karpinski's *Bibliography of mathematical works printed in America to 1850*, Ann Arbor and London, 1940. In the latter, most of the major items have reproductions of their title-pages in facsimile, and there is an historical introduction, commencing with the British background. The work lists 1,092 titles printed in the Americas up to 1850, with a total of 2,998 titles and subsequent editions. Entries are arranged chronologically by first editions, subsequent editions being listed with these. Locations of items are given, but full bibliographical descriptions are not provided. Indexes of authors, subjects, non-English and Canadian works, and of printers and publishers, complete this remarkable piece of bibliographical research.

Collections of classic papers are particularly useful when they contain translations of items hitherto available only in foreign languages. Although not to be classified as strictly bibliographical, they sometimes contain valuable information on the texts included. One such selection is D. J. Struik's *A source book in mathematics, 1200–1800*, Cambridge, Mass., 1969, and a similar work on a larger scale has been provided by James R. Newman in *The world of mathematics: a small library of the literature of mathematics from A'h-mose to Albert Einstein*, four volumes, New York, 1956.

Periodical bibliographies of mathematics include *Bulletin de Bibliographie, d'Histoire et de Biographie mathématiques*, eight volumes, 1855–62; *Bullettino di Bibliografia e di Storia delle Scienze Matematiche e Fisiche*, founded and edited by Baldassare Boncompagni (1831–94) and published at Rome from 1868 to 1887, a reprint of which was announced in 1964; *Revue semestrielle des Publications Mathématiques*, Amsterdam and Leipzig, 1893–1934; *Jahrbuch über die Fortschritte der Mathematik*, issued from Berlin from 1868 to 1934; *Zentralblatt für Mathematik*, Berlin, 1931 to date; and

Mathematical Reviews, published by the American Mathematical Society since 1940, an international organ of high repute, and the most significant reference tool devoted to the subject.

Geology and mineralogy are catered for by the following bibliographical tools, listed briefly: Brian Mason's *The literature of geology*, New York, 1953, is divided into two parts, the first being a classified list of the major general publications, reference books and serials, followed by entries arranged under the main subdivisions of geology. Part 2 lists by country the major official publications, geological surveys, maps, and publications of the international geological congresses from 1878 to 1952. Richard M. Pearl's *Guide to geologic literature*, New York, [*etc.*], 1951, contains information on libraries, library catalogues, indexing, bibliographies, abstracts, periodicals, books, theses, manuscripts, maps and other material. Carl Friedrich Wilhelm Schall's *Oryktologische Bibliothek*, Weimar, 1787, lists 1,500 items, and was followed by a second edition bearing the title *Anleitung zur Kenntniss der besten Bücher in der Mineralogie und physikalischen Erdbeschreibung*, Weimar, 1789 which contains about seventy extra items. The following bibliographies are also useful guides to the literature: Christian Keferstein's *Geschichte und Literatur der Geognosis*, Halle, 1840, and J. D. Dana's *Bibliography of mineralogy*, published in 1881.

An important list of nearly 4,000 bibliographies of geology published between 1726 and 1895 exists in Emmanuel de Margerie's *Catalogue des bibliographies géologiques*, [*etc.*], Paris, 1896. This in three parts, the first dealing with general bibliographies, the second with bibliographies of special subjects, while the third lists personal bibliographies and obituaries. An author index is provided. This work was supplemented by Edward B. Mathews' *Catalogue of published bibliographies in geology*, 1896–1920, Washington, 1923 (*Bulletin of the National Research Council*, 6, v, No. 36). This is arranged in exactly the same way as Margerie's *Catalogue*, and contains 3,699 titles. Michel Felix Mourlon was responsible for the *Bibliographia geologica*, [*etc.*], Brussels, of which Series A, 1–9, covers the years 1896–1906, and Series B, 1–7, the years 1897–1904. A guide to geological maps was published by A. Morley Davies as *Local geology; a guide to sources of information on the geology of the British Isles*, London, 1927.

Periodical bibliographies are represented by the *Revue de Géologie et des Sciences Connexes*, Liége, published monthly by the Société Géologique de Belgique since 1920. This contains signed abstracts of books and articles in periodicals, with author and subject indexes. The *Bibliographie des Sciences géologiques* has appeared since 1923, and is published in Paris. An important bibliography of periodicals in the geological sciences has been published by Josef Lomský as *Periodica geologica palaeontologica et mineralogica*, Prague, 1959, which lists 3,582 serial titles, and also contains a classified survey by subject matter.

An international *Annotated Bibliography of Economic Geology* is published half-yearly at Urbana, Illinois, and the first issue in 1929 covers the literature of the previous year. The Geological Society of America published at Washington a *Bibliography and Index of Geology exclusive of North America* from 1934-68, but from 1969 this was entitled *Bibliography and Index of Geology*, covering the world literature. This followed upon *Geological literature added to the Geological Society's Library*, [London], published from 1894 to 1934. The literature of North America is covered by John Milton Nickles' *Geological literature of North America, 1785–1928*, three volumes (*U.S. Geological Survey Bulletins* 746, 747 and 823), 1923–31, continued in biennial supplements as *Bibliography of North American geology, 1929–30*, [*etc.*] (*U.S. Geological Survey Bulletins* 834, *etc.*).[1]

The Mineralogical Society, London, has published since 1922 *Mineralogical Abstracts*, the first volume covering the year 1920. Entries are in classified order.

It has been suggested by Stella Keenan[2] that physics literature had increased by 77 per cent between 1960 and 1968, and that it doubles every seven years. A survey of physics literature is provided by Robert H. Whitford in *Physics literature: a reference manual. Second edition*, Metuchen, 1968, which provides information on bibliographies; journals; the history of the subject, including biographical sources; followed by experimental mathematical, educational and terminological approaches. A shorter and more selective book intended as an instruction manual for beginners was published by B. Yates as *How to find out about physics. A guide to sources of information arranged by the Decimal Classification*, Oxford, [*etc.*], 1965. This covers careers in physics, books, documents, periodicals, abstracts, societies and other subjects. Bibliographies of interest to physicists are innumerable, and the more general ones include Gustav Theodor Fechner's *Repertorium der Experimentalphysik, enthaltend eine vollständige Zusamenstellung der neuern Fortschritte dieser Wissenschaft*, three parts, Leipzig, 1832, listing 2,500 items; Sir Francis Ronalds' *Catalogue of books and papers relating to electricity, magnetism and the electric telegraph*, London, 1880; *A bibliography of electricity and magnetism, 1860–1883, with special reference to electro-technics*, London 1884, by Gustav May, of which a German edition was published in Vienna in the same year. Paul Fleury Mottelay's *Bibliographical history of electricity and magnetism, chronologically arranged. Researches into the domain of the early sciences, especially from the period of the revival of scholasticism, with biographical and other accounts of the most distinguished natural philosophers throughout the Middle Ages*, [*etc.*],

[1] See also Hawkes, H. E. Geology. *Library Trends*, 15, 1967, pp. 816–828.

[2] Keenan, Stella. Abstracting and indexing services in the physical sciences. *Library Trends*, 16, iii, 1968, pp. 329–336. See also Atherton, Pauline. Physics. *Library Trends*, 15, 1967, pp. 847–851; and Urquhart, D. J. Physics abstracting—use and users. *J. Document.*, 21, 1965, pp. 113–121.

London, 1922, commences at 2637 B.C. and finishes with Michael Faraday, and is a chronological history, with annotations, numerous references, illustrations and a very full index. A list of more recent bibliographies in physics has been compiled by Karl K. Darrow as *Classified list of published bibliographies in physics, 1910–1922*, 1924 (*Bulletin of the National Research Council*, Vol. 8, v, No. 47). This is classified, with a subject but no author index. Dwight E. Gray and Robert S. Bray[1] have provided details of abstracting and indexing services of physics interest, revealing a wide range of journals in this sphere. The major, general periodicals are *Physics Abstracts*, issued monthly by the Institution of Electrical Engineers as Section A of *Science Abstracts*, London, and published since 1898; *Die Fortschritte der Physik*, Brunswick, published between 1845 and 1918 and continued since 1920 in conjunction with *Beiblätter zu den Annalen der Physik*, Leipzig, 1877–1919, as *Physikalische Berichte*; and since 1920 the Société Française de Physique has published *Le Journal de Physique et le Radium: Revue Bibliographique*, which appears from Paris. *Current Papers in Physics* has been published twice-monthly since 1966, and *Current Contents: Physical Sciences* has been issued weekly by the Institute for Scientific Information, Philadelphia since 1961. *Nuclear Science Abstracts*, published by the Atomic Energy Commission since 1948, appears monthly with cumulative quarterly indexes, and a Report Number Index issued quarterly. *Electrical and Electronics Abstracts*, published monthly since 1965 by the Institution of Electrical Engineers, London, with a companion serial *Current Papers in Electrotechnology* (1967–68), published since 1969 in conjunction with the Institute of Electrical and Electronics Engineers, New York as *Current Papers in Electrical and Electronic Engineering*.

The American Institute of Physics has published *The periodical literature of physics: a handbook for graduate students*, New York, [1961?], a brief but useful guide covering mainly English language material, but also mentioning the most important journals in other languages. Although borderline in scope, and restricted in coverage, Charles Kenneth Moore and Kenneth John Spencer's *Electronics: a bibliographical guide*, London, 1961, contains much of interest to physicists. It covers the literature of 1945 to 1959, with some historical items, standard textbooks and bibliographies of earlier periods. Over 3,300 references are included in the sixty-eight sections into which this admirable bibliography is divided. Volume 2 of this work covers 1959 to 1964 and was published in 1965. Reprints of outstanding papers in the history of physics have appeared in W. F. Magie's *A source book in physics*, Cambridge, Mass., 1963, containing extracts covering 1600 to 1900, with biographical and explanatory notes; and in *Classical scientific papers. Physics. Facsimile*

[1] Gray, Dwight E., and Bray, Robert S. Abstracting and indexing services of physics interest. *Amer. J. Physics.*, 18, 1950, pp. 274–299. See also Keenan, Stella. *Op. cit.*

reproductions of famous scientific papers, with an introduction by Stephen Wright, London, [1964].

Chemistry is extremely rich in valuable bibliographical tools, but first we must mention several modern guides to the literature of the subject. The first, by E. J. Crane, Austin M. Patterson and Eleanor B. Marr,[1] was originally written as the result of the two first-named authors' experiences as editors of *Chemical Abstracts*, and contains numerous features of special interest. *A guide to the literature of chemistry* contains sections devoted to books; periodicals; patents; government publications; trade literature; other sources, including biographies, bibliographies, museum collections, reviews, theses, unpublished material, etc.; indexes; libraries; procedure in literature searches; appendixes on literature relating to chemical literature; symbols, abbreviations and standards; libraries; a bibliography of lists of periodicals; scientific and technical organizations; periodicals, including a list of periodicals discontinued before 1910; dealers and publishers.

Byron Amery Soule's *Library guide for the chemist*[2] deals with cataloguing and classification, periodicals and abstracts, dictionaries and encyclopaedias, patents and similar subjects. Melvin Guy Mellon's[3] *Chemical publications, their nature and use* also covers periodicals, institutional publications, patents, bibliographies, reference books and textbooks, libraries and indexes, etc., and provides lists of general scientific journals, of general chemical journals and of specialized periodicals. A British guide to the literature of chemistry by G. Malcolm Dyson[4] contains details of dictionaries, encyclopaedias, periodicals, abstract journals and reference books. An appendix mentions some old and obsolete journals (pp. 94–104), giving details of the number of volumes published, with dates and other information, and guidance is provided in tracing required literature.

The American Chemical Society has published *Searching the chemical literature. . . . Based on papers presented . . . at national meetings from 1947 to 1956*, Washington, 1961, consisting of thirty-one papers covering subjects including the use of *Chemical Abstracts*; language problems in literature searching; searching the older chemical literature, which contains a valuable list of obsolete chemical journals of the nineteenth century; searching less familiar periodicals, including a selective chronological list of early periodicals; abbreviations; theses; house organs; and patents.

[1] Crane, E. J., Patterson, Austin M., and Marr, Eleanor B. *A guide to the literature of chemistry. Second edition*, New York, London, 1957.
[2] Soule, Byron Amery. *Library guide for the chemist*, New York, 1938.
[3] Mellon, Melvin Guy. *Chemical publications, their nature and use. . . . Third edition*, New York, 1965. See also the same authors' *Searching the chemical literature*, Washington, 1964; and (with Power, Ruth T.) Chemistry. *Library Trends*, 15, 1967, pp. 836–846.
[4] Dyson, G. Malcolm. *A short guide to chemical literature, 2nd edition*, London, [*etc.*], 1958.

Harold Robert Malinowsky's *Science and engineering reference sources. A guide for students and librarians*, Rochester, New York, 1967 includes chemistry, and the following represent examples of reproductions of historical texts. A. J. Ihde and W. F. Kieffer, *Selected readings in the history of chemistry*, Easton, Penn., 1965, published by the *Journal of Chemical Education*; and Henry M. Leicester and Herbert M. Klickstein, *A source book in chemistry*, Cambridge, Mass., 1952. A survey of existing histories of chemistry is provided by Eduard Farber,[1] and John L. Thornton and R. I. J. Tully[2] have written a chapter on the history and biography of chemistry. This is contained in *The use of chemical literature. Second edition. Edited by R. T. Bottle*, London, 1969, a multiple author guide to the use of libraries; primary sources; abstracts and information retrieval; translations and their sources; background information; standard tables of physical data; patents; and Government and trade publications. Sections are also devoted to inorganic chemistry; nuclear chemistry; Beilstein's *Handbuch*; polymer science; and the practical use of chemical literature. This excellent volume in the Information Sources for Research and Development series provides an invaluable guide for the chemist requiring a guide to the literature of his subject both ancient and modern. Many of the chapters are based on lectures to graduates as courses of formal instruction in the use of chemical literature.

Our earliest bibliography of chemical interest is that by Elias Ashmole (1617–92),[3] the antiquary, astrologer and alchemist. His *Theatrum chemicum Britannicum; containing severall poeticall pieces of our famous English philosophers*, [*etc.*], London, 1652, consists of a number of old English poems on alchemy, with notes by Ashmole. Only the first part was published, and the work is very rare,[4] but it was reprinted with a new introduction by Allen G. Debus, New York, 1967; and with a preface by G. Heyer and J. M. Watkins, Hildesheim, 1967. Biographical information regarding Ashmole is provided in *Elias Ashmole (1617–1692). His autobiographical and historical notes, his correspondence, and other contemporary sources relating to his life and work. Edited, with a biographical introduction by C. H. Josten*, five volumes, 1966. The *Theatrum* also contains the first printed version of Thomas Norton's *Ordinall of alchimy*, which was reproduced in facsimile with an introduction by E. J. Holmyard in 1928. There has been some doubt about the authorship of the *Ordinall*, but J. Reidy[5] concludes that the book was written by "Thomas Norton of

[1] Farber, Eduard. Historiography of chemistry. *J. chem. Educ.*, 42, 1965, pp. 120–126.
[2] Thornton, John L., and Tully, R. I. J., History and biography of chemistry. *In* Bottle, R. T., *ed. The use of chemical literature. Second edition*, 1969, pp. 235–250.
[3] See Josten, C. H. Elias Ashmole, F.R.S. (1617–1692). *Notes Rec. Roy. Soc. Lond.*, 15,1960, pp. 221–230.
[4] Ferguson, Vol. 1, pp. 52–53.
[5] Reidy, J. Thomas Norton and the *Ordinall of alchemy*. *Ambix*, 6, 1957, pp. 59–85.

Bristol, gentleman of the king's privy chamber, and great-grandfather of Samuel Norton". Thomas Norton (*c.* 1437–1513 or 1514) began this book in 1477.

The next chemical bibliography emanated from France, where its author, a medical man, was physician to the King. Pierre Borel (*c.* 1620–71 or 1689) was the author of numerous books, many of which he left in manuscript, and he was keenly interested in natural history, astronomy, bibliography and antiquities. His *Bibliotheca chemica, seu catalogus librorum philosophicorum hermeticorum in quo quatuor millia circiter authorum chimicorum . . . quam in lucem editorum, cum eorum editionibus jusque ad annum 1653 continentur,* [*etc.*], Paris, 1654, lists 4,000 items devoted to alchemy, and was reprinted at Heidelberg in 1656.[1]

William Cooper, a bookseller at the Pelican in Little Britain, London during the latter half of the seventeenth century, specialized in alchemical literature, writing and publishing several works on the subject. He compiled *A catalogue of chymicall books . . . written originally, or translated into English*, London, 1675, which has been described as "an advance in detail and precise information on Borel's *Bibliotheca*".[2]

As an example of the *Bibliothecae*, which generally consist of reproductions of texts rather than mere lists, we mention one of the many similar works compiled by Jean Jacques Manget (1652–1742). His *Bibliotheca chemica curiosa, seu rerum ad alchemiam pertinentium thesaurus instructissimus,* [*etc.*], two volumes, Geneva, 1702, reproduces 140 tracts on the subject, many of these being extremely rare. Manget was a native of Geneva, and was a medical man.[3] A further bibliography of alchemical literature is contained in Nicolas Lenglet Du Fresnoy's *Histoire de la philosophie hermétique. Accompagnée d'un catalogue raisonné des écrivains de cette science,* [*etc.*], Volume 3, Paris, 1742 (also The Hague, 1744; and Paris, 1744), listing 1,500 items. Du Fresnoy (1674–1752) wrote extensively on historical subjects, the first two volumes of this work being of that nature. He disbelieved in alchemy, and his work is severely critical.[4] Works on alchemy contained in the British Museum have been listed by Kurt Karl Doberer in *A bibliography of books on alchemy in the British Museum*, 1946.

Johann Wilhelm Baumer (1719–88) was the author of *Bibliotheca chemica,* [*etc.*], Giessen, 1782, a "brief but useful bibliography", containing some 750 entries. Baumer was a professor of medicine who wrote extensively on his own subject and on minerals and geology, while he was also the author of *Elements of chemistry*, Giessen, 1783.[5] Another

[1] Ferguson, Vol. 1, pp. 116–117.
[2] Ferguson, Vol. 1, p. 135.
[3] Ferguson, Vol. 2, pp. 68–71; also Thornton, pp. 243–244.
[4] Ferguson, Vol. 2, p. 25.
[5] Ferguson, Vol. 1, p. 84.

professor of medicine, Georg Friedrich Christian Fuchs (1760–1813), a native of Jena, published numerous chemical papers, and was the author of two valuable histories of chemistry. These are *Versuch einer Uebersicht der chymischen Literatur und ihrer Branchen*, Altenburg, 1785; and *Repertorium der chemischen Literatur von 494 vor Christi geburt bis 1806 . . . von den Verfassern des systematischen Beschreibung aller Gesundbrunnen und Bäder in und ausser Europa*, four volumes, Jena and Leipzig, 1806–12.[1]

One of the most comprehensive bibliographies of chemistry is that by Henry Carrington Bolton (1843–1903) entitled *A select bibliography of chemistry, 1492–1892*, Washington, 1893, and published by the Smithsonian Institution. A reprint was published in 1966. The main work contains 12,031 titles, and there are two supplements, the first, published in 1899, covering the years 1492–1897, and listing 5,554 entries; the second, covering 1492–1892, was published in 1904. The work is devoted to independent works, but there is a biographical section, and lists of obituaries from certain periodicals. It is divided into seven sections: bibliography; dictionaries; history; biography; chemistry, pure and applied; alchemy; and periodicals; there is also a subject index. Section 8 of the work, devoted to academic dissertations, was separately published in 1901.

The best-known bibliography devoted to chemical literature is the catalogue of a private collection, compiled by a chemist who himself acquired a notable library, and of which a catalogue has been published in recent years. James Young (1811–83) was a native of Glasgow, and became the founder of the paraffin oil industry in Scotland. His collection of books is in the University of Strathclyde, Glasgow, and the catalogue of it was compiled by John Ferguson as *Bibliotheca chemica; a catalogue of the alchemical, chemical and pharmaceutical books in the collection of the late James Young*, [*etc.*], two volumes, Glasgow, 1906. This catalogue was printed for private distribution by the family of James Young, in accordance with his instructions, and was reprinted in 1954. It is in alphabetical order by author, full annotations providing valuable biographical information and references as sources of additional information. There are no indexes, but the work is an invaluable contribution to the history of chemistry. John Ferguson (1837–1916) was Regius Professor of Chemistry in the University of Glasgow from 1874 to 1915, and was the author of several other works, including *Bibliographia Paracelsica*, six parts, 1877–96, and a series of papers prepared for the Glasgow Archaeological Society and published in Glasgow, 1896–1915, which were reprinted in facsimile as *Bibliographical notes on histories of inventions and books of secrets*, two volumes, London, 1959. He amassed a fine collection of books, many of which were sold in Glasgow (June, 1920), and at Sotheby's (November,

[1] Ferguson, Vol. 1, p. 295.

1920), the remainder being purchased by the University of Glasgow. These number about 6,000, and a two-volume catalogue was printed as *Catalogue of the Ferguson collection of books, mainly relating to alchemy, chemistry, witchcraft and gipsies, in the Library of the University of Glasgow*, [By Katherine R. Thomson and Mary Margaret Service. Edited by William Ross Cunningham], [Glasgow], 1943. Only forty copies of this were issued.[1]

The Young collection is rich in sixteenth- and seventeenth-century items, and a companion bibliography supplementing this by its eighteenth- and nineteenth-century material must be mentioned here. This is Denis I. Duveen's *Bibliotheca alchemica et chemica. An annotated catalogue of printed books on alchemy, chemistry and cognate subjects in the library of Denis I. Duveen*, London, 1949, of which a second edition was published in 1965. The collection, minus the Lavoisier items, is now in the University of Wisconsin. This *Bibliotheca* was first printed in a limited edition of 200 copies, of which ten were on hand-made paper and not for sale. It is arranged alphabetically by authors, entries being annotated. There is no bibliographical information, or even bare dates of birth and death of authors, but under authors' names entries are chronologically arranged. Sixteen collotype plates are included, 3,000 items being listed. This collection is included, with additions in *Chemical, medical and pharmaceutical books printed before 1800. In the collection of the University of Wisconsin libraries. Edited by John Neu. Compiled by Samuel Ives, Reese Jenkins and John Neu*, Madison, Milwaukee, 1965.

Booksellers' catalogues are frequently of bibliographical interest, but few assume the role of permanent bibliographies of a subject. One such, however, is Henry Sotheran's *Bibliotheca chemico-mathematica: a catalogue of works in many tongues on exact and applied science, with a subject index. Compiled and annotated by H. Z[eitlinger] and H. C. S[otheran]*, two volumes, London, 1921. This was begun in 1906 as the catalogue of a large collection of books for sale, and grew into five volumes containing 47,490 items. It covers astronomy, chemistry, mathematics, physics and allied subjects. Volume 1 is alphabetical, with a supplement A–GILL; Volume 2 continues this alphabet, and after a brief "further addenda", proceeds to a further "final supplement" which is classified, and is completed by a subject index. The first supplement (1932), the second (two volumes), and the third (1952), are classified under broad headings, and then arranged alphabetically by authors. Throughout the work bibliographical and historical notes, facsimiles, portraits and prices are provided.

A list of bibliographies on chemistry was first published in 1925 by Clarence J. West and D. D. Berolzheimer as *Bibliography of bibliographies*

[1] See Sarton, George. John Ferguson (1837–1916). (Note.) *Isis*, 39, 1948, pp. 60–61.

on chemistry and chemical technology, 1900–1924, Washington, 1925.[1] Supplements published in 1929 and 1932 respectively cover 1924–28 and 1929–31. Part 1 of the work lists general bibliographies; Part 2 is concerned with abstract journals; Part 3 deals with general indexes of serials; Part 4 contains 9,000 bibliographies on special subjects, arranged by subjects; and Part 5 lists personal bibliographies of chemists.

Periodically published bibliographies of chemistry include Jacob Berzelius' *Jahresbericht über die Fortschritte der physischen Wissenschaften*, twenty volumes, Tübingen, 1822–41,[2] which was continued as *Jahresbericht über die Fortschritte der Chemie und Mineralogie*, Volumes 21–29, 1842–50. Since 1897 the Deutsche Chemische Gesellschaft has published *Chemisches Zentralblatt: vollständiger Repertorium für aller Zweige der reinen und angewandten Chemie*, published at Berlin. This began in 1830 as the *Pharmaceutisches Zentralblatt*, the name being changed in 1849 to *Chemisch-Pharmaceutisches Centralblatt*, in 1856 to *Chemisches Centralblatt*, and to its last title in 1897. It ceased publication in 1969, but was continued as *Chemische Information Dienst*, in two sections, *Organische Chemie* and *Anorganische und physikalische Chemie*. A list of periodicals examined for *Chemisches Zentralblatt*, with international standard abbreviations, was published by E. H. Maximilian Pflücke and Alice Hawelek as *Periodica chimica. Verzeichnis der im Chemischen Zentralblatt referierten Zeitschriften mit den entsprechenden genormten Titelabkurzungen*, Berlin, [*etc.*], 1961. This was followed chronologically by *Jahresbericht über die Fortschritte der reinen, pharmaceutischen und technischen Chemie, Physik, Mineralogie und Geologie*, covering the years 1847–48–56, and published at Giessen, 1849–57; this was continued as *Jahresbericht über die Fortschritte der Chemie und verwandter Theile anderer Wissenschaften* for the years 1847–1910 (1886 onwards published at Brunswick), 1848–1912–13. There was also the *Répertoire de Chimie pure et appliquée. Société Chimique de Paris*, published from 1858–63, continued as *Bulletin de la Société chimique de Paris* from 1863 to 1906, and since published in the *Bulletin de la Société chimique de France*, Paris.

The Chemical Society published the *Journal of the Chemical Society Abstracts* (afterwards *Papers*) from 1878 to 1923 (the Abstracts having been included in the *Journal* from 1871), which was continued as *Abstracts of chemical papers. . . . A. Pure Chemistry*, issued by the Bureau of Chemical Abstracts from 1924 to 1925. This was continued as the *British Chemical Abstracts*, 1926–37, and as *British Chemical and Physiological Abstracts*, 1938–44, and was taken over by the Bureau of Abstracts in 1937 and published from 1945 to 1953 as *British Abstracts* in the following series:

[1] *Bulletin of the National Research Council*, No. 50; the supplements are Nos. 71 and 86.

[2] See Ostwald, Wilhelm, Berzelius' "Jahresbericht" and the international organization of chemists. Translated by Ralph E. Oesper, *J. chem. Educ.*, 32, 1955, pp. 373–375.

A. Section I. General physical and inorganic chemistry; A. Section II. Organic chemistry; A. Section III. Physiology, biochemistry, [*etc.*]; B.I. Chemical engineering and industrial inorganic chemistry, [*etc.*]; B.II. Industrial inorganic chemistry; B.III. Agriculture, food, sanitation; and C. Analysis and apparatus. From 1954 series B.I. and B.II. appear in the *Journal of Applied Chemistry*, and B.III. in the *Journal of the Science of Food and Agriculture*, while A.III. was covered by *British Abstracts of Medical Science*, the title of which was changed to *International Abstracts of Biological Sciences* in 1956, while C. appears in *Analytical Abstracts* issued by the Society for Analytical Chemistry. From 1954 until 1969 the Chemical Society published each month *Current Chemical Papers* listing the titles in English of papers from some 400 journals. In that year the Chemical Society, acting for a group of British chemical societies and Aslib, linked with the American Chemical Society to market the American society's publications in the United Kingdom, to expand the Chemical Abstracts Service (C.A.S.) computer-based information sources in the U.K., to develop specialized publications and services of its own, and ultimately to supply C.A.S. with data from British journals. Some of the first fruits of this development were seen in new current-awareness bulletins entitled *Chemscan*, the first two of which were devoted to *Radiation and Photochemistry* and *Steroids*. The basis of this service is as mentioned, the Chemical Abstracts Service, which has grown from the monumental *Chemical Abstracts*, published by the American Chemical Society since 1907. *Chemical Abstracts* is prepared by examining some 12,000 journals and patents issued by twenty-five countries. By means of its computer-base, many other services to chemists are generated, such as *Chemical Titles*, issued every two weeks and covering over 700 journals, each issue of which contains up to 5,000 titles, and being divided into three parts, the keyword index, the bibliography, and the author index. The keyword index is permuted according to important words in the title, these words appearing in the centre of each column. Titles may appear several times under different keywords, and the reference code number at the extreme right of each entry refers one to the bibliography section. This provides the names of the authors of papers, full titles, and sources, with volume and inclusive page-numbers. The author index also provides reference code numbers. The keyword index entries are limited to a certain number of letters, and in order to provide the keyword in the appropriate position, entries are beheaded, reversed and otherwise mutilated. However, most of them are understandable; and since the material is produced by machine techniques, much time is saved. In fact, entries are recorded within a few weeks of their appearance in journals. Other C.A.S. services which must be mentioned are *Chemical-Biological Activities*, concerned with the interaction of organic compounds with biological systems, and *ACCESS* which in 1969 contained entries for over 10,000 currently published serials, and 6,000 entries for serials either no longer current or appearing under different titles. Library locations in several countries are also provided.[1] The extensive activity in chemical bibliography is shown by the appearance in 1961 of the first issue of a periodical devoted to this alone. Issued every six months, the *Journal of Chemical Documentation*, published by the American Chemical Society, contains papers on such matters as nomenclature, availability of chemical literature, chemical indexing, punched cards and mechanized searching techniques, and technical writing.

The *Handbuch der organischen Chemie* founded by Friedrich Konrad Beilstein has been described as the largest compilation of information on organic chemistry, and its function is detailed by T. C. Owen and R. M. W. Rickett.[2] The third edition consists of 27 volumes in the main work, covering the literature to 1910, 1918–37; 31 volumes in the first supplementary series covering the literature 1910–19; 29 volumes in the second supplementary series covering the literature 1920–29; and a third supplementary series covering the literature 1930–49 commenced publication in 1958.

Biology is a vast subject, and for the purpose of this chapter we have grouped together bibliographies devoted to general biology, zoology, and natural history, followed by separate sections devoted to ornithology, ichthyology, entomology and finally botany. A useful reference book for both the librarian and the scientist interested in biological literature is provided by *The use of biological literature. Editors R. T. Bottle; H. V. Wyatt*, 1966, a second edition of which is in preparation. Chapters are devoted to libraries and classification; primary sources of information; abstracts, reviews and bibliographies; foreign serials and translations; quick reference sources; taxonomy, treatises and museums; patents; Government publications and trade literature; zoology; anatomy, physiology and pathology; microbiology; and applied biology. H. V. Wyatt and R. T. Bottle[3] have also described courses for the instruction of post-graduate students in the use of libraries and of sources of information. The need for this instruction is also stressed by D. N. Wood and K. P. Barr,[4] and it is anticipated that similar courses will be organized in more

[1] Further information on this subject is available in *The use of chemical literature. Second edition. Edited by R. T. Bottle*, 1969, particularly Chapter 4, pp. 45–79; and Chapter 6, pp. 91–101.

[2] Owen, T. C., and Rickett, R. M. W. Beilstein's 'Handbuch' as a source of information on organic chemistry. In, Bottle, R. T., *ed. The use of chemical literature. Second edition*, 1969, pp. 142–154.

[3] Wyatt, H. V., and Bottle, R. T. Training in the use of biological literature. *Aslib Proc.*, 19, 1967, pp. 107–110.

[4] Wood, D. N., and Barr, K. P. Courses on the structure and use of scientific literature. *J. Document.*, 22, 1966, pp. 22–23.

British universities and similar institutes. It is better to educate the scientist to help himself than to rely upon mechanized retrieval, and the services of limited library staffs and information scientists.

A general guide to the literature is provided by Roger C. Smith in his *Guide to the literature of the zoological sciences*, Minneapolis, a seventh edition of which was published in 1966. This deals with the use of libraries; review journals; scientific reading and writing; library classification; bibliographies of zoology, journals on zoology and allied subjects; abstract journals; types of bibliographies, with examples; the preparation of scientific papers; and taxonomic indexes and literature.[1]

Although not strictly speaking classed as bibliographies, catalogues of extensive libraries are of vital importance to those searching the literature of a subject. For example, the British Museum (Natural History) *Catalogue of the books, MSS., maps and drawings*, [*etc.*], eight volumes, Vols. 1–5, 1903–15, and 6–8, Supplement, 1922–40, and the *Catalogue of the Library of the Zoological Society of London*, fifth edition, 1902, and the same Society's *List of the periodicals and serials in the Library*, [*etc.*], London, 1949, immediately suggest themselves as being of great significance to students of biology and its branches.

Two guides to the early literature of natural history are of special interest. Firstly Johann Ludwig Choulant (1791–1861), the author of several medical bibliographies,[2] compiled *Graphische Incunabeln für Naturgeschichte und Medizin*, [*etc.*], Leipzig, 1858 (facsimile reprint, Munich, 1924). This list of illustrated incunabula is not exhaustive, but provides complete bibliographical details, with particulars of place of publication, date, size (i.e., folio, *etc.*), name of printer, month, and annotation. The second item, although only a typescript thesis,[3] is worthy of mention, and might well have been completed and printed. It is by Catherine H. W. Bickle, and entitled *A bibliography of zoological works, by British authors or printed in Great Britain, 1477–1550*, 1948. The first twenty-three items (A–F) are fully collated, the remainder having only abbreviated entries, but there are useful annotations, locations of copies, S.T.C. references, an index of printers and a general index.

Eighteenth-century bibliographies include Johann Jacob Scheuchzer's (1672–1733) *Bibliotheca scriptorum historiae naturalis omnium terrae regionum inservientium. . . . Accessit . . . Jacobi Le Long . . . de scriptoribus historiae naturalis Galliae*, Zürich, 1716 (reissued in 1751). Scheuchzer was also the author of *Helvetiae historica naturalis*, three parts, Zürich, 1716–18, which is in Latin and German. F. C. Bruckmann's *Bibliotheca animalis* was published

[1] See also Bamber, Lyle E. Biology. *Library Trends*, 15, 1967, pp. 829–835.
[2] See Thornton, p. 248.
[3] University of London, School of Librarianship thesis; a copy is available at the School at University College, London.

at Wolfenbüttel in 1743, and was followed by Laurenz Theodor Gronovius' (1730–77) *Bibliotheca regni animalis atque lapidei, seu recensio auctorum et librorum, qui de regno animali et lapideo . . . tractant*, Leyden, 1760, a rare work listing 5,000 entries. Gronovius published his *Zoophylacium* in parts, 1763–81,[1] and was also the author of *Museum ichthyologicum*, [*etc.*], two volumes, Leyden, 1754. Georg Rudolph Boehmer compiled the *Bibliotheca scriptorum historiae naturalis, oeconomiae aliarumque artium ac scientiarum ad illam pertinentium realis systematica*, nine volumes, Leipzig, 1785–89, which has 65,000 entries.

The greatest bibliography of natural history published in the nineteenth century was commenced by Wilhelm Engelmann (1808–78). His *Bibliotheca historico-naturalis. Verzeichniss der Bücher über Naturgeschichte welche in Deutschland, Scandinavien, Holland, England, Frankreich, Italien und Spanien in den Jahren 1700–1846 erschienen sind*, [*etc.*], was published at Leipzig in 1846,[2] and reissued the following year with an English title-page. This first volume contains 10,000 entries, and was followed by a supplementary volume by J. Victor Carus and Wilhelm Engelmann, bearing the title *Bibliotheca zoologica. Verzeichniss der Schriften über Zoologie welche in den periodischen Werken enthalten und vom Jahre 1846–1860 selbständig erschienen sind*, two volumes, Leipzig, 1861. This lists 40,000 items, and in turn was continued in a work by Ernst Otto Wilhelm Taschenberg (1854–1928) entitled *Bibliotheca zoologica II. Verzeichniss der Schriften über Zoologie welche in den periodischen Werken enthalten, und vom Jahre 1861–1880 Selbständig erschienen sind*, [*etc.*], seven volumes [in eight], Leipzig, 1886–1930. The volumes forming this exhaustive bibliography are classified, with both author and subject indexes.

Louis Jean Rodolphe Agassiz (1807–73) was the author of numerous books on zoological topics, but his bibliographical effort is not as valuable as the names of the author and the publishers, the Ray Society, might suggest. It is not selective, the entries are very brief, and there are no annotations. Entitled *Bibliographia zoologiae et geologiae. A general catalogue of all books tracts and memoirs on zoology and geology. Corrected, enlarged and edited by H. E. Strickland* (*volume four . . . and Sir William Jardine*), four volumes, London, 1848–54 (reprinted New York, 1968), this includes articles in journals. Volume 1 lists periodicals by country, and includes authors A–B; the three other volumes complete the alphabet, listing a total of 40,000 items. Despite its inadequacies this is an important source-book for tracing the early literature. It lists the various editions and translations of items recorded, and the first volume contains an extensive list of sources consulted in the preparation of the bibliography.

[1] Wheeler, Alwyne C. The *Zoophylacium* of Laurens Theodore Gronovius. *J. Soc. Bib. Nat. Hist.*, 3, 1956, pp. 152–157.

[2] Reprinted in 1960 as Historia Naturalis Classica XIV.

Many of the following are particularly specialized, but are of outstanding importance in their respective classes. Max Meisel's *A bibliography of American natural history. The pioneer century, 1769–1865*, three volumes, Brooklyn, 1924–29 (reprinted New York, 1967), covers almost a century of important literature. It is the best source of information on the history and early publications of state geological surveys, of scientific societies and journals, and contains a bibliography of biographies. *A bibliography of eugenics*, by Samuel Jackson Holmes (born 1868), was issued in the University of California Publications in Zoology series (Volume 25, 1924, pp. 1–514), and is exhaustive within its field. John Charles Phillips (born 1876) was the author of *American game mammals and birds: a catalogue of books, 1582–1925; sports, natural history and conversation*, New York, 1930, covering an extensive period and an interesting subject.

One of the most important bibliographies of zoology is *An introduction to the literature of vertebrate zoology based chiefly on the titles in the Blacker Library of Zoology, the Emma Shearer Wood Library of Ornithology, the Bibliotheca Osleriana, and other libraries of McGill University, Montreal*, London, 1931, compiled and edited by Casey Albert Wood (1856–1942). This contains an extensive historical introduction consisting of chapters devoted to chronological periods, to particular branches of the subject, and to specialist material, such as periodicals, rare books, manuscripts, etc.; a chronological list of publications; and an alphabetical catalogue, arranged chronologically under names of authors. This bibliography lists about 15,000 works, and contains useful annotations. A dictionary catalogue recording 60,000 volumes was published thirty-five years after Casey Wood's *Introduction* as *A dictionary catalogue of the Blacker-Wood Library of Zoology and Ornithology*, nine volumes, Boston, 1966. Casey Wood was the author of numerous books and articles on ornithology, his most important monograph being *The fundus oculi of birds, especially as viewed by the ophthalmoscope*, [*etc.*], Chicago, 1917. Claus Nissen has published several parts of a work to appear in two volumes with the title *Die zoologische Buchillustration: ihre Bibliographie und Geschichte*, Stuttgart, 1966 onwards. Nissen has also contributed an historical and bibliographical text to an edition limited to one hundred copies of *Tierbuecher aus 5 Jahrhunderten. 58 Originalblätter aus zoologischen Prachtwerken von 1491 bis 1905*, Munich, 1967.

F. C. Sawyer[1] has provided a useful guide to reference books in zoology, citing 212 publications arranged under generalia, and the various phyla and classes. Brief annotations and a topographical index are provided.

[1] Sawyer, F. C. Books of reference in zoology, chiefly bibliographical. *J. Soc. Bib. Nat. Hist.*, 3, 1955, pp. 72–91.

Useful bibliographical information is also available in a series of articles[1] listing papers concerning the dates of publication of natural history books.

Brent Altsheler's *Natural history index-guide. An index to 3,365 books and periodicals in libraries*, was published in a second edition, New York, 1940, and is classified, with author and subject indexes. Compiled by Jessie Craft Ellis, *Nature and its applications: over 200,000 selected references to native forms and illustrations of nature as used in every way*, Boston, 1949, enables one to trace references to more popular articles on natural history, and to find requisite illustrations. Although the title of the next work suggests that it is confined to the literature of the primates, many of the entries are concerned with zoology in general. Theodore C. Ruch's *Bibliographia primatologica: a classified bibliography of primates other than man. Part 1. Anatomy, embryology & quantitative morphology; physiology, pharmacology & psychobiology; primate phylogeny & miscellanea*, [*etc.*], Springfield, Baltimore, (1941), was issued as Publication No. 4 of the Historical Library, Yale Medical Library, and lists material published up to 1939. The first three sections deal with the knowledge of the primates in the ancient world and the Middle Ages, the sixteenth and seventeenth centuries, and the eighteenth century respectively, these sections being arranged alphabetically by authors, The remainder of the work is classified, with an author index. Full names of authors, with dates of birth and death, are provided, and there are 4,630 entries.

The Association for the Study of Systematics in Relation to General Biology has published an extremely useful *Bibliography of key works for the identification of the British fauna and flora*, edited by John Smart and published in 1942. A second edition, edited by John Smart and George Taylor, London, 1953, was issued by the Systematics Association, with a third edition edited for the Association by G. J. Kerrick, R. D. Meikle and N. Tebble, London, 1967. This was much enlarged, containing some 2,363 titles, and is a very valuable work. We must also mention here the fact that the *Journal of the Society for the Bibliography of Natural History* contains several valuable bibliographies of specialist subjects, and note as an example the bibliography of whaling by James Travis Jenkins which occupies a complete number of the *Journal*.[2]

Periodical bibliographies covering this field include *Jahresbericht über die Fortschritte in der Biologie*, published at Erlangen from 1843 to 1850,

[1] Griffin, F. J., Sherborn, C. Davies, and Marshall, H. S. A catalogue of papers concerning the dates of publication of natural history books. *J. Soc. Bib. Nat. Hist.*, 1, 1936, pp. 1–30; first supplement, by F. J. Griffin. *Ibid.*, 2, 1943, pp. 1–17; second supplement, by W. T. Stearn and A. C. Townsend. *Ibid.*, 3, 1953, pp. 5–12; third supplement, by G. H. Goodwin, *Ibid.*, 3, 1957, pp. 165–174; fourth supplement, by G. H. Goodwin, W. T. Stearn and A. C. Townsend. *Ibid.*, 4, 1962, pp. 1–19.

[2] Jenkins, James Travis. Bibliography of whaling. *J. Soc. Bib. Nat. Hist.*, 2, iv, 1948, pp. 71–166.

which was known as C. Constatt's *Jahresbericht* during its last two years of publication; *The Record of Zoological Literature*, Volumes 1–6, 1864–99, which was continued as *The Zoological Record*, Volume 7, 1870 onwards, and is published by the Zoological Society of London, is an invaluable classified bibliography. The volumes of the latter for the years 1906–14 were issued as the Zoological volumes of the *International catalogue of scientific literature* (*q.v.*). *The Zoological Record* is now available in twenty separate sections, and G. D. R. Bridson[1] has provided a centennial history of this invaluable bibliographical tool. Other periodical bibliographies include the *Zoologische Jahresbericht. . . . Allgemeine Biologie*, Berlin, 1879–1913; *L'Année Biologique: Comptes Rendus Annuels des Travaux de Biologie Générale*, Paris, first published in 1895, and still current; the *Bibliographia Zoologica*, Leipzig, 1896–1934, which was edited by J. Victor Carus until 1913, and afterwards by H. H. Field, and was issued as a supplement to *Zoologischer Anzeiger* 1878–1934, published by the Deutsche Zoologische Gesellschaft, Leipzig. An index covering the period 1758 to 1850 was compiled by Charles Davies Sherborn as *Index animalium*, the first section of which covers 1758–1800, Cambridge, 1902; and the second section, covering 1801–50, thirty-three parts, London, 1922–33. This contains 440,000 references, and is implicitly reliable. The work was continued by Sheffield A. Neave in *Nomenclator Zoologicus*, covering the period from the tenth edition of Linnaeus, 1758, to the end of 1955, in six volumes published by the Zoological Society of London, 1939–66, and still in course of publication.

Probably the most important current abstracting periodical devoted to biology is *Biological Abstracts. A comprehensive abstracting and indexing journal of the world's literature in theoretical and applied biology, exclusive of clinical medicine*, of which Volume 1, 1926, was published at Menasha, Wisconsin, under the auspices of the Union of American Biological Societies. It is now published at the University of Pennsylvania, and is also available in the following sections: Section A, Abstracts of General Biology; Section B, Abstracts of Basic Medical Sciences; Section C, Abstracts of Microbiology, Immunology, and Parasitology; Section D, Abstracts of Plant Sciences; and Section E, Abstracts of Animal Sciences. Sections F–H and J, devoted to Abstracts of Animal Production and Veterinary Science, Food and Nutrition Research, Human Biology, and Cereals and Cereal Products respectively, were discontinued at the end of 1953. The indexes cover all the sections. Since 1965 *Biological Abstracts* has published annually a *List of serials with abbreviations*, that for 1969 comprising 7,593 titles, with American Standard for Periodical Title Abbreviations. There is also an individual country list giving percentage of total number of

[1] Bridson, G. D. R. The Zoological Record—a centenary appraisal. *J. Soc. Bib. Nat. Hist.*, 5, 1968, pp. 23–34.

serials for each. They also publish *B.A.S.I.C.*, 1959 onwards, and *BioResearch Titles*, 1965 onwards. *B.A.S.I.C.* is a subject in context index covering terms used by the authors of papers and additional terms provided by the scientific editorial staff. From 1970 it appears in each issue of *Biological Abstracts*. *BioResearch Titles* provides access to material not conveniently reported in *Biological Abstracts*. It covers symposia, reviews, letters, notes, bibliographies, reports, semi-popular journals, trade journals, institutional reports, selected government reports, and secondary sources. In 1969 it reported some 85,000 research papers additional to those in *Biological Abstracts*. The German abstracting periodical of a similar nature is *Berichte über die wissenschaftliche Biologie*, issued as Abteilung A of the *Berichte über die gesamte Biologie*, Berlin, 1926 onwards.[1]

A list of 3,500 serial publications devoted to the biological sciences has been edited by John Henry Richter and Charles P. Daly, and published by *Biological Abstracts*. Entitled *Biological sciences; serial publications. A world list, 1950–54. Prepared under the sponsorship of the National Science Foundation by the Science Division, Reference Department, Library of Congress*, Philadelphia, 1955, this is arranged in broad subject classes, within which there are subdivisions, entries then being arranged by country of origin. Details of title, issuing agency and publisher, date of first issue, frequency, annotations, and bibliographical data are included, and there are also indexes of titles, societies and institutions, and of subjects. The alphabetically arranged *List of serials in the British Museum (Natural History) Library*, London, 1968 (Publication No. 673) contains about 12,500 entries representing serials held in the general library and in the departments of botany, entomology, mineralogy, palaeontology and zoology.

Comparatively speaking, ornithology is rich in bibliographical tools, several of which are of outstanding importance. Elliott Coues (1842–99) intended to compile a "universal bibliography of ornithology", but only four parts of it were published as *Bibliography of ornithology*, four volumes, Washington, 1878–80. The first three volumes are devoted to American ornithology, and the fourth deals with faunal publications relating to British birds. Coues was the author of numerous other valuable works on ornithology.

Although restricted to a single species, we must mention the comprehensive bibliography of falconry compiled by James Edmund Harting (1841–1913?). Entitled *Bibliotheca accipitraria: a catalogue of books ancient and modern relating to falconry. With notes, glossary and vocabulary*, London, 1891 (reprinted London, 1964), this is arranged by countries, and entries are recorded in chronological order under these headings. The work contains a glossary of technical terms, an index to authors, printers and

[1] See also Schultz, Louise. New developments in biological abstracting and indexing. *Library Trends*, 16, 1968, pp. 337–352.

engravers, and the entries are usefully annotated. Numerous illustrations are included. Harting was also the author of *The birds of Middlesex*, London, 1866; *The birds of Shakespeare*, London, 1871; *Essays on sport and natural history*, London, 1883; *Recreations of a naturalist*, London, 1906; and several other similar books.

Two British ornithological bibliographies follow, the first by William Herbert Mullens (born 1866) and H. Kirke Swann (1871–1926), is entitled *A bibliography of British ornithology from the earliest times to the end of 1912, including biographical accounts of the principal writers and bibliographies of their published works*, London, 1917. This was published in six parts from 20th June, 1916, to 29th June, 1917, and after a preliminary list of bibliographies, is arranged alphabetically by authors, with valuable bibliographies and biographical details. A supplement by H. Kirke Swann was published in 1923. Mullens and Swann were joined by Francis Charles Robert Jourdain in compiling *A geographical bibliography of British ornithology from the earliest times to the end of 1918, arranged under counties: being a record of printed books, published articles, notes and records relating to local avifauna*, London, 1920. This also appeared in six bi-monthly parts from 7th November, 1919, to 1st September, 1920. It is arranged by countries, subdivided by counties, and entries are then chronological by date of publication. The work includes articles published in periodicals, but there are no indexes.

John Henry Gurney the Younger (1848–1922) was the author of *Early annals of ornithology*, London, 1921, which contains ancient passages relating to ornithology, particularly British birds, and is invaluable as a history of ornithology. The author wrote several other books on this subject.

Three books on the iconography of ornithology deserve mention. Claus Nissen's *Schöne Vogelbücher; ein Überblick der ornithologischen Illustration nebst Bibliographie*, Vienna, 1936; his *Die illustrierten Vogelbücher*, [*etc.*], Stuttgart, 1953; and Jean Anker's *Bird books and bird art: an outline of the literary history and iconography of descriptive ornithology, based principally on the collection of books containing figures of birds and their eggs now in the University Library at Copenhagen, and including a catalogue of these works*, Copenhagen, 1938. This latter describes 540 illustrated books on ornithology.

Reuben Myron Strong's *A bibliography of birds with special reference to anatomy, behaviour, biochemistry, embryology, pathology, physiology, genetics, ecology, aviculture, economic ornithology, poultry culture, evolution and related subjects*, four volumes, Chicago, 1939–59,[1] is mainly confined to material published up to 1926. It includes articles in periodicals, and pamphlets,

[1] *Publications of the Field Museum of Natural History, Zoology Series*, 25, Parts 1–3; Chicago Natural History Museum, *Fieldiana: Zoology*, 25, Part 4.

the first two volumes being arranged alphabetically by authors, while the third volume is devoted to a subject index. Volume 4 is described as a "finding index". The work also contains a list of bibliographies of ornithology, and of general bibliographies containing material on that subject, but it has been largely replaced by R. Berndt and W. Meise's *Naturgeschichte der Vogel. Dritter Band. Bibliographie und Register*, Stuttgart, 1966. This separately available volume contains a bibliography of about 11,500 entries representing some 5,000 authors. It is classified, covers the literature up to 1966, and has an author index, an index to general subjects, and lists of scientific names and of German names of birds.

An outstanding bibliography of British ornithology by Raymond Irwin is of particular importance on account of its arrangement and execution. Entitled *British birds books: an index to British ornithology A.D. 1481 to A.D. 1948*, London, 1951, this contains a supplementary list covering the years 1949 to May, 1950. Full bibliographical descriptions are not provided, and the work is not exhaustive in respect of articles in periodicals. It is divided into five parts, Part 1, Section 1, being devoted to bibliography and history, while Section 2 consists of a list of ornithological periodicals. Section 3 contains a check list of early works to A.D. 1800, arranged in chronological order; and Section 4 covers works published after 1800. Part 2 contains regional lists, books being arranged alphabetically by authors' names under each region, and these being prefaced by information on local societies, observatories, libraries, museums and nature reserves to be found in the respective localities. Part 3 comprises a systematic list of British birds, and contains details of books, etc., on each species, the locations of illustrations in more general works being indicated. Part 4, an index of authors, is actually an alphabetical list of books, providing names of author, dates of birth and death, title of book, size (8vo etc.), dates of various editions, and information on illustrations, biographies and bibliographies. Part 5 consists of an index of subjects, species and places, and is followed by two appendices: the first containing addresses of national societies, unions, periodicals, etc., the second being the supplementary list.

Covering a period of two hundred years, and restricted to items showing a high standard of book production, *Fine bird books 1700–1900*, London, New York, 1953, is more correctly termed a list than a bibliography. It reproduces thirty-six plates from bird books published between those dates, contains a text on the subject of the title by Sacheverell Sitwell, and a bibliography by Handasyde Buchanan and James Fisher. Items are noted for 'fineness' by means of asterisks, and there are interesting annotations to some of the items.

Ichthyology has been the subject of two conspicuous bibliographies, the first, by Thomas Westwood (died 1888) and Thomas Satchell, being entitled *Bibliotheca piscatoria. A catalogue of books on angling, the fisheries, and fish-culture, with bibliographical notes, and an appendix of citations touching on angling and fishing from old English authors*, London, 1883. The first edition of this appeared in 1861, with a supplement in 1869. This work records 3,158 editions and reprints of 2,148 separate works. There are sections on angling, the fisheries and fish culture, entries being arranged alphabetically under these headings; brief annotations are appended. In 1801 Robert Bright Marston published a supplement to Westwood and Satchell as an appendix to the *English Catalogue of Books* for 1900, listing an additional 1,250 titles. Both the 1883 edition and Marston's supplement have been reprinted together, London, 1966. The second work lists articles in periodicals in addition to books, and is a monumental bibliography by an eminent ichthyologist. Bashford Dean (1869–1928) collected the material for *A bibliography of fishes*, three volumes New York, 1916–23 (reprinted New York, 1962), which was published by the American Museum of Natural History. Volumes 1 and 2 were enlarged and edited by Charles Rochester Eastman (1868–1918), and the third volume was extended and edited by Eugene Willis Gudger, with the co-operation of Arthur Wilbur Henn. Entries are arranged alphabetically by authors, and there is a subject index in systematic order, with an alphabetical list of subjects. The first two volumes list about 50,000 items published between 1758 and 1914, Volume 3 bringing the material down to 1919. Brief biographical and bibliographical notes and annotations are provided, and there are lists of general bibliographies containing references to fishes, of voyages and expeditions relating to fish, and of periodicals.[1]

Entomological literature possesses a useful guide to W. J. Chamberlin's *Entomological nomenclature and literature. (Second edition)*, (Ann Arbor), 1946, which contains in chronological order a list of bibliographical works and a list of serial publications. General works on entomology from 1602 onwards are then dealt with, followed by those on single orders and lesser groups. Information on preparing articles is also provided. There also exists a useful article by John D. Sherman[2] dealing with entomological bibliographies.

The earliest bibliography of entomology is rare, and it is believed that the author, Jean Charles Emmanuel Nodier (born 1783), attempted to recall copies. Entitled *Bibliographie entomologique; ou catalogue raisonné des ouvrages relatifs a l'entomologie et aux insectes*, Paris, 1793, this lists about 400 items. It was followed by Johann Nepomuk Eiselt's *Geschichte*,

[1] See also *The Bashford Dean memorial volume. Archaic fishes. Edited by Eugene Willis Gudger. Article 1. Memorial of Bashford Dean, by William K. Gregory*, New York, 1930, published by the American Museum of Natural History. This contains a bibliography of 315 items by Bashford Dean.

[2] Sherman, John D. Some entomological and other bibliographies, *J. N.Y. Entomol. Soc.*, 32, 1924, pp. 206–215.

Systematik und Literatur der Insectenkunde, von der ältesten Zeiten bis auf die Gegenwart, Leipzig, 1836, which contains 1,500 entries. In the following year appeared Achille Percheron's *Bibliographie entomologique, comprenant l'indication par l'ordre alphabétique de noms d'auteurs.* 1. *Des ouvrages les plus reculés jusques et y compris l'année, 1834*; 2. *Des monographies et mémoires contenus dans les recueils, journaux et collections académiques françaises et étrangères*, [*etc.*], two volumes, Paris, London, 1837. This is arranged alphabetically by authors, with a subject index, a seventy-five-page list of anonymous titles, chronological tables, lists of periodicals and publications of societies. This is less useful than Hermann August Hagen's *Bibliotheca entomologica. The complete literature of entomology up to the year 1862*, two volumes, Leipzig, [*etc.*], 1862–63 (reprinted Weinheim, 1960), Volume 1 of which contains title-pages in English and German. The work is arranged by authors, with a subject index, and supplements i–iii by Albert Müller were published in 1873.[1] From 1928–29 appeared *Index literaturae entomologicae, Serie* 1, [*etc.*], four volumes, Berlin, by Walther Horn and Sigmund Schenkling, containing 8,000 titles not previously given by Hagen. Band 1, A-E of Serie II . . . 1864–1900, by Walter Derksen and Ursala Scheiding was published by the Academy of Agricultural Sciences in the German Democratic Republic, Berlin in 1963. Three more volumes are planned.

Sherman[2] gives a list of sources providing additions and corrections to Hagen. Several United States governmental departments have published official bibliographies, usually of individual species or groups. Two more general official items follow. The first, by Samuel Henshaw and Nathan Banks, bears the title *Bibliography of American economic entomology*, and was published in eight parts between 1889 and 1905. It was continued by the American Association of Economic Entomologists in its *Index*, No. 1, by Nathan Banks, covering 1905–14; and No. 2, by Mabel Colcord, covering 1915–19. The second item is Josephine Clark's *Catalogue of publications relating to entomology in the Library of the United States Department of Agriculture*, 1906.[3] This contains full bibliographical details of recorded items, useful annotations and valuable indexes. Although only devoted to termites, mention must be made of Francis J. Griffin's selective bibliography published in the *Journal of the Society for the Bibliography of Natural History*[4] as an example of an important tool devoted to the literature of a single species. It is arranged alphabetically by authors, and chronologically under these headings. Although selective, it contains all the important

[1] The first supplement appeared in *Trans. Entomol. Soc.*, 2, 1873, pp. 207–217; the others were separately issued. A reprint was published in 1960 as *Historiae naturalis classica*, XIII.
[2] Sherman, John D. *J. N.Y. Entomol. Soc.*, 32, 1924, pp. 210–211.
[3] Published as Bulletin No. 55 of the Library, United States Department of Agriculture.
[4] Griffin, Francis J. A bibliography of the isoptera (termites), 1758–1949. *J. Soc. Bib. Nat. Hist.*, 2, 1951, pp. 261–368.

books and articles on the subject. A more recent bibliography of this subject is by Thomas E. Snyder, bearing the title *Annotated, subject-heading bibliography of termites, 1350 B.C. to A.D. 1954*,[1] and with a supplement bringing it down to 1960. Entries are arranged alphabetically by authors under some seventy-five subject headings, with an alphabetical list under authors, and an index of subjects.

Arthur A. Lisney's *A bibliography of British lepidoptera, 1608–1799*, London, 1960, was privately printed for the author by the Chiswick Press, and contains a wealth of information based largely upon the works in the author's library. There is a brief biographical introduction to each writer, and each item included is given a detailed description.

Periodically published literature on entomological literature includes *Bericht über die wissenschaftlichen Leistungen im Gebiete der Entomologie*, Berlin, 1838 onwards: *Entomologische Literaturblätter: Repertorium der neuesten Arbeiten auf dem Gesamtgebeit der Entomologie*, fourteen volumes, Berlin, 1901–14, which was continued as *Repertorium Entomologicum. . . . Deutsche entomologische Gesellschaft*, ten volumes, 1924–33; and the *Review of Applied Entomology*, published since 1913 by the Commonwealth Bureau of Entomology, London, as a digest of the contents of most entomological publications.

A guide to the history of bacteriology by Thomas H. Grainger, New York, 1958, *Chronica Botanica No. 18* provides a guide to the literature of bacteriology. Part I in thirty-one sections covers very widely the bibliography of science and medicine inasmuch as bacteriology is included. Part II, History of bacteriology; with special reference to specific areas, is a more detailed bibliography of books, chapters in books, and papers. Part III, Biographical references in general. Part IV, Selective guide to biographies of selected bacteriologists, is arranged alphabetically under the name of the bacteriologist and includes details of biographies, collected works, and letters, with very brief biographical details and annotations.

A comprehensive bibliography of bibliographies of botany up to 1909, compiled and annotated by Jens Christian Bay,[2] was published with the title "Bibliographies of botany. A contribution toward a bibliotheca bibliographica".[3] Entries are arranged in twelve groups, viz.: 1. Methodology; 2. Periodicals and reviews; 2a. Collective indexes to periodicals; 3. General and comprehensive bibliographies; Supplement 1. Bibliographical catalogues of private collections; Supplement 2. Bibliographical catalogues of public collections; 3a. National (regional) bibliographies; 4. Morphology and anatomy, Microtechnique, Teratology; 5. Plant

[1] *Smithsonian Miscellaneous Collections*, Vol. 130, Washington, 1956; and Vol. 143, iii, 1961.
[2] See Thompson, Lawrence S. Jens Christian Bay, bibliologist. *Libri*, 12, 1962, pp. 320–330.
[3] *Progressus rei botanicae*, 3, 1910, pp. 331–456.

geography, Systematic botany, Ecology, Nomenclature; 5a. Phaneragamae; 5b. Cryptogamae; 6. Plant physiology, Phenology, Biology; 7. Palaeobotany; 8. Economic botany; 9. Bibliographies of individual works: (i) Collective, (ii) Individual; 10. Libraries of institutions, Lists of publications of institutions; 11. Auction and sales catalogues of private libraries; 12. Booksellers' catalogues, Trade lists.

The first attempt at a complete bibliography of botany, both of manuscripts and printed books, was the *Bibliotheca botanica seu Herboristarum Scriptorum promota synodia, cui accessit individualis graminum omnium ab auctoribus hujusque observationum numerosissima nomenclatura. Io. Antonio Bumaldo collectore*, Bologna, 1657. Bumaldus is an anagram of the name of the author, the work being by Ovidio Montalbani. It is chronologically arranged, and the book is very rare in this edition. However, it was contained in another bibliography published in the following century by Jean François Seguier as *Bibliotheca botanica, sive catalogus auctorum et librorum omnium qui de re botanica . . . tractant . . . Accessit Bibliotheca botanica Jo. Ant. Bumaldi, seu potius Ovidii Montalbani. Bononiensis nec non auctuarium . . . opera Laur. Theod. Gronovii*, The Hague, 1740 (Plate 16). A supplement was published at [Verona, 1745], and a further supplement at Leyden, 1760.

Albrecht von Haller (1708–77) has been hailed as the founder of medical and scientific bibliography, and Sir William Osler called him "the greatest bibliographer in our ranks". He was professor of anatomy, botany and surgery at the University of Göttingen, and wrote extensively on these and other subjects. Haller was also responsible for several extensive bibliographies,[1] including *Bibliotheca anatomica*, [*etc.*], two volumes, Zürich, 1774–77, arranged in sections representing the historical periods of medicine, and subdivided by authors. Information on their lives, writings and various editions, and notes on the contents of the books are provided. His *Bibliotheca botanica*, [*etc.*], two volumes, Zürich, 1771–72, is similarly arranged, and these, together with his other bibliographies, represent exhaustive tools that have proved invaluable to later bibliographers. An *Index emendatus* to *Bibliotheca botanica* was prepared by Jens Christian Bay, and published in Berne, 1908. A reprint of the 1771–72 work was published in two volumes, Bologna, 1966.[2]

Several bibliographies of botany were published between the date of the appearance of Haller's work and Pritzel's *Thesaurus*. These include Ernst Gottfried Baldinger's (1738–1804) *Über Literargeschichte der theoretischen und praktischen Botanik*, Marburg, 1794. Baldinger was also the author

[1] See Thornton, pp. 139–140, 244–245.

[2] See also Fulton, John F. Haller and the humanization of bibliography. *New Engl. J. Med.*, 206, 1932, pp. 323–328; Fulton, John F. *The great medical bibliographers: a study in humanism*, Philadelphia, 1951, pp. 38–45, 93–94, etc.; and Klotz, Oskar. Albrecht von Haller (1708–1777). *Ann. med. Hist.*, N.S. 8, 1936, pp. 10–26.

of numerous medical books, including *Litteratura universa materiae medicae, alimentariae, toxicologiae, pharmaciae, et therapiae generalis,* [*etc.*], Marburg, 1793. J. A. Schultes' *Grundriss einer Geschichte und Literatur der Botanik von Theophrastos Eresios bis auf die neusten Zeiten*, Vienna, 1817, which lists 2,500 items; Friedrich von Miltitz's *Handbuch der botanischen Literatur für Botaniker, Buchhändler, und Auctionatoren*, Berlin, 1829, containing twice that number; and Marcus Salomon Krüger's *Bibliographia botanica. Handbuch der botanischen Literatur*, Berlin, 1841, containing 10,000 entries.

Georg August Pritzel's *Thesaurus literaturae botanicae omnium gentium, inde a rerum botanicarum initiis ad nostra jusque tempora, quindecim millia operum recens*, Leipzig, [1847]–1851, was issued in parts. Part 1 is arranged alphabetically by authors, with an appendix of anonymous works, and Part 2 is classified, with an author index. Additions to this bibliography were published in 1853, 1859, with an index 1862–64, and a new edition was published at Leipzig, 1872–77. This was reprinted at Milan in 1950. With Pritzel we must mention Benjamin Daydon Jackson's *Guide to the literature of botany; being a classified selection of botanical works, including nearly 6,000 titles not given in Pritzel's "Thesaurus"*, London, 1881 (reprinted New York, 1964), originally published for the Index Society.[1] This contains over 9,000 short-title entries, with an author index arranged under bibliography; history; biography; encyclopaedias; systems; pre-Linnean botany; introductory works; and then systematically, followed by serial publications, etc. B. D. Jackson was also responsible for *Vegetable technology: a contribution towards a bibliography of economic botany, with a comprehensive subject index. . . . Founded upon the collections of G. J. Symons*, London, 1882.[2]

The following bibliographies are also worthy of note: Ernestus de Berg's *Catalogus systematicus bibliothecae horti imperialis botanici petropolitani*, St. Petersburg, 1852, of which a new edition by Ferdinand von Herder was published in 1886; H. Trimen's *Botanical bibliography of the British counties*, London, 1874; *Index Kewensis: an enumeration of the genera and species of flowering plants, from the time of Linnaeus to the year 1885 inclusive, together with their authors' names, the works in which they were first published, their native countries and their synonyms. Compiled at the expense of the late Charles Robert Darwin under the direction of Sir J. D. Hooker. By B. D. Jackson*, two volumes, 1895, the title of which is explanatory, and to which fourteen supplements were published between 1901 and 1969; Alfred Rehder's *The Bradley bibliography: a guide to the literature of the woody plants of the world published before the beginning of the twentieth century*, five volumes, Cambridge, Mass., 1911–18, was compiled at the

[1] Index Society Publication, Volume 8.
[2] Index Society Publication, Volume 11.

Arnold Aboretum of Harvard University. This covers arboriculture, forestry, etc., and there are indexes of authors and titles. Volume 5 is a supplement containing additions and corrections to the first four volumes. A list of over 20,000 figures of flowers is contained in *Index Londinensis, to illustrations of flowering plants, ferns and fern allies. Being an amended and enlarged edition continued up to the end of the year 1920 of Pritzel's Alphabetical register of representations of flowering plants and ferns. Compiled from botanical and horticultural publications of the XVIIIth and XIXth centuries. Prepared under the auspices of the Royal Horticultural Society of London at the London Botanic Gardens, Kew. By O. Stapf*, six volumes, 1929–31, to which a two-volume supplement covering the years 1931 to 1935 was published in 1941.

Two books deal in particular with botanical illustrations; Claus Nissen's *Die botanische Buchillustration. Ihre Geschichte und Bibliographie. Zweite Auflage. Durchgesehener und verbesserter Abdruck der Erstauflage ergänzt durch ein Supplement*, three volumes in one, Stuttgart, 1966. Volume 3, a supplement to the original edition (1951), is available separately. The second work is Sacheverell Sitwell and Wilfred Blunt's *Great flower books, 1700–1900. . . . The bibliography edited by Patrick M. Synge*, London, 1956. A guide to old herbals was published by Alfred Schmid as *Ueber alte Kräuterbücher*, Berne, 1939.

The United States Department of Agriculture has published Sidney Fox Blake and Alice C. Atwood's *Geographical guide to floras of the world: an annotated list with special reference to useful plants and common plant names*, Washington. Part 1[1] of this deals with Africa, Australasia, North America, South America and the islands of the Atlantic, Indian and Pacific Oceans, and contains about 2,597 primary and 428 subsidiary titles. Part 2 by S. F. Blake[2] covers Western Europe, Finland, Sweden, Norway, Denmark, Iceland, Great Britain with Ireland, Netherlands, Belgium, Luxembourg, France, Spain, Portugal, Andorra, Monaco, Italy, San Marino and Switzerland, and contains about 3,757 primary and 3,084 subsidiary titles. This work provides an annotated, geographically and alphabetically arranged catalogue of all the useful floras and floristic works, including those in periodicals or serial publications that list or describe the complete vascular flora (or the phanerogams or pteridophytes only), of any region or locality. Countries are dealt with in alphabetical order, and material is arranged by general floras, then large regions, followed by localities such as counties, and then by smaller areas within counties. There are author and geographical indexes.

N. Douglas Simpson has compiled *A bibliographical index of the British*

[1] United States, Department of Agriculture. Miscellaneous Publication No. 401, Part 1, 1942, reprinted 1963 and 1967.
[2] United States, Department of Agriculture. Miscellaneous Publication No. 797.

flora, including floras, herbals, periodicals, societies, etc., privately printed, 1960. Based largely on the compiler's library of some 5,400 books on the British flora and on his collection of periodicals; the work was completed in 1950, although additions were made whilst the book was in the press. Foreign works which include references to British plant records are included. Part I, containing Catalogues; Distribution; Economic; Families, genera, and species; Floras; Herbals; Periodicals; Plant lore; Reference; Trees, is arranged alphabetically within the sections. Part II covering plant records is arranged chronologically. Items without dates are placed at the end unless external evidence of date has been found. A brief but interesting survey of British floras was made by G. A. Nelson.[1]

A list of almost 1,900 references exists in Douglas H. Kent's *Index to botanical monographs. A guide to monographic and taxonomic papers relating to Phanerogams and Vascular Cryptogams found growing wild in the British Isles*, London, 1967, published for the Botanical Society of the British Isles. This is arranged alphabetically under authors within each family and genera in systematic order, with an index to families and genera. Frans A. Stafleu's *Taxonomic literature. A selective guide to botanical publications with dates, commentaries and types*, Utrecht, 1967, in the *Regnum Vegetabile* series is arranged alphabetically by authors or editors, there is an index of personal names and of titles, and although not a true bibliography, it is of great value to bibliographers. The annually published *Index to European taxonomic literature* is also issued in the *Regnum Vegetabile* series, published by the International Bureau for Plant Taxonomy and Nomenclature, Utrecht. The issues of the *Index* are edited by R. K. Brummit and I. K. Ferguson. P. J. Wanstall has edited *Local floras*, London, 1963, the report of the 1961 conference arranged by the Botanical Society of the British Isles (Conference Reports, No. 7), which provides a useful history of the local flora in Britain, and contains several other papers of bibliographical interest.

An attempt to enumerate the publications covering the history of botany in Russia has been made by Vladimir C. Asmous in *Fontes historiae botanicae Rossicae*,[2] in which entries are arranged alphabetically by authors or by titles of institutions responsible for publication. Biographical and bibliographical material is also included, the Russian titles are transliterated and translated, and there is also a useful subject index.

A guide to the literature of agriculture is provided by J. Richard Blanchard and Harald Ostvold's *Literature of agricultural research* (University of California Bibliographic Guides), Berkeley, 1958, in which entries are arranged under broad headings, and then subdivided into: Bibliographies

[1] Nelson, G. A. British floras. *Proc. Leeds Phil. & Lit. Soc., Sci. Sect.*, 7, iii, 1957, pp. 17–28.
[2] *Chron. Bot.*, 11, ii, 1947, pp. 87–118.

of bibliographies and general works; abstracting journals; bibliographies and indexes; encyclopaedias; dictionaries; handbooks; yearbooks; history and biography; geography; abbreviations; periodical lists; societies and organizations; tables; miscellaneous. Although chiefly devoted to American publications, certain important foreign works are included, and many of the references are annotated. J. Richard Blanchard[1] has also contributed an article on the current state and future trends of agricultural bibliography.

Catalogues of libraries, while not themselves bibliographies, are valuable guides to bibliographical research. We mention four as examples, the first, *Catalogue of the printed books and pamphlets in the Library of the Linnean Society of London. New edition.* London, 1925. This is an alphabetical list by authors; a previous edition was published in 1896. The Royal Horticultural Society's *The Lindley Library: catalogue of books, pamphlets, manuscripts and drawings*, London, 1927, is maintained up to date by lists of additions in the Society's *Journal.* Mary S. Aslin and E. J. Russell compiled the catalogue of the collection at the Rothamsted Experimental Station, Harpenden, published in a second edition as *Library catalogue of printed books and pamphlets on agriculture, published between 1471 and 1840*, 1940. The Forestry Commission's *Library catalogue of books, with lists of periodicals and Forestry Commission publications*, London, 1957, covers horticulture, botany, soil science, etc., and is particularly useful for details of conferences.

The first international bibliography of agriculture is an annotated classified list, which also includes forestry and veterinary science. Edited by Rudolph Lauche, the *Internationales Handbuch der Bibliographien des Landbaues. World Bibliography of Agricultural Bibliographies*, Munich, 1957 lists 4,157 items between 1596 and 1957.

A bibliography devoted to agriculture was compiled by Walter Frank Perkins as *British and Irish writers on agriculture. Third edition*, Lymington, 1939, previous editions having been published in 1929 and 1932. This includes books on agricultural chemistry, botany, entomology, grasses, weeds, etc., and the third edition contains a list of serials, journals and newspapers, with a subject index to the entries numbering about 2,000, and dating from 1600 to about 1900. Perkins' collection of books is housed in Southampton University Library, which has issued a *Catalogue of the Walter Frank Perkins Agricultural Library*, Southampton, 1961. This also contains a memoir on Perkins, and is a useful bibliographical tool.

Periodically published guides to botanical literature include the *Botanischer Jahresbericht*, [*etc.*], ten volumes, Berlin, 1873–82, which was continued as *Just's botanischer Jahresbericht*, [*etc.*], 1883; *Botanisches Zentralblatt: referierendes Organ für de Gesamtgebeit der Botanik des In-und Auslandes*,

[1] Blanchard, J. Richard. Agriculture. *Library Trends*, 15, 1967, pp. 880–891.

forty volumes, Cassel, 1880–1919, of which a new series commenced in 1920. A cumulative index covering the years 1922–32 was published in two volumes, 1927–32; and *Botanical Abstracts*, [*etc.*], is a classified list of abstracts with author and subject indexes, and was published in fifteen volumes, Baltimore, 1918–26. It was then continued in *Biological Abstracts*. The United Nations, Food and Agriculture Organization, Documentation Centre, publishes *FAO Documentation—current index*, which appears monthly with an annual cumulation. The 1969 cumulative issue is in two volumes, the first being an annotated bibliography of items arranged in accession number order, and the second an author and subject index.

A recent additional source for botanical literature is *Excerpta Botanica*, published in two sections from Stuttgart since 1959. Section A, *Taxonomica et Chorologica*, provides abstracts in the fields of systematic botany and plant distribution in English, French or German and covers the literature since 1955. Section B, *Sociologica*, aims to list, but not abstract, all new work, and to publish a retrospective bibliography of all material in the field, country by country. The arrangement is classified and an examination has shown that considerable material is included which had not been noticed in *Biological Abstracts* or the *Berichte über die wissenschaftliche Biologie*.

Although not issued for sale, and printed in an edition limited to 750 copies, the catalogue of the Rachel McMasters Miller Hunt Botanical Library, Carnegie Institute of Technology, Pittsburgh, must be mentioned for the wealth of its contents. Compiled by Jane Quinby and entitled *Catalogue of botanical books in the collection of Rachel McMasters Miller Hunt*, Volume 1, published in 1958, covers printed books, 1477 to 1700. The second volume, compiled by Allan Stevenson, 1961, covers 1701 to 1800; the third volume will cover 1801 to 1850, and there will be two supplementary volumes. Over 400 items are described in the first volume, which also contains essays on the history of botany, and a survey of early printing as 1epresented in this remarkable collection. The Hunt Botanical Library is almost an international institute for bibliographical studies of botanical and horticultural literature. It publishes the *Hunt Facsimile Series; Hunt Monograph Series; Bibliographia Huntiana; Huntia*, (1964), a yearbook devoted to studies in botanical bibliography; and *Botanico-periodicum huntianum*, edited by George H. M. Lawrence, A. F. Günther Buchheim, Gilbert S. Daniels and Helmut Dolezal, Pittsburgh, Pa., 1968. This is a compendium of all serial publications including articles dealing with the plant sciences, and is extremely useful and well executed.

A guide to the literature of the history of science and technology has been compiled by K. J. Rider as *The history of science and technology: a select bibliography for students*, published by the Library Association, 2nd edition, 1970. This covers bibliographies; museums and exhibitions; societies and institutions; periodicals; collective biographies; and history, by periods and by subjects. An excellent survey of bibliographies of interest to the historian of science has been provided by John Neu,[1] and the history and biography of biology have been the subject of a chapter by John L. Thornton and R. I. J. Tully.[2] A select list of journals devoted to the history of science and its specialist branches is provided in the Bibliography (p. 381). Catalogues of early printed books in libraries are also useful, and some of these are recorded in Chapter 12, and elsewhere in this Chapter. Richard J. Durling's *A catalogue of sixteenth century printed books in the National Library of Medicine*, Bethesda, Maryland, 1967, contains many items of interest to the historian of science, alchemy, for example, being well represented. Certain biographical sources have already been mentioned (e.g., Poggendorff, p. 297), and several other general biographical sources are worthy of notice. Anthony à Wood's *Athenae Oxoniensis: an exact history of all the writers and bishops who have had their education in the University of Oxford*, [*etc.*], which covers the period 1500 to 1695, London, 1813–20, published as five volumes in four; and Joseph François Michaud's *Biographie universelle, ancienne et moderne*, the second edition of which was in forty-five volumes, Paris, 1843–65, and was reprinted in 1966. The German equivalent *Allgemeine Deutsche Biographie*, 56 volumes, Leipzig, 1875–1912 is appearing in a new edition as *Neue Deutsche Biographie*, Berlin, 1953 onwards. The *Dictionary of National Biography* is an obvious source of information, and *Biography Index*, published by the H. W. Wilson Co., appears quarterly with an annual cumulation. In addition to biographies in book form, this includes biographical material from 1,600 periodicals, and also covers collective biographies. It also has an index to professions and occupations. *World who's who in science*, Chicago, 1968, and *A biographical dictionary of scientists*, London, 1969, are useful sources of information, as is an older work compiled by James Britten and George S. Boulger as *A biographical index of deceased British and Irish botanists. Second edition, revised and completed by A. B. Rendle*, London, 1931. The first edition was published in 1893, with supplements in 1899, 1905, and 1908. This provides a mass of information in concise entries alphabetically arranged, with dates and places of birth and death, details of career, publications and references to sources. Biographies in book form devoted to scientists with medical degrees are recorded in John L. Thornton's *A select bibliography of medical biography. . . . Second edition*, London, 1970.

A useful tool in tracing scientific directories relating to specific subjects

[1] Neu, John. The history of science. *Library Trends*, 15, 1967, pp. 776–792.

[2] Thornton, John L., and Tully, R. I. J. History and biography of biology. *In*, Bottle, R. T., and Wyatt, H. V., *eds. The use of biological literature*, 1966, pp. 230–245; (second edition in press).

or countries has been compiled by Anthony P. Harvey as *Directory of scientific directories. A world guide to scientific directories including medicine, agriculture, engineering, manufacturing and industrial directories*, London, 1969. This contains some 1,600 entries, and provides full information for each item, including author, editor or compiler; name and address of publisher; date of latest edition; pagination; price; and useful annotation.

A guide to modern scientific dictionaries is provided in William R. Turnbull's *Scientific and technical dictionaries. An annotated bibliography. Volume 1. Physical science and engineering*, San Bernardino, Cal., 1966 [1967]. Entries are arranged by subject, each division being provided with an introduction. Charles W. Rechenbach and Eugene R. Garnett have compiled *A bibliography of scientific, technical, and specialized dictionaries. Polyglot, bilingual, unilingual*, Washington, 1969, which lists dictionaries, mostly published since 1945, in subject groups, e.g., Bibliographical; General scientific and technical; Biology; Botany; Zoology; Palaeontology; Chemistry, etc., covering all the sciences, medicine, technology, engineering, agriculture, and also law, economics, library science, music, arts, etc. Where justified, the entries are further divided into A. Polyglot, B. Bilingual, C. Bilingual (without English), and D. Unilingual. Compilers, titles, place of publication, publisher and date of publication are given. Over 1,250 dictionaries are recorded, and there are indexes of languages, subjects and compilers.

The proceedings of conferences, symposia, congresses, etc., are constantly growing in number, and are very difficult to trace. The National Lending Library for Science and Technology has published monthly since 1964, *Index of conference proceedings received by the N.L.L.* Also published monthly (except July and August), but with annual cumulations INTERDOK's *Directory of published proceedings. Series SEMT,—Science/Engineering/Medicine/Technology* lists the proceedings of all conferences, meetings, seminars, symposia and congresses in chronological sequence. This covers all meetings held since January, 1964, and includes prices, together with editor, location and subject/sponsor indexes. A brief guide to locating the manuscript sources of science has been compiled by Alan Jeffreys.[1]

Books and articles in foreign languages present tremendous problems to both scientists and librarians. D. N. Wood[2] has published a report on the subject, suggesting among the other recommendations the publication of an international cumulated index of translations; the establishment of a centralized Japanese translation programme; and a national collection

[1] Jeffreys, Alan. Locating the manuscript sources of science. *Brit. J. Hist. Sci.*, 2, 1964, pp. 157–161.

[2] Wood, D. N. The foreign-language problem facing scientists and technologists in the United Kingdom—report of a recent survey. *J. Document.*, 23, 1967, pp. 117–130.

of translations from German. Among guides to translations and translators the following are noteworthy: Carl J. Himmelsbach and Grace E. Boyd's *A guide to scientific and technical journals in translation*, New York, 1968, published by the Special Libraries Association, which evaluates the translations, and provides full bibliographical details; The Special Libraries Association Translations Center, now known as the National Translations Centre, is situated in the John Crerar Library, and has published since 1967 the extremely valuable *Translations register index*; and *Directory of technical and scientific translators and services*, compiled by Patricia Millard, London, 1968. This includes some three hundred translators, covering about fifty languages, with a detailed subject breakdown. A guide to the recognition of foreign languages, and of particular interest to librarians is *A guide to foreign languages for science librarians and bibliographers, compiled by J. R. F. Piette, revised and enlarged by E. Horzelska. Published on behalf of the Welsh Plant Breeding Station, Aberystwyth by Aslib*, London, 1965.

In recent years there has been a growth of periodically published surveys of literature bearing titles such as *Annual Review*, *Advances in*, *Progress in*, and *Yearbook of*, which provide evaluations of recent work together with extensive references. These are invaluable sources of bibliographical information, and are extensively used by scientists. *Biological Reviews* is a good example of its type and H. Munro Fox[1] has outlined its development, stressing the value of authoritative review-articles, and indicating the growing popularity of the organ by means of a sales chart.

Abstracting and indexing services are the main keys to scientific literature, and are widely used by scientists and librarians. They should be as up to date as possible, comprehensive and readily understood by readers. Unfortunately, many of them are costly, and the smaller libraries find that they absorb a high percentage of their budgets. Guides to these services include the International Council of Scientific Unions, Abstracting Service's *Compared activities of the main abstracting and indexing services covering physics, chemistry and biology during the year 1965*, Paris, prepared mainly to provide information for the I.C.S.U./UNESCO Committee on the Feasibility of a World System for Scientific Information, and giving details of numbers of periodicals covered, and abstracts produced by the main abstracting services; Stella Keenan's "Abstracting and indexing services in science and technology",[2] which examines recent developments, and provides suggestions for the future; Calvin Mark Lee's *An inventory of some English language secondary information services in*

[1] Fox, H. Munro. The origin and development of "Biological Reviews". *Biol. Rev.*, 40, 1965, pp. 1–4.
[2] *Ann. Rev. Informn. Sci. Tech.*, 4, 1969, pp. 272–303.

science and technology, Paris, 1969, published by the Organisation for Economic Co-operation and Development, and providing factual details regarding coverage, countries represented, journals completely abstracted, indexes, etc.; *Abstracting services in science, technology, medicine, agriculture, social sciences and the humanities*, The Hague, 1965, issued by the International Federation for Documentation; and National Federation of Science Abstracting and Indexing Services, Report No. 102, *A guide to the world's abstracting and indexing services in science and technology*, Washington, 1963. Services are listed in U.D.C. order, and there are 1,855 titles from 40 countries.

Several "current title" type of bibliographical tools have been established in recent years to bridge the gap between the appearance of articles and the publication of abstracts. R. T. Bottle[1] has provided a survey of current awareness services, differentiating between those reproducing the contents pages of relevant journals (e.g., *Current Contents*), the Keyword in context (KWIC) services, and citation indexes. Bottle points out the drawbacks of any system based only on titles.

The Institute of Scientific Information, Philadelphia, publishes the *Science Citation Index*^{R}, (1961), an ordered list of cited articles each accompanied by a list of citing articles, and compiled by computer.[2] The early volumes of this have been the subject of some criticism, but more recently coverage has been greatly expanded, and the issues are increasingly used to supplement abstract and current awareness tools. John Martyn[3] has examined the principle of citation indexing, listing those indexes available in 1964, and suggesting that these will become more acceptable to users when they have become more firmly established. Efforts to control citation errors have been discussed by Irving H. Sher, Eugene Garfield and Arthur W. Elias.[4]

Before leaving bibliography it is necessary to note a guide to scientific periodicals which is most useful in tracing required literature, and provides the abbreviations of titles of periodicals looked upon as standard by many medical and scientific journals; the *World List of Scientific Periodicals*, fourth edition, three volumes, 1963–65 gives locations for journals in British libraries, as does the *British Union-Catalogue of Periodicals*, four volumes, 1955–58, with supplement, 1962. These are maintained up to date at the National Central Library by *British Union-Catalogue of Periodicals, incorporating World list of scientific periodicals, New periodical*

[1] Bottle, R. T. A user's assessment of current awareness services. *J. Document.*, 21, 1965, pp. 177–189.

[2] See Malin, Morton V. The Science Citation Index^{R}: a new concept in indexing. *Library Trends*, 16, 1968, pp. 374–387.

[3] Martyn, John. An examination of citation indexes. *Aslib. Proc.*, 17, 1965, pp. 184–196.

[4] Sher, Irving H., Garfield, Eugene, and Elias, Arthur W. Control and elimination of errors in ISI services. *J. chem. Document*,. 6, 1966, pp. 132–135.

titles, 1964, No. 1, March, 1964 onwards. Separate annual volumes containing the scientific, medical and technical entries are also published. The abbreviations in the *World List* are inconsistent and haphazard, and it is preferable to observe those published by the U.S.A. Standards Institute as *American Standard for Periodical Title Abbreviations*, New York, 1964. This has one abbreviation for each word, and is preferable to the British Standard, which was eventually aligned to the method adopted by the *World List* instead of the latter conforming to the more logical rules initially formulated by the committee devising the original standard for the British Standards Institute. Mary R. Kinney[1] has examined the problem of abbreviated citation to serial publications, reports and classic works, providing sources for the identification of abbreviated citations arranged by subjects arranged alphabetically.

[1] Kinney, Mary R. *The abbreviated citation—a bibliographical problem*, Chicago, 1967. ACRL Monograph No. 28, published by the American Library Association.

CHAPTER X

Private Scientific Libraries

"Show me the books he loves and I shall know
The man far better than through mortal friends."
S. WEIR MITCHELL

The ownership of books by scientists began at an early date, although large collections were rare until the seventeenth century. Previously, the high cost of books prohibited individuals from possessing large libraries, and even after the seventeenth century books were often collected rather as novelties than as items of educational value. Books and specimens were not used for teaching purposes until comparatively recent times, and it is sometimes impossible to prove that earlier scientists possessed even one volume, despite the fact that they may have owned very extensive libraries. Many large collections have been dispersed without the books bearing any signs of previous ownership. Others contain signatures or book-plates. Later, auction catalogues form monuments to one-time collections of books, but the absence of these indications of previous ownership makes it difficult to reconstruct many private scientific libraries. However, since it is impossible to record within a single chapter even the name of every scientist who has owned a library, we have selected the following, either for the special significance of the history of the books, or of their one-time owners.

Information on the more important book collectors is available in books devoted to that subject, the writings of the Eltons,[1] W. Y. Fletcher,[2] John Lawler[3] and Seymour De Ricci[4] being of particular significance. Certain of these great collections have found their way into university, national and other large libraries, and have had individual catalogues devoted to their contents. For example, the libraries of Carl Linnæus and Sir Joseph Banks have been thus honoured. Others have auction catalogues forming their sole monuments, many of these auction catalogues being contained in the British Museum and other large

[1] Elton, Charles Isaac, and Elton, Mary Augusta. *The great book collectors*, 1893.
[2] Fletcher, William Younger. *English book-collectors*, 1902.
[3] Lawler, John. *Book auctions in England in the seventeenth century, (1676–1700)*, [*etc.*], 1906.
[4] De Ricci, Seymour. *English collectors of books and manuscripts (1530–1930), and their marks of ownership*, [*etc.*], Cambridge, 1930.

libraries. Priced catalogues are of particular interest, especially if they record the names of purchasers.

There were few important private collections of scientific books during the sixteenth century, but we know that Conrad Gesner (1516–65) collected together an extensive library while preparing his incomparable bibliographies (see pp. 83, 178). A catalogue of his library was published in 1574.[1] George Hartgill (died *c.* 1597) was a parson and an astronomer, the author of *The prognostication for this year of our Lord God 1581*, London [S.T.C. 12896a], of which no copy is known, and *Generall calendars, or, most easie astronomicall tables* [*etc.*], 1594, published simultaneously in English and Latin. Only three copies of each edition of this are known to exist. A list of the books Hartgill owned in 1590 was discovered written on the fly-leaves of his copy of Johann Schöner's *Opera mathematica*, Nürnberg, 1551, now in Exeter College, Oxford. Listed in Hartgill's writing are 88 items, with 13 added in another hand, and the list is confined to his "Mathematicall, Astronomicall & Phisicall bookes", which he widely interpreted. Paul Morgan[2] has reproduced the list, with annotations, identification where possible, and an index of authors.

Dr. John Dee (1527–1608), another prominent sixteenth-century figure, acquired a very miscellaneous collection of books and manuscripts. Dee was born on 13th July, 1527, and became eminent as astrologer, astronomer and mathematician. In 1583, during his absence on the Continent, a mob raided his house at Mortlake, destroying or stealing much of his property, including his library. On Dee's return, however, he recovered about three-quarters of the total number of his books and many other of his possessions. Montague Rhodes James[3] has supplied interesting information on Dee's library, reprinting the three known catalogues, two of which list the manuscripts owned by Dee about 1556, while the third was compiled on 6th September, 1583. List A contains 45 items; list B, 32; and list C (1583) about 201, together with several other items probably belonging to Dee, but not mentioned in the catalogue. Dee informs us in his *Compendious rehearsal* that he possessed about 4,000 items, 1,000 of which were printed books, but M. R. James[4] doubts this, the catalogues listing nothing near this total. The manuscripts were mainly concerned with mediaeval science, alchemy, astrology, astronomy, physics, geometry, optics and mathematics, but Dee was frequently forced to dispose of books and manuscripts in order to pur-

[1] *Bibliotheca instituta et collecta primum a Conrado Gesner . . . in vera recognita . . . per Josiam Simlerum*, [*etc.*], Zürich, 1574.

[2] Morgan, Paul. George Hartgill: an Elizabethan parson-astronomer and his library. *Ann. Sci.*, 24, 1968 [1969], pp. 295–311.

[3] James, Montague Rhodes. Lists of manuscripts formerly owned by Dr. John Dee. With preface and identifications. *Supplement to Trans. Bib. Soc.*, 1921.

[4] James, Montague Rhodes, *Op. cit.*, p. 40.

chase food. The remains of his library were sold some years after his death, a few volumes finding their way to Corpus Christi College, Oxford, while the British Museum houses some of Dee's manuscripts and other relics. Dr. John Dee was consulted by Queen Elizabeth on several occasions, and despite his astrological and alchemical leanings, he was a brilliant scholar. Incidentally, he drew up a remarkable scheme for the foundation of a state national library,[1] which, if accepted, would have enhanced his prestige in the world of learning, and provided us with a forerunner of the Bodleian Library at Oxford. Dee was the author of many books, most of which remained in manuscript.[2]

Robert Burton (1577–1640), author of *The anatomy of melancholy*, and a Bachelor of Divinity, was also a mathematician. He left behind a very extensive library, but it was not selective. Burton appears to have purchased indiscriminately, but at his death bequeathed to the University Library at Oxford such books as it did not already possess, and the surveying books and instruments to "Mr. Jones, Chaplin and Chanter". After leaving certain other books to friends, the remainder were to be sold.[3]

The seventeenth century witnessed a decided increase in the number of private collectors, and it is of interest to record an early collection still extant in the United States. This belonged to John Winthrop, jr. (1606–76), first Governor of Connecticut, who assumed a prominent position in the early history of New England. Winthrop was a charter member of the Royal Society, and maintained a voluminous correspondence with other scientists. He began collecting alchemical books in England, and his collection was probably one of the largest in private hands in the seventeenth century. He owned a number of alchemical manuscripts, some from John Dee's collection. When Winthrop went to Boston in 1631 he possessed a thousand volumes, many of them being devoted to astronomy, chemistry, mathematics, natural history and physics, his chief interest being in chemistry. Winthrop imported books from England, many of those still extant containing his annotations, and the largest collection of his books is in the New York Academy of Medicine. Others are housed in the Society for the Preservation of New England Antiquities, and the New York Society Library contains 270 of his books, while the Massachusetts Historical Society, Boston, houses his papers and manuscripts. Ronald Sterne Wilkinson[4] suggests that John

[1] See Thornton, John L., Dr. John Dee and his scheme for a national library. *Med. Bookman & Historian*, 2, 1948, pp. 359–362.

[2] See Smith, Charlotte Fell. *John Dee (1527–1608)*, [*etc.*], 1909, Chapter XIX (pp. 236–246), Dee's Library. See also Fletcher, William Younger. *Op. cit.*, pp. 45–49.

[3] See Fletcher, William Younger, *Op. cit.*, pp. 72–75.

[4] Wilkinson, Ronald Sterne. The problem of the identity of Eirenaeus Philalethes. *Ambix*, 12, 1964, pp. 241–243.

Winthrop, jr. was possibly "Eirenaeus Philalethes", who has been described as "the most important expositor of alchemy in the second half of the seventeenth century". Wilkinson[1] has also written on Winthrop's library, and Robert C. Black[2] has published the first extensive biography of him.

Sir Charles Scarburgh, M.D. (1616–94) collected together a fine mathematical library, which was sold in 1698, the sale catalogue recording 1,097 lots. Yet another mathematical library was formed by Sir Jonas Moore (1617–79). Moore had intended to bequeath his remarkable library to the Royal Society, but their owner dying intestate the books were sold by auction. A catalogue of forty-one pages, and containing 1,560 lots, serves as their memorial.

Three eminent seventeenth-century figures who were not primarily scientists must be mentioned here. Samuel Pepys (1633–1703), the diarist, was keenly interested in science, attending the meetings of the Royal Society, of which he served as President. Pepys made an important collection of material on geography and hydrography, and compiled *Bibliotheca nautica*, which was first published in 1695. Pepys' library was bequeathed to his nephew, John Jackson, for life, on condition that it was kept intact, and then presented to a college, preferably Magdalene College, Cambridge. The latter received the collection in 1726, and it is still preserved in the original cases, arranged as Pepys left them. There are about 3,000 volumes in eleven mahogany cases, the books being in double rows on the shelves, with the smallest in the front. Some of the volumes were placed on pieces of wood, to make the tops level. The books are numbered and book-plated, three kinds of book-plates being employed. Two carry a portrait, while the third bears a monogram of S.P. and two anchors. The collection includes nine Caxtons, eight Pynsons, and nineteen items printed by Wynkyn de Worde, in addition to prints, broadsides, drawings and ballads.[3] The other famous seventeeth-century diarist, John Evelyn (1620–1706), was keenly interested in chemistry, botany and horticulture, and his scientific writings have been recorded elsewhere.[4] Evelyn's books and manuscripts have been placed on loan in Christ Church, Oxford, and W. G. Hiscock[5] has provided some information on this collection. There are about 2,000 books and 150 manu-

[1] Wilkinson, Ronald Sterne. The alchemical library of John Winthrop, jr. (1606–76) and his descendants in Colonial America. *Ambix*, 11, 1963, pp. 33–51; 13, 1966, pp. 139–186; see also Browne, C. A. Scientific notes from the books and letters of John Winthrop, jr. (1606–76), first Governor of Connecticut. *Isis*, 11, 1928, pp. 325–341.

[2] Black, Robert C. *The younger John Winthrop*, New York, London, 1966.

[3] See Savage, Ernest A. Samuel Pepys' library. *The Library*, N.S. 4, 1903, pp. 286–291; and Turner, F. McD. C. *The Pepys Library*, (Cambridge), [N.D.].

[4] See also pp. 127–128.

[5] Hiscock, W. G. John Evelyn's library at Christ Church. *Times Literary Supplement*, 6th April, 1951, p. 220.

scripts, compared with the 4,588 printed books mentioned in a catalogue prepared in 1687, and W. G. Hiscock records the history of the collection. The antiquarian John Aubrey (1626–97) possessed an extensive library, including a large number of mathematical books. He presented some 120 volumes to the Ashmolean Museum, these now being in the Bodleian Library, and also a collection of books to Gloucester Hall, now Worcester College, Oxford. A list of these latter has been published,[1] and includes items by Robert Recorde, John Dee, Thomas Digges and Richard Norwood. Many of these items are signed and annotated by Aubrey.

Robert Boyle (1627–91),[2] as befits a keen scientist and author, possessed an extensive library, but only a few books have been identified as having been in his possession.[3] The sale of his books was advertised in two issues of the *London Gazette* in 1692, but some of the volumes were apparently sold on bookstalls in Moorfields,[4] and the residue was sold with the library of Silvanus Morgan, the two collections being mixed together in the catalogue. The auction took place on 5th April, 1695, at Tom's Coffee House.

Robert Hooke (1635–1703) left a large and valuable library, which was sold by auction at Edward Millington's rooms on 29th April, 1703, and several following days. A copy of the sale catalogue is housed in the British Museum, and bears the title *Bibliotheca Hookiana sive catalogus diversorum librorum: . . . Plurimis facultatibus linguisque insignium quos Doct. R. Hooke, sibi congessit . . . quorum actio habenda est Londini, in edibus vulgo dictis Inner Lower-Walk in Exeter Exchange in the Strand, the 29th of April, 1703. Per Edoardum Millington, Bibliop. Londin.* The catalogues were sold at sixpence each, and listed over 3,250 titles. A few items inscribed by Hooke have been traced.

The astronomer Edward Bernard (1638–96) collected together many manuscripts and rare printed books, some of which he acquired in 1683 at the sale in Leyden of the library of Nicholas Heinsius. Bernard was Savile Professor of Astronomy at Oxford, and he initiated a scheme for re-editing the mathematical writings of the ancients. His library was collected with this in view, and at his death the more important items

[1] *Times Literary Supplement*, 13th January, 1950, p. 32; 20th January, 1950, p. 48.

[2] See pp. 101–106.

[3] See Fulton, John F. *A bibliography of the honourable Robert Boyle, . . . Second edition*, Oxford, 1961, pp. iv–vi, "The fate of Boyle's library". See also, Maddison, R. E. W. Robert Boyle's library. *Nature*, 165, 1950, p. 981; Fulton, John F. A book from Robert Boyle's library. *J. Hist. Med.*, 11, 1956, pp. 103–104; and Fulton, John F. Boyle and Sydenham. *J. Hist. Med.*, 11, 1956, pp. 351–352, noting a book of which the title-page bears the signature of both Boyle and Sydenham.

[4] Hooke acquired certain items from Boyle's library, and recorded in his Diary that he saw some of the books exposed for sale in Moorfields. See McKie, Douglas. Boyle's library. *Nature*, 163, 1949, pp. 627–628.

went to the Bodleian Library. The remainder was sold by auction on 25th October, 1697, the catalogue recording 1,462 lots.

Sir Isaac Newton (1642–1727) died intestate, and following the sale of his library in 1727, it was lost sight of for almost two hundred years. R. De Villamil[1] has shed much light on the history of Newton's collection, listing many of the items, and the following details are derived from De Villamil's book. He discovered at Somerset House an inventory of Newton's goods, showing that he left "362 books in folio, 477 in quarto, 1,057 in octavo . . . with above one hundredweight of pamphlets and wast [*sic*] books". John Huggins, a warden at the Fleet Prison, paid £300 for the collection in 1727, sending it to his son Charles, then rector of Chinnor, near Oxford. Charles Huggins added his bookplate to each volume, and when he died intestate his successor as rector, the Rev. Dr. James Musgrave, purchased the collection for £400. Musgrave pasted his book-plate over that of Charles Huggins, and all the books were catalogued and press-marked. De Villamil discovered this catalogue, which dates from about 1760. After 1778, when Musgrave died, the books were transferred to Barnsley Park, whence many were sent to Thame Park. There they were sold at an auction held in 1920, without any indication that the books had once belonged to Newton. Many were pulped, and others went to the United States, some of these being in the Babson Institute Library, Massachusetts.[2] In 1928 De Villamil catalogued the items remaining at Barnsley Park, of which there were only 858, and this collection was presented to Trinity College, Cambridge in 1943 through the generosity of the Pilgrim Trust. Newton's library contained a large number of Greek and Latin classics, French works, and books on travel, as revealed in the list provided by De Villamil in his book.[3] A. Rupert Hall and Marie Boas Hall have edited and translated *Unpublished scientific papers of Isaac Newton. A selection from the Portsmouth Collection in the University Library, Cambridge*, Cambridge, 1962.

The sale of the library of Martin Folkes (1690–1754) lasted well over a month,[4] fetching over £3,000.[5] It began on 2nd February, 1756, there being 5,126 lots. Martin Folkes was born in Lincoln's Inn Fields, and

[1] De Villamil, R. *Newton, the man*, [*etc.*], [1931].

[2] See *A descriptive catalogue of the Grace K. Babson Collection of the works of Sir Isaac Newton, and the material relating to him in the Babson Institute Library, Babson Park, Mass. With an introduction by Roger Babson Webber*, New York, 1950; and *A supplement to the Grace K. Babson Collection. . . . Compiled by Henry P. Macomber*, Babson Institute, 1955, listing 220 items added since the 1950 catalogue. See also *A catalogue of the Portsmouth Collection of books and papers written by or belonging to Sir Isaac Newton: the scientific portion of which has been presented by the Earl of Portsmouth to the University of Cambridge*, Cambridge, 1888.

[3] See also Feisenberger, H. A. The libraries of Newton, Hooke and Boyle. *Notes Rec. Roy. Soc. Lond.*, 21, 1966, pp. 42–55.

[4] One authority suggests 41 days, another 56 days!

[5] One authority gives £3,091 6s., another £3,090 5s.

educated at the University of Saumur and Clare Hall, Cambridge. He distinguished himself in mathematics and philosophy, being elected a Fellow of the Royal Society in 1714. Two years later he served on the Council of the Society, and was President from 1741 to 1753. Folkes received degrees from both Oxford and Cambridge Universities, and in 1792 a monument to his memory was erected in Westminster Abbey. He was the author of several papers published in the *Philosophical Transactions* and in *Archaeologia*, forming collections of prints, drawings, coins, gems and books, the latter being particularly rich in natural history.

The collections of Sir Hans Sloane (1660–1753) eventually went to form the nucleus of the British Museum, and a study of his career as physician, naturalist and collector is most fascinating. He was born at Killyleagh, County Down, and went to London to study medicine. Becoming friendly with Robert Boyle, his interest in chemistry was stimulated, while his botanical pursuits were encouraged by intimate contact with John Ray. Sloane went to Paris in 1683 to continue his medical studies, before proceeding to the University of Orange, where he acquired his M.D. On his return to London, Sloane assisted Thomas Sydenham in his practice, and was soon elected a Fellow of the College of Physicians and of the Royal Society. He went to Jamaica in 1687 as physician to the Duke of Albemarle, spending fifteen months in studying the natural history of the island, but on the death of the Duke, Sloane returned to England with an extensive natural history collection. He published *Catalogus plantarum quae in insula Jamaica sponte provenunt*, [*etc.*], 1696, in the preparation of which he was assisted by John Ray. This was followed by *A voyage to the islands of Madera* [*sic*], *Barbados, Nieves, S. Christophers and Jamaica, with the natural history of the herbs and trees, four-footed beasts, fishes, birds, insects, reptiles, &c. of the last of those islands*, [*etc.*], two volumes, 1707–25.[1] Sloane became physician to Christ's Hospital, and married an heiress, being created a baronet on the accession of George I. He served as President of the College of Physicians (1719–35) and of the Royal Society, in which latter appointment he succeeded Sir Isaac Newton. In 1712 Sloane bought the Manor of Chelsea for a country house, together with the freehold of the Physic Garden, which in 1722 he made over to the Society of Apothecaries in return for fifty dried specimens from the Garden to be sent annually to the Royal Society for forty years.[2] A catalogue of the Library at the Garden was published in 1956.[3] Sloane spent enormous sums of money in acquiring antiquities, coins, medals, crystals, seals and gems, as well as botanical

[1] An annotated copy of this, with the original drawings for the plates, is preserved in the British Museum (Natural History).

[2] See Copeman, W. S. C. Sir Hans Sloane and the Apothecaries Garden. *Med. Hist.*, 5, 1961, pp. 154–156.

[3] Chelsea Physic Garden. *Catalogue of the library at the Chelsea Physic Garden*, 1956.

and zoological specimens. He possessed between forty and fifty thousand volumes, in addition to prints, pictures and drawings; 32,000 medals; and nearly 4,000 manuscripts, containing letters and autograph writings by most of the significant people of his time. Sloane instructed his executors to offer the entire collection to the nation for £20,000. A lottery was held in order to raise this sum, and in 1754 Montague House was purchased, and the Sloane collection, with certain others, were placed therein to form the nucleus of the British Museum.[1] The Sloane manuscripts in the British Museum are of particular significance, and the printed books in the collection are rich in botanical literature and periodicals.[2]

Carl Linnæus (von Linné) (1707–78)[3] collected an enormous number of books and specimens, the subsequent history of which is of great interest. In his will Linnæus made provision for his scientific collections to be treated as was the remainder of his estate and shared among his family, but his son, Carl von Linné the younger, who succeeded his father as professor at Uppsala, took over the scientific material, renouncing his share in the other property. Sir Joseph Banks apparently made an offer for the collection, but it was declined. Unfortunately, von Linné the younger died in 1783, the collection reverting to his mother and sisters. Banks no longer required the material, but mentioned to friends that the collections of Linnæus were to be sold for 1,000 guineas. James Edward Smith, then still a student, sought further information, and discovered that there were about 19,000 plant sheets, 3,200 insects, 1,500 shells, 700–800 corals, 2,500 stones and minerals, together with the library of Linnæus, numbering 2,500 volumes. Smith purchased the complete collection, which arrived in London in October, 1784, packed in twenty-six large chests. The new owner found that he had also acquired numerous manuscripts of Linnæus (father and son), and of other scientists, together with about 3,000 letters. Two years later Smith and others founded the Linnean Society (1788), James Edward Smith, who was later knighted, serving as its President until his death in 1828. He published several previously unprinted manuscripts of Linnæus, including *A tour in Lapland*, 1811, and a selection from his correspondence, two volumes, 1821. After the death of Sir James Edward Smith his widow sold the collection to the Linnean Society for 3,000 guineas, and the Society still houses the material in a manner equal to its importance.

[1] See Francis, Sir Frank. Sir Hans Sloane, 1660–1753, as a collector. *Lib. Assn. Rec.*, 63, 1961, pp. 1–5; and Esdaile, Arundell, J. K. *The British Museum Library: a short history and survey*, [*etc.*], 1946.

[2] See Brooks, E. St. John. *Sir Hans Sloane. The great collector and his circle*, 1954; de Beer, G. R. *Sir Hans Sloane and the British Museum*, London, [*etc.*], 1953; Chance, Burton. Sketches of the life and interests of Sir Hans Sloane: naturalist, physician, collector and benefactor. *Ann. med. Hist.*, 2nd Ser., 10, 1938, pp. 390–404; and Thomson, W. W. D. Some aspects of the life and times of Sir Hans Sloane. *Ulster med. J.*, 7, 1938, pp. 1–17.

[3] See also pp. 193-194.

James West (1704?–72) was educated at Balliol College, Oxford, was called to the Bar, and sat in Parliament for St. Albans. West was keenly interested in science and antiquarian pursuits, being elected a Fellow of the Royal Society in 1726, and serving as President in 1768. He formed an extensive collection of books, manuscripts, paintings, prints, coins and medals. The books were auctioned in 1773, and totalled 4,653 lots, which included thirty-four Caxtons. Many of the manuscripts eventually found their way to the British Museum, and others are in the Bodleian Library at Oxford.

Benjamin Franklin (1706–90)[1] was an ardent book collector and avid reader all his life, having been a printer and a founder of several libraries. At his death he possessed 4,276 volumes, which were disposed of by bequest, sale and auction, and many of which are now housed in American libraries. Edwin Wolf[2] has attempted a reconstruction of Franklin's library, and Margaret B. Korty[3] has written on his bibliographical connections.

Albrecht von Haller (1708–77)[4] amassed a library of about 20,000 volumes which was acquired by the Emperor Joseph, and eventually went to Padua. About 13,000 letters addressed to Haller are preserved in the Stadtbibliothek in Berne.

The library of Sir Joseph Banks (1743–1820) has been described as "one of the finest libraries of books on natural history ever collected". Banks was born in London on 2nd February, 1743, and educated at Harrow, Eton and Christ Church, Oxford. His father died in 1761, leaving him a large estate, and five years later Joseph Banks went on a scientific expedition to Newfoundland, whence he brought back a large collection of plants and insects. In 1768 he equipped the *Endeavour* and sailed round the world with James Cook, the voyage occupying three years. The year 1772 saw Banks visiting Iceland, where he obtained botanical specimens, and acquired a large collection of Icelandic manuscripts and printed books, including the library of Halfdan Einarsson. This latter, Banks presented to the British Museum, and a few years later gave a further collection to the same foundation. Banks was elected a Fellow of the Royal Society in 1766, succeeding as President of the Society in 1778, which position he occupied for forty-one years.[5] Among other honours, Banks was created a baronet in 1781. His collections were always

[1] See also pp. 150-152.

[2] Wolf, Edwin. The reconstruction of Benjamin Franklin's library: an unorthodox jigsaw puzzle. *Papers Bib. Soc. Amer.*, 56, 1962, pp. 1–16.

[3] Korty, Margaret Barton. Franklin's world of books. *J. Lib. Hist.*, 2, 1967, pp. 271–328.

[4] See also pp. 178-179.

[5] See Cameron, Hector Charles. *Sir Joseph Banks, K.B., P.R.S., the autocrat of the philosophers*, 1952; and Arber, Agnes. Sir Joseph Banks and botany. *Chron. Bot.*, 9, 1945–46, pp. 94–106.

open to scholars. He bequeathed some of his manuscripts to the Royal Society and to the Mint, the remainder going to the British Museum, subject to a life interest in them by Robert Brown, the eminent botanist, who was librarian to Banks. The collection was transferred to the Museum in 1827, when Brown became Keeper of the Department of Botany. There were about 25,000 printed books mainly devoted to natural history, a catalogue of the library having been compiled by Jonas Dryander[1] when he was librarian to Banks. The Banks Library occupies a separate room in the British Museum, the scientific collections being housed at South Kensington. Banks was the author of several scientific articles and two books, *Short account of the cause of the disease in corn called by farmers blight, mildew and rust*, London, 1805, of which there were a few quarto copies printed before the octavo edition was published, and which went into several editions; and *Some circumstances relative to merino sheep*, 1809. He was also responsible for the publication of Kaempfer's *Icones plantarum*, 1791, and *Coromandel plants*, 1795–1819, by William Roxburgh (1751–1815). A 'grangerized' copy of E. Smith's *Life of Sir Joseph Banks*[2] acquired by the British Museum (Natural History) in 1963 contains several letters from Smith to Alexander Meyrick Broadley (1847–1916), some correspondence of Banks, and other interesting information detailed by Averil Lysaght.[3] An edition by Warren R. Dawson of Banks' correspondence was published in 1958.[4] It is also of interest to record that Sir Joseph's sister, Sarah Sophia Banks, had similar interests, and that when she died in 1818 much of her collection went to the British Museum.[5]

An American collector, Nathaniel Bowditch (1773–1838), the navigator and astronomer, collected about 2,500 volumes, mainly devoted to mathematics and astronomy. The collection includes many pamphlets, maps and charts, and is now housed in Boston Public Library.[6] Bowditch was born at Salem, Massachusetts, and became president of a marine insurance company after making five lengthy voyages. He revised J. H. Moore's *The practical navigator*, which was issued under Bowditch's name in the third American edition, and he also published many studies

[1] Dryander, Jonas. *Catalogus bibliothecae historico-naturalis Josephi Banks*, 5 vols., 1796–1800; reprinted 1965–66.

[2] Smith, E. *The life of Sir Joseph Banks*, 1911.

[3] Lysaght, Averil. A 'grangerized' copy of E. Smith's Life of Sir Joseph Banks. *J. Soc. Bib. Nat. Hist.*, 4, 1964, pp. 206–209.

[4] *Bank's letters: a calendar of M.SS. correspondence preserved in the British Museum, the British Museum (Natural History) and other collections in Great Britain, edited by Warren R. Dawson*, 1958. Supplementary letters in *Bull. Brit. Mus. (Nat. Hist.) Hist. Ser.*, 3, 1962, pp. 43–70; 3, 1965, pp. 71–93.

[5] See Fletcher, William Younger. *Op. cit.*, pp. 270–274; and Lysaght, Averil. Some eighteenth-century bird paintings in the Library of Sir Joseph Banks (1743–1820). *Bull. Brit. Mus. (Nat. Hist.)*, 1959, Hist. Ser. 1, No. 6, pp. 251–371.

[6] Munsterberg, Margaret. The Bowditch collection in the Boston Public Library. *Isis*, 34, 1942–43, pp. 140–142.

in astronomy, mathematics and navigation. Many of his MSS. are included in the Bowditch Collection, which contains many rare editions of the writings of Sacrobosco, Copernicus, Galileo, Tycho Brahe, Kepler, Euclid, Boyle and Newton.

The botanist and antiquary Dawson Turner (1773–1858) was the grandfather of Sir Joseph Hooker. In his younger days he was a keen field-botanist, making collecting expeditions to Scotland, Ireland, Wales and various parts of England. Dawson Turner compiled a great monograph on the *Fuci*, after which he relinquished the study of botany, and he gave his extensive herbarium to Sir William Hooker (Sir Joseph Hooker's father), when he was appointed professor at Glasgow University. Turner possessed a large collection of pictures, about 8,000 books, many of which contained drawings by himself, his wife and daughters, and 40,000 letters and manuscripts. Part of his library was sold in 1853, the remainder being dispersed at sales held in 1859, 1860 and 1869. Items from Dawson Turner's collection are now contained in the British Museum and the Bodleian Library.[1]

Peter Mark Roget (1779–1869) was a physician and the author of the remarkable *Thesaurus of English words and phrases*. He took part in the activities of the scientific societies of his period, was first Fullerian Professor at the Royal Institution (1834–36), and was Secretary of the Royal Society (1827–48). His library of over 4,000 items included 1,732 scientific books, and was sold at Sotheby's shortly after his death. D. L. Emblen has contributed an article on the collection, based on the sale catalogue, but only one item has since been identified.[2]

The Botany School at Cambridge houses 1,750 books that formerly belonged to Charles Darwin (1809–82),[3] which is remarkable considering Darwin's attitude towards printed material. It is recorded[4] that he never had books bound, cut heavy books in half, and it is said that Lyell published the second edition of his *Principles of geology* in two volumes because Darwin had told him that he was compelled to break the first into two parts, as it was unwieldy. Pamphlets and similar material he treated even more drastically, tearing out the pages that interested him, and throwing away the others. His library must have presented a picture

[1] See De Ricci, Seymour. *Op. cit.*, pp. 137–138; Dawson, Warren R. Sir Joseph Hooker and Dawson Turner, *J. Soc. Bib. Nat. Hist.*, 2, vi, 1950, pp. 218–222; Dawson, Warren R. Dawson Turner, F.R.S. (1755–1858). *J. Soc. Bib. Nat. Hist.*, 3, 1958, pp. 303–310; and Greene, S. W. The dates of publications of Smith's *Flora Britannica* and Turner's *Muscologia Hibernica*. *J. Soc. Bib. Nat. Hist.*, 3, 1957, pp. 281–290.

[2] Emblen, D. L. The library of Peter Mark Roget. *Book Collector*, 18, 1969, pp. 449–469.

[3] See Rutherford, H. W. *Catalogue of the library of Charles Darwin now in the Botany School, Cambridge*, Cambridge, 1908. The Darwin Library at Down House, Downe, Kent, contains about 5,000 volumes including notebooks and letters of Charles Darwin, and first editions of his works.

[4] *The life and letters of Charles Darwin.... Edited by Francis Darwin*, Vol. 1, 1888, pp. 150–151.

of a model working collection for individual use, but of little value to others.[1]

Although the greater part of the library of Baron Friedrich Heinrich Alexander von Humboldt was destroyed, it is of interest to record its fate. Alexander von Humboldt (1769–1859),[2] scientist, traveller and philosopher, by a codicil to his will left all his personal belongings, including his library, to his servant, Castellan Seifert, who was not latterly paid a salary. Humboldt had spent his inheritance on financing his 1799–1804 American expedition, and publishing his monumental account of it as *Voyage aux régions equinoxiales de Nouveau Continent fait en 1799–1804*, Paris, 1805–34. The library then contained 11,164 items, and was particularly strong in physics, botany, exploration and Americana, many of the books containing marginal notes by Humboldt. Seifert sold the collection to Asher & Co. of Berlin, and it was then acquired by Henry Stevens of London, who issued a catalogue of the library.[3] The library was to be auctioned by Messrs. Sotheby, and the sale began on 1st June, 1865. On the third day, however, the auction rooms caught fire, the majority of the books being destroyed. In 1871 Messrs. Sotheby issued a catalogue of the remaining items, which numbered only 574, and realized £134 19s., a few of the books going to the United States.

John Thomas Graves (1806–70), a native of Dublin, became an eminent lawyer, classical scholar and mathematician, and was professor of jurisprudence at University College, London from 1839 to 1843. Graves was a great friend of Sir William Rowan Hamilton, and himself wrote several important papers devoted to mathematics. He collected an extensive and very valuable mathematical library, which he bequeathed to University College. The Graves Library contains over 10,000 books and 4,600 pamphlets covering mathematical, astronomical and physical publications of all countries and periods. Probably the most important private mathematical collection ever made, it contains many rare and possibly unique items, most of the volumes being handsomely bound.

We must here record two private libraries now housed in Scotland, the collectors of which are well known because of their association with the *Bibliotheca chemica*.[4] James Young (1811–83) studied chemistry under Thomas Graham at Anderson's College, Glasgow, and became the founder of the shale oil industry in Scotland. About 1850 he began

[1] See also pp. 238–242.

[2] See also pp. 185–187.

[3] Stevens, Henry. *The Humboldt Library. A catalogue of the library of Alexander von Humboldt, with a bibliographical and biographical memoir*, 1863; reprint announced 1966.

[4] See Ferguson, John. *Bibliotheca chemica: a catalogue of the alchemical, chemical and pharmaceutical books in the collection of the late James Young of Kelly and Durris, Esq., LL.D., F.R.S., F.R.S.E.*, 2 vols., Glasgow, 1906, reprinted 1954; see also pp. 312–313.

to collect items of interest in the history of chemistry, in which he was later aided by John Ferguson (1838–1916), professor of chemistry at Glasgow University from 1874 to 1915. Young acquired about 1,400 volumes, many of them being very rare, and bequeathed the collection to the Chair of Technical Chemistry at Anderson's College, which is now the University of Strathclyde. The Ferguson Collection, now at the University of Glasgow, is more extensive, and the catalogue of the collection is a virtual bibliography of alchemy.[1] It was published in a limited edition of forty copies as *Catalogue of the Ferguson collection of books mainly relating to alchemy, chemistry, witchcraft and gipsies, in the Library of the University of Glasgow*, two volumes, Glasgow, 1943, with supplement, 1955.

Modern collections of rare books in the hands of private collectors are becoming extremely scarce, largely because of the vastly increased prices of old books. One of the most important collections of this century devoted to the history of zoology was that of Francis Joseph Cole (1872–1959), formerly professor of zoology in the University of Reading. Cole was born in London on 3rd February, 1872, at Clapham, and became interested in natural history at an early age. On leaving school he took up journalism, but prepared himself for university entrance, and went to Oxford. He left there to go to Edinburgh as private assistant to Professor Cossar Ewart, but in 1894 went to Liverpool as lecturer in zoology. During vacations he conducted research at Oxford, where he obtained a B.Sc. In 1906 Cole was appointed lecturer in zoology at Reading, and the following year became the first occupant of its chair of zoology, remaining there until his retirement in 1939. F. J. Cole founded the Museum of Comparative Anatomy at Reading, which is now known by his name, and contributed numerous papers on the cranial nerves and sense organs of fishes. He was elected F.R.S. in 1926. Cole was keenly interested in the history of biology, and was the author of *The history of protozoology*, 1926, *Early theories of sexual generation*, 1930, and *A history of comparative anatomy from Aristotle to the eighteenth century*, 1944, which was reprinted with corrections in 1949. Cole[2] has described how he acquired some of the items in his library of 8,000 volumes, which includes two incunabula, and early editions of the writings of Galen, Vesalius, Harvey, Willis, Leeuwenhoek, Malpighi and others. There are also long runs of early scientific periodicals, and many rare items. Cole's library was bequeathed to Nellie B. Eales, and was acquired by Reading University, which has published a remarkable catalogue edited by Nellie B. Eales, and entitled

[1] See Read, John. Vignette of a master and his masterpiece. *Endeavour*, 1, 1942, pp. 36–37, 91.
[2] Cole, Francis Joseph. Obiter dicta bibliographica. *J. Hist. Med.*, 13, 1958, pp. 2–9; see also The Cole Library of Zoology and Early Medicine, University of Reading. *Nature*, 188, 1960, pp. 1148–1151.

The Cole Library of early medicine and zoology. Catalogue of books and pamphlets. Part 1. 1472 to 1800, Reading, 1969. This contains entries for about 8,000 volumes and is arranged chronologically by the year of the first edition of an author's book, providing pagination, number of plates, with information on engravers, portraits and other illustrations. A list of general biographical works and of individual biographies, and subject and author indexes are included. Engraved book plates, signatures and inscriptions are noted, and Nellie Eales[1] has described these in a separate article.[2]

Denis I. Duveen[3] has described the genesis and growth of his collection of alchemical and chemical literature. This collection was the basis of his *Bibliotheca alchemica et chemica*,[4] and some twenty-three of the more interesting items are described. A growing interest in Lavoisier[5] led to Duveen specializing in a collection devoted solely to this scientist, and to the original collection going in its entirety to the University of Wisconsin. The nature and extent of the collection at Wisconsin has been described in some detail.[6] It contains over 3,000 items, including 8 incunabula, 256 items of the sixteenth century, 984 of the seventeenth, 905 of the eighteenth, 554 of the nineteenth and 310 of the twentieth century. The Duveen collection is included in *Chemical, medical and pharmaceutical books printed before 1800 in the collections of the University of Wisconsin libraries*, [*etc.*], Madison, Milwaukee, 1965, edited by John Neu. This contains 4,442 numbered entries arranged alphabetically by author, and lists several other special collections.

The classics of science in their original editions are finding their way into national, university and other libraries, whence they seldom emerge. The day of the private collector on a large scale is almost over, and one is delighted when these collections are housed in institutions where they will be cared for, catalogued, and exploited in a manner permitting the scholar to make use of them. Catalogues of these collections should be printed, research workers given access, and copies on microfilm or in photographic form made available, so that these treasures are made available for research purposes. Books were not written, printed, bound, purchased, and collected together as evidence of historical development, to

[1] Eales, Nellie B. On the provenance of some early medical and biological books. *J. Hist. Med.*, 24, 1969, pp. 183–192.

[2] See also Eales, Nellie B. Francis Joseph Cole, 1872–1959. *J. Hist. Med.*, 14, 1959, pp. 267–272; and Franklin, Kenneth James. Francis Joseph Cole, 1872–1959. *Biog. Mem. Fellows Roy. Soc.*, 5, 1960, pp. 37–47.

[3] Duveen, Denis I. The Duveen Alchemical and Chemical Collection. *Book Collector*, 5, 1956, pp. 331–342, 4 pl.

[4] See p. 313.

[5] See pp. 166–169.

[6] Ives, Samuel A., and Ihde, Aaron J. The Duveen Library. *J. chem. Educ.*, 29, 1952, pp. 244–247.

be gazed at through glass display cases at rare intervals. Facsimile reprints, although sometimes notoriously expensive, should be purchased even by libraries housing the originals, and the copies made available for loan. Scholars and students can seldom find enough time for study within libraries situated at a distance, and an interest in the classics of science should be encouraged at an early stage. A firm knowledge of the history of a subject ensures that research time and money is not wasted in duplicating the errors and successes of the past. Personal collections of books are said to reflect the personalities of their owners. Modern private libraries are limited in size by availability of requisite items, and by personal finance, but useful collections can still be acquired by selective individuals concentrating on particular fields of interest. Comparatively modern books can command high prices within a few years of publication, and collectors who specialize in books can serve their scientific interests and enjoy the pleasure of ownership, while investing their money in a highly desirable commodity.

CHAPTER XI

Scientific Publishing and Bookselling

> "And out of olde bokes, in good feith,
> Cometh al this newe science that men lere."
>
> CHAUCER

The development of science depends upon the rapid diffusion of scientific information throughout the world. Research is wasted if based on obsolete knowledge, and scientists should be prepared to spend a proportion of their time in the library keeping abreast of current developments. This suggests that scientists must rely heavily upon the book trade, and particularly upon the publishers of journals, for their basic materials embodying the results of work accomplished by workers in similar fields of research throughout the world. Without appreciating what has been done in the past, and what is being accomplished at the present time, it is almost impossible to plan for the future. Duplication of effort seldom results in progress.

Printers, publishers and booksellers are the main essentials constituting the book trade, and their work is vital in the development of science. Their products are the vehicles conveying eagerly awaited knowledge to the four corners of the world, and it is essential that they should be encouraged to produce their wares efficiently and economically. Speed in production without loss of accuracy, and economical marketing that will bring a profit to the publisher without pricing the book or periodical above the purse of intending purchasers, are necessary today more than at any period. Unfortunately, present-day trends are towards lengthy delays in production, due to various causes, and greatly enhanced prices that are constantly increasing, which tend towards making libraries and institutions the main purchasers of many highly-priced books and journals. The budgets of these libraries are not elastic, and there are indications that there will be drastic curtailments of expenditure on this type of literature in the near future.

A table based on an analysis of publications listed in the *British National Bibliography* from 1st July, 1967, to 30th June, 1968, provides information on book prices, and the details given below for certain scientific subjects are taken from fuller information available in the *Library Association Record*.[1]

	Per cent increase 1967–68 over 1966–67	*No. of volumes 1967–68*
General science	8·2	80
Mathematics	23·3	493
Astronomy	14·2	88
Physics	32·2	393
Chemistry	24·8	330
Geology, meteorology, palaeontology	19·8	147
General biology	13·1	213
Zoology	32·7	314

The book trade is as old as the book itself, but specialization is comparatively recent. Even today only a few publishers concentrate entirely on scientific literature, but a survey of the history of publishing and bookselling reveals that most of the early eminent printers issued some scientific books.

We have no knowledge of organized bookselling prior to the Greek and Roman civilizations, but in Alexandria the best editions of established writers were reproduced from texts in the Alexandrian Library.[2] We know little of the relationship between the Greek author and his publisher, but there was neither copyright protection nor royalties, and authors were content with the patronage of influential figures. Indeed, the remuneration of authors by means of royalties is a comparatively recent innovation. Rome had newspapers and public libraries, and every respectable house contained a library, the wealthier classes also having slave-readers (*anagnostae*) and transcribers (*librarii*). Atticus employed slaves to duplicate from dictation the writings of favourite authors, this system of multiplication being the foundation of the publishing trade. The bookseller was at first the manufacturer, publisher and retailer. Mainly owing to the employment of slave labour, prices were low and books plentiful. In Rome, the bookseller's haunt was in the Argiletum, the pillars outside the shops advertising the titles of the books on sale within. Authors brought their manuscripts to the publisher, who passed them on to the slave copyists, and a complete edition was

[1] Average book prices. *Lib. Assn. Rec.*, 70, 1968, pp. 233–238.
[2] Additional information on the early history of the subject is available in Mumby, Frank Arthur. *Publishing and bookselling. . . . With a bibliography by W. H. Peet*, 1934; and Pinner, H. L. *The world of books in classical antiquity*. Leyden, 1948.

produced within a very short period. Books were still in roll form, and consisted of parchment or papyrus, but after the third century A.D. the papyrus roll was largely displaced by parchment, and this being heavier led to the development of the codex form of book.

Titus Pomponius Atticus, the most prominent publisher of ancient Rome, was Cicero's literary adviser, and himself both scholar and author. He owned retail branches in the provinces. Those persons unable to afford rare or valuable books were permitted to inspect them upon payment of a fee,[1] and this scheme is similar to that adopted in the Middle Ages, when booksellers loaned texts by the sheet to students, charging high fees for the privilege.[2]

We have little knowledge of book distribution during the Middle Ages, but books were copied in the monasteries, the actual transcribing being confined to a few monks. Books became elaborate works of art, with beautiful illustrations and bindings, and were correspondingly expensive. There was no book-buying public, but monasteries exchanged volumes or loaned them to each other for copying. Most monasteries possessed a *scriptorium*, which was sometimes divided into carrels, in each of which worked a scribe. Needless to say, many of the books copied were theological.

The growth of universities as seats of learning necessitated the supply of texts for students. Booksellers (*stationarii*) were established, and these sold manuscripts, and also loaned them by the sheet. Sometimes the university controlled the book market, as at Paris in the fourteenth century. Copies of the texts for sale had to be submitted to the university, which examined them for accuracy, and also fixed their prices. In this country, Oxford and Cambridge Universities at one period acted as censors of the book trade.

Without doubt, the invention of printing was the most important factor in the diffusion of learning, and the spread of learning obviously increased the demand for books. The early booksellers made their own type, printed, bound and sold books, and also stocked stationery and newspapers. In this country, the stationers were first formed into a guild in 1403, but had no control over printed books until their first charter was granted in 1557. This gave them power to search for and destroy seditious and heretical literature. The entry of copies at Stationers' Hall commenced in 1558, but books themselves were not delivered until the year 1663.

The advent of printing did little to repay authors by means of royalties or other methods. Sometimes they lived at the houses of publishers,

[1] Pinner, H. L. *Op. cit.*, p. 48.

[2] See Chaytor, H. J. *From script to print: an introduction to medieval literature*, Cambridge, 1945, p. 136. This work is an invaluable study of the subject covered by the title.

enjoying hospitality, and perhaps editing or compiling future publications, but, generally speaking, any reward an author received was the result of patronage.

Chapter II provides additional information on scientific books printed in the fifteenth century, but it is of interest to record the names of the printers of certain of these works. Conrad Sweynheym and Arnold Pannartz, who carried the art of printing into Italy, were responsible for an edition of Strabo's *Geographia*, Rome, [1469], and also printed editions of Pliny the Elder's *Historia naturalis* at Rome in 1470 and 1473. Johannes de Spira had previously published an edition of the latter at Venice, 1469. The famous Venetian printer, Nicolas Jenson, was responsible for Macrobius' *In somnium Scipionis expositio Saturnalia*, Venice, 1472; Pliny the Elder's *Historia naturalis*, Venice, 1472, with an Italian version in 1476; and *Scriptores rei rusticae*, Venice, 1472, which contains Varro's *Rei rusticae libri III.* Venice was the home of numerous early printers, and scientific writings predominate among their publications. The great Aldus Manutius was born in 1450 at Sermoneta, near Rome, and it was not until he was forty-five that he went to Venice. His first publication appeared in 1495, to be followed by the five-volume edition of Aristotle's *Opera* in Greek, issued from 1495 to 1498. Aldus introduced the small 'pocket' classics, known as 'Aldines', and he issued 120 editions, both Greek and Latin, in twenty years. His device, the anchor and dolphin, is recognized as a sign of good printing. After the death of Aldus in 1515, his descendants carried on the business until 1597. Other scientific books from the press include Greek versions of Theophrastus' *De historia plantarum*, 1497, and Dioscorides' *De materia medica*, Venice, 1499; and Titus Lucretius Carus' *De rerum natura*, Venice, 1500,

Joannes and Gregorius de Gregoriis de Forlivio printed numerous scientific works at Venice, including Albertus Magnus' *De anima*, 1494; *De animalibus*, 1495; *De coelo et mundo*, 1490 and 1495; *De generatione et corruptione*, 1495; *Metaphysica*, 1494; *Meteorum libri IV*, 1494; *De mineralibus*, 1495; and *Physica sive de physico auditu*, 1488 and 1494. They printed editions of Alchabitius' *Liber isagogicus ad scientiam iudicialem astronomiae*, 1491; Aristotle's *De animalibus*, 1492, *De coelo et mundo*, 1495, and *Meteorum libri IV*, 1491; and Boëthius' *Opera*, 1491–92, and 1497[8]–99.

The work of Erhard Ratdolt as a printer of books on astronomy and mathematics is of particular significance. He was a native of Augsburg, but printed at Venice from 1476 to 1486, when he returned to Augsburg, resuming work in that town. In the year 1476, with Bernard Maler and Peter Löslein, Ratdolt printed in Latin and Italian the *Kalendarium* of Johann Müller (Regiomontanus).[1] This was followed in 1482 by an edition of Euclid's *Elementa geometriae*, a folio containing fine diagrams

[1] Ratdolt also printed editions of this in 1478, 1483 and 1485.

consisting of about 420 wood engravings. Ratdolt was the first printer to use several coloured inks simultaneously, his edition of Sacrobosco's *Sphaericum opusculum*, printed in 1485, containing coloured astronomical diagrams.[1] Other scientific works from his press at Venice include Alchabitius' *Liber isagogicus ad scientiam iudicialem astronomiae*, 1482 and 1485; Mela's *Cosmographia*, 1482; Alfonso's *Tabulae astronomicae*, [1482 or 1483]; Ptolemy's *Quadripartitum opus*, 1484; Pietro Borgo's *Arithmetica*, 1484; Hyginus' *Poeticon astronomicon liber*, 1485; and Michael Scot's *Physiognomia* [N.D.]. The following bear Ratdolt's Augsburg imprint: Albumasar's *Flores astrologiae*, 1488 and 1495; *Introductorium in astronomiam*, 1489; and his *De magnis coniunctionibus*, 1489; Boëthius' *Arithmetica*, 1488; and Johann Engel's *Astrolabium*, 1488. Erhard Ratdolt died in 1527 or 1528.[2]

William Caxton (*c.* 1422–*c.* 1491) introduced the art of printing into England when he set up his press at Westminster in 1476. He printed two editions of the *Myrrour of the worlde*, a folio [Westminster, 1481] [S.T.C. 24762], with a second edition, [Westminster, 1490] [S.T.C. 24763]. Another edition of this was printed by Laurence Andrewe, [1529?] [S.T.C. 24764]. William Caxton edited and translated some of his publications, and was responsible for printing over one hundred books. Nellie J. Kerling[3] has also connected him with an extensive trade importing books from the Continent, and H. S. Bennett[4] has published a study of the book trade from Caxton to 1557. Caxton's successor, Wynkyn de Worde, printed editions of Bartholomaeus Anglicus' *De proprietatibus rerum*, [Westminster, 1495][5] [S.T.C. 1536], and of Dame Juliana Berne's *Boke of St. Albans, containing treatises of hawking, hunting and cote armour*, 1496 [S.T.C. 3309], of which an edition had previously appeared [St. Albans, 1486] [S.T.C. 3308]. Another fifteenth-century London printer, William Machlinia, printed several scientific books, including Albertus Magnus' *Liber aggregationis*, 1485 [S.T.C. 258], and *Secreta mulierum*, 1485 [S.T.C. 273].

It has been estimated by Robert Steele[6] that of the 30,000 incunabula printed, about 1,800 are scientific, and from Procter's list of about 10,000 incunabula, 838 are devoted to science, using the term in a broad sense.

[1] Ratdolt had printed an earlier edition of this in 1482.

[2] See Redgrave, Gilbert R. *Erhardt Ratdolt and his work at Venice. A paper read before the Bibliographical Society, November 20, 1893*, 1894. (Illustrated Monographs issued by the Bibliographical Society, No. 1.)

[3] Kerling, Nellie J. M. Caxton and the trade in printed books. *Book Collector*, 4, Autumn, 1955, pp. 190–199.

[4] Bennett, H. S. *English books and readers 1475 to 1557: being a study in the history of the book trade from Caxton to the incorporation of the Stationers' Company*, Cambridge, 1952.

[5] See Mitchner, Robert W. Wynkyn de Worde's use of the Plimpton Manuscript of *De proprietatibus rerum. Lib. Trans. Bib. Soc.*, 5th Ser., 6, 1951–52, pp. 7–18.

[6] Steele, Robert. What fifteenth century books are about. 1. Scientific books. *The Library*, N.S. 4, 1903, pp. 337–354.

Mathematics has 24; astronomy 81; cosmography 76; natural history 33; physics 77; occult science 93; and medicine 247.

London became an important centre of the book trade, and St. Paul's, Little Britain, and later Paternoster Row, were the main centres of the printing, publishing and bookselling industry. Travelling booksellers took their wares to the fairs, the chief of which was at Stourbridge, while the stationers attended the world fair at Frankfurt, held twice annually, but which was later superseded by the Leipzig fair. The trade flourished, the output of books increasing every year, as did also the numbers of printers. It is impossible to record even the names of all those who have published scientific books, but it is of interest to note certain publishing houses founded at an early date which continue to serve scholarship. The University Presses of Oxford and Cambridge date from the fifteenth and sixteenth centuries respectively. The first book from Oxford was probably printed in 1478. The earliest printer there was Theodoric Rood, who went to Oxford from Cologne, continuing his work until 1485, when the first press came to an end. The second press was active between 1517 and 1520, but the output from these two was mainly classical or theological. After the year 1520 there is a gap in the history of the press, which was resumed in 1585, and in 1636 a Royal Charter was granted permitting the University to print all kinds of books. There is scant printed information on the early books printed by the Clarendon Press, Oxford, or its associated house, the Oxford University Press, but among early issues are numbered Captain John Smith's *Map of Virginia*, several editions of Burton's *Anatomy of melancholy*, and Bacon's *Advancement of learning*. The press maintains an ideal appropriate to its distinguished history, and the standard of its scientific publications is unsurpassed.[1]

The first press in Cambridge was set up by Joseph Laer, or Lair, of Siegburg,[2] generally known as John Siberch, in 1521, but he printed only ten books, and it was not until 1582, when Thomas Thomas was appointed University printer, that it began a continuous history. A history of the Cambridge University Press has been published,[3] and a separate list of publications[4] contains the following scientific items, selected for their historical value: Fr. Anthonie's *Medicinea, chymicae, et veri potabilis auri assertio*, [*etc.*], 1610; Isaac Barrow's *Euclid*, 1655; John Ray's *Catalogus plantarum circa Cantabrigiam nascentium*, 1660 (two states);

[1] See *Some account of the Oxford University Press, 1468–1921*, Oxford, 1922.

[2] Goldschmidt, E. P. *The first Cambridge University press in its European setting. . . . The Sanders Lectures in Bibliography, 1953*, Cambridge, 1955.

[3] Roberts S. C. *A history of the Cambridge University Press*, 1521–1921, Cambridge, 1921. Illustrations; bibliography; Appendix I. University printers, 1521–1921; Appendix II. Cambridge books, 1521–1750.

[4] *A list of books printed in Cambridge at the University Press, 1521–1800*, Cambridge, 1935.

G. Fournier's *Euclid*, 1665; P. Galtruchius' *Mathematicae totius institutio*, 1668; B. Varenius' *Geographia generalis*, edited by Isaac Newton, 1672, 1681, 1712; Sir Isaac Newton's *Arithmetica universalis*, 1707 and 1713; N. Saunderson's *Elements of algebra*, two volumes, 1740; R. Watson's *Chemical essays*, four volumes, 1781–86; J. Wood's *Elements of algebra*, 1795 and 1798; S. Vince's *Conic sections*, 1781; and several other of this author's astronomical writings. Cambridge University Press publishes numerous scientific journals, series of monographs, and academic paper-backs, maintaining a world-wide reputation for a high standard of scholarship and production.

In addition to these university presses several of our great publishing houses have a lengthy, distinguished history behind them. The firm of Longmans, Green & Co., for example, was founded by Thomas Longman, who in 1716 was apprenticed to John Osborn, a bookseller trading in Lombard Street. In 1724 Longman took over the *Black Swan* and the *Ship*, from which the present publishing house has grown. The firm was part-publisher of Boyle's *Works*, issued in 1725, and the following scientific publications have also been issued with the Longman imprint: a complete issue of Bacon's *Works*; Herschel's *Outlines of astronomy*; Watt's *Dictionary of chemistry*; Sir Edward Thorpe's *Dictionary of applied chemistry*; Mellor's *Inorganic and physical chemistry*; Thorburn's *British birds* and *British mammals*; J. G. Millais's two books on British ducks; and Maxwell's *Heat*.[1]

The firm of John Murray holds a reputation unrivalled throughout the world, its high ideals and standards being unique. The firm was founded by John McMurray, who in 1768 succeeded William Sandby, a Fleet Street bookseller. The medical portion of the stock was taken over in 1803 by Samuel Highley, a partner, who set up in business for himself. Among nineteenth-century scientific classics published by John Murray we mention Sir Charles Lyell's *Principles of geology*; Sir Roderick Murchison's *Silurian system*; and Paul B. Du Chaillu's *Explorations in Equatorial Africa*. Several of Charles Darwin's publications were handled by the firm. In 1845 Murray bought the copyright of Darwin's *Voyage of the Beagle*, which was first published by Colburn. Darwin's *Origin of species by means of natural selection* was published in November, 1859, the entire edition being sold at Murray's annual sale, and a reprint was immediately called for. Darwin's *Fertilization of orchids*, his *Descent of man*, and also his treatise on *Earthworms* were first published by John Murray.[2]

[1] See Cox, Harold, and Chandler, John E. *The house of Longman, with a record of their bicentenary celebrations . . . 1724–1924*, London, [etc.], *for private circulation*, 1925; Blagden, Cyprian. *Fire more than water: notes for the story of a Ship*, London, [etc.], *for private circulation*, 1949.

[2] See Paston, George. *At John Murray's; records of a literary circle, 1843–1892*, [etc.], 1932.

Taylor & Francis Ltd. of London have had almost two hundred years experience of printing scientific books and journals. In addition to their own periodicals they print for universities, societies and academic bodies, and they also issue reprints. *The Philosophical Magazine*, *Monthly Notices of the Royal Astronomical Society*, *The Ibis*, and *Annals of Science* are among the learned journals published by them, and their list of scientific monographs is equally impressive.

The House of Macmillan was founded in 1843 by Daniel Macmillan (1813–57) and Alexander Macmillan (1808–96), the first shop being opened in Aldersgate Street, London.[1] Before the end of the year, however, the business moved to Cambridge. A branch was opened in Henrietta Street, London in 1858, to cope with increasing business, and five years later saw the return of the business headquarters to London. The present premises in St. Martin's Street were erected in 1897, the Macmillan Company of New York having been founded by the British firm in 1869, while the Macmillan Company of Canada was established at Toronto in 1906. The publishing interests of the firm cover the whole realm of knowledge, including the production of many books devoted to mathematics, astronomy, physics, chemistry, and science in general. Among authors whose books have been handled by the firm are numbered Clerk Maxwell, Oliver Lodge, Watts, Thorpe, Francis Galton, Sedgwick, T. H. Huxley, Michael Foster, Geikie, J. D. Hooker, Ray Lancaster and Norman Lockyer. But the chief scientific publication for which the House of Macmillan is renowned is *Nature* (Plate 15). This weekly periodical[2], devoted to science in general, is the main vehicle for the prompt publication of recent advances in science. Its reputation is the envy of all similar publications. *Nature* was first edited by Sir Norman Lockyer, who was succeeded by Sir Richard Gregory in 1919. A. J. V. Gale and L. J. F. Brimble jointly assumed editorial leadership in 1938.[3]

The early booksellers of the United States merely imported books from Europe, but were soon acting as agents for publishers on this side of the Atlantic. Later they added cancel title-pages bearing their own imprints to books imported in sheets, and this is still an extensive two-way trade item between Great Britain and the States. Few publishers have specialized in scientific publications until recent years, and although most American publishers have handled this type of literature from time to time, it is impossible to consider many purely as publishers of scientific books. However, to take the firm of D. Appleton and Company as an example, we find that Daniel Appleton was importing books from Europe for sale in his general stores. In 1831 the first book was published

[1] See, Centenary of the House of Macmillan. *Nature*, 151, 1943, pp. 231–235.
[2] Beginning from January, 1971, *Nature* is published as three separate issues per week.
[3] See Morgan, Charles. *The House of Macmillan*, (*1843–1943*), 1943.

bearing the imprint "D. Appleton", and William Henry Appleton, son of Daniel, was in Europe purchasing books for the American market. Appleton became the American agent for the publications of Darwin, Huxley, Herbert Spencer and John Tyndall. The following were distributed in the United States by D. Appleton and Company: Darwin's *Origin of species*, 1859, his *Descent of man*, 1871, and certain other of his works; Herbert Spencer's *Education—intellectual, moral and physical*, 1860, his *First principles*, 1864, and his *Principles of biology*, 1867; Thomas Huxley's *Man's place in nature*, 1863; his *Lay sermons and addresses*, 1870, and his *Collected essays*, nine volumes, 1893; John Tyndall's *The forms of water*, 1862, his *Lectures on light*, 1873, and certain other of his writings; George Catlin's *Life among the Indians*, 1867, and *Last rambles among the Indians*; and Frank M. Chapman's *Handbook of birds of eastern North America*, together with several other books by this author.[1]

The firm of D. Appleton and Company sprang from the sale of books in a general store, and it is from such beginnings that the gigantic publishing houses of the United States have grown to their present proportions. The McGraw-Hill Book Company, for example, formed in 1909 by the amalgamation of the McGraw Publishing Company and the Hill Publishing Company, has reached the forefront of scientific and technical publishing. Its imprint is a guarantee of integrity and premier workmanship.

Information regarding scientific publishing in France is provided by Dominick Coppola,[2] who discusses three of the outstanding publishing houses in detail. Gauthier-Villars, a leading publisher in mathematics, assumed this name in 1864, having been founded in 1790. They started publishing the *Comptes rendus hebdomadaires de l'Académie de Sciences* in 1835, and in the following year launched *Le Journal de mathématiques pures et appliqués*, which they still publish. This firm has published the complete works of several eminent scientists, including Lagrange, La Place, Cauchy, and Poincaré, and also several notable series such as *Collection des monographies sur la théorie des fonctions*, edited by Émile Borel, of which fifty-three volumes have so far appeared; *Mémorial des sciences mathématiques*, of which over 160 facsimiles have been published, and *Mémorial des sciences physiques*, which has already reached fascicule sixty-seven. Masson et Cie publishes both books and periodicals devoted to science and medicine, and was founded in 1804 by Nicholas Crochard. They took over the *Annales de physique et de chimie*, and in 1824 began publishing the *Annales des sciences naturelles*. Victor Masson joined the

[1] See Overton, Grant. *Portrait of a publisher, . . . and the first hundred years of the House of Appleton, 1825–1925*, [*etc.*], New York, London, 1925.

[2] Coppola, Dominick. French scientific publishing. *Stechert-Hafner Book News*, 11, 1936, pp. 13–15.

firm in 1838, and published Cuvier's *Le règne animale* in twenty volumes. Masson publishes over fifty current journals, and has been responsible for the outstanding works of Poirier and Charpy, Charcot, Perrier, Grignard, Piveteau and Grassé. Louis Goevry organized the Librairie pour les Mathématiques et l'Architecture in 1791, which adopted the name of Dunod in 1860 when Charles Dunod became the proprietor. Comte, Carnot and Berthelot are among the outstanding names on their list, and Dunod publishes several important series and perodicals.

It is of interest to trace the development of publishing and bookselling from the earliest times, and the writings of Charles Knight[1] and Henry Curwen[2] provide us with additional material. The development of the economic history of the book trade has been the subject of a book by Marjorie Plant,[3] and it is significant to study the present trends in publishing. A recent survey devoted to book production in Great Britain supplies figures for the year 1969,[4] showing that 32,393 items were published, of which 9,106 were reprints and new editions. Figures for scientific subjects were: Aeronautics 132; Agriculture and Forestry 265; Astronomy 94; Chemistry and Physics 786; Geology and Meteorology 178; Mathematics 484; Natural Sciences 750; Science, General 99.

Publishing and bookselling suffered severely during and immediately following the war. Paper was scarce and expensive, and printing and binding costs increased considerably. One of the major casualties was the individual purchaser of books, who turned to libraries for his requirements, and made this a regular habit. Generations of students have grown up who have not been encouraged to own textbooks and standard works of reference. This decrease in the demand for such books led to the necessity for printing smaller editions, and the products were therefore more expensive. This further discouraged the intending purchaser, resulting in a vicious circle. Books and journals are more expensive because they are produced in smaller quantities, and the general public cannot afford the high prices prevalent today. A modern trend is towards publishing cheaper editions through book clubs, or paper-back editions, both of which ventures appear to prosper. But many valuable pieces of research remain unpublished because of limited demand, and many books issued through the normal channels have been heavily subsidized by the author, who cannot hope to receive his money back, or by institutions or research organizations. Commercial publishers must cover their costs and make profits, certain of them using money made on textbooks, for example, to subsidize more slowly selling items. It

[1] Knight, Charles. *Shadows of old booksellers*, [N.D.].
[2] Curwen, Henry. *A history of booksellers, the old and the new*, [*etc.*], (1873).
[3] Plant, Marjorie. *The English book trade; an economic history of the making and sale of books. Second edition*, 1965.
[4] *The Bookseller*, 3rd January, 1970, pp. 14–15.

may be that certain types of monographs, and highly specialized texts likely to have limited sales will eventually disappear from trade catalogues, to appear under the imprints of universities, institutions and research organizations, or possibly under the auspices of a national or international body incorporated to publish this type of material.

The universities have had presses since soon after the invention of printing, and even today some of the older ones still do some of their own printing. Some university presses are subsidized by their universities, but apparently none actually contributes to its parent body. They publish books which are intended to advance scholarship, education and religion, but which have small potential markets. Books are kept in print longer, out-of-print items are reprinted, and there is a growth in university presses throughout the world. Some of the newer ones use the older university presses for production, promotion and distribution. Seven are members of the Publishers Association, and in the U.S.A. there were sixty major university presses in 1967, issuing 2,300 titles in 1966, with a further thirty presses publishing less than five titles each year. John Brown[1] includes this information in an article on the subject, in which he suggests that "no commercial publishers can afford to issue a book which will not sell 3,000 copies in three years or to retain a book in print which is not selling 250 copies a year" (p. 314). There must be many highly specialized scientific books worth publishing that cannot hope to reach these figures.

Recent developments in copying and printing methods have caused rare books and out-of-print items to be made available quite cheaply. Facsimile reproductions of classics are common, and microcard, microfilm and micro-fiche copies are encountered in most libraries. These can take the place of originals for most purposes, and are certainly better than nothing. The prohibitive cost of rare items, and their growing scarceness, will obviously lead to an increasing use of photographic methods of reproduction, but it is pointless to produce so-called 'limited' editions at prices that defeat the whole object of attempting to make these works generally available. In certain instances the published price bears no obvious relationship to the cost of production. Scholarship certainly cannot profit when a much sought after item is reprinted in an edition of one thousand copies at ten guineas each, when half that price would have resulted in a handsome profit to the publisher, and more customers would have been able to afford it, without having to wait for it to be remaindered.

The price of books might be lowered by publishing them as separate monographs as parts of Transactions, such as the invaluable *Transactions of the American Philosophical Society* and the *Annals of the New York*

[1] Brown, John. University press publishing. *Lib. Assn. Rec.*, 70, 1968, pp. 310–314.

Academy of Science, so that they could be bought separately as well as included in orders for complete sets for the larger libraries. All libraries suffer from constantly rising prices of books and journals, and by the ever-increasing numbers of these. Certain publishers issue third-rate material and employ unethical devices to sell their products, imposing conditions of sale, increasing advertised prices on publication, and in general giving an inferior service to that enjoyed before this computer dominated age. This aspect has been detailed in connection with medical publishing,[1] but is also applicable to the sciences.

Second-hand booksellers have played an important part in building up many of the historical collections now housed in institutions, and some of their catalogues are of great interest historically.[2] Some of these contain interesting historical material and useful bibliographical information. References to standard bibliographies sometimes identify individual items, such as mention of the Garrison-Morton numbers, and certain booksellers quote "not in Garrison-Morton" as if criticizing that most useful tool, when the more likely explanation is that the item was not considered suitable for inclusion. One of the finest produced catalogues was that issued by Wm. Dawson & Sons Ltd. as Catalogue No. 91 with the title *Medicine and science: a bibliographical catalogue of historical and rare books from the 15th to the 20th century*, containing numerous plates, and a subject index. A catalogue such as this supplies invaluable historical information, and although one must in general approach these advertisements with caution (for, after all, the bookseller is trying to sell his wares), some of the booksellers display the wide scholarship that is acquired by many years of close acquaintance with rare books. As old books become scarcer, prices continue to soar, and libraries find that limited budgets make the acquisition of these desirable items increasingly difficult. Private collectors and donors must be their main sources of supply, and we trust that in the custody of libraries these volumes are cared for, yet made available to serious scholars.

Booksellers have distributed books since the invention of printing, since books were sold in sheets to retailers, and it was their responsibility to have them bound for sale to their own customers. Graham Pollard and Albert Ehrman[3] have prepared a fascinating study of the distribution of books by catalogue, giving details of broadsides and advertisements

[1] See Thornton, John L. British medical publishers 1868–1968. *Practitioner*, 201, 1968, pp. 231–237.

[2] See Zeitlin, Jake. Bookselling among the sciences. *College and Research Libraries*, 21, 1960, pp. 453–457; and Swann, C. Kirke. Natural history books from a bookseller's point of view. *J. Soc. Bib. Nat. Hist.*, 3, 1956, pp. 117–126.

[3] Pollard, Graham, and Ehrman, Albert. *The distribution of books by catalogue from the invention of printing to A.D. 1800. Based on material in the Broxbourne Library*, Cambridge, *Printed for presentation to members of the Roxburghe Club*, 1965.

concerning the book trade; the catalogues of sixteenth-century printers; the book fairs; the importation of books and their sale by auction; and the catalogues of books on pages at the end of monographs, and other interesting material. Unfortunately Pollard and Ehrman's book is too highly priced to reach all the readers it deserves.

CHAPTER XII

Scientific Libraries of Today

"Libraries are used by the scholar and author, and for such are the true universities."

J. S. BILLINGS

The increasing number of scientific books published each year, and the multitude of journals devoted to science and related subjects make it impossible for any individual to acquire all the literature he may desire to consult. The space required to house this overwhelming spate of printed matter, and the higher costs of both books and periodicals, have also sounded the death knell of the private collector on a large scale. Scientists must use libraries to satisfy their requirements, and those having access to comparatively small collections must tap the resources of the more comprehensive libraries, either through inter-library co-operation, or by means of personal visits. Some areas are rich in libraries, so that little hardship is caused by having to trace material outside the institution with which one is connected. In other instances local resources are meagre, and one must take full advantage of all the facilities available for inter-library lending, for the procurement of photo- and micro-copies, and also be able to trace references and their location in libraries with the minimum of trouble.

There are many thousands of libraries throughout the world containing scientific collections, and in this chapter we can mention only a few of the most important in Great Britain and in the United States. The *American Library Directory*[1] contains a classified list of over 21,000 libraries in the U.S.A. and in regions under its administration, and 2,000 Canadian libraries. Arranged by States, entries include details of date of foundation, statistics and special collections. The *International Library Directory*, second edition, edited by A. P. Wales, 1966–67, the *World Guide to Libraries*, second edition, three volumes, Munich, 1968, and the *World of Learning*[2] also contain information on libraries, and there are several other useful guides to libraries of various types or in certain areas.[3]

[1] *American Library Directory, 1968–1969. 26th edition. A classified list of libraries in the United States and Canada with personnel and statistical data. Compiled biennially by Eleanor F. Steiner-Prag*, New York, London, 1968.

[2] *The World of Learning, 1969–70. Twentieth edition*, 1970.

[3] See, for example, Spratt, H. Philip. *Libraries for scientific research in Europe and America*, 1936; and Irwin, Raymond, and Staveley, Ronald, *eds. The libraries of London. Second, revised edition*, 1961.

The World Directory of Agricultural Libraries & Documentation Centres, Harpenden, 1960, edited by D. H. Boalch for the International Association of Agricultural Librarians and Documentalists, for example, aims at recording all libraries and document centres in agriculture and kindred sciences throughout the world. It covers agriculture, forestry, horticulture, plant pathology and protection, veterinary medicine, animal and crop husbandry, dairy science, bee research, poultry, fisheries, food and nutrition. Botany, zoology and entomology in their applied aspects are included. Entries are arranged geographically by continents, then alphabetically by countries under the English form of their names, and then by towns in their vernacular forms. There is a place-name index, and a subject index. Information provided includes date of foundation; number of volumes; number of current periodicals; staff; classification; and publications issued.

Lee Ash and Denis Lorenz have edited *Subject collections: a guide to special book collections and subject emphases as reported by university, college, public and special libraries in the United States and Canada*, third edition, New York, London, 1967, containing 35,000 entries for collections in libraries in the U.S.A. and Canada; and Anthony T. Kruzas has compiled a *Directory of special libraries and information centres*, second edition, Detroit, Michigan, 1968, containing 13,000 entries. Unesco has published *World guide to science information and documentation services*, Paris, 1965, which lists under alphabetically arranged countries the major information and documentation services. Addresses, dates of foundation, names of librarians, subjects covered, stock and bibliographical facilities are recorded. R. T. Bottle and H. V. Wyatt's *The use of biological literature*, 1966, contains an appendix to Chapter 2 (pp. 18–26), giving information on thirty-eight British libraries covering biological subjects. A directory of libraries and information services in Great Britain and Ireland was published as *Aslib Directory. Vol. 1. Information sources in science, technology and commerce*, edited by Brian J. Wilson, London, 1968, and includes information on 2,800 research institutes, libraries and similar organizations.

The British Museum, the Library of Congress, the Bibliothèque Nationale and other national libraries[1] are very rich in scientific material. In several instances they house extensive collections of scientific books and manuscripts that have been acquired by purchase or bequest, and too often these holdings are not appreciated by the scientist, who may be unaware of their existence. Large, unwieldy general collections tend to swamp specialist literature, which may be difficult not only to trace but to consult. Furthermore, this type of material is not always administered by librarians having expert knowledge of the literature involved, and the

[1] See Esdaile, Arundell J. K. *National libraries of the world: their history, administration and public services. . . . Second edition, completely revised by F. J. Hill*, 1957.

necessity for the employment of suitably trained personnel cannot be over-emphasized.[1] Copyright libraries are other sources of large scientific collections, but are, for most purposes, reference libraries only. Most of the special collections housed in the British Museum are recorded in the history of that library by A. J. K. Esdaile,[2] but a directory recording details of specialist collections, particularly of those housed in more general libraries, remains urgently desirable.

University and college libraries present another useful source of scientific literature, certain of them housing very extensive stocks. It is also important to consider the departmental libraries existing within these institutions. D. W. Butcher[3] has published information on departmental libraries of the University of Cambridge, listing among others libraries devoted to anatomy, biochemistry, botany, chemistry, mathematics, physics, physiology and zoology. These are selected from over fifty libraries housed in the University, and most large institutions possess similar departmental libraries.

Government departmental libraries, such as that at the Ministry of Agriculture, Fisheries and Food (see p. 374), which has numerous branch libraries, are additional sources of scientific literature, and certain large public library systems possess important scientific collections. The libraries belonging to scientific societies and institutions, both national and local, present our major source, for many of these make their stocks available to members for home reading. Research institution libraries, particularly those maintained by commercial firms, have become of increasing importance during recent years, and, as information bureaux and sources of scientific literature, are invaluable in the development of industry. Many of these libraries are linked together in this country by Aslib (the Association of Special Libraries and Information Bureaux), which renders an important service to both personal and institutional members, and by the Reference and Special Libraries Section of the Library Association. In the United States, the Special Libraries Association performs a similar function.

Many of the libraries classified under the above headings are outlier libraries of the National Central Library, making their stocks available to borrowers through public and special libraries, and themselves obtaining literature through this channel. Much of the science material previously borrowed through the National Central Library is now acquired through the National Lending Library of Science and Technology (see p. 373).

[1] See Hunt, Judith Wallen. Science librarianship. *Science*, 104, 1946, pp. 171–173.
[2] Esdaile, Arundell J. K. *The British Museum Library: a short history and survey*, [*etc.*], 1946.
[3] Butcher, D. W. The departmental libraries of the University of Cambridge. *J. Document.*, 7, 1951, pp. 221–243.

The Royal Society of London[1] originated in meetings of scientists at Gresham College in 1645, where they were known as the Philosophical or Invisible College. About 1648 some of the members transferred to Oxford, eventually to form the Oxford Philosophical Society. After the Restoration, regular meetings were resumed in London, the first meeting of the new society being held in November, 1660. Charles II became its patron, and granted a charter of incorporation in 1662. The library began in 1660, and Hooke purchased books for the collection, while Fellows have always been expected to present their publications to the Society. Many eminent scientists have been numbered among the Fellows of the Royal Society, and its publications record the epoch-making discoveries of the past three hundred years. The library houses over 150,000 volumes, more than 100,000 manuscripts, and over 200,000 original letters. About fifty incunabula are among its treasures, which include manuscripts of Newton, Malpighi, Leewenhoek, Boyle, and many other eminent persons. The Royal Society is located in Carlton House, and it administers a grant on behalf of the Treasury to assist the publication of material by learned and scientific societies.[2]

The Linnean Society of London was founded in 1788, and is at present in Burlington House, the headquarters of so many prominent scientific societies. The library consists of 70,000 books, manuscripts and other material, including the library of Linnæus amounting to 1,600 volumes.[3] The stock is devoted to natural history, and several catalogues have been published.

The Royal Veterinary College, London was founded in 1791 and became a school of the University of London in 1949. It contains several important collections of historical material which has been described by R. Catton.[4] The library houses 22,000 volumes, and takes 400 current journals.[5] The Royal College of Veterinary Surgeons, London, was founded in 1844, and the library dates from 1853. Lists of acquisitions are published in *Veterinary Record*, and this library also houses a good historical collection.

The Royal Institution of Great Britain dates from 1799, when it was founded by Benjamin Thompson, Count Rumford. Michael Faraday

[1] See also pp. 262-264.

[2] See Martin, D. C. The Royal Society's interest in scientific publications and the dissemination of knowledge. *Aslib Proc.*, 9, 1957, pp. 127–141; and Bluhm, R. K. A guide to the archives of the Royal Society and to other manuscripts in its possession. *Notes Rec. Roy. Soc Lond.*, 12, 1956, pp. 21–39.

[3] See also p. 346.

[4] Catton, R. The historical collection in the Royal Veterinary College. *Veterinary Record*, 77, 1965, pp. 503–506; and Catton, R. Some association copies in the historical collection of the Royal Veterinary College Library. *Brit. vet. J.*, 120, 1964, pp. 583–586.

[5] See also Royal Veterinary College. *A catalogue of the books, pamphlets and periodicals up to 1950*, [*etc.*], Supplement to *Veterinary Record*, 1st May, 1965.

became the first Fullerian Professor of Chemistry, and many eminent scientists have been associated with the Institution. The library was inaugurated by the purchase of the library of Thomas Astle in 1804 for 1,000 guineas. The first librarian, William Harris, was responsible for the first catalogue published in 1809, containing a description of his interesting scheme of classification, which was divided into six main classes. Harris also compiled the second catalogue, published in 1821, and the third, issued in 1857, was the responsibility of Benjamin Vincent. Supplements appeared up to 1914. The Old Library contains books published up to 1857, and the Modern Library is devoted to literature issued since then, including a fine collection of scientific periodicals. The library is particularly rich in early scientific works, and contains 70,000 items, including books, incunabula, maps, pamphlets and other material. Manuscripts of Michael Faraday, Sir Humphry Davy, John Tyndall and others are among its treasures.[1]

Burlington House is the headquarters of the Geological Society of London, which was founded in 1807.[2] Two years later the formation of a library was commenced, and by 1828 it contained 1,000 volumes. This has grown into a collection of over 100,000 books and some 3,000 pamphlets, while about 500 current periodicals are taken. The library contains the original drawings and maps of William Smith (1769–1839) and the collections of G. B. Greenough and Sir John Prestwich. George Bellas Greenough (1778–1856) was the first President of the Geological Society of London, and toured Scotland in 1805, leaving two manuscript journals written during the journey. He published very little, but his scientific papers and correspondence are extensive.[3]

Although relatively small, the library of the Royal Astronomical Society, Burlington House, established in 1825, contains many incunabula and other rare items devoted to astronomy and mathematics. It houses about 10,000 books, in addition to pamphlets, charts, photographs and lantern slides.

The library of the Royal Entomological Society of London dates from 1833, and contains many rare items, including the entomological library of H. T. Stainton. The stock consists of 13,000 books and 34,000 pamphlets, and about 180 periodicals are regularly received. Several editions of the library catalogue have been published, and new additions are listed annually in the Society's *Proceedings*.

[1] See Vernon, K. D. C. The Royal Institution and its Library. *Lib. Assn. Rec.*, 61, 1959, pp. 283–289; and Anthony, H. D. Scientific books of the 18th century and the emergence of modern science. *Proc. Roy. Instn. G.B.*, 37, 1959, pp. 334–343.

[2] See Woodward, Horace B. *The history of the Geological Society of London*, London, [*etc.*], *1950*.

[3] See Rudwick, M. J. S. Hutton and Werner compared: George Greenough's geological tour of Scotland in 1805. *Brit. J. Hist. Sci.*, 1, 1962, pp. 117–135.

Although established in 1826, the Zoological Society of London did not possess a library until ten years later, when the collection was commenced with a gift of 218 volumes. The library now contains about 100,000 books and periodicals, 4,000 animal photographs, and a large collection of paintings of birds and animals.[1]

The Chemical Society was formed as the Chemical Society of London in 1841, and two years later certain journals were subscribed to for the use of members. In 1850 the library contained less than 400 volumes, but has now grown to about 65,000 volumes, and 10,000 pamphlets and reports. It takes 700 current periodicals. The Roscoe Collection, devoted to alchemy and early chemistry, and the Nathan Collection of material on firearms and explosives are both contained in the Chemical Society Library. Other libraries in the United Kingdom are permitted to become subscribers to this most important collection of chemical literature.[2]

In 1843 Sir Henry de la Beche presented his scientific books to the nation. These, together with the Education Library of South Kensington Museum (formed in 1857) and the books from the Museum of Practical Geology, were amalgamated in 1883 to form the Science Museum Library at South Kensington. Several other collections of books were subsequently added, and the library now covers all branches of pure and applied science. Over 10,500 current periodicals are received, and the library contains some 400,000 books, pamphlets and other material. The library is open to the public, lends material to approved institutions in the British Isles, and supplies photostat copies of requisite items at low cost. A *Monthly list of accessions* is published, and bibliographies of special subjects are issued at intervals. Some of the stock of the Science Library has been taken over by the National Lending Library for Science and Technology, and is available for loan through that library.[3] Another South Kensington library, that of the Geological Survey and Museum, was founded in 1851, and contains about 70,000 books and 28,000 maps.

The library of the Royal Botanic Gardens at Kew was founded in 1853, and now contains over 55,000 volumes, 80,000 pamphlets, about 150,000 paintings, drawings and photographs, and 6,500 maps. It houses many early books on botany, a large collection of prints and drawings of plants, manuscripts and letters of Darwin, Sir Joseph and Sir William Hooker, Sir Joseph Banks and George Bentham. There are two branch

[1] See Stratton, G. Burder. The Zoological Society of London. *Brit. Book News*, 95, 1948, pp. 358–360.

[2] See Moore, Tom Sidney, and Philip, James Charles. *The Chemical Society, 1841–1941: a historical review*, 1947; and Gibson, C. S. The Chemical Society (of London) after one hundred years. *Endeavour*, 6, 1947, pp. 63–68.

[3] See Pledge, H. T. The Science Library. *In*, Irwin, Raymond, and Staveley, Ronald, *eds*. *The Libraries of London. Second, revised edition*, 1961, Chapter III, pp. 48–53; Parker, Hannah J. Science Museum Library. *Libri*, 3, 1954, pp. 326–330; and Greenaway F., *A short history of the Science Museum* (*Ministry of Education, Science Museum*), 1951.

libraries at Kew; the Gardeners' Library devoted to gardening, and the Museum's library on economic botany. The *Index Kewensis* is prepared at Kew.[1] A similar collection, the Lindley Library, housed by the Royal Horticultural Society, although primarily devoted to horticulture, contains much material on systematic botany. The Horticultural Society of London had collected together a large library between 1804 and 1859, but in the latter year it was sold by auction owing to financial circumstances. In 1862 donations were received towards a new collection of books, and in 1868 the library of John Lindley (1799–1865) was acquired as the substantial nucleus of the Lindley Library, now containing about 33,000 books and pamphlets. A catalogue of the collection was published in 1898, with a revised edition in 1927 and lists of additions to the library are published in the Society's *Journal*. The Reginald Cory bequest, received in 1936, greatly enhanced the value of the collection, which probably represents the largest horticultural library in the world. It contains many unpublished coloured drawings of plants and other rare works.[2]

The National Lending Library for Science and Technology established at Thorp Arch, near Boston Spa in 1961, has proved an enormous boon to librarians in search of scientific material. A comprehensive stock of scientific journals, and an increasing coverage of books has made this library the primary source of borrowing for scientific literature. Taking over some of the facilities formerly provided by the Science Museum Library, the N.L.L. provides a prompt service, together with certain translation and copying facilities. D. J. Urquhart[3] has described the services available, coverage having been extended to medicine, and in 1966 the library was receiving 22,000 current journals. About 2,800 organizations were then approved borrowers, and the library received 12,000 loan requests per week.[4]

Planning a complementary reference service for scientists resulted in the planning of the National Reference Library of Science and Invention, originally to be established on the South Bank, London. It will now be established elsewhere in London, possibly near the British Museum, within whose structure it functions. Combining the more recent scientific literature received by that Library and by H.M. Patent Office, established in 1852, the N.R.L. now functions as the Holborn Division, formerly the

[1] See Davidge, R. The library of the Royal Botanic Gardens, Kew, *London Librarian*, 6, October, 1960, pp. 3–9; Hubbard, C. E. Bicentenary of the Royal Botanic Gardens, Kew. *Endeavour*, 18, 1959, pp. 156–160; and Turrill, W. B. *The Royal Botanic Gardens, Kew, past and present*, 1959.

[2] See Chittenden, F. J. The Lindley Library. *J. Roy. Hort. Soc.*, 65, 1939, pp. 350–352.

[3] Urquhart, D. J. Some recent developments at the National Lending Library for Science and Technology, Boston Spa, Yorkshire. *Proc. Roy. Soc. Med.*, 59, 1966, pp. 267–268.

[4] See also Barr, K. P. The National Lending Library for Science and Technology. *Postgrad. med. J.*, 42, 1966, pp. 695–697.

Patent Office Library, and the Bayswater Division. The following catalogues are guides to their respective holdings of current journals: *Periodical publications in the Patent Office Library. List of current titles. Third edition*, London, 1965; and *Periodical publications in the National Reference Library of Science and Invention. Part I. List of non-Slavonic titles in the Bayswater Division*, London, 1969.[1]

The British Museum (Natural History) at South Kensington was opened in 1881, having been separated from the British Museum at Bloomsbury. The former includes a General Library and five departmental libraries devoted to Zoology, Entomology, Geology, Botany and Mineralogy, with a sectional library devoted to anthropology. The stock numbers about 300,000 volumes, with large numbers of manuscripts, letters, original drawings and maps, over 1,000 current periodicals being taken. Special collections include the ornithological library of the ninth Marquess of Tweeddale, added in 1887; the entomological library of Lord Walsingham, presented in 1910; Sir John Murray's collection on oceanography, acquired in 1921; the Tring Museum Library of 40,000 volumes, bequeathed by Lord Rothschild in 1939; and the Owen MSS., consisting of twenty-seven volumes of letters to Sir Richard Owen. A collection of Linnæana (1,226 volumes) has a separate catalogue, and there are also catalogues of the main library, published as *Catalogue of the books, manuscripts, maps and drawings in the British Museum (Natural History)*, compiled by B. B. Woodward and A. C. Townsend, eight volumes, 1903–40, reprinted 1964. Also *List of serials in the British Museum (Natural History) Library*, London, 1968 (Publication No. 673), which contains about 12,500 entries. The British Museum (Natural History) Library is not generally available to the public, except by permission, and it is of interest to record that the *Zoological Record* is largely produced in the library.[2]

The Ministry of Agriculture, Fisheries and Food originated with the Board of Education about 1889, the Ministry of Food being merged in 1955. The library now contains 120,000 volumes, about 10,000 of these dealing with food technology and nutrition. The stock is devoted to agriculture and related subjects, with special collections on apiculture,

[1] See Webb, Maysie. The National Reference Library for Science and Invention: a review of progress. *J. Document.*, 22, 1966, pp. 1–12; National Reference Library of Science and Invention—a symposium, with contributions by R. S. Hutton, E. M. Nicholson, Maysie Webb, and A. H. Chaplin. *J. Document.*, 17, 1961, pp. 1–39; and Gravell, F. W. The Patent Office Library. *In.* Irwin, Ronald, and Staveley, Ronald, *eds.*, *The libraries of London. Second, revised edition*, 1961. Chapter V, pp. 60–73.

[2] See Townsend, A. C. The Library of the British Museum (Natural History) and some other libraries of natural history. *In* Irwin, Raymond, and Staveley, Ronald, *eds. Libraries of London. Second, revised edition*, 1961. Chapter II, pp. 31–47; see also Townsend, A. C., and Stratton, G. B. *Library resources in the Greater London area. No. 6. Zoological libraries*, 1957.

and several branch libraries are maintained throughout the country. The Horniman Museum and Library at Forest Hill, London, was founded in 1890, and is mainly devoted to zoology and anthropology. In addition to drawings, prints, manuscripts and lantern slides, the library houses about 23,000 volumes, and is open to the public for reference purposes.

In 1913 the Commonwealth Institute of Entomology, London, was founded as the Imperial Bureau of Entomology. It publishes the monthly *Review of Applied Entomology*, and the library contains about 14,000 volumes and 47,000 pamphlets, while over 700 current periodicals are taken. The Rothamsted Experimental Station at Harpenden was also founded in 1913, being financed by the Ministry of Agriculture, and by private benefactions. The library of 13,000 volumes is devoted to agriculture, botany, chemistry and biology, and contains maps, pamphlets manuscripts, prints and portraits. Eleven hundred current periodicals are taken; the library possesses fourteen incunabula, and several editions of a catalogue of printed books in the library have been issued, and a catalogue of serial publications.[1]

The Nature Conservancy in Belgrave Square, London, was set up in 1949 and its library covers all aspects of natural history, with special emphasis on British fauna and flora. Over 500 current periodicals are received, and a list of accessions is published monthly. Research stations are maintained in various parts of the country.

Although mainly devoted to medical subjects, the Wellcome Institute of the History of Medicine Library contains many alchemical and astrological works, and among other collections houses that of Ernest Darmstaedter, which includes much Paracelsian material and alchemical writings. Published catalogues include *Catalogue of the Wellcome Historical Medical Library. I. Books printed before 1641*, [*etc.*], 1962; *II. Printed books between 1641 and 1850: A–E*, 1966; S. A. J. Moorat's *Catalogue of Western manuscripts on medicine and science in the Wellcome Historical Medical Library. I. MSS. written before 1650 A.D.*, 1962; and F. N. L. Poynter's *A catalogue of incunabula in the Wellcome Historical Medical Library*, 1954. This Library also publishes *Current Work in the History of Medicine*, which is issued quarterly and contains a selection of scientific material.

The Crawford Library of the Royal Observatory, Edinburgh, was founded by James Ludovic Lindsay (1847–1913), 26th Earl of Crawford and 9th Earl of Balcarres, who in 1872 built a private observatory at Dun Echt, Aberdeenshire. There he built up a remarkable library, and presented the books and instruments to the nation on condition that a new Royal Observatory was built in Edinburgh. The present buildings on Blackford Hill were opened in 1896, and the library then contained

[1] Rothamsted Experimental Station. *Catalogue of the serial publications in the Library, 1953. Edited by D. H. Boalch*, Harpenden, 1954.

11,000 printed books and some manuscripts. The nucleus of this library was a collection from the library of Charles Babbage, and it now contains over 25,000 items, including incunabula, some thirty editions of Euclid, early editions of Copernicus, Galileo, Newton, Napier, and many other valuable items. Many of the books in this remarkable collection are in beautiful bindings, and the collection is maintained in a manner appropriate to its value as a unique scientific library specializing in astronomy. *The Catalogue of the Crawford Library of the Royal Observatory, Edinburgh*, published in 1890 is obviously not a complete record of the holdings of the Library, which acquired many rare items when books from Calton Hill Observatory joined the original collection on Blackford Hill.[1]

The United States of America contains a large number of outstanding scientific libraries, and a few of those of particular interest are here considered. Like most national libraries, the Library of Congress at Washington contains a large proportion of scientific literature. In 1815 Thomas Jefferson's collection, which was particularly rich in books on "natural philosophy", was purchased by Congress, and in 1866 the Smithsonian Institution's collection was transferred to the same location. Material is acquired by means of copyright deposit, and exchanges are arranged on a large scale with libraries abroad. On 30th June, 1957, there were 1,447,000 volumes devoted to science and technology in the collection, which took about 15,000 current journals in these fields. There were also 2,500,000 maps and 20,000 atlases, and the various services to science of the Library, including the issue of catalogues and bibliographies, are outlined in an article by John Sherrod.[2] The Smithsonian Institution, Washington, was founded in 1846 by the bequest of James Smithson (1765–1829), an Englishman, "for the increase and diffusion of knowledge among men". It finances expeditions to remote parts of the world, and acts as a central depot for the distribution of publications throughout the world. As previously mentioned, the books belonging to the Institution were deposited in the Library of Congress, the science section at the latter becoming known as the Smithsonian Division in 1900.[3]

The Department of Agriculture Library, renamed National Agricultural Library in 1962, was established in 1862, and was responsible for the publication of the *Index to the Literature of Economic Entomology;* the *Plant Science Catalog*, of which the subject catalogue cards were issued

[1] See Kemp, D. Alasdair. The Crawford Library of the Royal Observatory, Edinburgh. *Isis*, 54, 1963, pp. 481–483.

[2] Sherrod, John. The Library of Congress. *Science*, 127, 1958, pp. 958–959; see also Abrahams, Harold J. The chemical library of Thomas Jefferson. *J. chem. Educ.*, 37, 1960, pp. 357–360.

[3] See Oehser, Paul H. *Sons of science. The story of the Smithsonian Institution and its leaders*, New York, 1949; and True, Webster Prentiss. *The Smithsonian, America's treasure house*, [*etc.*], New York, 1950.

in book form in 1958;[1] and the monthly *Bibliography of Agriculture*. The two former titles ceased publication when the last started in 1942, but this was discontinued at the end of 1969. The Library is publishing *Dictionary Catalog of the National Agricultural Library, 1862–1965*, New York, 1967 onwards, volume 65 being reached in 1970, covering up to U.S. Forest Service (F). This is supplemented by *National Agricultural Library Catalog*, issued monthly since January, 1966, which is intended to have annual and quinquennial cumulations. The Department also published a list of current serials.[2,3]

The Academy of Natural Sciences, Philadelphia, was founded in 1812. The library housed there is devoted to natural history, representing the finest in the United States dealing with that subject. It is open to the public, as is the museum of natural history, and public lectures are another feature of its educational work. The library is particularly strong in periodicals, and many of its accessions are received in exchange for the Academy's own valuable publications.

The John Crerar Library, Chicago was opened to the public in 1897 with a stock of 22,000 volumes, and is mainly devoted to pure and applied science, including medicine. It concentrates on world coverage of chemical literature, housing over a million items, with current subscriptions to more than 11,000 periodicals. Membership is open to organizations and individuals interested in the subjects covered, and the Library serves as a Federal Technical Report Centre. There is a Research Information Service, and a translation centre of over 85,000 translations. There is a staff of "literature chemists" and bibliographers.[4]

The Chicago Natural History Museum, formerly the Field Museum of Natural History, Chicago, was established in 1893, and promotes research, finances expeditions, and conducts an information bureau, among other educational activities. The library stocks over 95,000 books and pamphlets, which are available to the public, and this remarkable natural history collection contains many rare items. The Museum of Science and Industry, Chicago, was opened in 1933, and is similar in scope to the Science Library, London. It is open to the public, and has a special room for children. The library contains about 15,000 volumes, and conducts an information service for the use of the public. There is a similar

[1] U.S. Department of Agriculture, Library. *Plant science catalog: botany subject index*, 15 vols., Boston, 1958.

[2] U.S. Department of Agriculture. *List of serials currently received in the United States Department of Agriculture as of July 1, 1950*, 1958.

[3] See Bercaw, Louise O. The National Agricultural Library. *Yearbook of Agriculture*, 1962, pp. 634–637.

[4] See Henkle, Herman H. Chemical literature in the John Crerar Library and other Chicago libraries. *Adv. Chem. Ser.*, 30, 1961, pp. 297–304.

Museum of Science and Industry at New York, which is also open to the public for reference purposes.

The Niels Bohr Library for the History and Philosophy of Physics is run by the Centre for History and Philosophy of Physics and is situated in the American Institute of Physics headquarters in New York. It contains biographies and autobiographies, histories, collected reference works and other materials devoted to the history of physics. A *Newsletter* is issued four times per annum. The Chemists' Club Library is also situated in New York, and contains some 60,000 volumes restricted mainly to chemical literature. Biographical material and portraits are well represented, and the Library is open to the public for reference purposes.[1]

The Kresge Science Library, Detroit, was formed at Wayne State University by the amalgamation of the Wayne science collection and the Samuel C. Hooker science library. The collection is particularly rich in chemical literature, both ancient and modern, and the collection of complete sets of periodicals is outstanding.[2]

The University of Wisconsin libraries are particularly rich in historical material, and several important collections have been added in recent years. These include most of Denis I. Duveen's chemical library, the Boyle and Priestley collection of Hugh Sinclair, the medical collections of William Snow Miller and Byron Robinson, and the pharmacy collection of Frederick Belding Power. John Neu has edited *Chemical, medical and pharmaceutical books printed before 1800. In the collections of the University of Wisconsin libraries. . . . Compiled by Samuel Ives, Reese Jenkins and John Neu*, Madison, Milwaukee, 1965. This contains 4,442 numbered entries arranged alphabetically by authors.

In addition to its outstanding medical collections[3] Yale University houses many early manuscripts and printed books dealing with alchemy and the occult. Several manuscripts of Sir Isaac Newton are also housed there. A catalogue of the Mellon collection of alchemy and the occult was announced for publication in 1968.[4]

The Massachusetts Horticultural Society is maintained by members of the Society, but the Library is open to the public for reference purposes. It stocks books, journals, reports and trade catalogues devoted to horticulture and landscaping, but some botanical material is also included. A printed catalogue was issued in four volumes, 1963. The Library of the Horticultural Society of New York (founded 1924) has published a *Catalog of printed books 1481–1900*, compiled by Elizabeth C. Hall, 1970, and contains 3,549 entries.

[1] See Ducca, Anne D. The Chemists' Club Library. *Adv. Chem. Ser.*, 30, 1961, pp. 282–284.

[2] See Powers, Wendell H. The Kresge-Hooker Science Library. *Adv. Chem. Ser.*, 30, 1961, pp. 282–284.

[3] See Thornton, p. 347.

[4] MacPhail, Ian. The Mellon collection of alchemy and the occult. *Ambix*, 14, 1967, pp. 198–202.

Three libraries of particular significance in the history of science merit special mention on account of their valuable collections of early scientific works. The Burndy Library at Norwalk, Connecticut, was founded by Bern Dibner in 1936, and issues an interesting series of publications on the history of science, including facsimiles and translations. A beautifully produced catalogue of part of the collection published in 1955[1] contains descriptions of 200 items, with numerous facsimiles, and includes first editions of the writings of Vesalius, Harvey, Copernicus, Kepler, Galileo, Newton, Gilbert and many other eminent scientists. The Linda Hall Library, Kansas City, Missouri, was founded by Herbert F. Hall, and is devoted to science and technology. It was opened to the public in 1946, and a new building was opened in March, 1956, when it contained almost 200,000 volumes. It is designed to house half a million books. In 1956 the library published a survey of some of the great classics of science,[2] containing a bibliography listing some of the treasures contained in this remarkable collection. The E. DeGolyer Collection in the History of Science and Technology was established at the University of Oklahoma in December, 1949, with the donation of 600 rare volumes. The collection has since grown considerably, and in 1954 housed about 5,000 books. A check list published in that year[3] reveals a valuable collection devoted to the history of science and technology, including first editions of many of the classics.

The value of scientific libraries in the development of modern science cannot be over-emphasized, for all research should originate in the study of work accomplished by predecessors in the field. The foundations of original work must be based firmly upon the literature representing previous studies. Depositories of printed literature must be administered as living collections, having in mind the needs of users, and trained staff should be employed in order to exploit the resources to the utmost capacity. Statistics of stock have little more significance than the weight of the volumes housed. The largest collections are not necessarily the best; in fact their value to scientists may be inversely proportionate to their size. General collections tend to smother the specialist material housed, and the small, highly-specialized libraries are better geared to serve their readers. Staffed by librarians who appreciate the literature of the subject, they can exploit this for the benefit of readers. Information officers generally have a thorough knowledge of a very limited field, and

[1] *Heralds of science as represented by two hundred epochal books and pamphlets from the Burndy Library. With notes by Bern Dibner*, Norwalk, Conn., 1955.

[2] Linda Hall Library, Kansas City. Missouri. *Some milestones in the history of science to 1800. Based upon selections of printed books in the collection of the Linda Hall Library*, Kansas City, 1956.

[3] See McAnally, Arthur, and Roller, Duane H. D. *A check list of the E. DeGolyer Collection in the History of Science and Technology as of August 1, 1950*. Norman, Oklahoma, 1954.

in common with the scientists they serve, are lost outside the confines of that closely defined area. Chemists no longer speak a common language, and the specialists within the numerous branches of the subject find it difficult to communicate with fellow research workers. This is applicable to other sciences, and where general scientific literature is gathered together there might well seemingly be the need for an array of information scientists, in fact one per reader. The only person capable of gleaning the information he requires is the research worker himself. He can sift the wheat from the chaff at source, and in searching for his material can sometimes be usefully sidetracked. Obviously he will need guidance in the use of bibliographical tools and reference books, and in the use of libraries. This can be provided by experienced librarians who have had lengthy acquaintance with the specific subject literature. Library administration and routine are only the mechanical aspects of librarianship; true librarianship is the employment of these to exploit the literature for the benefit of readers. Subject knowledge is acquired by finding out exactly what each reader requires, and making sure that either he receives it, or that he is convinced he is asking the impossible. Experience with the literature is more important than theoretical knowledge gained from lectures, for librarianship is a practical profession. A thorough knowledge of the literature of any scientific subject is gleaned only by a lifetime of experience, for science does not stand still, literature continues to be produced, and libraries must continue to flourish.

BIBLIOGRAPHY

This makes no attempt to be exhaustive, but lists the general histories consulted, together with the more important of the footnote references. Additional material will be found in the "Critical bibliography of the history of science and its cultural influences" contained in the issues of *Isis*; in *Current Work in the History of Medicine*, published by the Wellcome Institute of the History of Medicine; and in *Library & Information Science Abstracts* (formerly *Library Science Abstracts*), published by the Library Association and Aslib.

Books are published in London unless otherwise indicated.

Select List of Journals devoted to the History of Science:

Ambix, 1937–.
Annals of Science, 1936–.
Archives for History of Exact Sciences, 1960–.
Archives internationales d'Histoire des Sciences, 1947–.
British Journal for the History of Science, 1962–.
British Journal for the Philosophy of Science, 1950–.
Centaurus, 1950–.
Chymia, 1948–.
Discovery, 1920–66.
Endeavour, 1942–.
History of Science, 1962–.
Isis, 1912–.
Journal for the History of Astronomy, 1970–.
Journal of the History of Biology, 1968–.
Journal of the History of Ideas, 1940–.
Lychnos, 1936–.
Notes and Records of the Royal Society of London, 1938–.
Osiris, 1936–.
Physis. Rivista di Storia della Scienze, 1959–.
Revue d'Histoire de Science, 1947–.
Scientiarium Historia, 1959–.
Studies in History and Philosophy of Science, 1970–.
Zeitschrift für Geschichte der Naturwissenschaften, Technik und Medizin, 1960–.

Abbatt, William. Dr. Erasmus Darwin, the author of "Zoonomia". *Ann. med. Hist.*, 3, 1921, pp. 387–390.
Abetti, Giorgio. *The history of astronomy.... Translated from the Italian* Storia dell' astronomia, *by Betty Burr Abetti*, New York, 1952; London, 1954.
Abrahams, Harold J. The chemical library of Thomas Jefferson. *J. chem. Educ.*, 37, 1960, pp. 357–360.
——. Priestley answers the proponents of abiogenesis. *Ambix*, 12, 1964, pp. 44–71.
Adams, Alexander B. *John James Audubon: a biography*, New York, 1966; London, 1967.
Adams, Frank Dawson. *The birth and development of the geological sciences*, New York, 1954; reprint of 1938 edition.
Adelmann, Howard B. *Marcello Malpighi and the evolution of embryology*, 5 vols., Ithaca, New York; London, 1966.
Afnan, Soheil M. *Avicenna: his life and works*, 1958.
Africa, Thomas W. Copernicus' relation to Aristarchus and Pythagoras. *Isis*, 52, 1961, pp. 403–409.
Aiton, E. J. The Cartesian theory of gravity. *Ann. Sci.*, 15, 1959, pp. 27–49.
——. The celestial mechanics of Leibniz. *Ann. Sci.*, 16, 1960 [1962], pp. 65–82.
——. The celestial mechanics of Leibniz in the light of Newtonian criticism. *Ann. Sci.*, 18, 1962 [1964], pp. 31–41.
——. The celestial mechanics of Leibniz: a new interpretation. *Ann. Sci.*, 20, 1964, pp. 111–123.
——. An imaginary error in the celestial mechanics of Leibniz. *Ann. Sci.*, 21, 1965 [1966], pp. 169–173.
Aldis, Harry G. *The printed book. . . . Revised and brought up to date by John Carter and Brooke Crutchley. Third edition*, Cambridge, 1951.
Allan, Mea. *The Hookers of Kew, 1785–1911*, 1967.
Allen, D. E. John Martyn's Botanical Society: a bibliographical analysis of the membership. *Proc. Bot. Soc. Brit. Isles*, 6, 1967, pp. 305–324.
Allen, Elsa G. New light on Mark Catesby. *Auk*, 54, 1937, pp. 349–363.
American Chemical Society. *Searching the chemical literature. . . . Based on papers presented . . . at national meetings from 1947 to 1956*, Washington, 1961. (Advances in Chemistry Series, No. 30.)
Anawati, G. C. *Essai de bibliographie avicennienne*, Cairo, 1950.
Andrade, E. N. da C. Benjamin Franklin in London. *Journal of the Royal Society of Arts*, 104, 1956, pp. 216–234.
——. The birth and early days of the "Philosophical Transactions". *Notes Rec. Roy. Soc. Lond.*, 20, 1965, pp. 9–27.
——. *A brief history of the Royal Society*, 1960.
——. The early history of the vacuum pump. *Endeavour*, 16, 1957, pp. 29–35.

Andrade, E. N. da C. Galileo. *Notes Rec. Roy. Soc. Lond.*, 19, 1964, pp. 120–130.
——. *Isaac Newton*, [*etc.*], 1950. (Personal Portraits Series.)
——. A Newton collection. *Endeavour*, 12, 1953, pp. 68–75.
——. The presentation of scientific information. *Proc. Roy. Soc.*, A, 197, 1949–50, pp. 1–17; also in, *Proc. Roy. Soc.*, B, 136, 1949–50, pp. 317–333; and *The Royal Society Scientific Information Conference. 21 June–2 July, 1948. Report and papers submitted*, 1948, pp. 26–44.
——. Robert Hooke, F.R.S. (1635–1703). *Notes Rec. Roy. Soc. Lond.*, 15, 1960, pp. 137–145.
——. Samuel Pepys and the Royal Society. *Notes Rec. Roy. Soc. Lond.*, 18, 1963, pp. 82–93.
——. The scientific work of Benjamin Franklin. *Nature*, 177, 1956, pp. 60–61.
——. Wilkins Lecture. Robert Hooke, *Proc. Roy. Soc.*, A, 201, 1950, pp. 439–473.
——, and Martin, D. C. The Royal Society and its foreign relations. *Endeavour*, 19, 1960, pp. 72–80.
Andrews, John S. Philip Henry Gosse, F.R.S. (1810–88). *Lib. Assn. Rec.*, 63, 1961, pp. 197–201.
Anker, Jean. From the early history of the Flora Danica. *Libri*, 1, 1951, pp. 334–350.
——. *Otto Friderich Müller's "Zoologia Danica"*, Copenhagen, 1950. (Library Research Monographs, published by the University Library, Scientific and Medical Department, Vol. 1.)
Anthony, H. D. *Science and its background. . . . Third edition*, London, New York, 1957.
——. Scientific books of the 18th century and the emergence of modern science. *Proc. Roy. Inst. G.B.*, 37, 1959, pp. 334–343.
——. *Sir Isaac Newton*, London, [*etc.*], 1960.
Anthony, L. J., East, H., and Slater, M. J. The growth of the literature of physics. *Reports on Progress in Physics*, 32, 1969, pp. 709–767.
Apostle, Hippocrates George. *Aristotle's philosophy of mathematics*, Chicago, 1952.
Arber, Agnes. From medieval herbalism to the birth of modern botany. In, *Science, medicine and history*, Vol. 1, 1953, pp. 317–330.
——. *Herbals, their origin and evolution in the history of botany, 1470–1670. . . . A new edition*, [*etc.*], Cambridge, 1938.
——. *The natural philosophy of plant form*, Cambridge 1950.
——. Nehemiah Grew and Marcello Malpighi. *Proc. Linn. Soc.*, 153rd session, (1940–41), Part 2, pp. 218–238.
——. Nehemiah Grew (1641–1712) and Marcello Malpighi (1628–1694): an essay in comparison. *Isis*, 34, 1942–43, pp. 7–16.

Arber, Agnes. Robert Sharrock (1630–84): a precursor of Nehemiah Grew (1641–1712) and an exponent of "Natural law" in the plant world. *Isis*, 51, 1960, pp. 3–8.
Archibald, Raymond Clare. *Bibliography of Egyptian mathematics with special references to the Rhind Mathematical Papyrus, and sources in its study*, Oberlin, 1927.
——. A rare pamphlet of Moivre and some of his discoveries. *Isis*, 8, 1926, pp. 671–684.
Ardern, L. L. The Manchester Literary and Philosophical Society. *J. chem. Educ.*, 39, 1962, pp. 264–265.
Armitage, Angus. The astronomical work of Nicholas Louis de Lacaille. *Ann. Sci.*, 12, 1956, pp. 163–191.
——. *A century of astronomy*, 1950.
——. *Copernicus, the founder of modern astronomy. Revised edition*, 1957.
——. The cosmology of Giordano Bruno. *Ann. Sci.*, 6, 1948–50, pp. 24–31.
——. *Edmond Halley*, London, [*etc.*], 1966. (British Men of Science Series.)
——. *John Kepler*, 1966.
——. René Descartes (1596–1650) and the early Royal Society. *Notes Rec. Roy. Soc. Lond.*, 8, 1950–51, pp. 1–19.
——. *Sun, stand thou still: the life and work of Copernicus the astronomer*, New York, 1947.
——. *William Herschel*, London, 1962; New York, 1963.
Armstrong, Eva V., and Lukens, Hiram S. Lazarus Ercker and his "Probierbuch". Sir John Pettus and his "Fleta minor". *J. chem. Educ.*, 16, 1939, pp. 553–562.
Armytage, W. H. G. Coffee houses and science. *Brit. med. J.*, 1960, II, p. 213.
——. Giambattista della Porta and the Segreti. *Brit. med. J.*, 1960, I, pp. 1129–1130.
Ashmole, Elias. *His autobiographical and historical notes, his correspondence, and other contemporary sources relating to his life and work. Edited, with biographical introduction by C. H. Josten*, 5 vols., 1966 [1967].
Asimov, Isaac. *Asimov's Biographical encyclopedia of science and technology: the living stories of more than 100 great scientists from the age of Greece to the space age chronologically arranged*, New York, 1964.
——. *A short history of biology*, 1965.
Atherton, Pauline. Physics. *Library Trends*, 15, 1967, pp. 847–851.
Atkins, Hedley. The Darwin tradition. Thomas Vicary Lecture delivered at the Royal College of Surgeons of England on 29th October, 1964. *Ann. Roy. Coll. Surg. Engl.*, 36, 1965, pp. 1–25.
Aulie, Richard P. Caspar Friedrich Wolff and his "Theoria generationis", 1759. *J. Hist. Med.*, 16, 1961, pp. 124–144.

Ayres, Clarence, *Huxley*, New York, 1932.

Bacon, Egbert K. A precursor of the American Chemical Society—Chandler and the Society of Union College. *Chymia*, 10, 1965, pp. 183–197.

The Baldianus manuscript (*Codex Barberini Latin 241*), *Vatican Library: an Aztec herbal of 1552. Introduction, translation and annotations by Emily Walcott Emmart. With a foreword by Henry E. Sigerist*, Baltimore, 1940.

Baehni, Charles. Les grands systèmes botaniques depuis Linné. *Gesnerus*, 14, 1957, pp. 83–93.

——. M. de Goethe, botaniste. *Gesnerus*, 6, 1949, pp. 110–128.

Bailey, Sir Edward Battersby. Charles Lyell, F.R.S. (1797–1875). *Notes Rec. Roy. Soc. Lond.*, 14, 1959, pp. 121–138.

——. *James Hutton—founder of modern geology*, Amsterdam, [*etc.*], 1967.

Baker, John R. *Abraham Trembley of Geneva: scientist and philosopher, 1710–1784*, 1952.

Ball, W. W. Rouse. *An essay on Newton's "Principia"*, London, New York, 1893.

——. *A short account of the history of mathematics. Fourth edition*, 1908 and 1935.

Balss, Heinrich. Die Tausendfüssler, Insekten und Spinnen bei Albertus Magnus. *Sudhoffs Arch. Gesch. Med.*, 38, 1954, pp. 303–322.

Bamber, Lyle E. Biology. *Library Trends*, 15, 1967, pp. 829–835.

Baranowski, Henryk. *Bibliografia kopernikowska, 1509–1955*, Warsaw, 1958.

Barbensi, Gustavo. Marcello Malpighi (1628–1694). *Sci. med. ital.*, 2nd ser., 3, 1954, pp. 3–13.

Barlow, Horace Mallinson. Old English herbals, 1525–1640. *Proc. Roy. Soc. Med.*, 6, 1913, Section of Hist. of Med., pp. 108–149.

Barlow, Nora. Darwin's ornithological notes. [Cambridge University Handlist (1960), No. 29, (ii).] Edited with an introduction, notes and appendix. *Bull. Brit. Mus.* (*Nat. Hist.*) *Hist. Ser.*, 2, 1963, pp. 201–278.

——. Erasmus Darwin, F.R.S. (1731–1802). *Notes Rec. Roy. Soc. Lond.*, 1959, pp. 85–89.

Barnes, Sherman B. The editing of early learned journals. *Osiris*, 1, 1936, pp. 155–172.

Barnett, S. A., *ed. A century of Darwin*, London, [*etc.*], 1958.

Baron, John. *The life of Edward Jenner*, 2 vols., 1827–38.

Baron, Walter. Zu Louis Agassiz's Beurteilung des Darwinismus. *Sudhoffs Arch. Gesch. Med.*, 40, 1956, pp. 259–277.

Barr, E. Scott. Biographical material in the first series of the *Physical Review*. *Isis*, 58, 1967, pp. 245–246.

——. Biographical material in the *Philosophical Magazine* to 1900. *Isis*, 55, 1964, pp. 88–90.

Barr, E. Scott. *Nature*'s "Scientific worthies". *Isis*, 56, 1965, pp. 354–356.

Barr, K. P. Estimates of the number of currently available scientific and technical periodicals. *J. Document.*, 23, 1967, pp. 110–116.

——. The National Lending Library for Science and Technology. *Postgrad. med. J.*, 42, 1966, pp. 695–697.

Bates, Ralph S. *Scientific societies in the United States. . . . Third edition*, Oxford, [*etc.*], 1965.

Baumgardt, Carola. *Johannes Kepler: life and letters*, [*etc.*], 1952.

Baumgartner, Leona. Leonardo da Vinci as a physiologist. *Ann. med. Hist.*, N.S. 4, 1932, pp. 155–171.

Baur, Ludwig. *Dir Philosophie des Robert Grosseteste, Bischofs von Lincoln* (†*1253*). (*Beiträge zur Geschichte des Mittelalters*, 1917.)

——. *Die philosophischen Werke des Robert Grosseteste.* (*Beiträge zur Geschichte der Philosophie des Mittelalters*, 1912.)

Bay, Jens Christian. Bibliographies of botany. A contribution towards a bibliotheca bibliographica compiled and annotated by J. Christian Bay. *Progressus rei botanicae*, 3, 1910, pp. 331–456.

——. Conrad Gesner (1516–1565), the father of bibliography: an appreciation. *Pap. Bib. Soc. Amer.*, 10, 1916, pp. 53–86.

Bayon, H. P. The authorship of Carlo Ruini's "Anatomia del Cavallo". *J. comp. Path.*, 48, 1935, pp. 138–148.

——. William Harvey (1578–1657), his application of biological experiment, clinical observation, and comparative anatomy to the problems of generation. *J. Hist. Med.*, 2, 1947, pp. 51–96.

Beck, Curt W. Georg Ernst Stahl, 1660–1734. *J. chem. Educ.*, 37, 1960, pp. 506–509.

Beck, Hanno. Das literarische Testament Alexander von Humboldts, 1799. *Forschungen und Fortschritte*, 31, 1957, pp. 65–70.

Beck, L. J. *The method of Descartes: a study of the Regulae*, Oxford, 1952.

Beddall, Barbara G. Wallace, Darwin, and the theory of natural selection: a study in the development of ideas and attitudes. *J. Hist. Biol.*, 1, 1968, pp. 261–323.

Beer, John J. A. W. Hofmann and the founding of the Royal College of Chemistry. *J. chem. Educ.*, 37, 1960, pp. 248–251.

Beierlein, P. R. *Lazarus Ercker*, Berlin, 1955.

Bell, A. E. *Christian Huygens and the development of science in the seventeenth century*, 1947.

Belloni, Luigi. Francesco Redi als Vertreter der italienischen Biologie des XVII Jahrhunderts. *Münch. med. Wschr.*, 101, 1959, pp. 1617–1624.

Belt, Elmer. Leonardo da Vinci, medical illustrator. *Postgrad. Med.*, 16, 1954, pp. 150–157.

——. *Leonardo the anatomist.* (*Logan Clendening Lectures on the History and Philosophy of Medicine. Fourth series*), Lawrence, Kansas, 1955.

Bence-Jones, Henry. *The life and letters of Faraday*, 2 vols., 1870.

Bennett, H. S. *English books & readers 1475 to 1557: being a study in the history of the book trade from Caxton to the incorporation of the Stationers' Company*, Cambridge, 1952.

Bercaw, Louise O. The National Agricultural Library. *Yearbook of Agriculture*, 1962, pp. 634–637.

Bernal, J. D. *History of physics*, 1967.

——. *Science in history. Second edition*, 1957.

Bernstein, Henry T. J. Clerk Maxwell on the history of the kinetic theory of gases, 1871. *Isis*, 54, 1963, pp. 206–216.

Berry, Arthur John. *From classical to modern chemistry: some historical sketches*, Cambridge, 1954.

——. *Henry Cavendish: his life and scientific work*, 1960.

Berthelot, Marcelin. *La chimie au moyen âge*, [*etc.*], 3 vols., Paris, 1893; reprinted Amsterdam, 1967.

——. *Collection des anciens alchimistes grecs*, [*etc.*], 3 vols., Paris, 1887–88; reprinted 1963.

——. *Introduction a l'étude de la chimie des anciens et du moyen âge*, [*etc.*], Paris, 1889.

——. *La révolution chimique. Lavoisier. Ouvrage suivi de notices et extraits des registres inédits de laboratoire de Lavoisier*, Paris, 1890. (Bibliothèque scientifique internationale, LXIX.)

——. *Les origins de l'alchimie*, Paris, 1885.

Bettany, G. T. *Life of Charles Darwin*, 1887. (Great Writers, series.)

Bibby, Cyril. Huxley and the reception of the "Origin". *Vict. Stud.*, 3, 1959, pp. 76–86.

——. *T. H. Huxley: scientist, humanist and educator*, 1959.

Bicknell, P. J. Did Anaxagoras observe a sunspot in 467 B.C.? *Isis*, 59, 1968, pp. 87–90.

Bidez, Joseph, and others. *Catalogue des manuscrits alchimiques grecs*, 8 vols., Brussels, 1924–32.

Biermann, Kurt R., and Lange, Fritz G. Die Alexander von Humboldt-Briefausgabe. *Forsch. und Fortschr.*, 36, 1962, pp. 225–230.

Birembaut, Arthur. Fontonelle, Réaumur et le gaz naturel. *Rev. Hist. Sci.*, (*Paris*), 11, 1958, pp. 82–84.

——. Réaumur et l'élaboration des produits ferreux. *Rev. Hist. Sci.*, (*Paris*), 11, 1958, pp. 138–166.

Birkhoff, Garrett. Galois and group theory. *Osiris*, 3, 1938, pp. 260–268.

Biswas, Asit K. The automatic rain-gauge of Sir Christopher Wren, F.R.S. *Notes Rec. Roy. Soc. Lond.*, 22, 1967, pp. 94–104.

Black, Robert C. *The younger John Winthrop*, New York, London, 1966.

Blackwell, Richard J. Descartes' laws of motion. *Isis*, 57, 1966, pp. 220–234.

Blagden, Cyprian. *Fire more than water: notes for the story of a Ship*, London, [*etc.*], *for private circulation*, 1949.
Blake, Ralph M., and others. *Theories of scientific method. The Renaissance through the nineteenth century*, Seattle, 1960.
Blanchard, J. Richard. Agriculture. *Library Trends*, 15, 1967, pp. 880–895.
Blaschke, Wilhelm, and Schoppe, Günther. Regiomontanus: Commensurator. *Akad. Wiss. Lit. Abh. math.-nat. Klasse*, Jhg. 1956, pp. 445–529.
Bluhm, R. K. A guide to the archives of the Royal Society and to other manuscripts in its possession. *Notes Rec. Roy. Soc. Lond.*, 12, 1956, pp. 21–39.
——. Henry Oldenburg, F.R.S. (*c* 1615–1677). *Notes Rec. Roy. Soc. Lond.*, 15, 1960, pp. 183–197.
Blunt, Wilfrid. *The art of botanical illustration, by Wilfrid Blunt, with the assistance of William T. Stearn*, 1950.
Boas, Marie. Boyle as a theoretical scientist. *Isis*, 41, 1950, pp. 261–268.
——. An early version of Boyle's *Sceptical chymist*. *Isis*, 45, 1954, pp. 153–168.
——. *Robert Boyle and seventeenth-century chemistry*, Cambridge, 1958.
Bockstaele, P. Notes on the first arithmetics printed in Dutch and English. *Isis*, 51, 1960, pp. 315–321.
Bodenheimer, F. S. *The history of biology: an introduction*, 1958.
Boehm, W. John Mayow and his contemporaries. *Ambix*, 11, 1963, pp. 105–120.
Boerhaave, Herman. *Boerhaave's Correspondence. Edited by G. A. Lindeboom*, 2 vols., Leyden, 1962–64. (Analecta Boerhaaviana, Vols. 3, 5.)
Böttger, Herbert. Die Embryologie Leonardos da Vinci. *Centaurus*, 3, 1953–54, pp. 222–235.
Boissier, Gaston. *Étude sur la vie et les ouvrages de Varron*, Paris, 1861.
Bolton, Henry Carrington. *A catalogue of scientific and technical periodicals, 1665–1895. Together with chronological tables and a library check-list. . . . Second edition*, Washington, 1897. (Smithsonian Miscellaneous Collections, Vol. 40, 1898.) Reprinted New York, 1966.
——. *Chemical societies of the nineteenth century*, Washington, 1902.
Bonney, Thomas George. *Charles Lyell and modern geology*, 1895.
Bork, Alfred M. Maxwell and the vector potential. *Isis*, 58, 1967, pp. 210–222.
Bosmans, H. André Tacquet (S.J.) et son traité d "Arithmétique théorique et pratique". *Isis*, 9, 1927, pp. 66–82.
Bosworth, C. E. A pioneer Arabic encyclopedia of the sciences: al Khāwrizmī's Keys of the sciences. *Isis*, 54, 1963, pp. 97–111.
Bottle, R. T., *ed. The use of chemical literature. Second edition*, 1969.
——. A user's assessment of current awareness services. *J. Document.*, 21, 1965, pp. 177–189.

Bottle, R. T., and Wyatt, H. V., *eds. The use of biological literature*, 1966. (Second edition in press.)

Boulter, Ben Consitt. *Robert Grossetête. The Defender of our Church and our liberties*, 1936.

Boutroux, E. *Pascal*, Paris, 1919.

Bowen, E. J., and Hartley, Sir Harold. The Right Reverend John Wilkins, F.R.S. (1614–72). *Notes Rec. Roy. Soc. Lond.*, 15, 1960, pp. 47–56.

Bowley, Donovan R., and Smith, Hobart M. The dates of publication of Louis Agassiz's *Nomenclator zoologicus. J. Soc. Bib. Nat. Hist.*, 5, 1968, pp. 35–36.

Boyer, Carl B. *A history of mathematics*, New York, [*etc.*], 1968.

Bradley, D. *Count Rumford*, Princeton, N.J., 1967.

Bradley, John. On the operational interpretation of classical chemistry. *Brit. J. Phil. Sci.*, 5, 1955, pp. 32–42.

Brasch, Frederick E. A survey of the number of copies of Newton's *Principia* in the United States, Canada, and Mexico. *Scripta Mathematica*, 18, 1952, pp. 53–67.

Brewster, Sir David. *Memoirs of the life, writings and discoveries of Sir Isaac Newton*, 2 vols., Edinburgh, 1855.

Bridges, John Henry. *The life and work of Roger Bacon, with additions and notes by H. Gordon Jones*, 1914.

Bridson, G. D. R. The Zoological Record—a centenary appraisal. *J. Soc. Bib. Nat. Hist.*, 5, 1968, pp. 23–34.

British Council. *Scientific and learned societies of Great Britain. A handbook compiled from official sources. 61st edition (1964)*, 1964.

British Museum. *A catalogue of the works of Linnæus (and publications more immediately relating thereto) preserved in the libraries of the British Museum (Bloomsbury) and the British Museum (Natural History) (South Kensington). Second edition*, 1933.

Britten, James, and Boulger, George S., *compilers. A biographical index of deceased British and Irish botanists. Second edition. Revised and completed by A. B. Rendle*, 1931.

Brock, Thomas D. *Milestones in microbiology, translated and edited by Thomas D. Brock*, London, (printed in U.S.A.), 1961.

Brock, W. H. The London Chemical Society 1824. *Ambix*, 14, 1967, pp. 133–139.

Brockbank, E. M. *John Dalton as experimental physiologist*, Manchester, 1929.

——. *John Dalton: some unpublished letters of personal and scientific interest, with additional information about his colour-vision & atomic theories*, Manchester, 1944. (Publications of the University of Manchester, No. CCLXXXVII.)

Brodetsky, Selig. *Sir Isaac Newton: a brief account of his life and work*, 1927.
Brodrick, James. *Galileo: the man, his work, his misfortunes*, 1964.
Brooks, E. St. John. *Sir Hans Sloane. The great collector and his circle*, 1954.
Brown, Charles Harvey. *Scientific serials: characteristics and lists of most cited publications in mathematics, chemistry, geology, physiology, physics, botany, zoology, and entomology*, [*etc.*], Chicago, 1956.
Brown, Harcourt. *Scientific organizations in seventeenth century France, 1620–1680*, Baltimore, 1934.
Brown, Herbert C. Foundations of the structural theory. *J. chem. Educ.*, 36, 1959, pp. 104–110.
Brown, John. University press publishing. *Lib. Assn. Rec.*, 70, 1968, pp. 310–314.
Brown, John R., and Thornton, John L. Physiology before William Harvey. *St. Bart's Hosp. J.*, 63, 1959, pp. 116–124.
—— and ——. William Odling as Medical Officer of Health to Lambeth 1856–60. *Med. Officer*, 102, 1959, pp. 77–78.
Brown, Sanborn C. Count Rumford as a scientist. *Proc. Roy. Instn. G.B.*, 39, 1963, pp. 583–620.
Brown, W. S., Pierce, J. R., and Traub, J. F. The future of scientific journals. A computer-based system will enable a subscriber to receive a personalized stream of papers. *Science*, 158, 1967, pp. 1153–1159.
Browne, Charles Albert. Scientific notes from the books and letters of John Winthrop, jr. (1606–76), first Governor of Connecticut. *Isis*, 11, 1928, pp. 325–341.
——, and Weeks, Mary Elvira. *A history of the American Chemical Society: seventy-five eventful years*, Washington, 1952.
Browne, G. F. *The Venerable Bede, his life and writings*, 1919.
Brush, S. G. The development of the kinetic theory of gases. I. Herapath. *Ann. Sci.*, 13, 1957 [1959], pp. 188–198; III. Clausius. *Ann. Sci.*, 14, 1958, pp. 185–196; IV. Maxwell. *Ann. Sci.*, 14, 1958, pp. 243–255.
Buck, R. W. Doctors afield—Nicolaus Coppernic (1473–1543). *New Engl. J. Med.*, 250, 1954, pp. 954–955.
Bühler, Curt F. *The fifteenth century book. The scribes; the printers; the decorators*, Philadelphia, 1960.
——. Scientific and medical incunabula in American libraries. *Isis*, 35, 1944, pp. 173–175.
——. The statistics of scientific incunabula. *Isis*, 39, 1949, pp. 163–168.
Buess, Heinrich. Albrecht von Haller and his "Elementa physiologiae" as the beginning of pathological physiology. *Med. Hist.*, 3, 1959, pp. 123–131.
——. Zur Entstehung der *Elementa physiologiae* Albrecht Hallers (1708–1777). *Gesnerus*, 15, 1958, pp. 17–35.

Bullard, Sir Edward. Edmond Halley (1656–1741). *Endeavour*, 15, 1956, pp. 189–199.

Burgess, G. H. O. *The curious world of Frank Buckland*, 1967.

Burget, G. E. Stephen Hales (1677–1761). *Ann. med. Hist.*, 7, 1925, pp. 109–116.

Burrow, J. W. *Evolution and society: a study in Victorian social theory*, Cambridge, 1966.

Butcher, D. W. The departmental libraries of the University of Cambridge. *J. Document.*, 7, 1951, pp. 221–243.

Butterfield, Herbert. The history of science and the study of history. *Harvard Lib. Bull.*, 13, 1959, pp. 329–347.

——. *The origins of modern science, 1300–1800. New edition.* 1957.

Cahn, R. S. *Survey of chemical publications and report to the Chemical Society*, 1965.

Cajori, Florian. Johannes Kepler, 1571–1630. *Sci. Mon.*, 30, 1930, pp. 385–394.

——. Leibniz, the master-builder of mathematical notations. *Isis*, 7, 1925, pp. 412–429.

——. Notes on the Fahrenheit scale. *Isis*, 4, 1921–22, pp. 17–22.

——. A revaluation of Harriot's Artis analyticae praxis. *Isis*, 11, 1928, pp. 316–324.

Caley, Earle R. The Leyden Papyrus X: an English translation with brief notes. *J. chem. Educ.*, 3, 1926, pp. 1149–1166.

——. The Stockholm Papyrus: an English translation with brief notes. *J. chem. Educ.*, 4, 1927, pp. 979–1002.

Calinger, Ronald S. Frederick the Great and the Berlin Academy of Sciences (1740–1766). *Ann. Sci.*, 24, 1968, pp. 239–249.

Callus, Daniel A., *ed. Robert Grosseteste: scholar and bishop*, [*etc.*], Oxford, London, 1955.

Cameron, Sir Gordon Roy. Edward Jenner, F.R.S., 1749–1823. *Notes Rec. Roy. Soc. Lond.*, 7, 1949–50, pp. 43–53.

Cameron, Hector Charles. *Sir Joseph Banks, K.B., P.R.S., the autocrat of the philosophers*, 1952.

Campaigne, Ernest J. Wöhler and the overthrow of vitalism. *J. chem. Educ.*, 32, 1955, p. 403.

Campbell, L., and Garnett, W. *The life of James Clerk Maxwell, with a selection from his correspondence and occasional writings*, [*etc.*], 1882.

Cannon, H. Graham. *Lamarck and modern genetics*, Manchester, 1959.

Cardew, F. A note on the number of plates in Curtis's "Flora Londinensis", 1777, and Hooker's enlarged edition, 1817–28. *J. Soc. Bib. Nat. Hist.*, 2, vi, 1950, pp. 223–224.

Cardwell, D. S. L., *ed. John Dalton & the progress of science. Papers presented to a conference of historians of science held in Manchester September 19–24, 1966 to mark the bicentenary of Dalton's birth*, Manchester, New York, 1968.

Carli, A., and Favaro, Antonio. *Bibliographia Galileiana (1568–1895)*, Rome, 1896.

Carmody, Francis J. *Arabic astronomical and astrological sciences in Latin translation. A critical bibliography*, Berkeley, California, 1956.

Carnegie Institute of Technology. Hunt Botanical Library. *Adanson. The bicentennial of Michel Adanson's "Familles des plantes"*, Parts 1–2, Pittsburgh, 1963–64.

Carozzi, Albert V. Lamarck's theory of the earth: *Hydrogéologie. Isis*, 55, 1964, pp. 293–307.

Carter, Thomas Francis. *The invention of printing in China and its spread westward. . . . Revised by L. Carrington Goodrich. Second edition*, New York, 1955.

Caspar, Max. *Bibliographia Kepleriana. Ein Führer durch das gedruckte Schrifttum von Johannes Kepler*, Munich, 1936; reprinted 1968.

—— *Kepler. . . . Translated and edited by C. Doris Hellman*, London, New York, 1959.

Catton, R. The historical collection in the Royal Veterinary College. *Veterinary Record*, 77, 1965, pp. 503–506.

——. Some association copies in the historical collection in the Royal Veterinary College Library. *Brit. vet. J.*, 120, 1964, pp. 583–586.

Causey, Gilbert. *The cell of Schwann*, [*etc.*], Edinburgh, London, 1960.

Cave, A. J. E. The glands of Owen. *St Bart's Hosp. J.*, 57, 1953, pp. 131–133.

——. The muscles of Owen. *St Bart's Hosp. J.*, 61, 1957, pp. 138–140.

——. Sir Richard Owen. *St Bart's Hosp. J.*, 68, 1964, pp. 71–73, 76.

Centenaire de la création d'un enseignement de chimie organique au Collège de France. [Marcelin Berthelot.] *Bull. Soc. chim. Française*, 1964, pp. 1681–1689.

Centenary of the House of Macmillan. *Nature*, 151, 1943, pp. 231–235.

Centre International de Synthèse. *Pierre Gassendi, 1592–1655, sa vie et son oeuvre*, Paris, 1955.

Centre National de la Recherche Scientifique. *Tricentenaire de Pierre Gassendi, 1655–1955*, Paris, 1957.

Chaïa, Jean. Sur une correspondence inédite de Réaumur avec Artur, premier médecin du Roi à Cayenne. *Episteme*, 2, 1968, pp. 36–57, 121–138.

Chaldecott, J. A. Scientific activities in Paris in 1791. Evidence from the diaries of Sir James Hall for 1791 and other contemporary records. *Ann. Sci.*, 24, 1968, pp. 21–52.

Challinor, J. The early progress of British geology. I. From Leland to Woodward, 1538–1728. [II. From Strachey to Michell, 1719–1788. III. From Hutton to Playfair, 1788–1802.] *Ann. Sci.*, 9, 1953, pp. 124–153; 10, 1954, pp. 1–19, 107–148.

Chalmers, Gordon Keith. Sir Thomas Browne, true scientist. *Osiris*, 2, 1936, pp. 28–79.

Chalstrey, John. The life and works of John Freke (1688–1756). *St Bart's Hosp. J.*, 61, 1957, pp. 85–89, 108–112.

Chance, Burton. Sketches of the life and interests of Sir Hans Sloane: naturalist, physician, collector and benefactor. *Ann. med. Hist.*, 2nd Ser., 10, 1938, pp. 390–404.

Chapman, J. S. The strange books of Athanasius Kircher. *Proc. Roy. Inst. G.B.*, 38, 1960, pp. 259–268.

Chaytor, H. J. *From script to print: an introduction to medieval literature*, Cambridge, 1945.

Chipman, Robert A. The manuscript letters of Stephen Gray, F.R.S. (1666/7–1736). *Isis*, 49, 1958, pp. 414–433.

——. An unpublished letter of Stephen Gray on electrical experiments, 1707–1708. *Isis*, 45, 1954, pp. 33–40.

Chittenden, F. J. The Lindley Library. *J. Roy. Hort. Soc.*, 65, 1939, pp. 350–352.

Christianson, John. Tycho Brahe at the University of Copenhagen, 1559–1562. *Isis*, 58, 1967, pp. 198–203.

Clagett, Marshall. *Archimedes in the Middle Ages. Volume 1. The Arabo-Latin tradition*, Madison, 1964.

——. *Greek science in antiquity*, New York, 1955.

——. The impact of Archimedes on medieval science. *Isis*, 50, 1959, pp. 419–429.

——. A note on the *Commensurator* falsely attributed to Regiomontanus. *Isis*, 60, 1969, pp. 383–384.

Clark, Joseph T. Pierre Gassendi and the physics of Galileo. *Isis*, 54, 1963, pp. 352–370.

Clark, Sir Kenneth. *A catalogue of the drawings of Leonardo da Vinci in the collection of His Majesty the King at Windsor Castle*, 2 vols., Cambridge, 1935.

Clark, R. W. *The Huxleys*, 1968.

Clarke, Edwin, and Bearn, J. G. The brain "glands" of Malpighi elucidated by practical history. *J. Hist. Med.*, 23, 1968, pp. 309–330.

——, and O'Malley, C. D. *The human brain and spinal cord. A historical study illustrated by writings from antiquity to the twentieth century*, Berkeley, Los Angeles, 1968.

Clark-Kennedy, A. E. *Stephen Hales, D.D., F.R.S.: an eighteenth century biography*, Cambridge, 1929.

Clemens, F. J. *Giordano Bruno und Nicolaus von Cusa*, Bonn, 1847.

Clement, A. G., and Robertson, Robert H. S. *Scotland's scientific heritage*, Edinburgh, London, 1961.

Cochrane, Thomas. *Notes from Doctor Black's lectures on chemistry 1767–8. Edited with an introduction by Douglas McKie*, [*etc.*], (Wilmslow), Cheshire, 1966.

Cockayne, Oswald. *Leechdoms, wort-cunning and starcraft of early England, being a collection of documents, for the most part never before printed, illustrating the history of science in this country before the Norman Conquest.* 3 vols., 1864–66; reprinted 1961.

Cohen, Lord Henry. Erasmus Darwin. *Univ. Birmingham hist. J.*, 11, 1967, pp. 17–40.

Cohen, I. Bernard. *Franklin and Newton: an inquiry into speculative Newtonian experimental science and Franklin's work in electricity as an example thereof*, Boston, 1966.

——. Leibniz on elliptical orbits: as seen in his correspondence with the Académie Royale des Sciences in 1700. *J. Hist. Med.*, 17, 1962, pp. 72–82.

——. Neglected sources for the life of Stephen Gray (1666 or 1667–1736). *Isis*, 45, 1954, pp. 41–50.

——. Newton, Hooke, and "Boyle's law" (discovered by Power and Towneley). *Nature*, 204, 1064, pp. 618–621.

——. Newton in the light of recent scholarship. *Isis*, 51, 1960, pp. 489–514.

——. A note concerning Diderot and Franklin. *Isis*, 46, 1955, pp. 268–272.

——. Pemberton's translation of Newton's *Principia*, with notes on Motte's translation. *Isis*, 54, 1963, pp. 319–351.

——. Roemer and the determination of the velocity of light. *Isis*, 31, 1939–40, pp. 327–370.

——. Versions of Isaac Newton's first published paper. With remarks on the question of whether Newton planned to publish an edition of his early papers on light and color. *Arch. int. Hist. Sci.*, 11, 1958, pp. 357–375.

Cohen, John, Hansel, C. E. M., and May, Edith F. Natural history of learned and scientific societies. *Nature*, 173, 1954, pp. 328–333.

Cohen, Morris R., and Drabkin, I. E. *A source book in Greek science*, New York, [*etc.*], 1948; reissued 1959.

Cole, Francis Joseph. The *Biblia naturae* of Swammerdam. *Nature*, 165, 1950, p. 511.

——. Bibliographical recollections of a biologist. *Osiris*, 8, 1948, pp. 289–315; also in *Proc. Pap. Oxford Bib. Soc.*, 5, iii, 1938, pp. 169–186.

——. Harvey's animals. *J. Hist. Med.*, 12, 1957, pp. 106–113.

——. Henry Power on the circulation of the blood. *J. Hist. Med.*, 12, 1957, pp. 291–324.

Cole, Francis Joseph. The history of Albrecht Dürer's rhinoceros in zoological literature. In, *Science, medicine and history*, Vol. 1, 1953, pp. 337–356.

——. *A history of comparative anatomy, from Aristotle to the eighteenth century*, 1944.

——. Leeuwenhoek's zoological researches. Part 1 [–2]. *Ann. Sci.*, 2, 1937, pp. 1–46, 185–235.

——. Obiter dicta bibliographica. *J. Hist. Med.*, 13, 1958, pp. 2–9.

——, and Eales, Nellie B. The history of comparative anatomy. Part 1. A statistical analysis of the literature. *Sci. Progr.*, 11, 1916–17, pp. 578–596.

The Cole Library of Zoology and Early Medicine, University of Reading, *Nature*, 188, 1960, pp. 1148–1151.

Cole, R. J. Friedrich Accum (1769–1838). A bibliographical study. *Ann. Sci.*, 7, 1951, pp. 128–143.

Cole, Rufus. Francesco Redi (1626–1697), physician, naturalist and poet. *Ann. med. Hist.*, 8, 1926, pp. 347–359.

Coleman, William. *Georges Cuvier, zoologist. A study in the history of evolution theory*, Cambridge, Mass., 1964.

——. Lyell and the "reality" of species: 1830–1833. *Isis*, 53, 1962, pp. 325–338.

——. A note on the early relationship between Georges Cuvier and Louis Agassiz. *J. Hist. Med.*, 18, 1963, pp. 51–63.

Coley, N. G. The Animal Chemistry Club: assistant society to the Royal Society. *Notes Rec. Roy. Soc. Lond.*, 22, 1967, pp. 173–185.

——. The physico-chemical studies of Amedeo Avogadro. *Ann. Sci.*, 20, 1964 [1965], pp. 195–210.

Columbia University, School of Library Service. *Guide to the literature of science: for use in connection with the course on science literature. Prepared by Thomas P. Fleming. Second edition*, New York, 1957.

Comrie, John D. Michael Scot: a thirteenth-century scientist and physician. *Edinb. med. J.*, N.S. 25, 1920, pp. 50–60.

Conant, James Bryant, *ed. Harvard case histories in experimental science*, 2 vols., Cambridge, Mass., 1957.

Conradi, Edward. Learned societies and academies in early times. *Pedagogical Seminary*, 12, 1905, pp. 384–426.

Cook, Sir James. The science information problem. *Adv. Sci.*, 23, 1966, pp. 305–309.

Coonen, L. P. Aristotle on biology. *Thomist Reader*, 1958, pp. 35–71.

——. Theophrastus revisited. *Centennial Rev.*, 1, 1957, pp. 404–418.

Copeman, W. S. C. Sir Hans Sloane and the Apothecaries Garden. *Med. Hist.*, 5, 1961, pp. 154–156.

Coppleson, V. M. The life and times of Dr. George Bennett. *Bull. Post. Grad. Comm. Med., Univ. Sydney*, 11, ix, 1955, pp. 207–264.

Coppola, Dominick. French scientific publishing. *Stechert-Hafner Book News*, 11, 1936, pp. 13–15.
Corti, Alfredo. Omero sapeva quel che il Redi dimostro. *Riv. Storia Sci. med. nat.*, 45, 1954, pp. 114–135.
Cowan, C. F. Notes on Griffiths' "Animal kingdom" of Cuvier [1824–1835]. *J. Soc. Bib. Nat. Hist.*, 5, 1969, pp. 137–140.
Cowles, Thomas, Dr. Henry Power's poem on the microscope. *Isis*, 21, 1934, pp. 71–80.
Cowley, J. D. *Bibliographical description and cataloguing*, 1939.
Cox, Harold, and Chandler, John E. *The House of Longman, with a record of their bicentenary celebrations. . . . 1724–1924*, London, [*etc.*], *for private circulation*, 1925.
Cragg, R. H. "Work, finish and publish". The chemistry of Michael Faraday 1791–1867. *Chemistry in Britain*, 3, 1967, pp. 482–486.
Crane, E. J., Patterson, Austin M., and Marr, Eleanor B. *A guide to the literature of chemistry. Second edition*, New York, London, 1957.
Crombie, A. C. *Augustine to Galileo. Second edition*, 2 vols., 1961.
——. Descartes. *Sci. Amer.*, 201, 1959, pp. 160–173.
——. Helmholtz. *Sci. Amer.*, 198, 1958, 94–102.
——. Historians and the scientific revolution. *Endeavour*, 19, 1960, pp. 9–13.
——. *Medieval and early modern science*, 2 vols., New York, 1959.
——. The *Opus maius* of Roger Bacon. *Endeavour*, 8, 1949, pp. 163–166.
——. *Robert Grosseteste and the origins of experimental science, 1100–1700*, Oxford, 1953.
Crommelin, C. A. The clocks of Christian Huygens. *Endeavour*, 9, 1950, pp. 64–69.
Crook, Ronald E. *A bibliography of Joseph Priestley, 1733–1804*, 1966.
Crosland, Maurice P. The origins of Gay-Lussac's law of combining volumes of gases. *Ann. Sci.*, 17, 1961 [1963], pp. 1–26.
——. *The Society of Arcueil. A view of French science at the time of Napoleon I*, London, 1967.
——. The use of diagrams as chemical 'equations' in the lecture notes of William Cullen and Joseph Black. *Ann. Sci.*, 15, 1959, pp. 75–90.
Crowley, Denis. The Royal Dublin Society. *J. Roy. Inst. Chem.*, 82, 1958, pp. 10–18.
Crowther, J. G. *Francis Bacon: the first statesman of science*, 1960.
Curle, Richard. *The Ray Society: a bibliographical history*, 1954.
Curry, Walter Clyde. *Chaucer and the mediaeval sciences. Revised and enlarged edition*, 1960.
Curtis, W. Hugh. *William Curtis, 1746–1799, Fellow of the Linnean Society, botanist and entomologist*, [*etc.*], Winchester, 1941.
Curwen, Henry. *A history of booksellers: the old and the new*, [*etc.*], 1873.

Cushing, H. B. Darwin centennial publications. *Catholic Library World*, 31, 1959–60, pp. 423–429, 456–457.

Dales, Richard C. The authorship of the *Questio de fluxu et refluxus maris*, attributed to Robert Grosseteste. *Speculum*, 37, 1962, pp. 582–588.

——. The authorship of the "Summa in Physica" attributed to Robert Grosseteste. *Isis*, 55, 1964, pp. 70–74.

——. Robert Grosseteste's *Commentarius in octo libros Physicorum Aristotelis. Medievalia et Humanistica*, 11, 1957, pp. 10–33.

——. Robert Grosseteste's scientific work. *Isis*, 52, 1961, pp. 381–402.

——. Robert Grosseteste's views on astrology. *Mediaeval Studies*, 29, 1967, pp. 357–363.

——. The text of Robert Grosseteste's *Questio de fluxu et refluxu maris* with an English translation. *Isis*, 57, 1966, pp. 455–474.

Dampier, Sir William Cecil. *A shorter history of science*, Cambridge, 1944.

——. *The recent development of physical science*, 1927.

——, and Whetham, Margaret Dampier. *Cambridge readings in the literature of science: being extracts from the writings of men of science to illustrate the development of scientific thought*, Cambridge, 1924.

Darlington, C. D. *Darwin's place in history*, Oxford, 1959.

Darmstaedter, Ernst. *Die Alchemie des Geber*, [*etc.*], Berlin, 1922.

Darmstaedter, Ludwig. *Handbuch zur Geschichte der Naturwissenschaften und der Technik. In chronologischer Darstellung. Zweite umgearbeitete und vermehrte Auflage. Unter Mitwirkung von R. Du Bois-Reymond und D. C. Schaefer*, Berlin, 1908; reprint announced.

Darwin, Charles. *Charles Darwin and the voyage of the Beagle. Edited with an introduction by Nora Barlow*, 1945.

——, and Henslow, John Stevens. *Darwin and Henslow: the growth of an idea; letters, 1831–1860, edited by Nora Barlow*, 1967.

——, and Wallace, Alfred Russel. *Evolution by natural selection*, Cambridge, 1958.

Davidge, R. The library of the Royal Botanic Gardens, Kew. *London Librarian*, 6, Oct., 1960, pp. 3–9.

Davidson, J. N. The Royal Society of Edinburgh. *J. Roy. Inst. Chem.*, Nov., 1954, pp. 562–566.

Davies, Gordon L. The eighteenth-century denudation dilemma and the Huttonian theory of the earth. *Ann. Sci.*, 22, 1966, pp. 129–138.

——. The tour of the British Isles made by Louis Agassiz in 1840. *Ann. Sci.*, 24, 1968, pp. 131–146.

Davis, Tenney L. An ancient Chinese treatise on alchemy entitled Ts'an T'ung Ch'i. Written by Wei Po-Yang about 142 A.D. Now translated from the Chinese into English by Lu-Ch'iang Wu. With an introduction and notes by Tenney L. Davis. *Isis*, 18, 1932, pp. 210–289.

Davis, Tenny L. The Chinese beginnings of alchemy. *Endeavour*, 2 , 1943 pp. 154–160.

——. The first edition of the Sceptical chymist. *Isis*, 8, 1926, pp. 71–76.

——. The vicissitudes of Boerhaave's Textbook of chemistry. *Isis*, 10, 1928, pp. 33–46.

——, and Chao Yün Ts'ung. An alchemical poem by Kao Hsiang-Hsien. *Isis*, 30, 1939, pp. 236–240.

Davy, John. *Memoirs of the life of Sir Humphrey Davy, Bart. By his brother*, 2 vols., 1836.

Davy, Norman, *ed. British scientific literature in the seventeenth century*, London, [*etc.*], 1953 [i.e. 1954].

Davy de Virville, Ad. Réaumur, botaniste. *Rev. Hist. Sci., Paris*, 11, 1958, pp. 134–137.

Dawes, Ben. *A hundred years of biology*, 1952.

Dawrant, A. G. William Gilbert of Colchester. *St Bart's Hosp. J.*, 59, 1955, pp. 47–53.

Dawson, Warren R. Dawson Turner, F.R.S. (1775–1858). *J. Soc. Bib. Nat. Hist.*, 3, 1958, pp. 303–310.

——. *The Huxley papers: a descriptive catalogue of the correspondence, manuscripts and miscellaneous papers of the Rt. Hon. Thomas Henry Huxley, P.C., D.C.L., F.R.S.: preserved in the Imperial College of Science and Technology*, 1946.

——. Sir Joseph Hooker and Dawson Turner, *J. Soc. Bib. Nat. Hist.*, 2, vi, 1950, pp. 218–222.

Deacon, Richard. *John Dee, scientist, geographer, astrologer and secret agent to Elizabeth I*, 1968.

de Beer, Sir Gavin R. *Charles Darwin. Evolution by natural selection*, 1963. (British Men of Science Series.)

——. The Darwin letters of Shrewsbury School. *Notes Rec. Roy. Soc. Lond.*, 23, 1968, pp. 68–95.

——. Darwin's journal. *Bull. Brit. Mus.* (*Nat. Hist.*), *Hist. Ser.*, 2, 1960, 3–21.

——. Darwin's notebooks on transmutation of species. Part 1 [–4 Addenda, etc.] [etc.]. *Bull. Brit. Mus.* (*Nat. Hist.*), *Hist. Ser.*, 2, 1960–61, pp. 25–73, 75–118, 153–183, 187–200.

——. Further unpublished letters of Charles Darwin. *Ann. Sci.*, 14, 1958, pp. 83–115.

——. Haller's *Historia stirpium. Ann. Sci.*, 9, 1953, pp. 1–46.

——. Mendel, Darwin, and Fisher (1865–1965). *Notes Rec. Roy. Soc. Lond.*, 19, 1964, pp. 192–226.

——. Mendel, Darwin, and Fisher: addendum. *Notes Rec. Roy. Soc. Lond.*, 21, 1966, pp. 64–71.

——. The origins of Darwin's ideas on evolution and natural selection. The Wilkins Lecture. *Proc. Roy. Soc.*, 155B, 1962, pp. 321–338.

de Beer, Sir Gavin R. *Sir Hans Sloane and the British Museum*, London, [*etc.*], 1953.

——. Some unpublished letters of Charles Darwin. *Notes Rec. Roy. Soc. Lond.*, 14, 1959, pp. 12–66.

——. The volcanoes of Auvergne. *Ann. Sci.*, 18, 1962 [1964], pp. 49–61.

——, Rowlands, M. J., and Skramovsky, B. M., *eds.* Darwin's "Notebooks on transmutation of species". Part 6. Pages excised by Darwin. *Bull. Brit. Mus.* (*Nat. Hist.*), *Hist. Ser.*, 3, 1967, pp. 129–176.

Debuch, H., and Dawson, R. M. C. Prof. J. L. W. Thudichum (1829–1901). *Nature*, 207, 1965, p. 814.

Debus, Allen G. *The English Paracelsians*, 1965.

——. A forgotten chapter in the introduction of the new chemistry in Italy. *Ambix*, 11, 1963, pp. 153–157.

——. Mathematics and nature in the chemical texts of the Renaissance. *Ambix*, 15, 1968, pp. 1–28.

——. Renaissance chemistry and the work of Robert Fludd. *Ambix*, 14, 1967, pp. 42–59.

——. Sir Thomas Browne and the study of colour indications. *Ambix*, 10, 1962, pp. 29–36.

——, *ed. World who's who in science: a biographical dictionary of notable scientists from antiquity to the present*, Chicago, 1968.

——, and Multhauf, Robert P. *Alchemy and chemistry in the seventeenth century. Papers read . . . at a Clark Library Seminar, March 12, 1966*, Los Angeles, 1966.

De Grood, David H. *Haeckel's theory of the unity of nature*, Boston, 1965.

Delépine, Marcel. Les *Annales de Chimie* de leur fondation à la 173e année de leur parution. *Ann. Chim.*, 13e Sér., 7, 1962, pp. 1–11.

De Ricci, Seymour. *English collectors of books and manuscripts (1530–1930), and their marks of ownership*, [*etc.*], Cambridge, 1930.

De Rueck, Anthony, and Knight, Julie, *eds. Communication in science: documentation and automation*, 1967.

De Villamil, R. *Newton, the man*, [*etc.*], [1931].

Dewhurst, Kenneth. Locke's contribution to Boyle's researches on the air and on human blood. *Notes Rec. Roy. Soc. Lond.*, 17, 1962, pp. 198–206.

——. Willis in Oxford: some new MSS. *Proc. Roy. Soc. Med.*, 57, 1964, pp. 682–687.

Dibner, Bern. *Darwin of the Beagle. Second edition*, New York, 1964.

——. *Oersted and the discovery of electromagnetism*, New York, 1963.

Dickinson, H. W. *James Watt, craftsman & engineer*, Cambridge, 1936.

——. Tercentenary of Denis Papin. *Nature*, 160, 1947, pp. 422–423.

Diels, H. Ueber die ältesten Philosophischenschulen der Griechen. *Philosophische Aufsätze*, Leipzig, 1887, pp. 241–260.

Dijgraaf, Sven. Spallanzani's unpublished experiments on the sensory basis of object perception in bats. *Isis*, 51, 1960, pp. 9–20.

Dijksterhuis, E. J. *Archimedes*, Copenhagen, 1957. (Acta Historica Scientiarum Naturalium et Medicinalium, Vol. 12.)

——. *The mechanization of the world picture. Translated by C. Dikshoorn*, Oxford, 1961.

Dingle, Herbert, Astronomy in the sixteenth and seventeenth centuries. In *Science, medicine and history*, Vol. 1, 1953, pp. 455–468.

——, *ed. A century of science, 1851–1951. Written by specialist authors under the editorship of Herbert Dingle*, London, [*etc.*], 1951.

——. The dependence of science on its history. *Bull. Brit. Soc. Hist. Sci.*, 2, 1957, pp. 65–73.

——. Edmond Halley: his times and ours. *The Observatory*, 76, 1956, pp. 117–131.

——. Nicolaus Copernicus, 1473–1543. *Endeavour*, 2, 1943, pp. 136–141.

——. Physics in the eighteenth century. *Phil. Mag.*, Commemoration issue, July, 1948, pp. 28–46.

——. Tycho Brahe, 1546–1601. *Endeavour*, 5, 1946, pp. 137–141.

——. The work of Copernicus. *Nature*, 151, 1943, pp. 576–577.

Diogenes Laertius. *Lives of eminent philosophers, ix, 30–39, translated by R. D. Hicks*, 1925. (Loeb Classical Library.)

Dioscorides. *The Greek herbal of Dioscorides, illustrated by a Byzantine A.D. 512. Englished by John Goodyer A.D. 1655, edited and first printed A.D. 1933 by Robert T. Gunther*, [*etc.*], *for the author*, 1934.

Directory of British associations. Interests, activities and publications of trade associations, scientific and technical societies, professional institutes, learned societies, research organisations, chambers of trade and commerce, agricultural societies, trade unions, cultural, sports and welfare organisations in the United Kingdom and in the Republic of Ireland. Edition 2, Beckenham, Kent, 1967–68.

Disney, Alfred N., *ed. Origin and development of the microscope, as illustrated by catalogues of the instruments and accessories, in the collections of the Royal Microscopical Society, together with bibliographies of original authorities*, [*etc.*], 1928.

Ditmas, E. M. R. The co-ordination of abstracting services. UNESCO's approach to the problem. *J. Document.*, 4, 1948–49, pp. 67–83.

Dobbin, Leonard. A Cullen chemical manuscript of 1753. *Ann. Sci.*, 1, 1936, pp. 138–156.

Dobell, Clifford. *Antony van Leeuwenhoek and his "little animals": being some account of the father of protozoology and bacteriology, and his multiferous discoveries in these disciplines*, [*etc.*], 1932; reprinted 1958 and 1960.

Dobson, Jessie. Conservators of the Hunterian Museum. IV. William Henry Flower. *Ann. Roy. Coll. Surg. Engl.*, 30, 1962, pp. 383–391.

——. *John Hunter*, Edinburgh, London, 1969.

——. John Hunter's animals. *J. Hist. Med.*, 17, 1962, pp. 479–486.

Dobson, Jessie. John Hunter's museum. *St Bart's Hosp. J.*, 66, 1962, pp. 146–150.

Dodds, Sir Charles. Christopher Merrett, F.R.C.P. (1614–1695), first Harveian Librarian. *Proc. Roy. Soc. Med.*, 47, 1954, pp. 1053–1055.

Doetsch, Raymond N., *ed. Microbiology: historical contributions from 1776 to 1908 by Spallanzani, Schwann, Pasteur, Cohn, Tyndall, Koch, Lister, Schloesing, Burrill, Ehrlich, Winogradsky, Warington, Beijerinck, Smith, Orla-Jensen*, New Brunswick, New Jersey, 1960.

Donnan, F. G., and Haas, Arthur, *eds. A commentary on the scientific writings of J. Willard Gibbs*, [*etc.*], 2 vols., New Haven, London, 1938.

Drabkin, David L. *Thudichum: chemist of the brain*, Philadelphia, 1958.

Drabkin, I. E. A note on Galileo's De motu. *Isis*, 51, 1960, pp. 271–277.

Drachman, Julian M. *Studies in the literature of natural science*, New York, 1930.

Drecker, J. Hermannus Contractus Über das Astrolab. *Isis*, 16, 1931, pp. 200–219.

Dreyer, John Louis Emil. *Tycho Brahe: a picture of scientific life and work in the sixteenth century*, Edinburgh, 1890; reprinted New York (Dover), 1963, and Gloucester, Mass., 1964.

——, and Turner, Herbert Hall, *eds. History of the Royal Astronomical Society, 1820–1920*, 1923.

Drummond, H. J. H. Records of medical and scientific societies in Scotland. I. Early medical and scientific societies of North-East Scotland. *Bibliotheck*, 1, ii, 1957, pp. 31–33.

Dryander, Jonas. *Catalogus bibliothecae historico-naturalis Josephi Banks*, 5 vols., 1798–1800.

Dubbey, J. M. Cauchy's contribution to the establishment of the calculus. *Ann. Sci.*, 22, 1966, pp. 61–67.

Dubos, René J. *Louis Pasteur, free lance of science*, 1951.

Dubs, Homer H. The beginnings of alchemy. *Isis*, 38, 1947–48, pp. 62–85.

——. The origin of alchemy. *Ambix*, 9, 1961, pp. 23–36.

Ducca, Anne D. The Chemists' Club Library. *Adv. Chem. Ser.*, 30, 1961, pp. 282–284.

Duff, E. Gordon. *Fifteenth century English books*, [*etc.*], 1917.

Duhem, Pierre. *Le système du monde. Histoire des doctrines cosmologiques de Platon à Copernic*, 10 vols., Paris, 1954–58.

Dunleavy, Gareth W. The Chaucer ascription in Trinity College, Dublin MS. D. 2. 8. *Ambix*, 13, 1965, pp. 2–21.

Dunsheath, Percy. *A history of electrical engineering*, 1962.

Duveen, Denis I. Antoine Laurent Lavoisier and the French Revolution. Part 1 [–4]. *J. chem. Educ.*, 31, 1954, pp. 60–65; 34, 1957, pp. 502–503; 35, 1958, pp. 233–234, 470–471.

——. Antoine Lavoisier's Traité de chimie: a bibliographical note. *Isis*, 41, 1950, pp. 168–171.

Duveen. Denis I. The Duveen alchemical and chemical collection. (Contemporary collectors, xi.) *Book Collector*, 5, 1946, pp. 331–342.

——, and Hahn, Roger. A note on some Lavoisiereana in the "Journal de Paris". *Isis*, 51, 1960, pp. 64–66.

——. *Supplement to a bibliography of the works of Antoine Laurent Lavoisier, 1743–1794*, 1964.

——, and Klickstein, Herbert S. The "American" edition of Lavoisier's *L'art de fabriquer le salin et la potasse*. *William & Mary Quart.*, 13, 1956, pp. 493–498.

——, and ——. Benjamin Franklin (1706–1790) and Antoine Laurent Lavoisier (1743–1794). Part I. Franklin and the new chemistry. [Part II. Joint investigations. Part III. Documentation.] *Ann. Sci.*, 11, 1955, pp. 103–128, 271–308; 13, 1957, pp. 30–46.

——, and ——. A bibliographical study of the introduction of Lavoisier's *Traité élémentaire de chimie* into Great Britain and America. *Ann. Sci.*, 10, 1954, pp. 321–338.

——, and ——. *A bibliography of the works of Antoine Laurent Lavoisier, 1743–1794*, [*etc.*], 1954.

——, and ——. John Dalton's autopsy. *J. Hist. Med.*, 9, 1954, pp. 360–362.

——, and ——. A letter from Guyton de Morveau to Marquart [*sic.* Macquer] relating to Lavoisier's attack against the *phlogiston* theory (1778); with an account of de Morveau's conversion to Lavoisier's doctrines in 1787. *Osiris*, 12, 1956, pp. 342–367; see also A case of mistaken identity: Macquer and not Marquart. *Isis*, 49, 1958, pp. 73–74.

——, and ——. Some new facts relating to the arrest of Antoine Laurent Lavoisier. *Isis*, 49, 1958, pp. 347–348.

Dyson, G. Malcolm. *A short guide to chemical literature. Second edition*, London, [*etc.*], 1959.

Eales, Nellie B. Francis Joseph Cole, 1872–1959. *J. Hist. Med.*, 14, 1959, pp. 267–272.

——. On the provenance of some early medical and biological books. *J. Hist. Med.*, 24, 1969, pp. 183–192.

[Eames, Wilberforce.] *A list of editions of Ptolemy's Geography, 1475–1730*, New York, 1886.

Easton, Joy B. The early editions of Robert Recorde's *Ground of artes*. *Isis*, 58, 1967, pp. 515–532.

——. On the date of Robert Recorde's birth. *Isis*, 57, 1966, p. 121.

Easton, Stewart, C. *Roger Bacon and his search for a universal science*, [*etc.*], Oxford, 1952.

Eastwood, Bruce S. Grosseteste's "quantitative" law of refraction: a chapter in the history of non-experimental science. *J. Hist. Ideas*, 28, 1967, pp. 403–414.

Eby, Clifford H. "Anatomia del cavallo". [By Carlo Ruini.] *Western Veterinarian*, 7, 1960, pp. 88–91.

Egerton, Frank N. Leeuwenhoek as a founder of animal demography. *J. Hist. Biol.*, 1, 1968, pp. 1–22.

Eiseley, Loren C. Alfred Russel Wallace. *Sci. Amer.*, 200, 1959, pp. 70–84.

——. Charles Lyell. *Sci. Amer.*, 201, 1959, pp. 98–106.

——. *Darwin's century. Evolution and the men who discovered it*, 1958.

Ellegård, Alvar. *Darwin and the general reader. The reception of Darwin's theory of evolution in the British periodical press, 1859–1872*, Göteborg, 1958. (Göteborg Studies in English, 8.)

Elton, Charles Isaac, and Elton, Mary Augusta. *The great book collectors*, 1893.

Emblen, D. L. The library of Peter Mark Roget. *Book Collector*, 18, 1969, pp. 449–469.

Emery, Clark. Scientific theory in Erasmus Darwin's The Botanic Garden, 1789–91. *Isis*, 33, 1941–42, pp. 313–325.

Emmart, Emily Walcott. Concerning the Badianus manuscript, an Aztec herbal, "Codex Barberini, Latin 241" (Vatican Library). *Smithsonian miscellaneous collections*, 94, ii, 1935.

Esdaile, Arundell J. K. *The British Museum Library: a short history and survey*, [*etc.*], 1946.

——. *National libraries of the world: their history, administration and public services. . . . Second edition, completely revised by F. J. Hill*, 1957.

'Espinasse, Margaret. *Robert Hooke*, London, [*etc.*], 1956.

Essig, E. O. *A history of entomology*, New York, 1965.

——. A sketch history of entomology. *Osiris*, 2, 1936, pp. 80–123 (pp. 113–123, selected bibliography).

Estes, J. Worth, and White, Paul Dudley. William Withering and the purple foxglove. *Sci. Amer.*, 212, 1965, pp. 110–116, 119.

Evans, Herbert, M., *ed. Men and moments in the history of science*, Seattle, 1939.

Eve, A. S., and Creasey, C. H. *Life and work of John Tyndall. . . . With a chapter on Tyndall as a mountaineer by Lord Schuster*, [*etc.*], 1945.

Eyles, Joan M. Georgius Agricola (1494–1555). *Nature*, 176, 1955, pp. 949–950.

——. James Parkinson (1755–1824). *Nature*, 176, 1955, pp. 580–581.

——. William Smith (1769–1839): a bibliography of his published writings, maps and geological sections, printed and lithographed. *J. Soc. Bib. Nat. Hist.*, 5, 1969, pp. 87–109.

Eyles, V. A. A bibliographical note on the earliest printed version of James Hutton's *Theory of the earth*, its form and date of publication. *J. Soc. Bib. Nat. Hist.*, 3, 1955, pp. 105–108.

Eyles, V. A. Bibliography and the history of science. *J. Soc. Bib. Nat. Hist.*, 3, ii, 1955, pp. 63–71.

——. The evolution of a chemist. Sir James Hall, Bt., F.R.S., P.R.S.E., of Dunglass, Haddingtonshire, (1761–1832), and his relations with Joseph Black, Antoine Lavoisier, and other scientists of the period. *Ann. Sci.*, 19, 1963 [1965], pp. 153–182.

——. The history of geology: suggestions for further research. *History of Science*, 5, 1966, pp. 77–86.

——. James Hutton, 1726–1797. Commemoration of the 150th anniversary of his death. Note on the original publication of Hutton's *Theory of the earth*, and on the subsequent forms in which it was issued. *Proc. Roy. Soc. Edinb.*, 63B, 1950, pp. 377–386.

——. John Woodward, F.R.S. (1665–1728), physician and geologist. *Nature*, 206, 1965, pp. 868–870.

——. Nicolaus Steno, seventeenth-century anatomist, geologist and ecclesiastic. *Nature*, 174, 1954, pp. 8–10.

——, and Eyles, Joan M. Some geological correspondence of James Hutton. *Ann. Sci.*, 7, 1951, pp. 316–339.

Fabian, Bernhard. David Skene and the Aberdeen Philosophical Society. *Bibliotheck*, 5, 1968, pp. 81–99.

Fabricius, Hieronymus. *De venarum ostiolis, 1603. . . . Facsimile edition with introduction and notes by K. J. Franklin*, Springfield, Baltimore, 1933.

——. *The embryological treatises of Hieronymus Fabricius of Aquapendente. The formation of the egg and the chick.* [*De formatione ovi et pulli.*] *The formed fetus.* [*De formato foetu.*] *A facsimile edition, and a commentary by Howard B. Adelmann*, Ithaca, New York, 1942.

Fahie, J. J. *Galileo, his life and work*, 1903.

Fairchild, H. L. The history of the American Association for the Advancement of Science. *Science*, N.S. 59, 1924, pp. 365–369, 385–390, 410–415.

Farber, Eduard. Historiography of chemistry. *J. chem. Educ.*, 42, 1965, pp. 120–126.

Farrington, Benjamin. *Francis Bacon, philosopher of industrial science*, New York, 1949.

——. *Greek science: its meaning for us*, 2 vols., 1944–49. (Pelican Books.)

——. *Science in antiquity*, Oxford, 1936; reprinted 1947.

Farrukh, O. A. *Arab genius in science and philosophy*, Washington, 1954.

Fauré-Fremiet, E. Les origines de l'Académie des Sciences de Paris. *Notes Rec. Roy. Soc. Lond.*, 21, 1966, pp. 20–31.

Feisenberger, H. A. The libraries of Newton, Hooke and Boyle. *Notes Rec. Roy. Soc. Lond.*, 21, 1966, pp. 42–55.

Ferguson, Allan, and Ferguson, James. The Philosophical Magazine, *Phil. Mag.*, Commemoration issue, July, 1948, pp. 1–9.

Ferguson, John. *Bibliotheca chemica: a catalogue of the alchemical, chemical and pharmaceutical books in the collection of the late James Young of Kelly and Durris*, 2 vols., Glasgow, 1906. Reprinted 1954.

——. *A short biography and bibliography of Michael Scotus*, Glasgow, 1931.

Fichtmer, Gerhard. Neues zu Leben und Werk von Leonhart Fuchs aus seinen Briefen an Joachim Camerarius I. und II. in der Trew-Sammlung. *Gesnerus*, 25, 1968, pp. 65–82.

Finch, Sir Ernest. The forerunner [Leonardo da Vinci]. Thomas Vicary lecture delivered at the Royal College of Surgeons of England on 29th October, 1953. *Ann. Roy. Coll. Surg. Engl.*, 14, 1954, pp. 71–91.

Findlay, Alexander. *A hundred years of chemistry. Third edition, revised by T. I. Williams*, 1965.

Finlayson, C. P. Records of medical and scientific societies in Scotland. II. Records of scientific and medical societies preserved in the University Library, Edinburgh. *Bibliotheck*, 1, iii, 1958, pp. 14–19.

Fischer, Hans. Conrad Gessner (1516–1565) as bibliographer and encyclopedist. *The Library*, 5th Ser., 21, 1966, pp. 269–281.

Fisher, Ronald A. Has Mendel's work been rediscovered? *Ann. Sci.*, 1, 1936, pp. 115–137.

Flamsteed, John. *An account of the Rev. John Flamsteed, first Astronomer Royal. Supplement to the Account of the Rev. John Flamsteed, . . . by Francis Bailey*, 1837; reprinted 1966.

Fleming, Donald. The centenary of the *Origin of species*. *J. Hist. Ideas*, 20, 1959, pp. 437–446.

——. Galen on the motions of the blood in the heart and lungs. *Isis*, 46, 1955, pp. 14–21.

Fletcher, John. Athanasius Kircher and the distribution of his books. *The Library*, 5th Ser., 23, 1968, pp. 108–117.

Fletcher, William Younger. *English book collectors*, 1902.

Florkin, Marcel. Un manuscript de Schwann montrant que les "Mikroskopische Untersuchungen" constituent le premier volume d'une oeuvre intitullée "Theorie der Zellen", dont le second volume est constitué par une série de manuscrits encore inédits. *Bull. Acad. roy. Belg.* (*Cl. Sci.*), Sér. 5, 48, 1962, pp. 1054–1061.

Folta, J., *et. al. History of exact sciences in Bohemian lands up to the end of the nineteenth century*, Prague, 1961.

Forbes, Eric Gay. The bicentenary of the *Nautical Almanac* (1767). *Brit. J. Hist. Sci.*, 3, 1967, pp. 393–394.

——. Dr. Bradley's astronomical observations. *Quart. J. Roy. Astron. Soc.*, 6, 1965, pp. 321–328.

——. The life and work of Tobias Mayer (1723–62). *Quart. J. Roy. Astron. Soc.*, 8, 1967, pp. 227–251.

Forbes, R. J. Was Newton an alchemist? *Chymia*, 2, 1949, pp. 27–36.

Forbes, Thomas R. John Hunter on spontaneous intersexuality. *Amer. J. Anat.*, 116, 1965, pp. 269–300.

——. William Yarrell, British naturalist. *Proc. Amer. Philos. Soc.*, 106, 1962, pp. 505–515.

Ford, A. E. *John James Audubon*, Norman, Okla., 1964.

Foregger, Richard. Jan Ingen Housz and Joseph Priestley on carbon dioxide absorption. *Anesthesiology*, 17, 1956, pp. 511–522.

——. Two types of respiratory apparatus of Stephen Hales. *Anaesthesia*, 11, 1956, pp. 235–240.

Foster, W. D. *A history of parasitology*, Edinburgh, 1965.

Fountain, R. B. George Stubbs (1724–1806) as an anatomist. *Proc. Roy. Soc. Med.*, 61, 1968, pp. 639–646.

Fowler, Maureen J. *Guides to scientific periodicals. Annotated bibliography*, 1966.

Fox, H. Munro. The origin and development of "Biological Reviews". *Biol. Rev.*, 40, 1965, pp. 1–4.

Francis, Sir Frank. Sir Hans Sloane, 1660–1753, as a collector. *Lib. Assn. Rec.*, 63, 1961, pp. 1–5.

Francis, W. W., and Stevenson, Lloyd G. Three unpublished letters of Edward Jenner. *J. Hist. Med.*, 10, 1955, pp. 359–368.

Frank, F. C. Reflections on Sadi Carnot. *Physics Education*, 1, 1966, pp. 11–18.

Franklin, Benjamin. *Benjamin Franklin's autobiographical writings. Selected and edited by Carl van Doren*, 1946.

——. *Benjamin Franklin's experiments. A new edition of Franklin's Experiments and observations on electricity. Edited, with a critical and historical introduction, by I. Bernard Cohen*, Cambridge, Mass., 1941.

Franklin Kenneth James. Francis Joseph Cole, 1872–1959. *Biog. Mem. Fellows Roy. Soc.*, 5, 1960, pp. 37–47.

——. History of physiology, 1851–1951. *Advanc. Sci.*, 8, 1951–52, pp. 293–302.

——. *A short history of physiology. . . . Second edition*, 1949.

——. Some textual changes in successive editions of Richard Lower's *Tractatus de corde item de motu & colore sanguinis et chyli in eum transitu. Ann. Sci.*, 4, 1939, pp. 283–294.

Frati, Carlo. *Bibliografia Malpighiana. Catalogo descrittivo delle opere a stampa di Marcello Malpighi e degli scritti che lo riguardano*, Milan, 1897: facsimile reprint, London, 1960.

Frederick II. *The art of falconry, being the De arte venandi cum avibus of Frederick II of Hohenstaufen. Translated and edited by Casey A. Wood and F. Marjorie Fyfe*, Stanford Univ., London, 1943.

Freeman, R. B. On the origin of species 1859. *Book Collector*, 16, 1967, pp. 340–344.

Freeman, R. B. *The works of Charles Darwin. An annotated bibliographical handlist*, 1965.
——, and Gautrey, P. J. Darwin's *Questions about the breeding of animals*, with a note on *Queries about expression. J. Soc. Bib. Nat. Hist.*, 5, 1969, pp. 220–225.
Freund, Hugo. 80 Jahre Zeitschrift für wissenschaftliche Mikroskopie und mikroskopische Technik. Ihr Begründer und seine Nachfolger. *Z. wissen. Mikr.*, 66, 1964, pp. 3–17.
Fric, René. Contribution à l'étude de l'évolution des idées de Lavoisier sur la nature de l'air et sur la calcination des métaux. *Arch. int. Hist. Sci.*, 12, 1959, pp. 137–168.
Frick, George Frederick. Mark Catesby: the discovery of a naturalist. *Pap. Bib. Soc. Amer.*, 54, 1960, pp. 163–175.
——, and Stearns, Raymond Phineas. *Mark Catesby: the colonial Audubon*, Urbana, 1961.
Friedman, Reuben. Thomas Moffet. The tercentenary of his contribution to scabies. *Med. Life*, 41, 1934, pp. 620–637.
Fullmer, June Z. Davy's biographers: notes on scientific biography. *Science*, 155, 1967, pp. 285–291.
——. Davy's sketches of his contemporaries. *Chymia*, 12, 1967, pp. 127–150.
——. Humphry Davy and the gunpowder manufactory. *Ann. Sci.*, 20, 1964 [1965], pp. 165–194.
——. Humphry Davy's adversaries. *Chymia*, 8, 1962, pp. 147–164.
——. *Sir Humphry Davy's published works*, Cambridge, Mass., 1969.
Fulton, John Farquhar. *A bibliography of the Honourable Robert Boyle, Fellow of the Royal Society. . . . Second edition*, Oxford, 1961.
——. A bibliography of two Oxford physiologists. Richard Lower, 1631–1691, John Mayow, 1643–1679. *Oxford Bib. Soc. Proc. Pap.*, 1, i, 1934, 62 pp. 1935.
——. A book from Robert Boyle's library. *J. Hist. Med.*, 11, 1956, pp. 103–104.
——. Boyle and Sydenham. *J. Hist. Med.*, 11, 1956, pp. 351–352.
——. *The great medical bibliographers: a study in humanism*, Philadelphia, 1951.
——. Haller and the humanization of bibliography. *New Engl. J. Med.*, 206, 1932, pp. 323–328.
——. The Honourable Robert Boyle, F.R.S. (1627–1692). *Notes Rec. Roy. Soc. Lond.*, 15, 1960, pp. 119–135.
——. Robert Boyle and his influence on thought in the seventeenth century. *Isis*, 18, 1932, pp. 77–102.
——. The scientific writings of Joseph Priestley (1733–1804). *Atti del XIV Congresso Internazionale di Storia della Medicina*, 1954, pp. 1–8.

Fulton, and Cushing, Harvey. A bibliographical study of the Galvani and the Aldini writings on animal electricity. *Ann. Sci.*, 1, 1936, pp. 239–268.

Furley, David J. *Two studies in the Greek atomists: Study I. Indivisible magnitudes. Study II. Aristotle and Epicurus on voluntary action*, Princeton, N.J., 1967.

Gade, John Allyne. *The life and times of Tycho Brahe*, Princeton, 1947.

Gage, A. T. *A history of the Linnean Society of London*, 1938.

Gandz, S. The sources of al-Khowarizmi's algebra. *Osiris*, 1, 1936, pp. 263–277.

Ganzinger, Kurt. Ein Kräuterbuchmanuskript des Leonhart Fuchs in der Wiener Nationalbibliothek. *Sudhoffs Arch. Gesch. Med.*, 43, 1959, pp. 213–224.

Gardiner, K. R., and Gardiner, D. L. André-Marie Ampère and his English acquaintances. *Brit. J. Hist. Sci.*, 2, 1965, pp. 235–245.

Garrison, Fielding Hudson. The medical and scientific periodicals of the 17th and 18th centuries. *Bull. Inst. Hist. Med.*, 2, 1934, pp. 285–343. (Supplement to *Bull. Johns Hopkins Hosp.*, 55, 1934.)

Gasking, Elizabeth B. Why was Mendel's work ignored? *J. Hist. Ideas*, 20, 1959, pp. 60–84.

Gates, William. *The De la Cruz-Badiano Aztec herbal of 1552. Translation and commentary*, Baltimore. 1939. (Maya Society Publication No. 23.)

Gebler, Karl von. *Galileo Galilei and the Roman curia. . . . Translated . . . by Mrs. George Sturge*, 1879.

Geikie, Sir Archibald. *The founders of geology*, London, New York, 1905.

——. *Life of Sir Roderick Impey Murchison*, [*etc.*], 2 vols., 1875.

George, Wilma. *Biologist philosopher; a study of the life and writings of Alfred Russel Wallace*, London, New York, 1964.

Geoghegan, D. Some indications of Newton's attitude towards alchemy. *Ambix*, 6, 1957, pp. 102–106.

George, Albert J. The genesis of the Académie des Sciences. *Ann. Sci.*, 3, 1938, pp. 372–401.

George Stubbs. Rediscovered anatomical drawings from the Free Public Library, Worcester, Massachusetts, 1958.

Gershenson, Daniel E., and Greenberg, Daniel A. *Anaxagoras and the birth of physics. Introduction by E. Nagel*, New York, 1964.

Gerstner, Patsy A. James Hutton's theory of the earth and his theory of matter. *Isis*, 59, 1968, pp. 26–31.

Getman, Frederick H. Sir Charles Blagden, F.R.S. *Osiris*, 3, 1938, pp. 69–87.

Geymonat, L. *Galileo Galilei: a biography and inquiry into his philosophy of science. . . . Text translated from the Italian with additional notes and appendix by Stillman Drake*, New York, 1965.

Gibbs, F. W. Boerhaave and the botanists. *Ann. Sci.*, 13, 1957, pp. 47–61.
——. Boerhaave's chemical writings. *Ambix*, 6, 1958, pp. 117–135.
——. Dr. Johnson's first published work? *Ambix*, 8, 1960, pp. 24–34.
——. *Joseph Priestley. Adventurer in science and champion of truth*, 1965.
——. A notebook of William Lewis and Alexander Chisholm. *Ann. Sci.*, 8, 1952, pp. 202–220.
——. Peter Shaw and the revival of chemistry. *Ann. Sci.*, 7, 1951, pp. 211–237.
Gibson, C. S. The Chemical Society (of London) after one hundred years. *Endeavour*, 6, 1947, pp. 63–68.
Gibson, R. W. *Francis Bacon: a bibliography of his works and of Baconiana to the year 1750*, Oxford, 1950; privately printed *Supplement*, Oxford, 1959.
Gibson, Tom. The first homografts: Trembley and the polyps. *Brit. J. plast. Surg.*, 19, 1966, pp. 301–307.
Gillispie, Charles Coulston. *The edge of objectivity: an essay in the history of scientific ideas*, Princeton, New Jersey, London, 1960.
——. The formation of Lamarck's evolutionary theory. *Arch. int. Hist. Sci.*, 9, 1956, pp. 323–338.
——. Lamarck and Darwin in the history of science. *Amer. Scient.*, 46, 1958, pp. 388–409.
Gilmour, J. S. L. A "Catalogue of London plants" attributed to William Curtis. *J. Soc. Bib. Nat. Hist.*, 2, v, 1949, pp. 181–182.
——. William Curtis. *Nature*, 157, 1946, pp. 14–15.
Glass, Hiram Bentley, Temkin, Owsei, and Straus, William L., *eds. Forerunners of Darwin, 1745–1859*, Baltimore, 1959.
Glazebrook, R. T. *James Clerk Maxwell and modern physics*, London, [*etc.*], 1896. (Century of Science Series.)
Gloyne, S. Roodhouse, *John Hunter*, Edinburgh, 1950.
Goldstein, Bernard R. The Arabic version of Ptolemy's *Planetary hypotheses*. *Trans. Amer. Philos. Soc.*, N.S., 57, iv, 1967.
Golino, Carlo L., *ed. Galileo re-appraised*, Berkeley, Los Angeles, 1966.
Goodfield-Toulmin, June. Blasphemy and biology. *Rockefeller Univ. Rev.*, Sept.–Oct., 1966, pp. 9–18.
Gordon, Mervyn. Memories of Frank Buckland and extracts from some of his records. *Ann. Roy. Coll. Surg. Engl.*, 10, 1952, pp. 133–139.
Gorman, Mel. Gassendi in America. *Isis*, 55, 1964, pp. 409–417.
——. Some copies of Jean Beguin's textbook of chemistry. *J. chem. Educ.*, 35, 1958, pp. 575–577.
Goss, Charles Mayo. On anatomy of nerves by Galen of Pergamon. *Amer. J. Anat.*, 118, 1966, pp. 327–335.
——. The precision of Galen's anatomical descriptions compared with Galenism. *Anat. Rec.*, 152, 1965, pp. 376–380.

Gough, J. B. Lavoisier's early career in science: an examination of some new evidence. *Brit. J. Hist. Sci.*, 4, 1968, pp. 52–57.

Gould, R. P. The place of historical statements in biology. *Brit. J. Phil. Sci.*, 8, 1957, pp. 192–210.

Gourlie, Norah. *The prince of botanists: Carl Linnæus*, 1953.

Grainger, Thomas H. *A guide to the history of bacteriology*, New York, 1958. (*Chronica Botanica*, No. 18.)

Grant, R. *Johann Kepler. A tercentenary commemoration of his life and work*, Baltimore, 1931.

Gray, Andrew. *Lord Kelvin: an account of his scientific life and work*, London, New York, 1908. (English Men of Science Series.)

Gray, Dwight E., and Bray, Robert S. Abstracting and indexing services of physics interest. *Amer. J. Phys.*, 18, 1950, pp. 274–299.

Gray, George J. *A bibliography of the works of Sir Isaac Newton. Together with a list of books illustrating his works. . . . Second edition*, [*etc.*], Cambridge, 1907.

Gray, Ronald D. *Goethe the alchemist: a study of alchemical symbolism in Goethe's literary and scientific works*, Cambridge, 1952.

Green, H. Gwynedd, and Winter, H. J. J. John Landen, F.R.S. (1719–1790)—mathematician. *Isis*, 35, 1944, pp. 6–10.

Green, J. H. S. Robert Brown (1773–1858) and the Brownian movement. *Research*, 11, 1958, pp. 290–291.

Greenaway, Frank. The biographical approach to John Dalton. *Mem. Proc. Manchester Lit. Phil. Soc.*, 100, 1958–59, pp. 1–98.

——. John Dalton. *Endeavour*, 25, 1966, pp. 73–78.

——. *John Dalton and the atom*, 1966.

——. John Dalton as a historical figure. *Nature*, 211, 1966, pp. 1013–1014.

——. John Dalton in London. *Proc. Roy. Instn.*, 41, 1966, pp. 162–177.

——. A pattern of chemistry. Hundred years of the periodic table. *Chemistry in Britain*, 5, 1969, pp. 97–99.

——. *A short history of the Science Museum.* (*Ministry of Education, Science Museum*), 1951.

Greene, S. W. The dates of publication of Smith's *Flora Britannica* and Turner's *Muscologia Hibernica. J. Soc. Bib. Nat. Hist.*, 3, 1957, pp. 281–290.

——. The publication date of William Withering's A systematic arrangement of British plants (Ed. 4), London, 1801. *J. Soc. Bib. Nat. Hist.*, 4, 1962, pp. 66–67.

Greenstreet, W. J., *ed. Isaac Newton, 1642–1727. A memorial volume edited for the Mathematical Association by W. J. Greenstreet*, 1927.

Gregory, Joshua C. *Combustion from Heracleitos to Lavoisier*, 1934.

——. *The scientific achievements of Sir Humphry Davy*, 1930.

Gregory, Sir Richard. News and views from the scientific front. *Nature*, 151, 1943, pp. 517–519.

Gregory, Tullio. *Scetticismo ed empirismo, studio su Gassendi*, Bari, 1961.

Griffin, Francis J. Charles Davies Sherborn (1861–1942): some personal recollections. *J. Soc. Bib. Nat. Hist.*, 3, 1953, pp. 1–4.

Grimsley, Ronald. *Jean D'Alembert (1717–1783)*, Oxford, 1963.

Gruber, Howard E., and Gruber, Valmai. The eye of reason: Darwin's development during the *Beagle* voyage. *Isis*, 53, 1962, pp. 186–200.

Gudger, E. W. The five great naturalists of the sixteenth century: Belon, Rondelet, Salviani, Gesner and Aldrovandi: a chapter in the history of ichthyology. *Isis*, 22, 1934–35, pp. 21–40.

——. Pliny's Historia naturalis: the most popular natural history ever published. *Isis*, 6, 1924, pp. 269–281.

Guédès, Michel. L'Édition originale de la Philosophia botanica de Linné (1751). *J. Soc. Bib. Nat. Hist.*, 4, 1968, pp. 385–389.

Guerlac, Henry. Joseph Black and fixed air: a bicentenary retrospective, with some new or little known material. *Isis*, 48, 1957, pp. 124–151, 433–456.

——. Joseph Priestley's first papers on gases and their reception in France. *J. Hist. Med.*, 12, 1957, pp. 1–12.

——. *Lavoisier—the crucial year. The background and origin of his first experiments in 1772*, Ithaca, New York, 1961.

——. Lavoisier and his biographers. *Isis*, 45, 1954, pp. 51–62.

——. Newton's optical aether. His draft of a proposed addition to his *Opticks*. *Notes Rec. Roy. Soc. Lond.*, 22, 1967, pp. 45–57.

——. Some Daltonian doubts. *Isis*, 52, 1961, pp. 544–554.

Gunther, R. T. *Early science in Oxford. Vol. 3. Part I. The biological sciences. Part II. The biological collections*, Oxford, *for the subscribers*, 1925.

——. *Early science in Oxford. Vol. 4. The Philosophical Society*, Oxford, *for the subscribers*, 1925.

——. *Early science in Oxford. Vol. 6–7. The life and work of Robert Hooke*, Oxford, 1930. (Complete set, 15 volumes, 1923–67.)

Guppy, Henry. *Rules for the cataloguing of incunabula. . . . Second edition, revised*, 1932.

Habakkuk, H. J. Thomas Robert Malthus, F.R.S. (1766–1834). *Notes Rec. Roy. Soc. Lond.*, 14, 1959, pp. 99–108.

Haber, Francis C. *The age of the world: Moses to Darwin*, Baltimore, 1959.

Haden, Russell L. Galileo and the compound microscope. *Bull. Hist. Med.*, 12, 1942, pp. 242–247.

Hadzsits, G. D. *Lucretius and his influence*, New York, 1963.

Hagberg, Knut. *Carl Linnæus. . . . Translated from the Swedish by Alan Blair*, 1952.

Hahn, Roger. The Boscovitch archives at Berkeley. *Isis*, 56, 1965, pp. 70–78.

Hall, Alfred Rupert. Galileo and the science of motion. *Brit. J. Hist. Med.*, 2, 1964–65, pp. 185–199.

——. Galileo's system of the world. *Quart. J. Roy. Astron. Soc.*, 5, 1964, pp. 304–317.

——. *Hooke's Micrographia 1665–1965. A lecture in commemoration of the tercentenary of the publication of Micrographia: or some physiological descriptions of minute bodies by Robert Hooke, delivered at the Middlesex Hospital Medical School on 25 November, 1965*, London, 1966.

——. *The scientific revolution 1500–1800. The formation of the modern scientific attitude*, London, [*etc.*], 1954.

——. Wren's problem. *Notes Rec. Roy. Soc. Lond.*, 20, 1965, pp. 140–144.

——, and Hall, Marie Boas. Further notes on Henry Oldenburg. *Notes Rec. Roy. Soc. Lond.*, 23, 1968, pp. 33–42.

——, and ——. The intellectual origins of the Royal Society—London and Oxford. *Notes Rec. Roy. Soc. Lond.*, 23, 1968, pp. 157–168.

——, and ——. Some hitherto unknown facts about the private career of Henry Oldenburg. *Notes Rec. Roy. Soc. Lond.*, 18, 1963, pp. 94–103.

——, and ——. Why blame Oldenburg? *Isis*, 53, 1962, pp. 482–491.

Hall, Marie Boas. Oldenburg and the art of scientific communication. *Brit. J. Hist. Sci.*, 2, 1965, pp. 277–290.

——. Robert Boyle. *Sci. Amer.*, 217, 1967, pp. 97–102.

——. *Robert Boyle on natural philosophy. An essay with selections from his writings*, Bloomington, 1965.

——. Sources for the history of the Royal Society in the seventeenth century. *History of Science*, 5, 1966, pp. 62–76.

——. What happened to the Latin edition of Boyle's *History of cold*? *Notes Rec. Roy. Soc. Lond.*, 17, 1962, pp. 32–35.

Hall, Thomas S. The biology of the "Timaeus" in historical perspective. *Arion*, 4, 1965, pp. 109–122.

——. *A source book in animal biology*, Cambridge, Mass., 1951.

Hamarneh, Sami. Al-Kindī, a ninth-century physician, philosopher, and scholar. *Med. Hist.*, 9, 1965, pp. 328–342.

Hammond, Kathleen Mary. *August Wilhelm von Hofmann: bibliography submitted in part requirement for University of London Diploma in Librarianship*, 1967. [Typescript.]

Hamy, E. T. William Davisson, Intendant au Jardin du Roi, et Professeur de Chemie (1647–51). *Nouvelle Archives du Museum d'Histoire Naturelle*, 3 me Série, 10, i, 1898.

Hanlon, C. Rollins. The decline and fall of scientific societies. *Surgery*, 46, 1959, pp. 1–8.

Haraszti, Zoltán. Dr. Sarton on scientific incunabula. *Isis*, 32, 1940 (1947), pp. 52–62.

Harris, D. Fraser. Stephen Hales, the pioneer in the hygiene of ventilation. *Sci. Mon.*, 1916, pp. 440–454.

Harrison, Thomas P. Longolius on birds. *Ann. Sci.*, 14, 1958, pp. 257–268.

Hart, Ivor B. *James Watt and the history of steam power*, New York, 1949.

——. *Makers of science: mathematics; physics; astronomy*, [*etc.*], 1924.

——. *The mechanical investigations of Leonardo da Vinci*, 1925.

——. *The world of Leonardo da Vinci, man of science, engineer and dreamer of flight*, 1961.

Hartley, Sir Harold. Antoine Laurent Lavoisier, 26 August, 1743–8 May, 1794. *Proc. Roy. Soc.*, A, 189, 1947, pp. 427–456.

——. *Humphry Davy*, 1966. (British Men of Science Series.)

——. Michael Faraday as a physical chemist. *Trans. Faraday Soc.*, **49**, 1953, pp. 473–488.

——. The place of Berzelius in the history of chemistry, [*etc.*], *K. Svenska Vetenskapsakademiens Årsbok för är 1948*, *Bilaga*, pp. 31–50.

——. *The Royal Society—its origins and founders*, 1960.

——. Stanislao Cannizzaro, F.R.S. (1826–1910) and the first International Chemical Conference at Karlsruhe in 1860. *Notes Rec. Roy. Soc. Lond.*, 21, 1966, pp. 56–63.

Hartman, L. Wöhler and the vital force. *J. chem. Educ.*, 34, 1957, pp. 141–142.

Hartog, Sir Philip J. The newer views of Priestley and Lavoisier. *Ann. Sci.*, 5, 1941–47, pp. 1–56.

Haschmi, Mohamed Yahia. The beginnings of Arab alchemy. *Ambix*, 9, 1961, pp. 155–161.

Haskins, Charles Homer. The "Alchemy" ascribed to Michael Scot. *Isis*, 10, 1928, pp. 350–359.

——. The "De arte venandi cum avibus" of the Emperor Frederick II. *Engl. hist. Rev.*, 36, 1921, pp. 334–335.

——. *Studies in the history of mediaeval science. . . . Second edition*, Cambridge, Mass., 1927.

Hassinger, H. J. J. Becher. Ein Beitrag zur Geschichte der Merkantilismus. *Veröffl. d. Kommission f. neuere Geschichte Österreichs*, 37, 1951.

Hatton, A. P., and Rosenfeld, L. An analysis of Joule's experiments on the expansion of air. (Including original results not previously published.) *Centaurus*, 4, 1955–56, pp. 311–318.

Hauser, Gaston. *Geometrie der Griechen von Thales bis Euklid, mit einem einleitenden Abschnitt über die vorgrieche Geometrie*, Lucerne, 1955.

Hawkes, H. E. Geology. *Library Trends*, 15, 1967, pp. 816–828.

Hawks, Ellison. *Pioneers of plant study*, London, [*etc.*], 1928.

Hayward, John. The location of copies of the first editions of Giordano Bruno. *Book Collector*, 5, 1956, pp. 152–157.

Heathcote, N. H. de V. Franklin's introduction to electricity. *Isis*, 46, 1955, pp. 29–35.

——. Guericke's sulphur globe. *Ann. Sci.*, 6, 1950, pp. 293–305.

Heiberg, J. L. *Mathematics and physical science in classical antiquity. Translated from the German . . . by D. C. Macgregor*, 1922. (Chapters in the History of Science, 2.)

Hellmann, C. Doris. Was Tycho Brahe as influential as he thought? *Brit. J. Hist. Sci.*, 1, 1963, pp. 295–324.

Hemmeter, John C. Leonardo da Vinci as a scientist. *Ann. med. Hist.*, 3, 1921, pp. 26–44.

Henderson, E. *Life of James Ferguson, F.R.S., in a brief autobiographical account, and further extended memoir*, [*etc.*], Edinburgh, [etc.], 1867.

Henkle, Herman H. Chemical literature in the John Crerar Library and other Chicago libraries. *Adv. Chem. Ser.*, 30, 1961, pp. 297–304.

Henry, William Charles. *Memoirs of the life and scientific researches of John Dalton*, 1854. (Works of the Cavendish Society.)

Herdan, G. Chaucer's authorship of *The equatorie of the planetis:* the use of Romance vocabulary as evidence. *Language*, 32, 1956, pp. 254–259.

Herivel, John W. *The background to Newton's* Principia. *A study of Newton's dynamical researches in the years 1664–84.* Oxford, 1965 [1966].

——. Newton's discovery of the law of centrifugal force. *Isis*, 51, 1960, pp. 546–553.

Herrick, Francis Hobart. *Audubon the naturalist: a history of his life and time. . . . Second edition. Two volumes in one*, New York, London, 1938.

Herrlinger, Robert. C. F. Wolff: "Theoria generationis" (1759). *Neue Z. ärztl. Fortbild.*, 48, 1959, pp. 954–955.

——. C. F. Wolff's "Theoria generationis" (1759). Die Geschichte einer epochemachenden Dissertation. *Z. Anat. EntwGesch.*, 121, 1959, pp. 245–270.

Hesse, Mary B. Gilbert and the historians. *Brit. J. Phil. Sci.*, 11, 1960, pp. 1–10, 130–142.

——. Hooke's philosophical algebra. *Isis*, 57, 1966, pp. 67–83.

——. Hooke's vibration theory and the isochromy of springs. *Isis*, 57, 1966, pp. 433–441.

Hewsen, Robert H. Science in seventh-century Armenia: Ananias of Sirak. *Isis*, 59, 1968, pp. 32–45.

Heydenreich, Ludwig H. *Leonardo da Vinci*, 2 vols., 1954.

Hierons, Raymond, and Meyer, Alfred. Willis' place in the history of muscle physiology. *Proc. Roy. Soc. Med.*, 57, 1964, pp. 687–692.

Hill, Brian. The best laid schemes: Bonnell Thornton, M.B. (1724–1768), and Robert James Thornton, M.D. (1768?–1837). *Practitioner*, 200, 1968, pp. 722–726.
——. Georgian careerist: John Frost (1803–1840). *Practitioner*, 188, 1962, pp. 262–266.
Hill, Christopher. The intellectual origins of the Royal Society—London or Oxford? *Notes Rec. Roy. Soc. Lond.*, 23, 1968, pp. 144–156.
Himmelfarb, Gertrude. *Darwin and the Darwinian revolution*, 1959.
Hirsch, Rudolf. The invention of printing and the diffusion of alchemical and chemical knowledge. *Chymia*, 3, 1950, pp. 115–141.
Hiscock, W. G. John Evelyn's library at Christ Church. *Times Literary Supplement*, 6 April, 1951, p. 220.
Historical Association. *The early history of science: a short handlist*, 1950. (Helps for Students of History, No. 50.)
Hoare, Cecil A. Erasmus Darwin in Russia. *Ann. Sci.*, 11, 1956, pp. 255–256.
Hoare, M. E. "Cook the discoverer": an essay by Georg Forster, 1787. *Rec. Austral. Acad. Sci.*, 1, iv, 1969, pp. 7–16.
Hobson, E. W. *John Napier and the invention of logarithms, 1614*, Cambridge, 1914.
Hoff, Hebbel F. Galvani and the pre-Galvanian electrophysiologists. *Ann. Sci.*, 1, 1936, pp. 157–172.
——. Nicolaus of Cusa, van Helmont and Boyle: the first experiment of the Renaissance in quantitative biology and medicine. *J. Hist. Med.*, 19, 1964, pp. 99–117.
——, and Geddes, L. A. The contributions of the horse to knowledge of the heart and circulation. I. Stephen Hales and the measurement of blood pressure. *Conn. Med.*, 29, 1965, pp. 795–800.
Hofmann, Joseph Ehrenfried. *Classical mathematics: a concise history of the classical era in mathematics*, 1960.
Holford, Lord. The new home of the Royal Society. *Notes Rec. Roy. Soc. Lond.*, 22, 1967, pp. 23–36.
Hollindale, S. H. Charles Babbage and Lady Lovelace—two 19th century mathematicians. *Bull. Inst. Math. Applic.*, 2, 1966, pp. 2–15.
Holmberg, Arne. *Bibliographie de J. J. Berzelius*, [*etc.*], 3 parts, Stockholm, 1933–36.
Holmes, Frederic L. Elementary analysis and the origins of physiological chemistry. *Isis*, 54, 1963, pp. 50–81.
——. From elective affinities to chemical equilibria: Berthollet's law of mass action. *Chymia*, 8, 1962, pp. 105–145.
Holmes, Sir Maurice Gerald. Captain James Cook, R.N., F.R.S. *Endeavour*, 8, 1949, pp. 11–17.
——. *Captain James Cook, R.N., F.R.S., a bibliographical excursion*, 1952; reprinted Franklin, New York, 1968.

Holmyard, Eric John. *Alchemy*, Harmondsworth, 1957. (Pelican Books.)
——. Alchemy in medieval Islam. *Endeavour*, 14, 1955, pp. 117–125.
——. *The great chemists. . . . Third edition*, 1929.
——. Jābir ibn Hayyān. *Proc. Roy. Soc. Med.*, 16, Parts 1–2, 1923, Section Hist. Med., pp. 46–57.
Holt, Anne. *A life of Joseph Priestley*, [*etc.*], 1931.
Holt-White, Rashleigh. *The life and letters of Gilbert White of Selborne. Written and edited by his great-grand-nephew*, [*etc.*], 2 vols., 1901.
Home, Roderick W. Francis Hauksbee's theory of electricity. *Archive for History of Exact Sciences*, 4, 1967, pp. 203–217.
Hopley, I. B. Clerk Maxwell's experiments on colour vision. *Sci. Progr.*, 48, 1960, pp. 46–66.
——. Maxwell's determination of the number of electrostatic units in one electromagnetic unit of electricity. *Ann. Sci.*, 15, 1959, pp. 91–108.
——. Maxwell's work on electrical resistance. *Ann. Sci.*, 13, 1957, pp. 265–272; 14, 1958, pp. 197–210; 15, 1959, pp. 51–55.
Hoppen, K. Theodore. The Royal Society and Ireland—William Molyneux, F.R.S. (1656–98). *Notes Rec. Roy. Soc. Lond.*, 18, 1963, pp. 125–135.
Horne, R. A. Aristotelian chemistry. *Chymia*, 11, 1966, pp. 21–27.
Hoskin, Michael A. *William Herschel and the construction of the heavens*, 1963.
Houstoun, R. A. Athanasius Kircher and the magic lantern. *Sci. Progr.*, 45, 1957, pp. 462–464.
——. Kepler and the law of refraction. *Institute of Physics Bulletin*, 9, 1958, pp. 3–6.
Howard, Arthur Vyvyan. *Chambers's dictionary of scientists*, 1951.
Howarth, O. J. R. The British Association. *Endeavour*, 3, 1944, pp. 57–61.
——. *The British Association for the Advancement of Science: a retrospect, 1831–1931. . . . Centenary* (*second*) *edition*, 1931.
Hubbard, C. E. Bicentenary of the Royal Botanic Gardens, Kew. *Endeavour*, 18, 1959, pp. 156–160.
Huber, Friedrich. *Daniel Bernouli* (*1700–1782*) *als Physiologer und Statistiker*, Basle, 1958. (*Basler Veröffentlichungen zur Geschichte der Medizin und der Biologie*, Fasc. VIII.)
Hudson, Noel. *An early English version of Hortus sanitatis: a recent bibliographical discovery by Noel Hudson*, 1954 [1955].
Hughes, Arthur. *A history of cytology*, London, New York, 1959.
——. Science in English encyclopaedias. *Ann. Sci.*, 7, 1951, pp. 340–370; 8, 1952, pp. 323–367; 9, 1953, pp. 233–264; 11, 1955, pp. 74–92.
Hull, Lewis William Halsey. *History and philosophy of science*, London, [*etc.*], 1959.
Hulls, L. G. Phil. Trans.—the first fifty years. *Discovery*, 21, 1960, pp. 236–240.

Hummel, A. W. The printed herbal of 1249 A.D. *Isis*, 33, 1941–42; also in *Ann. Rep. Library of Congress for 1940*, 1941, pp. 155–157.

Hunkin, J. W. William Curtis, founder of *The Botanical Magazine*. *Endeavour*, 5, 1946, pp. 13–17.

Hunt, Judith Wallen. Science librarianship. *Science*, 104, 1946, pp. 171–173.

Hunter, Richard A., and Cuttler, Emily. Walter Charleton's "Natural history of the passions" (1674) and J. F. Senault's "The use of passions" (1649). A case of mistaken identity. *J. Hist. Med.*, 13, 1958, pp. 87–92.

——, and Macalpine, Ida. William Harvey and Robert Boyle. *Notes Rec. Roy. Soc. Lond.*, 13, 1958, pp. 115–127.

——, and Rose, F. Clifford. Robert Boyle's "Uncommon observations about vitiated sight", (London, 1688). *Brit. J. Ophthal.*, 42, 1958, pp. 726–731.

Huntley, Frank Livingstone. *Sir Thomas Browne: a biographical and critical study*, Ann Arbor, 1962.

Huxley, Julian, and Kettlewell, H. B. D. *Charles Darwin and his world*, 1965.

Huxley, Leonard. *Life and letters of Sir Joseph Dalton Hooker*, [*etc.*], 2 vols., 1918.

——. *The life and letters of Thomas Henry Huxley*, 1900.

Ihde, A. J., and Kieffer, W. F. *Selected readings in the history of chemistry*, Easton, Penn., 1965.

Iltis, Anne. Gregor Mendel's autobiography. *J. Hered.*, 45, 1954, pp. 231–234.

Iltis, Hugo. *Life of Mendel. . . . Translated by Eden and Cedar Paul*, [*etc.*], 1932.

International Botanical Congress (VI), Amsterdam, 1935. Catalogue of the exhibition of books.

Ireland, Norma Ruth. *Index to scientists of the world, from ancient to modern times: biographies and portraits*, Boston, 1962.

Ironmonger, E. F. The Royal Institution and the teaching of science in the nineteenth century. *Proc. Roy. Inst. G.B.*, 37, 1958, pp. 139–158.

Irvine, William. *Apes, angels, and Victorians: a joint biography of Darwin and Huxley*, 1955.

Irwin, Raymond, and Staveley, Ronald, *eds. The libraries of London. Second, revised edition*, 1961.

Iskandar, A. Z. *A catalogue of Arabic manuscripts on medicine and science in the Wellcome Historical Medical Library*, 1967.

Isler, Hansruedi. *Thomas Willis, 1621–1675, doctor and scientist*, New York, London, 1968.

Ives, Samuel A., and Ihde, Aaron J. The Duveen Library. *J. chem. Educ.*, 29, 1952, pp. 244–247.

Jackson, Benjamin Daydon. *Linnæus (afterwards Carl von Linné): the story of his life, adapted from the Swedish of Theodor Magnus Fries, and brought down to the present time in the light of recent research*, 1923.
Jaeger, Werner. *Aristotle. Fundamentals of the history of his development: translated with the author's corrections and additions by Richard Robinson. Second edition*, Oxford, 1948.
Jakubíček, M., and Kubíček, J. *Bibliographia Mendeliana*, Brno, 1965.
James, Montague Rhodes. Lists of manuscripts formerly owned by Dr. John Dee. With preface and identifications. Supplement to *Trans. Bib. Soc.*, 1921.
James, R. R. The father of British optics. Roger Bacon, *c.* 1214–1294. *Brit. J. Ophthal.*, 12, 1928, pp. 50–60.
James, W. O. Linnæus (1707–1778). *Endeavour*, 16, 1957, pp. 107–112.
Jaquot, Jean. Thomas Harriott's reputation for impiety. *Notes Rec. Roy. Soc. Lond.*, 9, 1952, pp. 164–187.
Jeffers, Robert H. *The friends of John Gerard (1545–1612), surgeon and botanist*, Falls Village, Conn., 1967.
Jeffreys, Alan Edward. Locating the manuscript sources of science. *Brit. J. Hist. Sci.*, 2, 1964, pp. 157–161.
——. *Michael Faraday: a list of his lectures and published writings*, [*etc.*], 1960.
Jessup, Everett Colgate. Rabanus Maurus: "De sermonum proprietate, seu de universo". *Ann. med. Hist.*, N.S. 6, 1934, pp. 35–41.
Jevons, F. R. Boerhaave's biochemistry. *Med. Hist.*, 6, 1962, pp. 343–362.
Job of Edessa. *Encyclopaedia of philosophical and natural sciences as taught in Baghdad about A.D. 817; or, Book of treasures. . . . Syriac text edited and translated with a critical apparatus by A. Mingana*, [*etc.*], Cambridge, 1935.
Johnson, Francis R. Astronomical text-books in the sixteenth century. In, *Science, medicine and history*, Vol. 1, 1953, pp. 285–302.
——. The influence of Thomas Digges on the progress of modern astronomy in sixteenth-century England. *Osiris*, 1, 1936, pp. 390–410.
——, and Larkey, Sanford V. Robert Recorde's mathematical teaching and the anti-Aristotelian movement. *Huntington Lib. Bull.*, 7, April, 1935, pp. 59–87.
——, and ——. Thomas Digges, the Copernican system, and the idea of the infinity of the universe in 1576. *Huntington Lib. Bull.*, 5, April, 1934, pp. 69–117.
Johnson, Obed Simon. *A study of Chinese alchemy*, Shanghai, 1928.
Johnson, Walter. *Gilbert White, pioneer, poet, and stylist*, 1928.
Jones, Charles W. *Bedae Pseudoepigraphia. Scientific writings falsely attributed to Bede*, Ithaca, New York, 1939.
——. Manuscripts of Bede's De natura rerum. *Isis*, 27, 1937, pp. 430–440.
Jones, F. Wood. John Hunter as a geologist, [*etc.*], *Ann. Roy. Coll. Surg. Engl.*, 12, 1953, pp. 219–245.

Jones, Sir Harold Spencer. The Royal Observatory, Greenwich. *Endeavour*, 7, 1948, pp. 9–14.

——. Tycho Brahe (1546–1601). *Nature*, 158, 1946, pp. 856–861.

Jones, T. I. *The contribution of Welshmen to science*, 1934. (Cymmrodorion Society's publications. From the *Transactions of the Honourable Society of Cymmrodorion*, Session 1932–33).

Jorpes, J. E. *Jac. Berzelius: his life and work. Translated by B. Steele*, Stockholm, 1966. (*Bidrag till K. Svenska Vetenskapakademiens Historia*, VII.)

Joseph Black, *Endeavour*, 14, 1955, pp. 115–116.

Josten, C. H. Elias Ashmole, F.R.S. (1617–1692). *Notes Rec. Roy. Soc. Lond.*, 15, 1960, pp. 221–230.

——. Robert Fludd's "Philosophicall key" and his alchemical experiment on wheat. *Ambix*, 11, 1963, pp. 1–23.

——. A translation of John Dee's "Monas hieroglyphica" (Antwerp, 1564), with an introduction and annotations. *Ambix*, 12, 1964, pp. 84–221.

Judd, John W. *The coming of evolution: the story of a great revolution in science*, Cambridge, 1910. (Cambridge Manuals of Literature and Science.)

Kahn, Charles H. *Anaximander and the origins of Greek cosmology*, New York, 1960.

Kantor, J. R. Goethe's place in modern science. *Goethe bicentennial studies, by members of the Faculty of Indiana University. Edited by H. J. Meeson*, Bloomington, 1950, pp. 61–82.

Kaplon, Morton F., *ed. Homage to Galileo: papers presented at the Galileo quadricentennial, University of Rochester, October 8 and 9, 1964*, Cambridge, Mass., London, 1965.

Kapoor, S. C. Berthollet, Proust, and proportions. *Chymia*, 10, 1965, pp. 53–110.

Kargon, Robert Hugh. *Atomism in England from Hariot to Newton*, Oxford 1966.

——. Thomas Hariot, the Northumberland circle and early atomism in England. *J. Hist. Ideas*, 27, 1966, pp. 128–136.

——. Walter Charleton, Robert Boyle, and the acceptance of Epicurean atomism in England. *Isis*, 55, 1964, pp. 184–192.

Karpinski, Louis Charles. The first printed arithmetic of Spain: Francesch Sanct Climont, *Suma de la art de arismetica*, Barcelona, 1482. *Osiris*, 1, 1936, pp. 411–420.

——. *Robert of Chester's Latin translation of the Algebra of Al-Khowarizmi*, [*etc.*], New York, 1915. (University of Michigan Studies, Humanistic Series, Vol. IX.)

——. The whetstone of witte (1557). *Bibliotheca mathematica*, 13, 1913, pp. 223–228.

Keele, Kenneth David. *Leonardo da Vinci on movement of the heart and blood*, [*etc.*], 1952.

——. *William Harvey: the man, the physician and the scientist*, London, [*etc.*], 1965.

Keenan, Stella. Abstracting and indexing services in science and technology. *Ann. Rev. Informn Sci. Technol.*, 4, 1969, pp. 272–303.

——. Abstracting and indexing services in the physical sciences. *Library Trends*, 16, 1968, pp. 329–336.

——, and Atherton, Pauline. *The journal literature of physics. A comprehensive study based on* Physics Abstracts (Science Abstracts, Section A) *1961 issues*, New York, 1964.

Keevil, J. J. William Anderson, 1748–1778, master surgeon, Royal Navy. *Ann. med. Hist.*, 2nd Ser., 5, 1933, pp. 511–524.

Keith, Sir Arthur. *Darwin revalued*, 1955.

Kekulé centennial. A symposium co-sponsored by the Division of History of Chemistry, the Division of Organic Chemistry, and the Division of Chemical Education at the 150th meeting of the American Chemical Society, . . . 1965, Washington, 1966. (Advances in Chemistry, Series 61.)

Kellner, L. *Alexander von Humboldt*, 1963.

Kelly, Suzanne. The *De mundo* of William Gilbert. *In* Gilbert, William. *Guilielmi Gilberti Colcestrensis, medici regii, De mundo nostro sublunari philosophia nova. Facsimile of the first and only edition, published by Elzevier in Amsterdam in 1651*. Amsterdam, 1965.

——. Gilbert's influence on Bacon: a revaluation. *Physics* (*Firenze*), 5, 1963, pp. 249–258.

Kemp, D. Alasdair. *Astronomy and astrophysics. A bibliographical guide*, 1970.

——. The Crawford Library of the Royal Observatory, Edinburgh. *Isis*, 54, 1963, pp. 481–483.

Kendall, James. The Alembic Club. *Endeavour*, 13, 1954, pp. 94–96.

——. *Michael Faraday, man of simplicity*, 1955.

——. The Royal Society of Edinburgh. *Endeavour*, 5, 1946, pp. 54–57.

——. Some eighteenth-century chemical societies. *Endeavour*, 1, 1942, pp. 106–109.

Kent, Andrew. Thomas Thomson (1773–1852) historian of chemistry. *Brit. J. Hist. Sci.*, 2, 1964–65, pp. 59–63.

——, and Hannaway, Owen. Some new considerations on Beguin and Libavius. *Ann. Sci.*, 16, 1960 [1963], pp. 241–250.

Kerker, Milton. Herman Boerhaave and the development of pneumatic chemistry. *Isis*, 46, 1955, pp. 36–49.

——. Sadi Carnot. *Sci. Mon.*, 85, 1957, pp. 143–149.

Kersaint, G. Antoine François de Fourcroy (1755–1809). Sa vie et son oeuvre. *Mémoires du Muséum national d'Histoire naturelle*, N.S. série D, Sciences physico-chimiques, Tome II, 1966; abbreviated version in *Rev. d'Hist. Pharmacie*, 15, 1967, pp. 589–596.

Keston, Hermann. *Copernicus and his world. . . . Illustrated by Hugo Steiner-Prag.* (*Translated by E. B. Ashton and Norbert Guterman. Second edition*), 1946.

Kew, H. W., and Powell, H. E. *Thomas Johnson, botanist and royalist,* [*etc.*], 1932.

Keynes, Sir Geoffrey Langdon. *A bibliography of Dr. Robert Hooke*, Oxford, 1960.

——. *A bibliography of Sir Thomas Browne, Kt., M.D. . . . Second edition, revised and augmented*, Oxford, 1968.

——. *A bibliography of the writings of Dr. William Harvey, 1578–1657. . . . Second edition, revised*, Cambridge, 1953.

——. Harvey and his books. . . . Being the Harveian Lecture delivered to the Harveian Society of London, in March, 1953. *St Bart's Hosp. J.*, 57, 1953, pp. 177–182, 212–216.

——. Harvey through John Aubrey's eyes. *Lancet*, 1958, II. pp. 859–865.

——. *The Honourable Robert Boyle. A handlist of his works. G. L. K*[*eynes*]*; for & from J. F. F*[*ulton*], 1932.

——. *John Evelyn: a study in bibliophily with a bibliography of his writings.* (*Second edition*), Oxford, 1968.

——. *John Ray: a bibliography*, 1951.

——. *John Ray, F.R.S. A handlist of his works*, Cambridge, 1944.

——. *The life of William Harvey*, Oxford, 1966.

[Keynes, Sir Geoffrey Langdon.] Robert Boyle. *St Bart's Hosp. J.*, 39, 1931–32, pp. 184–189.

——. Sir Thomas Browne. *Brit. med. J.*, 1965, II, pp. 1505–1510.

[——, and Bartholomew, Augustus Theodore.] *A handlist of the works of John Evelyn . . . and of books connected with him*, Cambridge, 1916.

Keys, Thomas E. The "Salmonia" of Sir Humphry Davy. *Bull. Med. Lib. Ass.*, 44, 1956, pp. 431–442.

——. Sir Humphry Davy and his safety lamp for coal miners. *Mayo Clinic Proc.*, 43, 1968, pp. 865–891.

Khatchadourian, Haig, and Rescher, Nicholas. Al-Kindī's Epistle on the concentric structure of the universe. *Isis*, 56, 1965, pp. 190–195.

Kibre, Pearl. Albertus Magnus. *De occultis nature. Osiris*, 13, 1958, pp. 157–183.

——. An alchemical tract attributed to Albertus Magnus. *Isis*, 35, 1944, pp. 303–316.

——. Alchemical writings ascribed to Albertus Magnus. *Speculum*, 17, 1942, pp. 499–518.

Kibre, Pearl. The *Alkimia minor* ascribed to Albertus Magnus. *Isis*, 32, 1949, pp. 267–300.

——. The De occultis naturae attributed to Albertus Magnus. *Osiris*, 11, 1954, pp. 23–39.

——. Further manuscripts containing alchemical tracts attributed to Albertus Magnus. *Speculum*, 34, pp. 238–247.

Kilgour, Frederick G. Galen. *Sci. Amer.*, 196, 1957, pp. 105–114.

——. Harvey manuscripts. *Pap. Bib. Soc. Amer.*, 54, 1960, pp. 177–179.

——. William Harvey and his contributions. *Circulation*, 23, 1961, pp. 286–296.

King, Lester S. Stahl and Hoffmann: a study in eighteenth century animism. *J. Hist. Med.*, 19, 1964, pp. 118–130.

King-Hele, Desmond G. Dr. Erasmus Darwin and the theory of evolution. *Nature*, 200, 1963, pp. 304–306.

——. *Erasmus Darwin*, 1963.

——. The Lunar Society of Birmingham. *Nature*, 212, 1966, pp. 229–233.

Klebs, Arnold C. Desiderata in the cataloguing of incunabula: with a guide for catalogue entries. *Pap. Bib. Soc. Amer.*, 10, 1916, pp. 143–163.

——. Incunable editions of Pliny's Historia naturalis. *Isis*. 24, 1935–36, pp. 120–121.

——. Incunabula scientifica et medica. Short title list. *Osiris*, 4, 1938, pp. 1–359.

Klein, Jacob. *Greek mathematical thought and the origin of algebra. . . . Translated by Eva Brann. With an appendix containing Vieta's* Introduction to the analytical art. *Translated by J. Winfree Smith*, Cambridge, Mass.; London, 1968.

Klickstein, Herbert S. Thomas Thomson: pioneer historian of chemistry. *Chymia*, 1, 1948, pp. 37–53.

Kliem, Fritz, and Wolff, Georg. *Archimedes*, Berlin, 1927.

Klotz, Oskar. Albrecht von Haller (1708–1777). *Ann. med. Hist.*, N.S. 8, 1936, pp. 10–26.

Kneale, W. Boole and the algebra of logic. *Notes Rec. Roy. Soc. Lond.*, 12, 1956, pp. 53–63.

Knight, Charles. *Shadows of old booksellers*, [*N.D.*].

Knight, David M. *Atoms and elements: a study of theories of matter in England in the nineteenth century*, 1967.

Kobler, John. *The reluctant surgeon: the life of John Hunter*, London, [*etc.*], 1960.

Kopf, L. The zoological chapter of the *Kitāb al-Imta 'wal- Mu'-ānasa* of Abū Hayyān al-Tauhīdī (10th century). (Translated from the Arabic and annotated.) *Osiris*, 12, 1956, pp. 390–466.

Korty, Margaret Barton. Franklin's world of books. *J. Lib. Hist.*, 2, 1967, pp. 271–328.

Koster, C. J. A history of the British Union Catalogue of Periodicals. *Occasional News Letter. National Central Library*, No. 10, April, 1970, pp. 10–11.

Koyré, Alexandre. *Études galiléennes. I. A l'aube de la science classique*, Paris, 1939; . . . *II. La loi de la chute des corps. Descartes et Galilée*, Paris, 1939; . . . *III. Galilée et la loi d'inertie*, Paris, 1939. (Actualités scientifiques et industrielles, 852–854. Histoire de la pensée, I–III.)

——. *Newtonian studies*, 1965.

——, and Cohen, I. Bernard. The case of the missing *Tanquam*: Leibniz, Newton & Clarke. *Isis*, 52, 1961, pp. 555–566.

——, and ——. Newton & the Leibniz-Clarke correspondence, with notes on Newton, Conti, & Des Maizeaux. *Arch. int. Hist. Sci.*, 15, 1962, pp. 64–126.

Kraus, Paul. *Jabīr ibn Hayyān. Contribution à l'histoire des idées scientifiques dans l'Islam. . . . Vol. 1. Le corpus des écrits jabiriens.* (*Vol. 2. Jabir et la science greque.*), Cairo, 1942–43. (Mémoires presentés à l'Institut d'Égypte, Tomes 44–45.)

Kronick, David A. The Fielding H. Garrison list of medical and scientific periodicals of the 17th and 18th centuries; addenda et corrigenda. *Bull. Hist. Med.*, 32, 1958, pp. 456–474.

——. *A history of scientific and technical periodicals: their origins and development of the scientific and technological press, 1665–1790*, New York, 1962.

Krotikov, V. A. The Mendeleev Archives and Museum of the Leningrad University. *J. chem. Educ.*, 37, 1960, pp. 625–628.

Krumbhaar, E. B. The bicentenary of Erasmus Darwin and his relation to the doctrine of evolution. *Ann. med. Hist.*, N.S. 3, 1931, pp. 487–500.

——. Thoughts on bibliographies and Harvey's writings. *J. Hist. Med.*, 12, 1957, pp. 235–240.

Kubrin, David. Newton and the cyclical cosmos: providence and the mechanical philosophy. *J. Hist. Ideas*, 28, 1967, pp. 325–346.

Kuhn, Thomas S. Sadi Carnot and the Cagnard engine. *Isis*, 52, 1961, pp. 567–574.

Kurzer, Frederick, and Sanderson, Phyllis M. Urea in the history of organic chemistry. *J. chem. Educ.*, 33, 1956, pp. 452–459.

Lange, Erwin F. Alchemy and the sixteenth century metallurgists. *Ambix*, 13, 1966, pp. 92–95.

Lanham, Url. *Origins of modern biology*, New York, London, 1968.

Larder, David F. The editions of Cardanus' *De rerum varietate*. *Isis*, 59, 1968, pp. 74–77.

Larson, James L. Goethe and Linnaeus. *J. Hist. Ideas*, 28, 1967, pp. 590–596.

——. Linnaeus and the natural method. *Isis*, 58, 1967, pp. 304–320.

Larsson, B. Hjalmar. Carolus Linnæus, physician and botanist. *Ann. med. Hist.*, 2nd Ser., 10, 1938, pp. 197–214.

Lassere, François. *The birth of mathematics in the age of Plato*, 1964.
Laue, Max von. *History of physics, . . . translated by Ralph Oesper*, New York, 1950.
Lawler, John. *Book auctions in England in the seventeenth century, (1676–1700)*, [*etc.*], 1906.
Layton, David. Diction and dictionaries in the diffusion of scientific knowledge: an aspect of the history of the popularization of science in Great Britain. *Brit. J. Hist. Sci.*, 2, 1965, pp. 221–234.
Leake, Chauncey D. Valerius Cordus and the discovery of ether. *Isis*, 7, 1925, pp. 14–24.
Leclair, E. Matthias de Lobel, médecin et botaniste Lillois (1538–1616). *Monspeliensis Hippocrates*, 11, 1968, pp. 11–20; reprinted from *J. Sci. med. Lille*, No. 36, 1938.
LeFanu, William Richard. *A bio-bibliography of Edward Jenner, 1749–1823*, 1951.
——. *John Hunter: a list of his books*, for the Royal College of Surgeons of England, 1946.
——. Letters from Edward Jenner. *Ann. Roy. Coll. Surg. Engl.*, 39, 1966, pp. 370–372.
——. A volume associated with Leonhart Fuchs. *J. Hist. Med.*, 11, 1956, pp. 344–346.
Leibowitz, J. O. Manuscript notes in a Rhazes-Maimonides incunable, 1497. An appreciation of Rhazes and an exposition of the anatomy of the heart. *Bull. Hist. Med.*, 39, 1965, pp. 424–434.
Leicester, Henry M. Boyle, Lomonosov, Lavoisier, and the corpuscular theory of matter. *Isis*, 58, 1967, pp. 240–244.
——. Factors which led Mendeleev to the periodic law. *Chymia*, 1, 1948, pp. 67–74.
——. *The historical background to chemistry*, New York, 1956.
——, and Klickstein, Herbert S. *A source book in chemistry*, Cambridge, Mass., 1952.
Lerner, R. Marcello Malpighi: ein ikonographischer und bio-bibliographischer Überblick. *Mitteilungen des Instituts für die Geschichte der Medizin in Würzburg*, 6, 1960, pp. 1–4.
Levene, John R. Nevil Maskelyne, F.R.S., and the discovery of night myopia. *Notes Rec. Roy. Soc. Lond.*, 20, 1965, pp. 100–108.
Levere, T. H. Faraday, matter, and natural theology—reflections on an unpublished manuscript. *Brit. J. Hist. Sci.*, 4, 1968, pp. 95–107.
Levey, Martin. The Aqrābādhīn of al-Kindī and early Arabic chemistry. *Chymia*, 8, 1962, pp. 11–20.
——. Evidences of ancient distillation, sublimation and extraction in Mesopotamia. *Centaurus*, 4, 1955–56, pp. 23–33.

Levey, Martin. Mediaeval Arabic bookmaking and its relation to early chemistry and pharmacology. *Trans. Amer. Philos. Soc.*, 52, iv, 1962.

——. Research sources in ancient Mesopotamian chemistry. *Ambix*, 6, 1958, pp. 149–154.

Lewes, George Henry. *Aristotle: a chapter from the history of science, including analyses of Aristotle's scientific writings*, 1864.

Libby, Walter. *An introduction to the history of science*, London, [*etc.*], 1924.

Lilley, S. "Nicholson's Journal" (1797–1813). *Ann. Sci.*, 6, 1948–50, pp. 78–101.

Lindberg, David C. Alhazen's theory of vision and its reception in the West. *Isis*, 58, 1967, pp. 321–341.

——. Roger Bacon's theory of the rainbow: progress or regress? *Isis*, 57, 1966, pp. 235–248.

Lindeboom, Gerrit Arie. *Bibliographia Boerhaaviana: list of publications written by or provided by H. Boerhaave, or based on his works and teaching*, Leyden, 1959. (Analecta Boerhaaviana, Vol. 1.)

——. Herman Boerhaave (1668–1738), teacher of all Europe. *J. Amer. Med. Assoc.*, 206, 1968, pp. 2297–2301.

——. *Herman Boerhaave: the man and his work*, 1968.

——, *ed. Iconographia Boerhaavii*, Leyden, 1963. (Analecta Boerhaaviana, Vol. 4.)

——. Linnaeus and Boerhaave, *Janus*, 47, 1957, pp. 264–274.

Linder, Bertel, and Smeaton, W. A. Schwediauer, Bentham and Beddoes: translators of Bergman and Scheele. *Ann. Sci.*, 24, 1968 [1969], pp. 259–273.

Lipman, Timothy O. Vitalism and reductionism in Liebig's physiological thought. *Isis*, 58, 1967, pp. 167–185.

A List of books printed in Cambridge at the University Press, 1521–1800, Cambridge, 1935.

Little, A. G., *ed. Roger Bacon: essays contributed by various writers on the occasion of the commemoration of the seventh centenary of his birth. Collected and edited by A. G. Little*, Oxford, 1914.

Lloyd, G. E. R. *Aristotle: the growth and structure of his thought*, Cambridge, 1968.

Lockyer, T. Mary, and others. *Life and work of Sir Norman Lockyer. By T. Mary Lockyer and Winifred L. Lockyer, with the assistance of H. Dingle*, [*etc.*], 1928.

Locy, William A. *Biology and its makers. . . . Third edition, revised*, New York, 1915.

Loewenberg, Bert James. Darwin and Darwin studies, 1959–63. *History of Science*, 4, 1965, pp. 15–54.

Lohne, J. A. Isaac Newton: the rise of a scientist, 1661–1671. *Notes Rec. Roy. Soc. Lond.*, 20, 1965, pp. 125–139.

Lohne, Johs. Documente zur Revalidierung von Thomas Harriot als Algebraiker. *Archive for the History of the Exact Sciences*, 3, 1966–67, pp. 185–205.

——. The fair fame of Thomas Harriott. Rigaud versus Baron von Zach. *Centaurus*, 8, 1963, pp. 69–84.

——. Thomas Harriott (1560–1621). The Tycho Brahe of optics; preliminary notice. *Centaurus*, 6, 1959, pp. 113–121.

Lohwater, A. J. Mathematical literature. *Library Trends*, 15, 1967, pp. 852–867.

Lonie, I. M. Erasistratus, the Erasistrateans, and Aristotle. *Bull. Hist. Med.*, 38, 1964, pp. 426–443.

Lopez, Claude A. Saltpetre, tin and gunpowder: addenda to the correspondence of Lavoisier and Franklin. *Ann. Sci.*, 16, 1960 [1962], pp. 83–94.

Loria, Gino. Saggio di una bibliografia Lagrangiana. *Isis*, 40, 1949, pp. 112–117.

Lough, John. *Essays on the* Encyclopédie *of Diderot and D'Alembert*, London, [*etc.*], 1968.

Lovell, D. J. Herschel's dilemma in the interpretation of thermal radiation. *Isis*, 59, 1968, pp. 46–60.

Lowe, D. N. The British Association for the Advancement of Science. *British Book News*, 93, May, 1948, pp. 232–234.

Lubbock, Lady Constance A. *The Herschel chronicle: the life-story of William Herschel and his sister Caroline Herschel. Edited by his granddaughter Constance A. Lubbock*, Cambridge, 1933.

Lurie, Edward. *Louis Agassiz: a life in science*, Chicago, 1960; abridged edition Chicago, London, 1966.

Lynam, Edward. *The first engraved atlas of the world. The Cosmographia of Claudius Ptolemaeus, Bologna, 1477*, Jenkintown, 1941.

Lysaght, Averil, *ed. Directory of natural history and other field study societies in Great Britain, including societies for archaeology, astronomy, meteorology, geology and cognate subjects*, 1959.

——. A 'grangerized' copy of E. Smith's *Life of Sir Joseph Banks. J. Soc. Bib. Nat. Hist.*, 4, 1964, pp. 206–209.

——. Some eighteenth century bird paintings in the Library of Sir Joseph Banks (1743–1820). *Bull. Brit. Mus. (Nat. Hist.) Hist. ser.*, 1, 1959, pp. 251–371.

Macalister, Alexander. *James Macartney: . . . a memoir*, 1900.

McAtee, W. L. The North American birds of George Edwards, *J. Soc. Bib. Nat. Hist.*, 2, vi, 1950, pp. 194–205.

——. The North American birds of Mark Catesby and Eleazar Albin. *J. Soc. Bib. Nat. Hist.*, 3, 1957, pp. 177–194.

McAtee, W. L. The North American birds of Thomas Pennant. *J. Soc. Bib. Nat. Hist.*, 4, 1963, pp. 100–124.

McGuire, J. E. Transmutation and immutability: Newton's doctrine of physical qualities. *Ambix*, 14, 1967, pp. 69–95.

M'Kendrick, John Gray. *Hermann Ludwig Ferdinand von Helmholtz*, 1899.

McKenzie, A. E. E. *The major achievements of science*, 2 vols., Cambridge, 1960.

McKie, Douglas. Antoine Laurent Lavoisier, F.R.S., 1743–1794. *Notes Rec. Roy. Soc. Lond.*, 7, 1949–1950, pp. 1–41.

——. *Antoine Lavoisier, scientist, economist, social reformer*, 1952.

——. *Antoine Lavoisier, the father of modern chemistry.... With an introduction by F. G. Donnan*, 1935.

——. Bernard le Bovier de Fontenelle, F.R.S., 1657–1757. *Notes Rec. Roy. Soc. Lond.*, 12, 1957, pp. 193–200.

——. Books on the history of science and technology, 1945–1960. I [–III]. *British Book News*, Nos. 241–243, 1960, pp. 621–625, 699–703, 767–772.

——. Boyle's law. *Endeavour*, 7, 1948, pp. 148–151.

——. Boyle's library. *Nature*, 163, 1949, pp. 627–628.

——. The descent of Pierre Joseph Macquer (1718–1784). *Nature*, 163, 1948, p. 628.

——. The early years of the *Académie des Sciences*. *Endeavour*, 25, 1966, pp. 100–103.

——. English writers on atomism before Dalton. *Endeavour*, 25, 1966, pp. 13–15.

——. Fire and the flamma vitalis: Boyle, Hooke and Mayow. In *Science, medicine and history*, Vol. 1, 1953, pp. 469–488.

——. The Hon. Robert Boyle's *Essays of effluviums* (1673). *Sci. Progr.*, 29, 1934–35, pp. 253–256.

——. John Harris and his *Lexicon technicum*. *Endeavour*, 4, 1945, pp. 53–57.

——. Joseph Priestley and the Copley Medal. *Ambix*, 9, 1961, pp. 1–22.

——. Joseph Priestley (1733–1804), chemist. *Sci. Progr.*, 28, 1933, pp. 17–35.

——. Kekulé and the benzene ring. *Pharmaceut. J.*, 195, 1956, pp. 197–200.

——. Macquer, the first lexicographer of chemistry. *Endeavour*, 16, 1957, pp. 133–136.

——. Men and books in English science (1600–1700). Part 1. *Sci. Progr.*, 46, 1958, pp. 606–631.

——. Newton and chemistry. *Endeavour*, 1, 1942, pp. 141–144.

——. The *Observations* of the Abbé François Rozier (1734–93). *Ann. Sci.*, 13, 1957, pp. 73–89.

McKie, Douglas. On some MS. copies of Black's chemical lectures. II [-VI]. *Ann. Sci.*, 15, 1959, pp. 65–73; 16, 1960, pp. 1–9; 18, 1962 [1964], pp. 87–97; 21, 1965, pp. 209–255; 23, 1967, pp. 1–33.
——. On Thom. Cochrane's MS. notes of Black's chemical lectures, 1767–68. *Ann. Sci.*, 1, 1936, pp. 101–110.
——. The phlogiston theory. *Endeavour*, 18, 1959, pp. 144–147.
——. Priestley's laboratory and library and other of his effects. *Notes Rec. Roy. Soc. Lond.*, 12, 1956, pp. 114–136.
——. René Antoine Ferchault de Réaumur (1683–1757), the Pliny of the eighteenth century. *Sci. Progr.*, 45, 1957, pp. 619–627.
——. *Science and history. . . . An inaugural lecture delivered at University College, London, 22nd May, 1958*, 1958.
——. Science in France and Britain: a retrospect. *Sci. Progr.*, 42, 1954, pp. 569–586.
——. The scientific periodical from 1665–1798. *Phil. Mag.*, Commemoration issue, July, 1948, pp. 122–132.
——. Scientific societies to the end of the eighteenth century. *Phil. Mag.*, Commemoration issue, July, 1948, pp. 133–144.
——. Some notes on a students' scientific society in eighteenth-century Edinburgh. *Sci. Progr.*, 49, 1961, pp. 228–241.
——. Some notes on the history of science. *Aslib Proc.*, 8, 1956, pp. 73–80.
——. William Cleghorn's *De igne* (1779). *Ann. Sci.*, 14, 1958, pp. 1–82.
——. Wöhler's synthetic urea and the rejection of vitalism: a chemical legend. *Nature*, 153, 1944, pp. 608–610.
——, and Kennedy, David. On some letters of Joseph Black and others. *Ann. Sci.*, 16, 1960 [1962], pp. 129–170.
——, and Heathcote, Niels H. de V. *The discovery of specific and latent heats*, [*etc.*], 1935.
McKinney, H. Lewis. Alfred Russel Wallace and the discovery of natural selection. *J. Hist. Med.*, 21, 1966, pp. 333–357.
MacLeod, Roy M. Evolutionism and Richard Owen, 1830–1868: an episode in Darwin's century. *Isis*, 56, 1965, pp. 259–280.
McMichael, Sir John. William Withering in perspective. *Univ. Birmingham hist. J.*, 11, 1967, pp. 41–50.
McMillan, Nora F. William Swainson's *Exotic conchology*. *J. Soc. Bib. Nat. Hist.*, 4, 1963, pp. 198–199.
McMurrich, J. Playfair. *Leonardo da Vinci the anatomist, 1452–1519*, 1930.
MacPhail, Ian. The Mellon collection of alchemy and the occult. *Ambix*, 14, 1967, pp. 198–202.
MacPike, Eugene Fairfield. *Correspondence and papers of Edmond Halley*, Oxford, 1932.

MacPike, Eugene Fairfield. *Dr. Edmond Halley (1656–1742). A bibliographical guide to his life and work arranged chronologically preceded by a list of sources, including references to the history of the Halley family*, 1939.
——. *Hevelius, Flamsteed and Halley: three contemporary astronomers and their mutual relations*, 1937.
Maddison, R. E. W. The earliest published writing of Robert Boyle. *Ann. Sci.*, 17, 1961 [1963], pp. 165–173.
——. The first edition of Robert Boyle's *Medicinal experiments*. *Ann. Sci.*, 18, 1962 [1964], pp. 43–47.
——. *The life of the Honourable Robert Boyle, F.R.S.*, 1969.
——. The portraiture of the Honourable Robert Boyle, F.R.S. *Ann. Sci.*, 15, 1959, pp. 141–214.
——. Robert Boyle's library. *Nature*, 165, 1950, p. 981.
——. Studies in the life of Robert Boyle, F.R.S. *Notes Rec. Roy. Soc. Lond.*, 9, 1951–52, pp. 1–35, 196–219; 10, 1952, pp. 15–27; 11, 1954, pp. 38–53, 159–188; 18, 1963, pp. 104–124; 20, 1965, pp. 51–77.
——. A summary of former accounts of the life and work of Robert Boyle. *Ann. Sci.*, 13, 1957, pp. 90–108.
——. A tentative index of the correspondence of the Honourable Robert Boyle, F.R.S. *Notes Rec. Roy. Soc. Lond.*, 13, 1958, pp. 128–201.
——, and Maddison, Francis R. Joseph Priestley and the Birmingham Riots. *Notes Rec. Roy. Soc. Lond.*, 12, 1956, pp. 98–113.
Maddox, John. Journals and the literature explosion. *Nature*, 221, 1969, pp. 128–130.
——. Tape or type? *Lancet*, 1968, II, p. 1071.
Magie, W. F. *A source book in physics*, Cambridge, Mass., 1963.
Maire, Albert. *Bibliographie générale des oeuvres de Blaise Pascal*, 2 vols., Paris, 1925–26.
Major, Ralph H. Cl. Galen. *Int. Rec. Med.*, 172, 1959, pp. 37–43.
Malin, Morton V. The Science Citation Index[R]: a new concept in indexing. *Library Trends*, 16, 1968, pp. 374–387.
Malpas, J. S. The life and work of Louis Pasteur. Part I [–II]. *St Bart's Hosp. J.*, 59, 1955, pp. 150–159, 185–194.
Mani, Nikolaus. Die griechische Editio princeps des Galenos (1525), ihre Entstehung und ihre Wirkung. *Gesnerus*, 13, 1956, pp. 29–52.
Mann, J. H. *Louis Pasteur*, New York, 1964.
Manuel, Frank E. *A portrait of Isaac Newton*, Cambridge, Mass., 1968.
Marchant, James. *Alfred Russel Wallace: letters and reminiscences*, 2 vols., 1916.
Marcus, Margaret Fairbanks. The herbal as art. *Bull. Med. Libr. Ass.*, 32, 1944, pp. 376–384.
Martin, D. C. Former homes of the Royal Society. *Notes Rec. Roy. Soc. Lond.*, 22, 1967, pp. 12–19.

Martin, D. C. The Royal Society's interest in scientific publications and the dissemination of information. *Aslib Proc.*, 9, 1957, pp. 127–141.

Martin, Edward A. *A bibliography of Gilbert White, the naturalist and antiquarian of Selborne*, [*etc.*], 1934.

Martin, Thomas. Early years at the Royal Institution. *Brit. J. Hist. Sci.*, 2, 1964, pp. 99–115.

——. Origins of the Royal Institution. *Brit. J. Hist. Sci.*, 1, 1962, pp. 49–63.

Martyn, John. An examination of citation indexes. *Aslib Proc.*, 17, 1965, pp. 184–196.

——, and Gilchrist, Alan. *An evaluation of British scientific journals*, 1968 (Aslib Occasional Publication No. 1.)

Mason, S. F. *A history of the sciences. Main currents of scientific thought*, 1953.

Masson, Flora. *Robert Boyle: a biography*, 1914.

Masson, Irvine. *Three centuries of chemistry: phases in the growth of a science*, 1925.

Meiklejohn, M. F. M. The birds of Dante. *Ann. Sci.*, 10, 1954, pp. 33–43.

Meldrum, Andrew N. *Avogadro and Dalton. The standing in chemistry of their hypothesis*, Edinburgh, 1904.

——. Lavoisier's early work in science, 1763–1771. *Isis*, 19, 1933, pp. 330–363; 20, 1933, pp. 396–425.

Mellon, Melvin Guy. *Chemical publications, their nature and use. . . . Fourth edition*, New York, 1965.

——. *Searching the chemical literature*, Washington, 1964.

——, and Power, Ruth T. Chemistry. *Library Trends*, 15, 1967, pp. 836–846.

Melsen, Andrew G. van. *From atomos to atom. The history of the concept Atom. . . . Translated by Henry J. Koren*, Pittsburg, 1952. (Duquesne Studies, Philosophical Studies, 1.)

Mendelsohn, Everett. The biological sciences in the nineteenth century: some problems and sources. *History of Science*, 3, 1964, pp. 39–59.

Mendoza, Eric. Sadi Carnot and the Cagnard engine. *Isis*, 54, 1963, pp. 262–263.

——. The surprising history of the kinetic theory of gases. *Mem. Proc. Manchester Lit. Phil. Soc.*, 105, 1962, pp. 15–28.

Menshutkin, B. N. *Russia's Lomonosov*, Princeton, N.J., 1952.

Merrill, Elmer Drew. The botany of Cook's voyages, and its unexpected significance in relation to anthropology, biogeography and history. *Chron. Bot.*, 14, 1954, pp. 163–383.

Merton, E. Stephen. The botany of Sir Thomas Browne. *Isis*, 47, 1956, pp. 161–171.

——. Old and new physiology in Sir Thomas Browne: digestion and some other functions. *Isis*, 57, 1966, pp. 249–259.

Meyer, Alfred. Karl Friedrich Burdach on Thomas Willis. *J. neurol. Sci.*, 3, 1966, pp. 109–116.

——. Marcello Malpighi and the dawn of neurohistology. *J. neurol. Sci.*, 4, 1967, pp. 185–193.

——, and Hierons, Raymond. A note on Thomas Willis' views on the corpus striatum and the internal capsule. *J. neurol. Sci.*, 1, 1964, pp. 547–554.

Meyer, Arthur William. *An analysis of the De generatione of William Harvey*, Stanford, London, 1956.

——. Leeuwenhoek as an experimental biologist. *Osiris*, 3, 1938, pp. 103–122.

Meyer, Ernst von. *A history of chemistry from earliest times to the present day; being also an introduction to the study of science. Third edition*, 1906.

Meyer, Ernst Heinrich Friedrich. *Geschichte der Botanik*, 4 vols., Königsberg, 1854–57; reprinted Amsterdam, 1965.

Middleton, W. E. Knowles. Archimedes, Kircher, Buffon and the burning-mirrors. *Isis*, 52, 1961, pp. 533–543.

——. A footnote to the history of the barometer: an unpublished note by Robert Hooke, F.R.S. *Notes Rec. Roy. Soc. Lond.*, 20, 1965, pp. 145–151.

——. The place of Torricelli in the history of the barometer, *Isis*, 54, 1963, pp. 11–28.

Middleton, William Shainline. Joseph Leidy, scientist. *Ann. med. Hist.*, 5, 1923, pp. 100–112.

Mieli, A. *La science arabe et son rôle dans l'évolution scientifique mondiale. Réimpression anastatique augmentée d'une bibliographie avec index analytique par A. Mazahéri*, Leyden, 1966.

Miles, Wyndham D. The Columbian Chemical Society. *Chymia*, 5, 1959, pp. 145–154.

——. Early American chemical societies. 1. The 1789 Chemical Society of Philadelphia. 2. The Chemical Society of Philadelphia. *Chymia*, 3, 1950, pp. 95–113.

——. John Redman Coxe and the founding of the Chemical Society of Philadelphia in 1792. *Bull. Hist. Med.*, 30, 1956, pp. 469–472.

——. Joseph Black, Benjamin Rush and the teaching of chemistry at the University of Pennsylvania. *Lib. Chron.*, 22, 1956, pp. 9–18.

Millás-Vallicrosa, J. M. La autenticidad de comentario a las tablas astronómicas de Al-jwārizmī por Ahmad ibn al-muṯannā. *Isis*, 54, 1963, pp. 114–119.

Miller, G. A. *Historical introduction to mathematical literature*, New York, 1921.

Millhauser, M. *Just before Darwin: Robert Chambers and 'Vestiges'*, Middletown, 1968.

Millington, E. C. History of the Young-Helmholtz theory of colour vision. *Ann. Sci.*, 5, 1941–47, pp. 167–176.

Mitchner, Robert W. Wynkyn de Worde's use of the Plimpton Manuscript of *De proprietatibus rerum. Lib. Trans. Bib. Soc.*, 5th Ser., 6, 1951–52, pp. 7–18.

Mizwa, Stephen P. *Nicholas Copernicus. A tribute of the nations*, New York, 1945.

Mokrushin, S. G. Thomas Graham and the definition of colloids. *Nature*, 195, 1962, p. 861.

Montagu, M. F. Ashley. *Edward Tyson, M.D., F.R.S., 1650–1708, and the rise of human and comparative anatomy in England: a study in the history of science*, [*etc.*], Philadelphia, 1943. (Memoirs of the American Philosophical Society, Vol. XX.)

Moody, J. W. T. Erasmus Darwin, M.D., F.R.S.: a biographical and iconographical note. *J. Soc. Bib. Nat. Hist.*, 4, 1964, pp. 210–213.

Moore, A. G. N. A herbal found by chance [John Gerard's]. *Brit. med. J.*, 1962, I, pp. 1756–1757; II, p. 58.

Moore, Tom Sidney, and Philip, James Charles. *The Chemical Society, 1841–1941: a historical review*, 1947.

Moorehead, A. *Darwin and the Beagle*, 1969.

More, Louis Trenchard. *Isaac Newton: a biography. . . . 1642–1727*, New York, London, 1934.

——. *The life and works of the Honourable Robert Boyle*, London, [*etc.*], 1944.

——. Tysoniana. *Isis*, 36, 1945–46, pp. 105–108.

Morgan, Charles. *The House of Macmillan (1843–1943)*, 1943.

Morgan, Paul. George Hartgill: an Elizabethan parson-astronomer and his library. *Ann. Sci.*, 24, 1968 [1969], pp. 295–311.

Morley, F. V. Thomas Hariot—1560–1621. *Sci. Monthly*, 14, 1922, pp. 60–66.

Moseley, M. *Irascible genius: a life of Charles Babbage, inventor*, 1964.

Mudford, Peter G. William Lawrence and *The natural history of man. J. Hist. Ideas*, 29, 1968, pp. 430–436.

Muir, John Reid. *The life and achievements of Captain James Cook, R.N., F.R.S., explorer, navigator, surveyor, and physician*, London, Glasgow, 1939.

Multhauf, Robert P. *The origins of chemistry*, 1966.

Mumby, Frank Arthur. *Publishing and bookselling. . . . With a bibliography by W. H. Peet*, 1934.

Munby, A. N. L. *The history and bibliography of science in England: the first phase, 1833–1845. To which is added a reprint of a catalogue of scientific manuscripts in the possession of J. O. Halliwell, Esq.*, Berkeley, Calif., 1968.

Munby, A. N. L. The two title-pages of the "Principia". *Times Literary Supplement*, 21st December, 1951, p. 828; 28th March, 1952, p. 228.

Munsterberg, Margaret. The Bowditch collection in the Boston Public Library. *Isis*, 34, 1942–43, pp. 140–142.

Murray, P. D. F. The voyage of the "Beagle". *New Biology*, No. 28, 1959, pp. 7–24.

Murray, Robert H. *Science and scientists in the nineteenth century*, [*etc.*], 1925.

Myers, J. L. *Learned societies. A lecture delivered in the Library of the Department of Education at the University of Liverpool, 4th November, 1922.*

Napier Tercentenary Memorial Volume, Edinburgh, 1915.

Nash, Leonard K. The Carnot cycle and Maxwell's relations. *J. chem. Educ.*, 41, 1964, pp. 368–372.

——. The origin of Dalton's chemical atomic theory. *Isis*, 47, 1956, pp. 101–116.

Nasr, Seyyed Hossein. *Science and civilization in Islam*, Cambridge, Mass., 1968.

Nathanson, Leonard. *The strategy of truth. A study of Sir Thomas Browne*, Chicago, London, 1967.

National Academy of Sciences—National Research Council. *Scientific and technical societies of the United States and Canada. Sixth edition. Part I. Scientific and technical societies of the United States. . . . Part II. Scientific and technical societies of Canada*, [*etc.*], Washington, 1955.

National Book League. *Flower books and their illustrators. An exhibition arranged for the National Book League by Wilfrid Blunt*, Cambridge, 1950.

National Reference Library of Science and Invention—a symposium, with contributions by R. S. Hutton, E. M. Nicholson, Maysie Webb, and A. H. Chaplin, published in *J. Document.*, 17, 1961, pp. 1–39.

National Science Foundation, Office of Science Information Services. *Characteristics of scientific journals, 1949–1959*, Washington, 1964.

Neave, E. W. J. Joseph Black's Lectures on the elements of chemistry. *Isis*, 25, 1936, pp. 372–390.

Needham, Joseph. *A history of embryology. . . . Second edition. Revised with the assistance of Arthur Hughes*, Cambridge, 1959.

——. *Science and civilisation in China*, [*etc.*], Vols. 1–4, Cambridge, 1954–65.

Nelson, G. A. British floras. *Proceedings of the Leeds Philosophical and Literary Society, Scientific Section*, 7, iii, 1957, pp. 17–28.

—. William Turner's contribution to the first records of British plants. *Proceedings of the Leeds Philosophical and Literary Society, Scientific Section*, 8, 1959, pp. 109–138

Němec, B Julius Sachs in Prague. In *Science, medicine and history*, Vol. 2, 1953, pp. 211–216.

Nethery, Wallace, "On the origin of species", 1859. *Book Collector*, 17, 1968, p. 216.

Neu, John. The history of science. *Library Trends*, 15, 1967, pp. 776–792.

Neugebauer, Otto Eduard Hermann. *Astronomical cuneiform texts*, 3 vols., 1955.

——. *The exact sciences in antiquity. . . . Second edition*, Providence, R.I., 1957.

——. Ptolemy's Geography, Book VII, Chapters 6 and 7. *Isis*, 50, 1959, pp. 22–29.

——, and Parker, Richard A. *Egyptian astronomical texts. I. The early Decans*, Providence, R.I., London, 1960.

——, and Sachs, A. *Mathematical cuneiform texts*, New Haven, Conn., 1945. (American Oriental Series, Vol. 29.)

Nève de Mévergnies, P. *Jean Baptiste van Helmont, philosophe*, Paris, 1935.

Neville, Roy G. Christophle Glaser and the *Traité de la chymie*, 1663. *Chymia*, 10, 1965, pp. 25–52.

——. The discovery of Boyle's law, 1661–62. *J. chem. Educ.*, 39, 1962, pp. 356–359.

——. Macquer and the first chemical dictionary, 1766. A bicentennial tribute. *J. chem. Educ.*, 43, 1966, pp. 486–490.

——. Macquer's "Dictionnaire de chymie", 1766. *Book Collector*, 15, 1966, pp. 484–485.

——. "Observations sur la mine de fer de Bagory" (1767), an unpublished manuscript by P.-J. Macquer. *Chymia*, 8, 1962, pp. 89–96.

——. The "Pratique de chymie" of Sébastien Matte la Faveur. *Ambix*, 10, 1962, pp. 14–28.

——. "The sceptical chymist", 1661. A tercentenary tribute, *J. chem. Educ.*, 38, 1961, pp. 106–109.

——. Unrecorded Daltoniana: two letters to John Bostick and a prospectus to the "New system", 1808. *Ambix*, 8, 1960, pp. 42–45.

New South Wales Public Library. *Bibliography of Captain James Cook, R.N., F.R.S., circumnavigator*, Sydney, 1928.

Newth, D. R. Lamarck in 1800. A lecture on the invertebrate animals and a note on fossils taken from the *Système des animaux sans vertèbres*, by J. B. Lamarck. Translated and annotated by D. R. Newth. *Ann. Sci.*, 8, 1952, pp. 229–254.

Nicolle, Jacques. *Louis Pasteur: a master of scientific enquiry*, 1961.

Nielsen, Lauritz. *Tycho Brahes Bogtrykkeri. En bibliografisk-bokhistorisk Undersøgelse*, Copenhagen, 1946.

Nordenskiöld, Erik. *The history of biology: a survey. . . . Translated from the Swedish by Leonard Bucknall Eyre*, New York, 1946.

North, J. D. *Isaac Newton*, 1967.

North, J. D. Werner, Apian, Blagrave and the meteoroscope. *Brit. J. Hist. Sci.*, 3, 1966, pp. 57–65.

Nowicki, A. Early editions of Giordano Bruno in Poland. *Book Collector*, 13, 1964, pp. 342–345.

O'Brien, J. J. Samuel Hartlib's influence on Robert Boyle's scientific development. Part 1 [–2]. *Ann. Sci.*, 21, 1965, pp. 1–14, 257–276.

Oehser, Paul H. *Sons of science. The story of the Smithsonian Institution and its leaders*, New York, 1949.

Olby, Robert C. Charles Darwin's manuscript of "Pangenesis". *Brit. J. Hist. Sci.*, 1, 1963, pp. 251–263.

——. The Mendel centenary. *Brit. J. Hist. Sci.*, 2, 1965, pp. 343–349.

——. *Origins of Mendelism*, 1966.

——. and Gautrey, Peter. Eleven references to Mendel before 1900. *Ann. Sci.*, 24, 1968, pp. 7–20.

Oldfather, W. A., Ellis, C. A., and Brown, D. M. Leonhard Euler's elastic curves. (De curvis elasticis, Addimentum I to his Methodus inveniendi lineas curvas maximi minimive propriete guadentes, Lausanne and Geneva, 1744.) *Isis*, 20, 1933, pp. 72–160.

Oldham, Graham. Peter Shaw. *J. chem. Educ.*, 37, 1960, pp. 417–419.

O'Leary, De Lacy. *How Greek science passed to the Arabs*, 1948 [1949].

Olmsted, J. M. D. French contributions to physiology in the XIX century. *Tex. Rep. Biol. Med.*, 13, 1955, pp. 306–316.

Olschki, Leonardo. The scientific personality of Galileo. *Bull. Hist. Med.*, 12, 1942, pp. 248–273.

O'Malley, Charles Donald. *Andreas Vesalius of Brussels, 1514–1564*, Berkeley, Calif., 1964.

——. The evolution of physiology. *J. Int. Coll. Surg.*, 30, 1958, pp. 115–129.

——. A review of Vesalian literature. *History of Science*, 4, 1965, pp 1–14.

Ore, Oystein, *Niels Henrik Abel: mathematician extraordinary*, Minneapolis, London, 1957.

Orel, V., and Vávra, M. Mendel's program for the hybridization of apple trees. *J. Hist. Biol.*, 1, 1968, pp. 219–224.

Ornstein, Martha. *The rôle of scientific societies in the seventeenth century. Third edition*, Chicago, 1938.

Osborn, Henry Fairfield. Joseph Leidy, founder of vertebrate paleontology in America, [*etc.*], *Science*, 59, 1924, pp. 173–176.

Osman, W. A. Alessandro Volta and the inflammable air eudiometer. *Ann. Sci.*, 14, 1958, pp. 215–242.

Ostwald, Wilhelm. Berzelius' "Jahresbericht" and the international organization of chemists. Translated by Ralph E. Oesper. *J. chem. Educ.*, 32, 1955, pp. 373–375.

Overton, Grant. *Portrait of a publisher, . . . and the first hundred years of the House of Appleton, 1825–1925*, [*etc.*], New York, London, 1925.
Owen, Richard. *The life of Richard Owen. By his grandson Richard Owen. . . . Also an essay on Owen's position in anatomical science by T. H. Huxley*, [*etc.*], 2 vols., 1894.
Packard, Alpheus S. *Lamarck, the founder of evolution: his life and work. With translations of his writings on organic evolution*, New York, [*etc.*], 1901.
Pagel, Walter. J. B. van Helmont, De tempore, and biological time. *Osiris*, 8, 1948, pp. 346–417.
——. Paracelsus and the neo-Platonic and Gnostic tradition. *Ambix*, 8, 1960, pp. 125–166.
——. *Paracelsus: an introduction to philosophical medicine in the era of the Renaissance*, Basle, New York, 1958.
——. The Prime Matter of Paracelsus. *Ambix*, 9, 1961, pp. 117–135.
——. The reaction to Aristotle in seventeenth-century biological thought: Campanella, van Helmont, Glanvill, Charleton, Harvey, Glisson, Descartes. In *Science, medicine and history*, Vol. 1, 1953, pp. 489–509.
——. The religious and philosophical aspects of van Helmont's science and medicine, Baltimore, 1944. (Supplement No. 2 to *Bull. Hist. Med.*)
——. Van Helmont's ideas on gastric digestion and the gastric acid. *Bull. Hist. Med.*, 30, 1956, pp. 524–536.
——. The "Wild spirit" (Gas) of John Baptist van Helmont (1579–1644) and Paracelsus. *Ambix*, 10, 1962, pp. 1–13.
——. *William Harvey's biological ideas*, Basle, 1967.
Paneth, F. A. Thomas Wright of Durham. *Endeavour*, 9, 1950, pp. 117–125.
Pannekoek, Antonie. *A history of astronomy*, 1961.
Pantin, C. F. A. Alfred Russel Wallace, F.R.S., and his essays of 1858 and 1855. *Notes Rec. Roy. Soc. Lond.*, 14, 1959, pp. 67–84.
Paris, J. A. *The life of Sir Humphry Davy*, 1831.
Parke, Nathan Grier. *Guide to the literature of mathematics and physics, including related works on engineering science. Second revised edition*, New York, 1958.
Parker, Hannah J. Science Museum Library. *Libri*, 3, 1954, pp. 326–330.
Parsons, Edward Alexander. *The Alexandrian Library, glory of the Hellenic world: its rise, antiquities and destructions*, 1952; reprinted Amsterdam, 1967.
Partington, J. R. An ancient Chinese treatise on alchemy. *Nature*, 136, 1935, pp. 287–288.
——. Chemistry in the ancient world. In *Science, medicine and history*, Vol. 1, 1953, pp. 35–46.

Partington, J. R. Chemistry through the eighteenth century. *Phil. Mag.*, Commemoration issue, July, 1948, pp. 47–66.
——. The discovery of oxygen. *J. chem. Educ.*, 39, 1962, pp. 123–125.
——. Edmund O. von Lippman. *Osiris*, 3, 1938, pp. 5–21.
——. *A history of chemistry*, Vols. 2–4, 1961–64. (Vol. 1, Part 1, 1970.)
——. Joan Baptista van Helmont. *Ann. Sci.*, 1, 1936, pp. 359–384.
——. The life and work of John Mayow (1641–1679). Part I [–II]. *Isis*, 47, 1956, pp. 217–230, 405–417.
——. Lignum neplariticum. *Ann. Sci.*, 11, 1955, pp. 1–26.
——. Seventeenth century chemistry, the phlogiston theory and Dalton's atomic theory. *Nature*, 174, 1954, pp. 291–293.
——. *A short history of chemistry. . . . Third edition*, 1957.
——. Some early appraisals of the work of John Mayow. *Isis*, 50, 1959, pp. 211–226.
——. Thomas Thomson, 1773–1852. *Ann. Sci.*, 6, 1948–50, pp. 115–126.
——. William Higgins, chemist (1763–1825). *Nature*, 176, 1955, pp. 8–9.
Paston, George. *At John Murray's: records of a literary circle, 1843–1892*, [*etc.*], 1932.
Pastore, N. Samuel Bailey's critique of Berkeley's theory of vision. *J. Hist. behav. Sci.*, 1, 1965, pp. 321–337.
Patterson, Louise Diehl. Hooke's gravitation theory and its influence on Newton. I [–II]. *Isis*, 40, 1949, pp. 327–341; 41, 1950, pp. 32–45.
——. Robert Hooke and the conservation of energy. *Isis*, 38, 1947–48, pp. 151–156.
Patterson, T. S. Jean Beguin and his "Tyrocinium chymicum". *Ann. Sci.*, 2, 1937, pp. 243–248.
——. John Mayow in contemporary setting. A contribution to the history of respiration and combustion. *Isis*, 15, 1931, pp. 47–96, 504–546.
——. Van Helmont's ice and water experiments. *Ann. Sci.*, 1, 1936, pp. 462–467.
Pav, Peter Anton. Gassendi's statements of the principle of inertia. *Isis*, 57, 1966, pp. 24–34.
Payne, Joseph Frank. English herbals. [Summary.] *Trans. Bib. Soc.*, 9, 1906–08, pp. 120–123; *Ibid*, 11, 1909–11, pp. 299–310.
——. *The Fitz-Patrick Lectures for 1903. English medicine in the Anglo-Saxon times*, [*etc.*], Oxford, 1904.
Payne, Leonard M., Wilson, Leonard G., and Hartley, Sir Harold. William Croone, F.R.S. (1633–1684). *Notes Rec. Roy. Soc. Lond.*, 15, 1960, pp. 211–219.
Peachey, George C. *A memoir of William & John Hunter*, Plymouth, *for the author*, 1924.
Peacock, George. *The life of Thomas Young*, 1855.

Pearson, Hesketh. *Dr.* [*Erasmus*] *Darwin*, London, Toronto, 1930.

Pearson, Karl. *The life, letters and labours of Francis Galton*, 3 vols. [in 4], Cambridge, 1914–30.

Peck, Thomas Whitmore, and Wilkinson, Kenneth Douglas. *William Withering of Birmingham, M.D., F.R.S., F.L.S.*, Bristol, London, 1950.

Peddie, R. A. *Fifteenth-century books*, [*etc.*], 1913.

Pemberton, John E. *How to find out in mathematics: a guide to sources of mathematical information arranged according to the Dewey Decimal Classification*, Oxford, 1963.

Pepper, Jon V. The study of Thomas Harriot's manuscripts. II. Harriot's unpublished papers. *History of Science*, 6, 1957, pp. 17–40.

Peset, V. Spanish version of Dioscorides' "Materia medica". *J. Hist. Med.*, 9, 1954, pp. 49–58.

Peterson, Houston. *Huxley, prophet of science*, 1932.

Petit, Georges, and Théodorides, Jean. Les cahiers de notes de Georges Cuvier et son premier *Diarium zoologicum*. *Comptes rendus des séances de l'Académie des Sciences*, 246, 1958, pp. 352–354.

Phelps, Ralph H., and Herlin, John P. Alternatives to the scientific periodical: a report and bibliography. *Unesco Bull. Lib.*, 14, 1960, pp. 61–75.

Phillips, J. P. Liebig and Kolbe, critical editors. *Chymia*, 11, 1966, pp. 89–97.

Picken, Laurence. The fate of Wilhelm His. *Nature*, 178, 1956, pp. 1162–1165.

Pingree, David. Astronomy and astrology in India and Iran. *Isis*, 54, 1963, pp. 229–246.

——. Sanskrit astronomical tables in the United States. *Trans. Amer. Philos. Soc.*, 58, iii, 1968.

——, J. *Thomas Henry Huxley: a list of his scientific notebooks, drawings and other papers, preserved in the College Archives*, London, *Imperial College*, 1968.

Pinner, H. L. *The world of books in classical antiquity*, Leyden, 1948.

Plant, Marjorie. *The English book trade: an economic history of the making and sale of books. Second edition*, 1965.

Platt, Sir Robert. Darwin, Mendel and Galton. *Med. Hist.*, 3, 1959, pp. 87–99.

Pledge, H. T. *Science since 1500: a short history of mathematics, physics, chemistry, biology. Second edition*, 1966.

Pliny the Elder. *The Elder Pliny's chapters on chemical subjects. . . . Edited, with translation and notes, by Kenneth C. Bailey*, 2 vols., 1929–32.

Pollard, Graham, and Ehrman, Albert. *The distribution of books by catalogue from the invention of printing to A.D. 1800. Based on material in the Broxbourne Library*, Cambridge, *Printed for presentation to members of the Roxburghe Club*, 1965.

Pomper, Philip. Lomonosov and the discovery of the law of the conservation of matter in chemical transformations. *Ambix*, 10, 1962, pp. 119–127.

Ponsonby, Arthur, Lord Ponsonby of Shulbrede. *John Evelyn, Fellow of the Royal Society, author of Sylva*, 1933.

Porter, J. R. The scientific journal—300th anniversary. *Bact. Rev.*, 28, 1964, pp. 211–230.

Portmann, Georges. Louis Pasteur (1822–1895). *Arch. Laryngol.*, 90, 1969, pp. 800–812.

Posin, Daniel Q. *Mendeleyev. The story of a great scientist*, New York, 1948.

Posner, E. The enigmatic Mendel. *Bull. Hist. Med.*, 40, 1966, pp. 430–440.

——, and Skutil, J. Darwin and Mendel. *Midland med. Rev.*, 2, 1967, pp. 112–118.

Potter, Owen. August Laurent's contributions to chemistry. *Ann. Sci.*, 9, 1953, 271–286.

Power, Sir D'Arcy. Thomas Johnson (1597–1644), botanist and barber surgeon. *Glasg. med. J.*, 133, 1940, pp. 201–203.

Powers, Wendell H. The Kresge-Hooker Science Library. *Adv. Chem. Ser.*, 30, 1961, pp. 288–296.

Poynter, F. N. L. *A catalogue of incunabula in the Wellcome Historical Medical Library*, London, [*etc.*], 1954.

——. Centenary of the Darwin-Wallace paper on natural selection. *Brit. med. J.*, 1958, I, pp. 1538–1540.

——. Linnæus, naturalist and doctor, 1707–78. *Brit. med. J.*, 1957, I, pp. 1359–1361.

Prance, Claude A. Some uncollected authors. XLIII. Gilbert White, 1720–1793. *Book Collector*, 17, 1968, pp. 300–321.

Prandi, Dino. *Bibliografici delle Opere di Lazzaro Spallanzani: delle traduzione e degli scritti su di lui*, [*etc.*], Florence, 1951.

Price, Derek J. de Solla. Chaucer's astronomy. *Nature*, 170, 1952, pp. 474–475.

——. *Little science, big science*, New York, London, 1963.

——. Networks of scientific papers: the pattern of bibliographic references indicates the nature of the scientific research front. *Science*, 149 1965, pp. 510–515.

——. *Science since Babylon*, New Haven, London, 1961.

Priestley, Joseph. *Memoirs of Dr. Joseph Priestley, to the year 1795, written by himself*, [*etc.*], 2 vols, 1806–07.

Primer, Irwin. Erasmus Darwin's *Temple of nature:* progress, evolution, and the Eleusinian mysteries. *J. Hist. Ideas*, 25, 1964, pp. 58–76.

Prowe, Leopold. *Nicolaus Coppernicus*, 2 vols., Berlin, 1883–1884.

Pupilli, Giulio Cesare. Luigi Galvani. *Sci. med. ital.*, 4, 1955, pp. 5–24.

Purver, Margery. *The Royal Society: concept and creation*, [*etc.*], 1967.

Rádl, Em. Paracelsus. Eine Skizze seines Lebens. *Isis*, 1913, pp. 62–94.
Raikov, B. E. *Karl Ernst von Baer, 1792–1876. Sein Leben und Werk. Deutsche Übersetzung mit Anmerkungen von H. von Knorre*, Leipzig, 1969. (Acta Historica Leopoldina, Nr. 5.)
Ramsay, Sir William. *Essays biographical and chemical*, 1909.
——. *The life and letters of Joseph Black*, [*etc.*], 1918.
Rappaport, Rhoda. Lavoisier's geologic activities, 1763–1792. *Isis*, 58, 1967, pp. 375–384.
——. Problems and sources in the history of geology, 1749–1810. *History of Science*, 3, 1964, pp. 60–77.
——. Rouelle and Stahl—the phlogistic revolution in France. *Chymia*, 7, 1961, pp. 73–102.
Rattansi, P. M. The intellectual origins of the Royal Society. *Notes Rec. Roy. Soc. Lond.*, 23, 1968, pp. 129–143.
Rauschenberg, Roy Anthony. Daniel Carl Solander, naturalist on the "Endeavour". *Trans. Amer. Philos. Soc.*, N.S. 58, viii, 1968.
——. Daniel Carl Solander, the naturalist on the *Endeavour* voyage. *Isis*, 58, 1967, pp. 367–374.
——. A letter of Sir Joseph Banks describing the life of Daniel Solander. *Isis*, 55, 1964, pp. 62–67.
Raven, Charles E. *English naturalists from Neckam to Ray: a study of the making of the modern world*, Cambridge, 1947.
——. *John Ray, naturalist: his life and works. Second edition*, Cambridge, 1950.
Ravetz, J. R. *Astronomy and cosmology in the achievement of Nicolaus Copernicus*, Warsaw, 1968. (Polish Academy of Sciences, Research Centre for the History of Science and Technics. Monographs on the History of Science and Technics, Vol. 30.)
Ravier, Émile. *Bibliographie des oeuvres de Leibniz*, Paris, 1957.
Rây, Priyadaranjan, *ed. History of chemistry in ancient and medieval India, incorporating the History of Hindu chemistry by Acharya Prafulla Chandra Rây. Edited by P. Rây*, Calcutta, 1956.
Read, John. *Prelude to chemistry; an outline of alchemy, its literature and relationships. Second edition*, 1939; reprinted 1961.
——. Vignette of a master and his masterpiece. *Endeavour*, 1, 1942, pp. 36–37, 91.
——. William Davidson of Aberdeen. The first British professor of chemistry. *Ambix*, 9, 1961, pp. 70–101; also *Aberdeen University Studies*, No. 129, 1951.
Redgrave, Gilbert R. *Erhard Ratdolt and his work at Venice*, [*etc.*], 1894. (Illustrated Monographs issued by the Bibliographical Society, No. 1.)
Redgrove, H. Stanley. *Alchemy: ancient and modern. . . . Second . . . edition*, 1922.

Reed, Howard S. Jan Ingenhousz, plant physiologist. With a history of the discovery of photo-synthesis. *Chron. Bot.*, 11, 1949, pp. 285–393.

Reidy, J. Thomas Norton and the Ordinall of alchimy. *Ambix*, 6, 1957, pp. 59–85.

Reilly, J., and O'Flynn, N. Richard Kirwan, an Irish chemist of the eighteenth century. *Isis*, 13, 1929–30, pp. 298–319.

Renan, Ernest. *Averroës et l'averroisme. Troisième édition*, 1869.

Rescher, Nicholas. *Al-Kindī: an annotated bibliography*, Pittsburgh, 1965.

——, and Khatchadourian, Haig. Al-Kindī's Epistle on the finitude of the universe. *Isis*, 56, 1965, pp. 426–433.

Reti, Ladislao. The Leonardo da Vinci codices in the Biblioteca Nacional of Madrid. *Technology and Culture*, 8, 1967, pp. 437–455.

——. Leonardo da Vinci's experiments on combustion. *J. chem. Educ.*, 29, 1952, pp. 590–596.

Riccardi, Pietro. *Saggio di una bibliografia Euclidea*, 4 parts, Bologna, 1887–93.

Richeson, A. W. The first arithmetic printed in English. *Isis*, 37, 1947, pp. 47–56.

Rider, K. J. *The history of science and technology: a select bibliography for students. Second edition*, 1970.

Ritchie, Arthur David. *Studies in the history and methods of science*, Edinburgh, 1958.

Robbins, Rossell Hope. Alchemical texts in middle English verse: corrigenda and addenda. *Ambix*, 13, 1966, pp. 62–73.

Roberts, Ffrangcon. Alfred Russel Wallace (1823–1913). *Lancet*, 1958, II, pp. 580–581.

Roberts, Stanley. Captain Cook's voyages: a bibliography of the French translations, 1772–1800. *J. Document.*, 3, 1947–48, pp. 160–176.

Roberts, Sydney C. *A history of the Cambridge University Press, 1521–1921*, Cambridge, 1921.

Robinson, Eric. The Derby Philosophical Society. [*c.* 1783?–1857.] *Ann. Sci.*, 9, 1953, pp. 359–367.

——. Erasmus Darwin's Botanic Garden and contemporary opinion. *Ann. Sci.*, 10, 1954, pp. 314–320.

——, and McKie, Douglas, *eds. Partners in science. Letters of James Watt and Joseph Black. Edited with introductions and notes by Eric Robinson and Douglas McKie*, 1970.

Robinson, Henry W. Robert Hooke, M.D., F.R.S., with special reference to his work in biology. *Proc. Roy. Soc. Med.*, 38, 1944–45, pp. 485–489.

Robinson, Herbert Spencer. Thomas Young: a chronology and a bibliography, with estimates of his work and character. *Med. Life*, 36, 1929, pp. 227–540.

Robinson, Mabel L. *Runner of the mountain tops. The life of Louis Agassiz*, New York, 1939.

Robinson, Victor. Jons Jakob Berzelius (1779–1848). *Med. Life*, 35, 1928, pp. 165–185.

Rohde, Eleanor Sinclair. *The old English herbals*, [*etc.*], 1922.

Roller, Duane H. D. *The De magnete of William Gilbert*, Amsterdam, 1959.

Rolleston, Sir Humphry Davy. Walter Charleton, D.M., F.R.C.P., F.R.S. *Bull. Hist. Med.*, 8, 1940, pp. 403–416.

Rome, Remacle. Nicolas Sténon et la Royal Society of London. *Osiris*, 12, 1956, pp. 244–268.

Ronchi, Vasco. From seventeenth-century to twentieth-century optics. *Internat. Council of Sci. Unions Review*, 4, 1962, pp. 145–156.

Rooseboom, Maria. Christian Huygens et la microscopie. *Arch. néerl. Zool.*, 13, 1958, Suppl., pp. 59–73.

Rose, J. John Dalton the atomist. *Nature*, 211, 1966, pp. 1015–1016.

Rosen, Edward. The authentic title of Copernicus's major work. *J. Hist. Ideas*, 4, 1943, pp. 457–474.

——. The Commentariolus of Copernicus. *Osiris*, 3, 1938, pp. 123–141.

——. Galileo and Kepler: their first two contacts. *Isis*, 57, 1966, pp. 262–264.

——. Kepler's *Dream* in translation. *Isis*, 57, 1966, pp. 392–394.

——. *Three Copernican treatises* (*The* Commentariolus *of Copernicus, the* Letter against Werner, *the* Narratio prima *of Rheticus*, New York, 1939). (Records of Civilization, No. 30.) Reprinted, 1959.

——. Title of Galileo's *Sidereus nuncius*. *Isis*, 41, 1950, pp. 287–289.

Rosner, Fred. Moses Maimonides (1135 to 1204). *Ann. intern. Med.*, 62, 1965, pp. 372–375.

Ross, S. Faraday consults the scholars: the origins of the terms of electrochemistry. *Notes Rec. Roy. Soc. Lond.*, 16, 1961, pp. 187–220.

Ross, Sir W. D. *Aristotle, 5th edition, revised*, 1949; (reprinted, 1960).

Rossi, Paolo. *Francis Bacon: from magic to science. . . . Translated from the Italian by Sacha Rabinovitch*, 1968.

Rostand, Jean. Réaumur, embryologiste et geneticien. *Rev. Hist. Sci., Paris*, 11, 1958, pp. 34–50.

——. Réaumur et la resistance des insectes à la congélation. *Rev. Hist. Sci. Paris*, 15, 1962, pp. 71–72.

Rothschuh, K. E. Alexander von Humboldt und die Physiologie seiner Zeit. *Sudhoffs Arch. Gesch. Med.*, 43, 1959, pp. 97–113.

Roule, Louis. *Buffon et la description de la nature*, Paris, 1924.

Rourke, Constance. *Audubon. . . . With 12 plates from original Audubon prints*, [*etc.*], London, [*etc.*], 1936.

Rowbottom, Margaret E. The earliest published writings of Robert Boyle, *Ann. Sci.*, 6, 1948–50, pp. 376–389.

Royal Society. *Newton tercentenary celebrations, 15–19 July, 1946*, Cambridge, 1947.

Rudnicki, Józef. *Nicholas Copernicus (Mikolaj Kopernik), 1473–1543. . . . Translated from the Polish, by B. W. A. Massey,* [*etc.*], 1943.

Rudwick, M. J. S. The foundation of the Geological Society of London: its scheme for co-operative research and its struggle for independence. *Brit. J. Hist. Sci.*, 1, 1963, pp. 325–355.

——. Hutton and Werner compared: George Greenough's geological tour of Scotland in 1805. *Brit. J. Hist. Sci.*, 1, 1962, pp. 117–135.

Rukeyser, Muriel. *Willard Gibbs*, New York, 1942.

Ruska, Julius. Die Alchemie des Avicenna. *Isis*, 21, 1934, pp. 14–51.

——. Pseudepigraphe Rasis-Schriften. *Osiris*, 7, 1939, pp. 31–94.

——. Zur ältesten arabischen Algebra und Rechenkunst. *Sitzungsberichte der Heidelberger Akademie der Wissenschaften, Philosophischhistorische Klasse*, 1917, pp. 1–125.

Ruske, Walter. August Kekulé und die Entwicklung der chemischen Strukturtheorie. *Naturwissenschaften*, 52, 1965, pp. 485–489.

Russell, Colin A. The electrochemical theory of Berzelius. Part I [–II]. *Ann. Sci.*, 19, 1963, pp. 117–145.

——.The electrochemical theory of Sir Humphry Davy. Part I. The Voltaic pile and electrolysis. Part II. Electrical interpretations of chemistry. Part III. The evidence of the Royal Institution manuscripts. *Ann. Sci.*, 15, 1959, pp. 1–25; 19, 1963, pp. 255–271.

——. Kekulé and benzene. *Chemistry in Britain*, 1, 1965, pp. 141–142.

Russell, L. J. Kepler's laws of planetary motion, 1609–1666. *Brit. J. Hist. Sci.*, 2, 1964, pp. 1–24.

Rutherford, H. W. *Catalogue of the library of Charles Darwin now in the Botany School, Cambridge*, Cambridge, 1908.

Ryan, W. F. Science in medieval Russia: some reflections on a recent book. *History of Science*, 5, 1966, pp. 52–61.

Rytel, Alexander. Nicolaus Copernicus, physician and humanitarian. A new approach. *Bull. Pol. Med. Hist. Sci.*, 1, 1956, pp. 3–11.

Sabra, A. I. The authorship of the *Liber de crepusculis*, an eleventh-century work on atmospheric refraction. *Isis*, 58, 1957, pp. 77–85.

Sachs, Julius von. *History of botany, 1530–1860. Authorized translation by Henry E. F. Garnsey. Revised by Isaac Bayley Balfour*, Oxford, 1890.

Sadler, D. H. The bicentenary of the Nautical Almanac. *Quart. J. Roy. Astron. Soc.*, 8, 1967, pp. 161–171.

Saidan, A. S. The earliest extant Arabic arithmetic *Kitāb al-Fuṣūl fī al Hisāb al-Hindī* of Abū al Hasan, Aḥmad ibn Ibrāhīm al-Uqlīdisī. *Isis*, 57, 1966, pp. 475–490.

Salié, Hans. Poggendorff and *Poggendorff*. Translated by Ralph E. Oesper. *Isis*, 57, 1966, pp. 389–392.

Salisbury, Sir Edward. The Royal Botanic Gardens, Kew. *Endeavour*, 6, 1947, pp. 58–62.

Salvestrini, Virgilio. *Bibliografia de Giordano Bruno (1582–1950). 2nd ed. postuma, a cura di Luigi Firpo*, Florence, 1958.

Sambursky, S. Conceptual developments in Greek atomism. *Arch. int. Hist. Sci.*, 11, 1958, pp. 251–262.

——. *Physics of the Stoics*, 1959.

Santillana, Giorgio de. *The crime of Galileo*, Chicago, 1955.

Sarton, George. *Ancient science and modern civilization: Euclid and his time; Ptolemy and his time; the end of Greek science*, 1954.

——. *The appreciation of ancient and mediaeval science during the Renaissance (1450–1600)*, Philadelphia, 1955.

——. Discovery of the aberration of light. With facsimile reproduction (No. xii) of James Bradley's letter to Edmond Halley giving an account of his discovery, [*etc.*]. *Isis*, 16, 1931, pp. 233–265.

——. The discovery of the mammalian egg and the foundation of modern embryology. With a complete facsimile (No. xiii) of K. E. von Baer's fundamental memoir: De ovi mammalian et hominis genesi (Leipzig, 1827). *Isis*, 16, 1931, pp. 315–379.

——. Discovery of the theory of natural selection. With facsimile reproductions (Nos. xviii-ix) of Darwin's and Wallace's earliest publications on the subject (1859). *Isis*, 14, 1930, pp. 133–154.

——. Evariste Galois. *Osiris*, 3, 1938, , pp. 241–259.

——. *Galen of Pergamon*, Laurence, Kansas, 1954.

——. *A history of science: ancient science through the golden age of Greece*, 1953.

——. *A history of science (II). Hellenistic science and culture in the last three centuries B.C.*, Cambridge, Mass., London, 1959.

——. *Horus. A guide to the history of science. A first guide for the study of the history of science, with introductory essays on science and tradition*, Waltham, Mass., 1952.

——. Incunabula wrongly dated: fifteen examples with eighteen illustrations. *Isis*, 40, 1949, pp. 227–240.

——. *Introduction to the history of science*, 3 vols. [in 5], Baltimore, 1927–48.

——. John Ferguson (1837–1916). (Note). *Isis*, 39, 1948, pp. 60–61.

——. Montucla (1725–1799), his life and works. *Osiris*, 1, 1936, pp. 519–567.

——. The scientific literature transmitted through the incunabula. (An analysis and discussion illustrated with sixty facsimiles.) *Osiris*, 5, 1938, pp. 41–245.

Sarton, George. Simon Stevin of Bruges (1548–1620). *Isis*, 21, 1934, pp. 241–303; continued as, The first explanation of decimal fractions and measures (1585). Together with a history of the decimal idea and a facsimile (No. xvii) of Stevin's Disme. *Isis*, 23, 1935, pp. 153–244.

——. *Six wings: men of science in the Renaissance*, 1957.

——. Spinoza. *Isis*, 10, 1928, pp. 11–15.

——. The study of early scientific textbooks. *Isis*, 38, 1947-48, pp. 137–148.

Saunders, J. B. de C. M. Leonardo da Vinci as anatomist and physiologist: a critical evaluation. *Tex. Rep. Biol. Med.*, 13, 1955, pp. 1010–1026.

Savage, Ernest A. Samuel Pepys' library. *The Library*, N.S. 4, 1903, pp. 286–291.

Savage, S. Curtis as naturalist and humanist. *Nature*, 157, 1946, pp. 15–16.

Sawyer, F. C. John Ray (1627–1705)—a portrait in oils. *J. Soc. Bib. Nat. Hist.*, 4, 1963, pp. 97–99.

——. Some additions to P. Stageman's "Bibliography of the first editions of Philip Henry Gosse", 1955. *J. Soc. Bib. Nat. Hist.*, 3, 1957, pp. 221–222.

——. Some natural history drawings made during Captain Cook's first voyage round the world. *J. Soc. Bib. Nat. Hist.*, 2, vi, 1950, pp. 190–193.

Scherz, Gustav. Niels Stensens Smaragdreise. *Centaurus*, 4, 1955–56, pp. 51–57.

Schiek, O. Centenary of spectral analysis. In commemoration of its discovery by Robert Kirchhoff and Wilhelm Bunsen. *Jena Review*, 4, 1959, pp. 143–146.

Schierbeek, Abraham. *Jan Swammerdam (12 February 1637–17 February 1680). His life and works*, Amsterdam, 1967.

——. The main trends of zoology in the 17th century. *Janus*, 50, 1963, pp. 159–175.

——. *Measuring the invisible world: the life and works of Antoni van Leeuwenhoek, F.R.S. With a bibliographical chapter by Maria Rooseboom* London, New York, 1959.

Schiller, J. Physiology's struggle for independence in the first half of the nineteenth century. *History of Science*, 7, 1968 [1969], pp. 64–89.

Schlichtung. Th. H. Das Tagebuch von Niels Stenson. *Centaurus*, 3, 1953–54, pp. 305–310.

Schmitz, Rudolf. Zur Bibliographie der Erstausgabe des Dispensatoriums Valerii Cordi. *Sudhoffs Arch. Gesch. Med.*, 42, 1958, pp. 260–270.

Schneer, Cecil J. The rise of historical geology in the seventeenth century. *Isis*, 45, 1954, pp. 256–268.

Schneer, Cecil J. *The search for order: the development of the major ideas in the physical sciences from the earliest times to the present,* [*etc.*], 1960.

Schofield, Robert E. Electrical researches of Joseph Priestley. *Arch. int. Hist. Sci.*, 16, 1963, pp. 277–286.

——. Histories of scientific societies: needs and opportunities for research. *History of Science*, 2, 1963, pp. 70–83.

——. The industrial orientation of science in the Lunar Society of Birmingham. *Isis*, 48, 1957, pp. 408–415.

——. Joseph Priestley, natural philosopher. *Ambix*, 14, 1967, pp. 1–15.

——. Joseph Priestley, the theory of oxidation and the nature of matter. *J. Hist. Ideas*, 25, 1964, pp. 285–294.

——. The Lunar Society of Birmingham: a bicentenary appraisal. *Notes Rec. Roy. Soc. Lond.*, 21, 1966, pp. 144–161.

——. *The Lunar Society of Birmingham. A social history of provincial science and industry in eighteenth century England*, Oxford, 1963.

——. Membership of the Lunar Society of Birmingham. *Ann. Sci.*, 12, 1956, pp. 118–136.

——. The scientific background of Joseph Priestley. *Ann. Sci.*, 13, 1957, pp. 148–163.

Schramm, Gottfried. Über die Chemie im alten China. *Nova Acta Leopoldina*, N.F. 27, 1963, pp. 145–166.

Schramm, Matthias. *Ibn al-Haytham's Weg zur Physik*, Wiesbaden, 1963. (Boethius. Texte und Abhandlungen zur Geschichte der exakten Wissenschaften, Bd. 1.)

Schrecker, Paul. Une bibliographie de Leibniz. *Revue philosophique*, 63, 1938, pp. 324–346.

Schüepp, Otto. Goethe als Botaniker. *Gesnerus*, 6, 1949, pp. 144–158.

Schütte, Godmund. *Ptolemy's maps of northern Europe. A reconstruction of the prototypes*, Copenhagen, 1917.

Schullian, Dorothy M. New documents on Volcher Coiter. *J. Hist. Med.*, 6, 1951, pp. 176–194.

——, and Sommer, Francis E. *A catalogue of incunabula and manuscripts in the Army Medical Library*, New York, [1950].

Schultz, Louise. New developments in biological abstracting and indexing. *Library Trends*, 16, 1968, pp. 337–352.

Schulz, Wilhelm. Aimé Bonpland. Alexander von Humboldts Begleiter auf d. Amerikanreise, 1799–1804. Sein Leben und Wirken, besonders nach 1817, in Argentinien. *Akademie der Wissenschaft und Literatur. Abhandlungen der math.-nat. Klasse*, 1960, Nr. 9.

Scolari, Felici. *Guide bibliografiche, Alessandro Volta per incario delle R. Commissione per l'Edizione Nazionale delle Opere Voltiane*, [*etc.*], Rome, 1927.

Scott, Flora Murray. Marcello Malpighi, Doctor of Medicine. *Sci. Mon.*, 25, 1927, pp. 546–553.

Scott, J. F. *A history of mathematics from antiquity to the beginning of the nineteenth century*, [*etc.*], 1958 [on title-page; on verso, second edition, 1960].

——. John Wallis as a historian of mathematics. *Ann. Sci.*, 1, 1936, pp. 335–357.

——. *The mathematical work of John Wallis, D.D., F.R.S., 1616–1703*, [*etc.*], 1938.

——. Mathematics through the eighteenth century. *Phil. Mag.*, Commemoration issue, July, 1948, pp. 67–91.

——. *The scientific work of René Descartes (1596–1650)*, [*etc.*], 1952.

Scott, Walter S. Gilbert White's *The natural history of Selborne*. *Book Collector*, 18, 1969, pp. 89–90.

——. *White of Selborne*, 1950.

Scriba, Christoph J. John Wallis' *Treatise of angular sections* and Thâbit ibn Qurra's generalization of the Pythagorean theorem. *Isis*, 57, 1966, pp. 56–66.

——. A tentative index of the correspondence of John Wallis, F.R.S. *Notes Rec. Roy. Soc. Lond.*, 22, 1967, pp. 58–93.

Scurla, Herbert. *Alexander von Humboldt. Leben und Wirken, 2 Aufl.*, Berlin, 1959.

Seaton, E. Thomas Hariot's secret script. *Ambix*, 5, 1956, pp. 111–114.

Sebba, Gregor. *Bibliographia Cartesiana. A critical guide to the Descartes literature 1800–1960*, The Hague, 1964. (International Archives of the History of Ideas.)

Seeger, R. J. Michael Faraday: his scientific insights and philosophical outlook. *Physis*, 8, 1966, pp. 220–231.

Seifert, Hans. Nicolaus Steno als Bahnbrecher der modernen Krystallographie. *Sudhoffs Arch. Gesch. Med.*, 38, 1954, pp. 29–47.

Seligmen, Paul. *The "Apeiron" of Anaximander. A study in the origin and function of metaphysical ideas*, 1962.

Sen, S. N. *A bibliography of Sanskrit works on astronomy and mathematics. Part 1: Manuscripts, texts, translations, and studies*, Delhi, 1966.

Sher, Irving H., Garfield, Eugene, and Elias, Arthur W. Control and elimination of errors in ISI services. *J. chem. Document.*, 6, 1966, pp. 132–135.

Sherbo, Arthur. The translation of Boerhaave's *Elementa chemiae*. *Ambix*, 13, 1966, pp. 108–117.

Sherlock, T. P. The chemical works of Paracelsus. *Ambix*, 3, 1948, pp. 33–63.

Sherman, John D. Some entomological and other bibliographies. *J. N.Y. Entomol. Soc.*, 32, 1924, pp. 206–215.

Sherrington, Sir Charles. *Goethe on nature and on science*, Cambridge, 1942.

——. Language distribution of scientific periodicals. *Nature*, 134, 1934, p. 625.

Sherrod, John. The Library of Congress. *Science*, 127, 1958, pp. 958–959.

Sidgwick, J. B. *William Herschel, explorer of the heavens*, 1953.

Siegel, Rudolph E. *Galen's system of physiology and medicine. An analysis of his doctrines and observations on bloodflow, respiration, humors and internal diseases*, Basle, New York, 1968.

——. Theories of vision and color perception of Empedocles and Democritus: some similarities to the modern approach. *Bull. Hist. Med.*, 33, 1959, pp. 145–159.

——. Why Galen and Harvey did not compare the heart to a pump. *Amer. J. Cardiol.*, 20, 1967, pp. 117–121.

Siegfried, Robert. Boscovich and Davy: some cautionary remarks. *Isis*, 58, 1967, pp. 236–238.

——. Further Daltonian doubts. *Isis*, 54, 1963, pp. 480–481.

——. Sir Humphry Davy on the nature of the diamond. *Isis*, 57, 1966, pp. 325–335.

Silliman, Robert H. William Thomson: smoke rings and nineteenth-century atomism. *Isis*, 54, 1963, pp. 461–474.

Silverstein, Theodore. *Medieval Latin scientific writings in the Barberini Collection: a provisional catalogue*, Chicago, 1957.

Simpkins, Diana M. Early editions of Euclid in England. *Ann. Sci.*, 22, 1966 [1967], pp. 225–249.

Simpson, Thomas K. Maxwell and the direct experimental test of his electromagnetic theory. *Isis*, 57, 1966, pp. 411–432.

Singer, Charles, Dr. William Gilbert (1544–1603). *J. Roy. Nav. Med. Serv.*, 2, 1916, pp. 495–510.

——. *The evolution of anatomy*, [*etc.*], 1925; a slightly revised edition was published by Dover Books as *A short history of anatomy and physiology from the Greeks to Harvey*, New York, 1957.

——. The first English microscopist: Robert Hooke (1635–1703). *Endeavour*, 14, 1955, pp. 12–18.

——. Galen's Elementary course on bones. *Proc. Roy. Soc. Med.*, 45, 1952, pp. 767–776.

——. *Greek biology and Greek medicine*, Oxford, 1922. (Chapters in the History of Science, I.)

——. The herbal in antiquity. *J. Hellenic Stud.*, 47, 1927, pp. 1–52.

——. *History of biology to about the year 1900; a general introduction to the study of living things. Third revised edition*, New York, 1959.

——. *A short history of science to the nineteenth century*, Oxford, 1941.

——. *A short history of scientific ideas to 1900*, Oxford, 1959.

——. *Science, medicine and history. Essays on the evolution of scientific thought and medical practice: written in honour of Charles Singer. Collected and edited by E. Ashworth Underwood*, 2 vols., London, [*etc.*], 1953.

Singer, Charles. *ed. Studies in the history and method of science*, 2 vols,. Oxford, 1917–21.

Singer, Dorothea Waley. Alchemical writings attributed to Roger Bacon. *Speculum*, 7, 1932, pp. 80–86.

——. *Catalogue of Latin and vernacular alchemical manuscripts in Great Britain and Ireland dating from before the XVth century, by Dorothea Waley Singer assisted by Annie Anderson*, 3 vols., Brussels, 1928–31.

——. The cosmology of Giordano Bruno (1548–1600). *Isis*, 33, 1941–42, pp. 187–196.

——. *Giordano Bruno, his life and thought. With annotated translation of his work On the infinite universe and worlds*, New York, 1950.

Sivin, Nathan. William Lewis (1708–1781) as a chemist. *Chymia*, 8, 1962, pp. 63–88.

Smeaton, W. A. Antoine-François de Fourcroy (1755–1809). *Nature*, 175, 1955, pp. 1017–1018.

——. L'Avant-Coureur. The journal in which some of Lavoisier's earliest research was reported. *Ann. Sci.*, 13, 1957, pp. 219-234.

——. Centenary of the law of octaves: Döbereiner, Newlands, Odling and others. *Roy. Instn Chem. J.*, 88, 1964, pp. 271–274.

——. The contributions of P.-J. Macquer, T. O. Bergmann and L. B. Guyton de Morveau to the reform of chemical nomenclature. *Ann. Sci.*, 10, 1954, pp. 87–106.

——. The early years of the Lycée and the Lycée des Arts. A chapter in the lives of A. L. Lavoisier and A. F. de Fourcroy. 1. The Lycée of the Rue de Valois. [2. The Lycée des Arts.] *Ann. Sci.*, 11, 1956, pp. 257–267, 309–319.

——. F.-J. Bonjour and his translation of Bergman's "Disquisitio de attractionibus electivis". *Ambix*, 7, 1959, pp. 47–50.

——. Fourcroy and the anti-phlogistic theory. *Endeavour*, 18, 1959, pp. 70–74.

——. *Fourcroy, chemist and revolutionary, 1755–1809*, Cambridge, *for the author*, 1962.

——. Guyton de Morveau and chemical affinity. *Ambix*, 1, 1963, pp. 55–64.

——. Guyton de Morveau's course of chemistry in the Dijon Academy. *Ambix*, 9, 1961, pp. 53–69.

——. Lavoisier's membership of the Assembly of Representatives of the Commune of Paris, 1789–1790. *Ann. Sci.*, 13, 1957, pp. 235–248.

——. Lavoisier's membership of the Société Royale d'Agriculture and the Comité d'Agriculture. *Ann. Sci.*, 12, 1956, pp. 267–277.

——. Lavoisier's membership of the Société Royale de Médicine. *Ann. Sci.*, 12, 1956, pp. 228–244.

——. Louis Bernard Guyton de Morveau, F.R.S. (1737–1816) and his relations with British scientists. *Notes Rec. Roy. Soc. Lond.*, 22, 1967, pp. 113–130.

Smeaton, W. A. Macquer on the composition of metals and the artificial production of gold and silver. *Chymia*, 11, 1966, pp. 81–88.

——. New light on Lavoisier: the research of the last ten years. *History of Science*, 2, 1963, pp. 51–69.

——. Some unrecorded editions of Fourcroy's *Philosophie chimique*. *Ann. Sci.*, 23, 1967 [1968], pp. 295–298.

——. Two unrecorded publications of the "Régie des Poudres et Salpêtres" probably written by Lavoisier. *Ann. Sci.*, 12, 1956, pp. 157–159.

Smit, P. Ernst Haeckel and his "Generelle Morphologie": an evaluation. *Janus*, 1967, pp. 236–252.

Smith, Barbara M. D., and Moilliet, J. L. James Keir of the Lunar Society. *Notes Rec. Roy. Soc. Lond.*, 22, 1967, pp. 144–154.

Smith, Charlotte Fell. *John Dee (1527–1608)*, [*etc.*], 1909.

Smith, David Eugene. The first great commercial arithmetic. *Isis*, 8, 1926, pp. 41–49.

——. The first printed arithmetic (Treviso, 1478). *Isis*, 6, 1924, pp. 311–331.

Smith, E. *The life of Sir Joseph Banks*, 1911.

Smith, Henry Monmouth. *Torchbearers of chemistry: portraits and brief biographies of scientists who have contributed to the making of modern chemistry. . . . With bibliographies of biographies, by Ralph E. Oesper*, New York, 1949.

Smith, J. E. *A selection of the correspondence of Linnaeus and other naturalists*, [*etc.*], 2 vols., 1821.

Smith, Roger C. *Guide to the literature of the zoological sciences. Seventh edition*, Minneapolis, 1966.

Smith, Sydney. The origin of the "Origin" as discerned from Charles Darwin's notebooks and his annotations in the books he read between 1837 and 1842. *Advanc. Sci.*, 16, 1960, pp. 391–401.

Smyth, A. L. *John Dalton, 1766–1844: a bibliography of works by and about him*, Manchester, 1966.

Snelders, H. A. M. Reply to query, "Was Boyle the first to use spot test analysis?" *Isis*, 56, 1965, pp. 210–211.

Snell, William E. Frank Buckland—medical naturalist. *Proc. Roy. Soc. Med.*, 60, 1967, pp. 291–296.

Snyder, C. The eyes of John Dalton. *Archives of Ophthal.*, 67, 1962, pp. 671–673.

Snyder, Emily Eveleth. *Biology in the making*. New York, London, 1940.

Solmsen, Friedrich. *Aristotle's system of the physical world. A comparison with his predecessors*, Ithaca, [*etc.*], 1960.

Soloviev, I. I. New materials for the scientific biography of J. J. Berzelius—the scientific relations of Berzelius with Russian scholars, from unpublished letters. *Chymia*, 7, 1961, pp. 109–125.

Some account of the Oxford University Press, 1468–1921, Oxford, 1922.
Sorsby, Arnold. Gregor Mendel. *Brit. med. J.*, 1965, I, pp. 333–338.
Soubiran, André. *Avicenne, prince des médicins: sa vie et sa doctrine*, Paris, 1935.
Soule, Byron Amery. *Library guide for the chemist*, New York, 1938.
South African Council for Scientific and Industrial Research, Pretoria, Transvaal, Union of South Africa, *Directory of scientific, technical and medical libraries in the Union of South Africa*, 1949.
Sparrow, W. J. *Knight of the White Eagle. A biography of Sir Benjamin Thompson, Count Rumford (1753–1814)*, 1964.
Spence, Sydney Alfred. *Captain James Cook, R.N. (1728–1779): a bibliography of his voyages, to which is added other works relating to his life, conduct and nautical achievements*, Mitcham, 1960.
Spencer-Jones, Sir Harold. Astronomy through the eighteenth century. *Phil. Mag.*, Commemoration issue, July, 1948, pp. 10–27.
Spiers, C. H. Sir Humphry Davy and the leather industry. *Ann. Sci.*, 24, 1968, pp. 99–113.
Spooner, Roy C., and Wang, C. H. The Divine Nine Turn Tan Sha method: a Chinese alchemical recipe. *Isis*, 38, 1947–48, pp. 235–242.
Sprague, T. A., and Nelmes, E. The herbal of Leonhardt Fuchs. *J. Linn. Soc. (Bot.)*, 48, 1931, pp. 545–642.
——, and Sprague, M. S. The herbal of Valerius Cordus. *J. Linn. Soc. (Bot.)*, 52, 1939, pp. 1–113.
Sprat, Thomas. *The history of the Royal Society of London, for the improving of natural knowledge*, 1667.
Spratt, H. Philip. *Libraries for scientific research in Europe and America*, 1936.
Spronsen, J. W. van. William Higgins. *Arch. internat. d'Hist. des Sci.*, 19, 1966, pp. 263–270.
Stageman, Peter. *A bibliography of the first editions of Philip Henry Gosse, F.R.S. With introductory essays by Sacheverell Sitwell and Geoffrey Lapage*, Cambridge, 1955.
Stahl, William H. *Roman science. Origins, development and influence to the later Middle Ages*, Madison, 1962.
Stapleton, H. E. The antiquity of alchemy. *Ambix*, 5, 1953, pp. 1–43.
——, Azo, R. F., Husain, M. Hidāyat, and Lewis, G. L. Two alchemical treatises attributed to Avicenna. *Ambix*, 10, 1962, pp. 41–82.
Stauffer, Robert Clinton. Ecology in the long manuscript version of Darwin's *Origin of species* and Linnæus' *Oeconomy of nature*. *Proc. Amer. Philos. Soc.*, 104, 1960, pp. 235–241.
——. Haeckel, Darwin & ecology. *Quart. Rev. Biol.*, 32, 1957, pp. 138–144.
——. "On the origin of species": an unpublished version. *Science*, 130, 1959, pp. 1449–1452.

Stauffer, Robert Clinton. Speculation and experiment in the background of Oersted's discovery of electromagnetism, *Isis*, 48, 1957, pp. 33–50.

Stearn, William T. Alexander von Humboldt (1769–1859) and plant geography. *New Scientist*, 5, 1959, pp. 957–959.

——. The botanical results of the *Endeavour* voyage. *Endeavour*, 27, 1968, pp. 3–10.

——. Franz and Ferdinand Bauer, masters of botanical illustration. *Endeavour*, 19, 1960, pp. 27–35.

——. Humboldt's "Essai sur la géographie des plantes". *J. Soc. Bib. Nat. Hist.*, 3, 1960, pp. 351–357.

Stecher, R. M. The Darwin-Innes letters: the correspondence of an evolutionist with his vicar, 1848–1884. *Ann. Sci.*, 17, 1961 [1964], pp. 201–258.

Steele, Robert. Practical chemistry in the twelfth century. Rasis de aluminibus et salibus. Translated by Gerard of Cremona. *Isis*, 12, 1929, pp. 10–46.

——. What fifteenth century books are about. I. Scientific books. *The Library*, N.S. 4, 1903, pp. 337–354.

Steeves, George Walter. *Francis Bacon: a sketch of his life, works and literary remains, chiefly from a bibliographical point of view*, 1910.

Steiger, R. Erschliessung des Conrad-Gessner-Materials der Zentralbibliothek Zürich. *Gesnerus*, 25, 1968, pp. 29–64.

Steiner, Hans. Goethe und die vergleichende Anatomie. *Gesnerus*, 6, 1949, pp. 129–143.

Stern, Curt, and Sherwood, Eva R., *eds. The origin of genetics. A Mendel source book*, San Francisco, London, 1966.

Stevens, Henry. *The Humboldt library. A catalogue of the library of Alexander von Humboldt, with a bibliographical and biographical memoir*, 1863.

Stevens, Henry N. *Ptolemy's Geography. A brief account of all the printed editions down to 1730. . . . Second edition*, 1908.

Stevens, Neil E. Factors in botanical publication and other essays. *Chron. Bot.*, 11, iii, 1947, pp. 119–204.

Stevenson, I. P. John Ray and his contributions to plant and animal classification. *J. Hist. Med.*, 2, 1947, pp. 250–261.

Stillman, John Maxson. *The story of early chemistry*, New York, 1924; reissued with title *The story of alchemy and early chemistry*, New York, 1960. (Dover Books.)

Stillwell, Margaret Bingham, *ed. Incunabula in American libraries: a second census of fifteenth-century books owned in the United States, Mexico, and Canada*, New York, 1940.

Stoddart, Anna M. *The life of Paracelsus Theophrastus von Hohenheim, 1493–1541*, [*etc.*], 1911.

Stomps, Th. J. On the rediscovery of Mendel's work by Hugo de Vries. *J. Hered.*, 45, 1954, pp. 293–294.

Storrs, F. C. Lavoisier's technical reports: 1768–1794. Part I [–II]. *Ann. Sci.*, 22, 1966 [1967], pp. 251–275; 24, 1968, pp. 179–197.
Stratton, G. Burder. The Zoological Society of London. *Brit. Book News*, 95, July, 1948, pp. 358–360.
Strebel, J. Paracelsus als Chemiker und Verfasser des ersten deutschsprachigen Lehrbuches der Chemie. *Praxis*, 38, 1949, pp. 806–814.
Strowski, F. *Pascal et son temps.*, 3 vols., Paris, 1931.
Strube, Irene. Die Phlogistonlehre Georg Ernst Stahls (1659–1734) in ihrer historischen Bedeutung. *Z. Gesch. Naturw. Techn. Med.*, 1, 1961, pp. 27–51.
Strube, Wilhelm. Die Ausbreitung der Naturanschauung G. E. Stahls unter den deutschen Chemikern des 18. Jahrhunderts. *Z. Gesch. Naturw. Techn. Med.*, 1, 1961, pp. 52–61.
Struik, D. J. American science between 1780 and 1830. *Science*, 129, 1959, pp. 1100–1106.
——, *ed. A source book in mathematics, 1200–1800*, Cambridge, Mass., 1969.
Sturt, Mary. *Francis Bacon: a biography*, 1932.
Suchting, W. A. Berkeley's criticism of Newton on space and motion. *Isis*, 58, 1967, pp. 186–197.
Sudhoff, Karl. *Nachweise zur Paracelsus-Literatur*, Munich, 1932.
——. *Versuch einer Kritik der Echtheit der paracelsischen Schriften*, 2 vols., Berlin, 1894–99.
Süsskind, Charles. Hertz and the technological significance of electromagnetic waves. *Isis*, 56, 1965, pp. 342–345.
Sullivan, J. W. N. *The history of mathematics in Europe; from the fall of Greek science to the rise of the conception of mathematical rigour*, 1925. (Chapters in the History of Science, IV.)
——. *Isaac Newton, 1642–1727*, [*etc.*], 1938.
Summent, George A. A lost work on Euclid rediscovered. *Isis*, 48, 1957, pp. 66–68.
Swann, C. Kirke. Natural history books from a bookseller's point of view. *J. Soc. Bib. Nat. Hist.*, 3, 1956, pp. 117–126.
Swinburne, R. G. Galton's law-formulation and development. *Ann. Sci.*, 21, 1965, pp. 15–31.
Swinton, W. E. Early history of comparative anatomy. *Endeavour*, 19, 1960, pp. 209–214.
Swisher, Charles N. Charles Darwin on the origins of behaviour. *Bull. Hist. Med.*, 41, 1967, pp. 24–43.
Talbot, G. R., and Pacey, A. J. Some early kinetic theories of gases: Herapath and his predecessors. *Brit. J. Hist. Sci.*, 3, 1966, pp. 133–149.
Tallmadge, Guy K. The third part of the *De extractione* of Valerius Cordus. *Isis*, 7, 1925, pp. 394–395.

Tanner, Rosalind C. H. On the role of equality and inequality in the history of mathematics. *Brit. J. Hist. Sci.*, 1, 1962–63, pp. 159–169.

——. The study of Thomas Harriot's manuscripts. I. Harriot's will. *History of Science*, 6, 1957, pp. 1–16.

——. Thomas Harriot as mathematician—a legacy of hearsay. Part I. *Physis*, 9, 1967, pp. 235–247.

Taton, René. *A general history of the sciences. Translated by A. J. Pomerans*, 4 vols., 1963–65.

——. Réaumur, mathématicien. *Rev. Hist. Sci. Paris*, 11, 1958, pp. 130–133.

Taylor, A. E. *Aristotle*, London, [*etc.*], 1943. (Nelson's Discussion Books, No. 77.)

Taylor, D. W. Galen's physiology. *N.Z. med. J.*, 66, 1967, pp. 176–181.

Taylor, Eva Germaine Rimington. *The mathematical practitioners of Hanoverian England, 1714–1840*, Cambridge, 1966.

——. *The mathematical practitioners of Tudor & Stuart England*, Cambridge, 1954.

Taylor, F. Sherwood. An alchemical work of Sir Isaac Newton. *Ambix*, 5, 1956, pp. 59–84.

——. *Galileo and the freedom of thought*, 1938.

——. *A short history of science*, London, Toronto, [1939].

——. A survey of Greek alchemy. *J. Hellenic Stud.*, 15, 1930, pp. 109–139.

Taylor, Sir Geoffrey. George Boole, F.R.S., 1815–1864. *Notes Rec. Roy. Soc. Lond.*, 12, 1956, pp. 44–52.

Tchenpkal, V. L. M. V. Lomonosov and his astronomical works, in commemoration of the 250th anniversary of his birth. *Observatory*, 81, 1961, pp. 183–189.

Thackray, Arnold W. Documents relating to the origins of Dalton's chemical atomic theory. *Manchester Lit. Philos. Soc. Mem. Proc.*, 108, 1965–66, pp. 21–42.

——. The emergence of Dalton's chemical atomic theory: 1801–08. *Brit. J. Hist. Sci.*, 3, 1966, pp. 1–23.

——. Fragmentary remains of John Dalton. Part 1. Letters. *Ann. Sci.*, 22, 1966, pp. 145–174.

——. "In praise of famous men"—the John Dalton bicentenary celebrations, 1966. *Notes Rec. Roy. Soc. Lond.*, 22, 1967, pp. 40–44.

——. John Dalton—accidental atomist. *Discovery*, 27, Sept., 1966, pp. 28–33.

——. "Matter in a nut-shell": Newton's *Opticks* and eighteenth-century chemistry. *Ambix*, 15, 1968, pp. 29–53.

——. The origin of Dalton's chemical atomic theory: Daltonian doubts resolved. *Isis*, 57, 1966, pp. 35–55.

Thalamus, A. *Étude bibliographique de La géographie d'Eratosthène*, Paris, 1921.

Théodoridès, Jean. Conrad Gesner et la zoologie: les invertébrés. *Gesnerus*, 23, 1966, pp. 230–237.

——. Quelques documents inedits ou peu connus relatifs à Georges Cuvier, à sa famille et à son salon. *Stendahl Club*, No. 33, 1966, pp. 55–64; No. 34, pp. 179–188.

——. Réaumur (1683–1757) et les insectes sociaux. *Janus*, 48, 1959, pp. 62–76.

——. Les sciences naturelles et particulièrement la zoologie dans le "Traité des poisons" de Maimonide. *Rev. Hist. Med. hébr.*, No. 31, 1956, pp. 87–104.

Thomas, K. Bryn. A Jenner letter. *J. Hist. Med.*, 12, 1957, pp. 449–458.

Thomas-Sanford, Charles. *Early editions of Euclid's Elements*, 1926. (Bibliographical Society. Illustrated Monographs, No. XX.)

Thompson, D. John Tyndall and the Royal Institution. *Ann. Sci.*, 13, 1957, pp. 9–22.

Thompson, Lawrence S. Jens Christian Bay, bibliologist. *Libri*, 12, 1962, pp. 320–330.

Thompson, Reginald Campbell. *The Assyrian herbal*, [etc.], 1924.

——. *A dictionary of Assyrian botany*, 1949.

——. *A dictionary of Assyrian chemistry and geology*, Oxford, 1936.

Thompson, Silvanus P. *Life of Lord Kelvin*, 2 vols., 1910.

——. *Michael Faraday, his life and work*, London, [etc.], 1898. (The Century Science Series.)

Thomson, S. Harrison. Grosseteste's *Questio de calore, De cometis* and *De operacionibus solis. Medievalia et Humanistica*, 11, 1957, pp. 34–43.

——. The *Summa in viii libros physicorum* of Grosseteste. *Isis*, 22, 1934–35, pp. 12–18.

——. The text of Grosseteste's De cometis. *Isis*, 19, 1933, pp. 19–25.

——. The texts of Michael Scot's Ars alchemie. *Osiris*, 5, 1938, pp. 523–559

——. An unnoticed treatise of Roger Bacon on time and motion. *Isis* 27, 1937, pp. 219–224.

——. *The writings of Robert Grosseteste, Bishop of Lincoln, 1235–1253*, Cambridge, 1940.

Thomson, Thomas. *History of the Royal Society, from its institution to the end of the eighteenth century*, 1812.

Thomson, W. W. D. Some aspects of the life and times of Sir Hans Sloane. *Ulster med. J.*, 7, 1938, pp. 1–17.

Thoren, Victor E. An early instance of deductive discovery: Tycho Brahe's lunar theory. *Isis*, 58, 1967, pp. 19–36.

Thorndike, Lynn. Epitomes of Pliny's Natural history in the fifteenth century. *Isis*, 26, 1936, p. 39.
——. *History of magic and experimental science*, 8 vols., New York, 1923–58.
——. The Latin translations of astrological works of Messahala. *Osiris*, 12, 1956, pp. 49–72.
——. Mediaeval magic and science in the seventeenth century. *Speculum*, 28, 1953, pp. 692–704.
——. *Michael Scot*, London, [*etc.*], 1965.
——. Notes on some medieval astronomical, astrological and mathematical manuscripts at Florence. Milan, Bologna and Venice. *Isis*, 50, 1959, pp. 33–50.
——. Notes on some medieval Latin astronomical, astrological and mathematical manuscripts at the Vatican. Part I [–II]. *Isis*, 47, 1956, pp. 391–404; 49, 1958, pp. 34–49.
——. The Pseudo-Galen, *De plantis* (with Latin text of chapters on stones and those of chemical interest). *Ambix*, 11, 1963, pp. 87–96.
——. Robertus Anglicus. *Isis*, 34, 1942–43, pp. 467–469.
——. Rufinus: a forgotten botanist of the thirteenth century. *Isis*, 18, 1932, pp. 63–76.
——. *Science and thought in the fifteenth century: studies in the history of medicine and surgery, natural and mathematical science, philosophy and politics*, New York, 1929.
——. Some alchemical manuscripts at Bologna and Florence. *Ambix*, 5, 1956, pp. 85–110.
——. *The Sphere of Sacrobosco and its commentators*, Chicago, 1949.
——. Translations of works of Galen from the Greek by Peter of Abano. *Isis*, 33, 1941–42, pp. 649–653.
——. Vatican Latin manuscripts in the history of science and medicine. *Isis*, 13, 1929–30, pp. 51–102.
——, and Kibre, Pearl. *Incipits of mediaeval scientific writings in Latin*, 1963.
Thornton, John L. A diary of James Macartney (1770–1843), with notes on his writings. *Med. Hist.*, 12, 1968, pp. 164–173.
——. Dr. John Dee and his scheme for a national library. *Med. Bookman & Historian*, 2, 1948, pp. 359–362; also in Thornton, John L. *Selected readings in the history of librianship. Second edition*, 1966, pp. 26–29.
——. James Macartney (1770–1843). *St Bart's Hosp. J.*, 65, 1961, pp. 121–123.
——. *Medical books, libraries and collectors. . . . Second edition*, 1966.
——. *A select bibliography of medical biography. With an introductory essay on medical biography. Second edition*, 1970.
——, and Tully, R. I. J. History and biography of biology. *In*, Bottle, R. T., and Wyatt, H. V., *eds. The use of biological literature*, 1966, pp. 230–245. (Second edition in press.)

Thornton, John L. and ——. History and biography of chemistry. *In*, Bottle, R. T., *ed. The use of chemical literature. Second edition*, 1969, pp. 235–250.

——, and ——. Scientific societies. Their growth and contribution to scientific literature. *Lib. World*, 46, 1943–44, pp. 119–122, 136–139.

——, and Wiles, Anna. William Odling, 1829–1921. *Ann. Sci.*, 12, 1956 [1957], pp. 288–295.

Thorpe, Sir Thomas Edward. *Essays in historical chemistry*, London, New York, 1894.

——. *Humphry Davy*, 1896.

Thrower, W. R. Contributions to medicine of Captain James Cook, F.R.S., R.N. *Lancet*, 1951, II, pp. 215–219.

Tilden, Sir William A. *Chemical discovery and invention in the twentieth century. Second edition*, London, New York, 1917.

——. *Sir William Ramsay, K.C.B., F.R.S.: memorials of his life and work*, 1918.

——. A sketch of Berzelius. *Med. Life*, 35, 1928, pp. 187–209.

Times, The. *The Royal Society tercentenary. Compiled from a special supplement of The Times, July, 1960*, 1961.

Torlais, Jean. Chronologie de la vie et des oeuvres de René-Antoine Ferchault de Réaumur. *Rev. Hist. Sci., Paris*, 11, 1958, pp. 1–12.

——. *Réaumur d'après les documents inédits*, Paris, 1961.

——. Réaumur, de Geer et la création de l'entomologie. *Prog. méd., Paris*, 86, 1958, pp. 143–145.

——. Réaumur, philosophe. *Rev. Hist. Sci., Paris*, 11, 1958, pp. 13–33.

——. Une rivalité célèbre: Réaumur et Buffon. *Pr. méd.*, 66, 1958, pp. 1057–1058.

Tortonese, E. Lazzaro Spallanzani: founder of experimental physiology. *Endeavour*, 7, 1948, pp. 92–96.

Townsend, A. C. Guides to scientific literature. *J. Document.*, 11, 1955, pp. 73–78.

——, and Stratton, G. B. *Library resources in the Greater London area. No. 6. Zoological libraries*, 1957.

Trattner, Ernest R. *Architects of ideas. . . . The story of the great theories of mankind*, New York, 1938.

Travers, Morris W. *A life of Sir William Ramsay, K.C.B., F.R.S.*, 1956.

Treneer, Anne. *The mercurial chemist. A life of Sir Humphry Davy*, 1963.

Trengove, Leonard. Chemistry at the Royal Society of London in the eighteenth century. I [–III]. *Ann. Sci.*, 19, 1963 [1965], pp. 183–237; 20, 1964 [1965], pp. 1–57; 21, 1965 [1966], pp. 81–130.

True, Webster Prentiss. *The Smithsonian, America's treasure house*, [*etc.*], New York, 1950.

Truesdale, C. The new Bernouli edition. *Isis*, 49, 1958, pp. 54–62.

Tuge, Hideomi, *ed. Historical development of science and technology in Japan*, Tokyo, 1961.

Turbane, C. M. Grosseteste and an ancient optical principle. *Isis*, 50, 1959, pp. 467–472.

Turnbull, Herbert Westren. *Bi-centenary of the death of Colin Maclaurin (1698–1746)*, [*etc.*], Aberdeen, 1951. (Aberdeen University Studies, No. 127.)

——, *ed. James Gregory tercentenary memorial volume. Containing his correspondence with John Collins and his hitherto unpublished mathematical manuscripts, together with addresses and essays communicated to the Royal Society of Edinburgh, July, 1938*, 1939.

——. *The mathematical discoveries of Newton*, London, Glasgow, 1947.

——, and Bushnell, George Herbert. *University of St. Andrews: James Gregory Tercentenary. Record of the celebrations held in the University Library, July fifth, MCMXXXVIII*, St. Andrews, 1939.

Turner, F. McD. C. *The Pepys library*, Cambridge, [N.D.].

Turner, H. D. Robert Hooke and Boyle's air pump. *Nature*, 184, 1959, pp. 395–397.

Turrill, W. B. *Joseph Dalton Hooker*, 1964. (British Men of Science Series.)

——. Joseph Dalton Hooker, F.R.S. (1817–1911). *Notes Rec. Roy. Soc. Lond.*, 14, 1959, pp. 109–120.

——. Pioneer plant geography. The phytogeographical researches of Sir Joseph Dalton Hooker. *Lotsya*, 4, 1953.

——. *The Royal Botanic Gardens, Kew, past and present*, 1959.

Uggla, Arvid Hj. *Linnaeus*, Stockholm, 1957.

Underwood, E. Ashworth. Boerhaave after three hundred years. *Brit. med. J.*, 1968, 4, pp. 820–825.

——. Lavoisier and the history of respiration. *Proc. Roy. Soc. Med.*, 37, 1943–44, pp. 247–262.

Union Académique Internationale. *Katalog der arabischen alchimistischen Handschriften Deutschlands: Handschriften der Öffentlichen Wissenschaftlichen Bibliothek. Bearbeitet von Alfred Siggel*, Berlin, 1949; and, *Handschriften der Ehemals Herzoglichen Bibliothek zu Gotha. Bearbeitet von Alfred Siggel*, Berlin, 1950.

Union Internationale d'Histoire et de Philosophie des Sciences. *La science au seizième siècle. Colloque international de Royaumont 1–4 Juillet, 1957*, Paris, 1960.

Urquhart, D. J. How the science information problem concerns YOU. *Adv. Sci.*, 23, 1966–67, p. 496.

——. Physics abstracting—use and users. *J. Document.*, 21, 1965, pp. 113–121.

Urquhart, D. J. Some recent developments at the National Lending Library for Science and Technology, Boston Spa, Yorkshire. *Proc. Roy. Soc. Med.*, 59, 1966, pp. 267–268.

Vaccaro, Leopold. Galileo Galilei. *Ann. med. Hist.*, 2nd Ser., 7, 1935, pp. 372–384.

Vallery-Radot, René. *The life of Pasteur. . . . Translated from the French, by Mrs. R. L. Devonshire*, [*etc.*], 2 vols., 1911.

Valson, C. A. *Étude sur la vie et les ouvrages d'Ampère*, Lyons, 1885.

Van der Pas, Peter W. A note on the bibliography of Gregor Mendel. *Med. Hist.*, 3, 1959, pp. 331–333.

Van Klooster, H. S. The story of Liebig's Annalen der Chemie. *J. chem. Educ.*, 34, 1957, pp. 27–30.

Van Seters, W. H. *Pierre Lyonet, 1706–1789; sa vie, ses collections de coquillages et de tableaux, ses recherches entomologiques*, The Hague, 1962.

Varoissieau, W. W. A survey of scientific abstracting and indexing services. *Review of Documentation*, 16, 1949, pp. 25–46.

Vecerek, O. Johann Gregor Mendel as a beekeeper. *Bee World*, 46, 1965, pp. 86–96.

Verdoorn, Frans. The development of scientific publications, and their importance in the promotion of international scientific relations. *Science*, 107, 1948, pp. 492–497.

——. On the aims and methods of biological history and biography. *Chron. Bot.*, 8, 1944, pp. 425–448.

Verga, Ettore. *Bibliografia Vinciana, 1493–1930*, 2 parts, Bologna, 1931.

Vernon, K. D. C. The foundation and early years of the Royal Institution. *Proc. Roy. Instn. G.B.*, 39, 1963, pp. 364–462.

——. The Royal Institution and its library. *Lib. Assn. Rec.*, 61, 1959, pp. 283–289.

Vickery, B. C. Science, research and history. *Stechert-Hafner Book News*, 13, 1958–59, pp. 65–66.

——. Statistics of scientific and technical articles. *J. Document.*, 24, 1968, pp. 192–195.

——. The use of scientific literature. *Lib. Assn. Rec.*, 63, 1961, pp. 263–269.

Viets, Henry. De staticis experimentis of Nicolaus Cusanus. *Ann. med. Hist.*, 4, 1922, pp. 115–135.

Virtanen, R. *Marcelin Berthelot: a study of a scientist's public role*, Lincoln, Nebraska, 1965. (University of Nebraska Studies, N.S. No. 31.)

Vorzimmer, Peter J. Darwin and Mendel: the historical connection. *Isis*, 59, 1968, pp. 77–82.

Vucinich, Alexander. Mendeleev's views on science and society. *Isis*, 58, 1967, pp. 342–351.

——. *Science in Russian culture. Volume 1. A history to 1860*, 1965; Stanford, 1963.

Wächter, Otto, The "Vienna Dioskurides" and its restoration. *Libri*, 13, 1963, pp. 107–111.

Waerden, B. L. van der. *Science awakening. . . . English translation by Arnold Dresden, with additions of the author*, Groningen, 1961.

Waley, A. Notes on Chinese alchemy (supplementary to Johnson's A study of Chinese alchemy). *Bull. School Orient. Stud. Lond. Univ.*, 6, 1930–32, pp. 1–24.

Walker, Oswald J. August Kekulé and the benzene problem. *Ann. Sci.*, 4, 1939, pp. 34–36.

Walker, R. E. The veterinary papyrus of Kahun. A revised translation and interpretation of the ancient Egyptian treatise known as the veterinary papyrus of Kahun. *Vet. Rec.*, 76, 1964, pp. 198–200.

Walker, W. Cameron. The beginnings of the scientific career of Joseph Priestley. *Isis*, 21, 1934, pp. 81–97.

Wallace, Alfred Russel. *My life: a record of events and opinions*, [*etc.*], 2 vols., 1905.

Wallis, P. J. *A check list of British Euclids up to 1850*, Newcastle, 1967. [Typescript.]

——. Fun with figures. *Accountant's Magazine*, December 1968, pp. 1–4.

——. William Oughtred's 'Arithmeticae in numeris' or 'Clavis mathematicae', 1647: an unrecorded (unique?) edition in Aberdeen University Library. *Bibliotheck*, 5, 1968, pp. 147–148.

——. William Oughtred's 'Circles of proportion' and 'Trigonometries'. *Trans. Cambridge Bib. Soc.*, 4, 1968, pp. 372–382.

Walls, E. W. The Journal of Anatomy, 1867–1966. *J. Anat.*, 100, 1966, pp. 1–4.

Warren, R. M. *Helmholtz on perception, its physiology and development*, New York, 1968.

Waterfield, Reginald L. *A hundred years of astronomy*, 1938.

Watson, E. C. The early days of the Académie des Sciences as portrayed in the engravings of Sébastien Le Clerc. *Osiris*, 7, 1939, pp. 556–587.

Webb, K. R. Gay-Lussac (1778–1850) as chemist. *Endeavour*, 9, 1950, pp. 207–210.

Webb, Maysie. The National Reference Library for Science and Invention: a review of progress. *J. Document.*, 22, 1966, pp. 1–12.

Webster, Charles. The discovery of Boyle's law, and the concept of the elasticity of air in the seventeenth century. *Arch. Hist. exact. Sci.*, 2, 1965, pp. 441–502.

——. Harvey's De generatione: its origins and relevance to the theory of circulation. *Brit. J. Hist. Sci.*, 3, 1967, pp. 262–274.

——. Henry Power's experimental philosophy. *Ambix*, 14, 1967, pp. 150–178.

——. Richard Towneley and Boyle's law. *Nature*, 197, 1963, pp. 226–228.

Webster, Charles. *ed. Samuel Hartlib and the advancement of learning*, Cambridge, 1970.

——. Water as the ultimate principle of nature: the background to Boyle's Sceptical chymist. *Ambix*, 13, 1966, pp. 96–107.

Weil, E. An unpublished letter by Davy on the safety-lamp. *Ann. Sci.*, 6, 1948–50, pp. 306–307.

Weimann, Karl-Heinz. *Paracelsus-Bibliographie 1932–1960. Mit einem Verzeichnis neu entdeckter Paracelsus-Handschriften (1900–1960)*, Wiesbaden, 1963. (Kosmosophie. Forschungen und Texte zur Geschichte der Weltbildes, der Naturphilosophie, der Mystik und des Spiritulismus vom Spatmittelalter bis zur Romantik herausgegeben von Kurt Goldammer, Vol. II.)

Weir, J. A. Agassiz, Mendel, and heredity. *J. Hist. Biol.*, 1, 1968, pp. 179–203.

Weiss, Helene. Notes on the Greek ideas referred to in Van Helmont, De tempore. *Osiris*, 8, 1948, pp. 418–449.

Weld, Charles Richard. *A history of the Royal Society*, [*etc.*], 2 vols, 1848.

Wells, George A. Goethe and evolution. *J. Hist. Ideas*, 28, 1967, pp. 537–550.

——. Goethe and the intermaxillary bone. *Brit. J. Hist. Sci.*, 3, 1967, pp. 348–361.

Wendell-Smith, C. P. Harvey and embryology. *St Bart's Hosp. J.*, 61, 1957, pp. 180–183.

——. William Harvey, man-midwife. *St Bart's Hosp. J.*, 53, 1949, pp. 212–214.

West, Muriel. Notes on the importance of alchemy to modern science in the writings of Francis Bacon and Robert Boyle. *Ambix*, 9, 1961, pp. 102–114.

Westfall, Richard S. The development of Newton's theory of colour. *Isis*, 53, 1962, pp. 339–358.

——. Hooke and the law of universal gravitation. A reappraisal of a reappraisal. *Brit. J. Hist. Sci.*, 3, 1967, pp. 245–261.

——. Newton and his critics on the nature of colors. *Arch. internat. Hist. Sci.*, 15, 1962, pp. 47–58.

——. Newton defends his first publication: the Newton-Lucas correspondence. *Isis*, 57, 1966, pp. 299–314.

——. Newton's optics: the present state of research. *Isis*, 57, 1966, pp. 102–107.

——. Newton's reply to Hooke and the Theory of colors. *Isis*, 54, 1963, pp. 82–96.

——, and Thoren, Victor E., *eds. Steps in the scientific tradition: readings in the history of science*, New York, [*etc.*], 1968.

Wheeler, Alwyne C. The *Zoophylacium* of Laurens Theodore Gronovius. *J. Soc. Bib. Nat. Hist.*, 3, 1956, pp. 152–157.

Wheeler, T. S., and Partington, J. R. *The life and work of William Higgins, chemist (1763–1825). Including reprints of 'A comparative view of the phlogistic and antiphlogistic theories', and 'Observations on the atomic theory and electrical phenomena', by William Higgins*, Oxford, [*etc.*], 1960.

Wheeler, William Morton, and Barbour, Thomas, *eds. The Lamarck manuscripts at Harvard*, Cambridge, Mass., 1933.

Wheelwright, Philip. *Heraclitus*, Princeton, London, 1959.

White, T. H., *ed. The book of beasts. Being a translation from a Latin bestiary of the twelfth century, made and edited by T. H. White*, 1954.

Whitehead, E. S. *A short account of the life and work of John Joseph Fahie*, Liverpool, London, 1939.

Whitehead, P. J. P. The dating of the 1st edition of Cuvier's *Le règne animal distribué d'après son organisation. J. Soc. Bib. Nat. Hist.*, 4, 1967, pp. 300–301.

Whiteside, Derek Thomas. The expanding world of Newtonian research. *History of Science*, 1, 1962, pp. 16–29.

——. A face-lift for Newton: current facsimile reprints. *History of Science*, 6, 1967, pp. 59–68.

——. Isaac Newton: birth of a mathematician. *Notes Rec. Roy. Soc. Lond.*, 19, 1964, pp. 53–62.

——. Newton's early thoughts on planetary motion: a fresh look. *Brit. J. Hist. Sci.*, 2, 1964, pp. 117–137.

——. Newton's marvellous year: 1666 and all that. *Notes Rec. Roy. Soc. Lond.*, 21, 1966, pp. 32–41.

——. Patterns of mathematical thought in the later seventeenth century. *Archive for the History of Exact Sciences*, 1, 1961, pp. 179–388.

Whitford, Robert H. *Physics literature: a reference manual. Second edition*, Metuchen, 1968.

Whitrow, G. J. Galileo's significance in the history of astronomy. *Quart. J. Roy. Astron. Soc.*, 5, 1964, pp. 182–195.

——. Kant and the extragalactic nebulae. *Quart. J. Roy. Astron. Soc.*, 8, 1967, pp. 48–56.

Whitteridge, Gweneth. Harvey's Galen. *St Bart's Hosp. J.*, 6, 1957, pp. 174–175.

Whyte, Lancelot Law, *ed. Roger Joseph Boscovich, S.J., F.R.S., 1711–1787. Studies of his life and work on the 250th anniversary of his birth*, 1961.

Wichler, Gerhard. *Charles Darwin: the founder of the theory of evolution and natural selection*, Oxford, [*etc.*], 1961.

Wightman, William P. D. "Philosophical Transactions" of the Royal Society, *Nature*, 192, 1961, pp. 23–24.

——. Science and the Renaissance. *History of Science*, 3, 1964, pp. 1–19.

——. William Cullen and the teaching of chemistry. [1–2.] *Ann. Sci.*, 11, 1955, pp. 154–165; 12, 1956, pp. 192–205.

Wilkie, J. S. Galton's contribution to the theory of evolution, with special reference to his use of models and metaphors. *Ann. Sci.*, 11, 1956, pp. 194–205.
——. Some reasons for the rediscovery and appreciation of Mendel's work in the first years of the present century. *Brit. J. Hist. Sci.*, 1, 1962, pp. 5–17.
Wilkins, G. L. Notes on the "Historia conchyliorum" of Martin Lister (1683–1712). *J. Soc. Bib. Nat. Hist.*, 3, 1957, pp. 196–205.
Wilkinson, Ronald Sterne. The alchemical library of John Winthrop, Jr. (1606–1676) and his descendants in Colonial America. *Ambix*, 11, 1963, pp. 33–51; 13, 1966, pp. 139–186.
——. The Hartlib papers and seventeenth-century chemistry. *Ambix*, 15, 1968, pp. 54–69.
——. The problem of the identity of Eirenaeus Philalethes. *Ambix*, 12, 1964, pp. 241–243.
Williams, L. Pearce. *Michael Faraday*, 1965,
——. The physical sciences in the first half of the nineteenth century: problems and sources. *History of Science*, 1, 1962, pp. 1–15.
Williams, Terrence. *A checklist of Linneana 1735–1835 in the University of Kansas Libraries*, Lawrence, 1964.
Williams, Trevor I., *ed. A biographical dictionary of scientists*, 1969.
Williams-Ellis, Amabel. *Darwin's moon. A biography of Alfred Russel Wallace*, 1966.
Wilson, Curtis. Kepler's derivation of the elliptical path. *Isis*, 59, 1958, pp. 5–25.
Wilson, George. *The life of the Honble. Henry Cavendish, including abstracts of his more important scientific papers, and a critical enquiry into the claims of all the alleged discoverers of the composition of water*, 1851.
Wilson, Leonard G. Erasistratus, Galen, and the *Pneuma. Bull. Hist. Med.*, 33, 1959, pp. 293–314.
——. William Croone's theory of muscular contraction. *Notes Rec. Roy. Soc. Lond.*, 16, 1961, pp. 158–178.
Wilson, M. H., and Brocklebank, R. W. Goethe's colour experiments. *Yr Bk Phys. Soc.*, 1958, pp. 12–21.
Wilson, William. *A hundred years of physics*, 1950.
Wilson, William Jerome. An alchemical manuscript by Arnaldus de Bruxella. *Osiris*, 2, 1921, pp. 220–405.
——. Catalogue of Latin and vernacular alchemical manuscripts in the United States and Canada. *Osiris*, 7, 1939.
——. The origin and development of Greco-Egyptian alchemy. Special number of *Ciba Symposia*, 3, August, 1941.
Winder, Marianne. A bibliography of German astrological works printed between 1465 and 1600, with locations of those extant in London libraries. *Ann. Sci.*, 22, 1966 [1967], pp. 191–220.

Wingate, S. D. *The mediaeval Latin versions of the Aristotelian scientific corpus, with special reference to the biological works*, London, Leamington Spa, 1931.
Winsor, Justin. *A bibliography of Ptolemy's Geography*, Cambridge, Mass., 1884. (Library of Harvard University, Bib. Contrib.)
Winspear, A. D. *Lucretius and scientific thought*, Montreal, 1963.
Winter, H. J. J. The Arabic achievement in physics. *Endeavour*, 9, 1950, pp. 76–79.
——. *Eastern science: an outline of its scope and contribution*, 1952. (The Wisdom of the East Series.)
Wohl, Robert. Buffon and his project for a new science. *Isis*, 51, 1960, pp. 186–199.
Wolf, Abraham. *A history of science, technology, and philosophy in the eighteenth century. . . . Second edition revised by D. McKie*, [*etc.*], 1952.
——. *A history of science, technology, and philosophy in the 16th & 17th centuries, by A. Wolf with the co-operation of F. Dannemann and A. Armitage. New edition prepared by Douglas McKie*, [*etc.*], 1950.
Wolf, Edwin. The reconstruction of Benjamin Franklin's library: an unorthodox jigsaw puzzle. *Papers Bib. Soc. Amer.*, 56, 1962, pp. 1–16.
Wollaston, A. F. R. *Life of Alfred Newton*, [*etc.*], 1921.
Wood, Alexander. History of the Cavendish Laboratory, Cambridge. *Endeavour*, 4, 1945, pp. 131–135.
——, and Oldham, Frank. *Thomas Young, natural philosopher, 1773–1829*, [*etc.*], Cambridge, 1954.
Wood, Casey A. *An introduction to the literature of vertebrate zoology, based chiefly on the titles in the Blacker Library of Zoology, the Emma Shearer Wood Library of Ornithology, the Bibliotheca Osleriana and other libraries of McGill University, Montreal*, 1931.
Wood, D. N. The foreign-language problem facing scientists and technologists in the United Kingdom—report of a recent survey. *J. Document.*, 23, 1967, pp. 117–130.
——, and Barr, K. P. Courses on the structure and use of scientific literature. *J. Document.*, 22, 1966, pp. 22–32.
Wood, S. Martin Lister, zoologist and physician. *Ann. med. Hist.*, N.S. 1, 1929, pp. 87–104.
Woodruff, A. E. William Crookes and the radiometer. *Isis*, 57, 1966, pp. 188–198.
Woodward, Horace B. *The history of the Geological Society of London*, London, [*etc.*], 1908.
Wyatt, H. V. Research newsletters in the biological sciences—a neglected literature service. *J. Document.*, 23, 1967, pp. 321–325.
——, and Bottle, R. T. Training in the use of biological literature. *Aslib Proc.*, 19, 1967, pp. 107–110.

Wyckoff, Dorothy. Albertus Magnus on ore deposits. *Isis*, 49, 1958, pp. 109–122.

Wykes, Alan. *Doctor Cardano, physician extraordinary*, 1969.

Yagello, Virginia E. Early history of the chemical periodical. *J. chem. Educ.*, 45, 1968, pp. 426–429.

Yates, B. *How to find out about physics. A guide to sources of information arranged by the Decimal Classification*, Oxford, [*etc.*], 1965.

Yates, Frances A. *Giordano Bruno and the Hermetic tradition*, 1964.

Yeldham, Florence A. The alleged early English version of Euclid. *Isis*, 9, 1927, pp. 234–238.

Yonge, C. M. George Johnston and the Ray Society. *Endeavour*, 14, 1955, pp. 136–139.

[Young, H.] *A record of the scientific work of John Tyndall, D.C.L., LL.D., F.R.S., 1850–1888*, for private circulation, 1935.

Young, James. Malpighi. *N.Z. med. J.*, 20, 1921, pp. 1–19.

Youschkevitch, A. P. Recherches sur l'histoire des mathematiques au moyen-âge dans les pays d'orient: bilans et perspectives. *History of Science*, 6, 1967, pp. 41–58.

Zacharias, Procopios D. Chymeutike. The real Hellenic chemistry, *Ambix*, 5, 1956, pp. 116–128.

Zeitlin, Jake. Bookselling among the sciences. *College and Research Libraries*, 21, 1960, pp. 453–457.

Zietz, Joseph R. "The pirotechnia" of Vannocio Biringuccio. *J. chem. Educ.*, 29, 1952, pp. 507–510.

Zinner, Ernst. Cl. Ptolemaeus und das Astrolab. *Isis*, 41, 1950, pp. 286–287.

——. *Leben und Wirken des Johannes Müller von Königsberg genannt Regiomontanus*. (*Schriftenreihe zur bayerischen Landesgeschichte. Herausgegeben von den Kommission für bayerischen Landesgeschichte bei der Bayerischen Akademie der Wissenschaften*, Bd. 31), Munich, 1938.

——. Neuer Regiomontanus-Forschungen und ihre Ergebnisse. *Sudhoffs Arch. Gesch. Med.*, 37, 1953, pp. 104–108.

Zirkle, Conway. The death of Gaius Plinius Secundus (23–79 A.D.). *Isis*, 58, 1967, pp. 553–559.

——. Natural selection before the "Origin of species". *Proc. Amer. Philos. Soc.*, 84, 1941, pp. 71–123.

——. The role of Liberty Hyde Bailey and Hugo de Vries in the rediscovery of Mendelism. *J. Hist. Biol.*, 1, 1968, pp. 205–218.

——. Some oddities in the delayed discovery of Mendelism. *J. Heredity*, 55, 1964, pp. 65–72.

——. Species before Darwin. *Proc. Amer. Philos. Soc.*, 103, 1959, pp. 636–644.

Zubov, V. P. *Leonardo da Vinci. Translated from the Russian by David H. Kraus*, Cambridge, Mass., 1968.

Index

This Index has been compiled as a guide to the material contained in a reference book containing innumerable names of individuals, titles, etc., and it has been necessary to select the entries likely to be of particular significance to readers. Names of persons, societies and institutions, titles of journals, and subjects are included. Footnotes and the Bibliography are indexed, and the abbreviation *loc.* has been used to indicate the location of specific items, etc. in the possession of certain libraries and institutions. Cross-references have been cut to the minimum, it being advantageous as a rule to add page numbers instead of *see* . . .

Entries are brief, and details such as forenames and dates of birth and death are used only to distinguish between persons with identical names. Fuller information is readily available in the text.

Universities named after locations generally appear under University of . . .

Alphabetical arrangement is by the "all-through" or "letter-by-letter" system.

Bib. = Bibliography *Loc.* = Location *Note* = Footnote